Healthy Longevity in Ch

This is dedicated to:

Professor Thomas Pullum

Oct, 2009

THE SPRINGER SERIES ON
DEMOGRAPHIC METHODS AND POPULATION ANALYSIS

Series Editor
KENNETH C. LAND
Duke University

In recent decades, there has been a rapid development of demographic models and methods and an explosive growth in the range of applications of population analysis. This series seeks to provide a publication outlet both for high-quality textual and expository books on modern techniques of demographic analysis and for works that present exemplary applications of such techniques to various aspects of population analysis.

Topics appropriate for the series include:

- General demographic methods
- Techniques of standardization
- Life table models and methods
- Multistate and multiregional life tables, analyses and projections
- Demographic aspects of biostatistics and epidemiology
- Stable population theory and its extensions
- Methods of indirect estimation
- Stochastic population models
- Event history analysis, duration analysis, and hazard regression models
- Demographic projection methods and population forecasts
- Techniques of applied demographic analysis, regional and local population estimates and projections
- Methods of estimation and projection for business and health care applications
- Methods and estimates for unique populations such as schools and students

Volumes in the series are of interest to researchers, professionals, and students in demography, sociology, economics, statistics, geography and regional science, public health and health care management, epidemiology, biostatistics, actuarial science, business, and related fields.

For other titles published in this series, go to
www.springer.com/series/6449

Healthy Longevity in China

Demographic, Socioeconomic, and Psychological Dimensions

edited by

Zeng Yi
Duke University, Durham, NC, USA; Peking University, Beijing, China

Dudley L. Poston, Jr.
Texas A&M University, College Station, TX, USA

Denese Ashbaugh Vlosky
Louisiana State University, Baton Rouge, LA, USA

Danan Gu
Duke University, Durham, NC, USA

Editors

Zeng Yi
Duke University Durham, NC, USA
Peking University, Beijing,
China

Dudley L. Poston, Jr.
Texas A&M University,
College Station,
TX, USA

Denese Ashbaugh Vlosky
Louisiana State University,
Baton Rouge, LA,
USA

Danan Gu
Duke University,
Durham, NC,
USA

ISBN: 978-1-4020-9478-1 (PB)

ISBN: 978-1-4020-6751-8 (HB)   e-ISBN: 978-1-4020-6752-5

Library of Congress Control Number: 2008940442

© Springer Science + Business Media B.V. 2009
No part of this work may be reproduced, stored in a retrieval system, or transmitted
in any form or by any means, electronic, mechanical, photocopying, microfilming, recording
or otherwise, without written permission from the Publisher, with the exception
of any material supplied specifically for the purpose of being entered
and executed on a computer system, for exclusive use by the purchaser of the work.

Printed on acid-free paper.

9 8 7 6 5 4 3 2 1

springer.com

Preface

The Chinese Longitudinal Healthy Longevity Study (CLHLS) was launched a decade ago. Professor Zeng Yi, Professor Xiao Zhenyu and I, with help from various researchers at Peking University, the Chinese Academy of Social Science, Duke University, the Max Planck Institute for Demographic Research, and the University of Southern Denmark, coordinated the design of the study and the required fundraising. The Max Planck Institute for Demographic Research provided some modest but crucial seed money. A continuing research grant from the National Institute on Aging (NIA) of the U.S. National Institutes of Health covered international research expenditures as well as some expenses in China; I was the original director of this grant, but Zeng Yi has now taken on that responsibility. The bulk of the effort in carrying out the survey—a mammoth undertaking—as well as much of the analysis in China was provided by Peking University and the other involved Chinese institutions.

In our 1997 grant application to the NIA, we argued that it is important to study people 80 years old and older—the oldest-old—because this segment of the population is rapidly growing, not only in developed countries but also in China and many other developing countries. United Nations statistics indicate that the global population of the oldest-old is growing at 3.4 percent annually. Because of this rapid growth, the world's oldest-old may number almost 200 million in 2030 and almost 400 million in 2050, compared with less than 70 million in 2000. The oldest-old are especially significant because of the intensive health and social care they require—and because, compared with other segments of the population, so little is known about their needs and about opportunities to improve their health and wellbeing.

More specifically we argued in our 1997 grant application that the elderly in China were worthy of careful study. China is home to about a fifth of the world's population—and about a fifth of the world's elderly population. As described in the introductory chapter of this book, the pace of increase in the numbers and proportions of the elderly and the oldest-old in China is extraordinary. The numbers of the elderly in China are being swelled by the survivors of the huge cohorts born in the 1950s and 1960s. The rise in the percentage of people in China who are elderly is further fuelled by the rapid decline in fertility since the 1970s.

As demographers we were aware of the problems of age misreporting that make studies of the oldest-old problematic in many countries. For some of the minority populations of China this is a major concern, but for the Han Chinese, who make up the great bulk of the Chinese population, reports of age are generally accurate. This was documented by various scholars and confirmed by a pilot study reported in article by Wang Zhenglian, Zeng Yi, Bernard Jeune and me in *Genus*.

Thus we were convinced in 1997 that it was feasible and important to conduct a longitudinal study focusing on the determinants of healthy longevity among the oldest-old in the 22 provinces of China with overwhelmingly Han Chinese populations. Under the leadership of Professor Zeng Yi and co-investigator Professor Xiao Zhenyu and with the help of hundreds of survey enumerators and other scientists, the CLHLS was successfully launched and continued over the past decade. As summarized in Chapter 2 of this book, the first survey was undertaken in 1998 with follow-up surveys in 2000, 2002 and 2005. These four surveys included interviews with about 9,000, 1,1000, 1,6000 and another 16,000 elderly people. In total, face-to-face interviews have been conducted with nearly 11,000 centenarians, more than 14,000 nonagenarians, some 16,500 octogenarians and almost 10,000 people aged 65–79. At each wave, CLHLS re-interviewed survivors, and replaced deceased interviewees with additional participants. For the 12,500 people who died (at ages from 65 to more than 105) between waves, data on mortality and health status before death were collected in interviews with a close family member. No other study anywhere in the world has conducted detailed interviews with so many centenarians, nonagenarians and octogenarians. Furthermore, the study of 10,000 people aged 65–79 is one of the largest such studies ever undertaken and provides valuable information on the younger old in comparison with the oldest-old.

A decade ago little was known about the determinants of healthy longevity. We documented this in our 1997 grant application. The scarceness of knowledge was remarkable—given the interest people have shown in "secrets of long life" for thousands of years and given the rapidly growing numbers of the oldest-old. We now know much more, in important part due to research based on CLHLS data. As cited in various chapters of this book and in the Epilogue, research output includes 24 articles in international journals, 91 articles in Chinese journals, 80 papers presented at international conferences, at least 8 doctoral theses and at least 13 M.A. degree theses, 11 policy reports submitted to governmental agencies and four books published in China (two of them in both Chinese and English).

The research in this book represents an important milestone in our emerging understanding of how individuals can endeavor to live a long, healthy life and how societies can help them do so. The book captures highlights of a decade of effort—by the tens of thousands of Chinese who were interviewed, by the hundreds of survey enumerators who organized and carried out the interviews, and by the scores of scientists who have sifted through the data. We still have only a partial understanding of the determinants of healthy longevity, but we have a considerably better understanding than we had a decade ago—and we can look forward to prospects for deeper understanding based in significant measure on further analysis of data from the ongoing CLHLS endeavor.

Director, Max Planck Institute for Demographic Research   James W. Vaupel
Research Professor, Duke University

Acknowledgments

Early versions of almost all the chapters in this book were first presented at one of the four seminars, symposia, and conferences focusing on the analyses of data from the "Chinese Longitudinal Healthy Longevity Survey (CLHLS)." There were: (1) the "International Seminar on Determinants of Healthy Longevity in China," held at the Max Planck Institute for Demographic Research (MPIDR), Rostock, on August 2–4, 2004; (2) the "Conference on Chinese Healthy Aging and Socioeconomics: International Perspectives," held at Duke University, on August 20–21, 2004; (3) the "Symposium on the Chinese Longitudinal Healthy Longevity Survey: Unique Data Resource and Research Opportunities," at the 57th annual scientific meeting of GSA, Washington DC, November 19–23, 2004 and (4) the "International Conference on Healthy Longevity," held in Pengshan, Sichuan Province, China, on September 21–23, 2005.

We thank the several institutions and their personnel for supporting the academic activities which have led to the research reported this book. The National Institute on Aging has provided funding in support of the CLHLS (P01 AG 08761, 10/1998–8/2004, PI: Zeng Yi and P01 Program Director: James W. Vaupel; R01 AG023627–01, 9/2004–8/2009, PI: Zeng Yi). The United Nations Fund for Population Activities (UNFPA) and the China National Foundation for Social Sciences joined NIA to co-sponsor an expanded survey in 2002. Since 2005, the National Natural Science Foundation of China (grants 70440009 and 70533010) and the Hong Kong Research Grant Council joined the NIA to co-sponsor the expanded survey. The Academia Sinica in Taiwan has provided funding to support the sub-sample of the adult children of the elderly interviewees. Peking University, Duke University, and the China Academy of Social Sciences have all provided institutional support. Since 1997, the Max Planck Institute for Demographic Research (MPIDR) has provided support for international training as well as funding support for the 2004 seminar. Duke University's Vice Provost for International Affairs and Development and the Vice Provost for Interdisciplinary Research, along with the Asian/Pacific Studies Institute, provided funding to support the 2004 conference at Duke University. The Gerontology Society of America offered the opportunity of our symposium held in Washington DC. The governments of the Mei Shan Municipality and Peng Shan County of Sichuan Province, China, provided funding to support the 2005 international conference at Peng Shan. We also wish to thank the Centre for Health Aging and Family Studies/ China Center for Economic Research at Peking University, and the Center for Study of Aging and Human Development, Geriatrics Division, Population Research Institute and the Center for Chinese Population and Socioeconomic Studies at Duke University for their institutional supports and contributions as local hosts of the seminars and conferences.

We sincerely thank Peking University and the China Mainland Information Group for their valuable support and contributions in the CLHLS data collections. We are very grateful to all interviewees and interviewers who participated in the CLHLS surveys. Without their efforts and collaboration, this nationwide project could never have been conducted. We would like to sincerely thank the following individuals who participated in or provided advice or administrative support for the CLHLS pilot studies, questionnaire design, training, survey field work, data procession and archiving (alphabetically listed): Wenmei Cai, Huashuai Chen, Junhong Chu, Harvey Cohen, Qiushi Feng, Baochang Gu, Danan Gu, Ling Guan, Zhigang Guo, Cheng Jiang, Leiwen Jiang, Jianxin Li, Ling Li, Qiang Li, George Linda, Guiping Liu, Yuzhi Liu, Jiehua Lu, Cindy Owens, Kenneth Land, Diane Parham, Georgeanne E. Patmios, Dudley L. Poston, Jr., Ke Shen, Jacqui Smith, Richard Suzman, Liqun Tao, Becky Tesh, James W. Vaupel, Nancy Vaupel, Denese Ashbaugh Vlosky, Zhenglian Wang, Changping Wu, Deqin Wu, Zhenyu Xiao, Qin Xu, Ye Yuan, Xianxin Zeng, Zeng Yi, Jie Zhan, Chunyuan Zhang, Fengyu Zhang, Wenjuan Zhang, Zhen Zhang, Baohua Zhao, Zhenzhen Zheng, Yun Zhou.

We would like to express our special gratitude and respect to Professor Wenmei Cai who passed away at age 80 in 2004. Professor Cai was one of the initial members of the CLHLS research team, and she actively participated in the pilot studies and the field surveys in the first three waves of CLHLS in 1998, 2000, and 2002. We are deeply saddened by her death; her passing will be a loss for the CLHLS project. Her life was exemplarily active, productive, and fruitful. Our CLHLS study has benefited significantly from her anthropological field observations and contributions in sociological and demographic theories. We dedicate this volume to her.

Contents

1 Introduction: Aging and Aged Dependency in China 1
  Dudley L. Poston Jr. and Zeng Yi

Part I CLHLS and its Data Quality Assessment

2 Introduction to the Chinese Longitudinal Healthy Longevity Survey
  (CLHLS) 23
  Zeng Yi

3 General Data Quality Assessment of the CLHLS 39
  Danan Gu

4 Reliability of Age Reporting Among the Chinese Oldest-Old in the
  CLHLS Datasets 61
  Zeng Yi and Danan Gu

5 Age Reporting in the CLHLS: A Re-assessment 79
  Heather Booth and Zhongwei Zhao

6 Assessment of Reliability of Mortality and Morbidity in the
  1998–2002 CLHLS Waves 99
  Danan Gu and Matthew E. Dupre

Part II The Effects of Demographic and Socioeconomic Factors

7 The Effects of Sociodemographic Factors on the Hazard of Dying
  Among Chinese Oldest Old 121
  Dudley L. Poston Jr. and Hosik Min

8 When I'm 104: The Determinants of Healthy Longevity Among the
  Oldest-Old in China 133
  D.A. Ahlburg, E. Jensen and R. Liao

9 Association of Education with the Longevity of the Chinese Elderly .. 149
    Jianmin Li

10 Analysis of Health and Longevity in the Oldest-Old Population—A
    Health Capital Approach 157
    Zhong Zhao

11 The More Engagement, the Better? A Study of Mortality of the
    Oldest Old in China 177
    Rongjun Sun and Yuzhi Liu

Part III Living Arrangements and Elderly Care

12 Living Arrangements and Psychological Disposition of the Oldest
    Old Population in China 197
    Zheng Wu and Christoph M. Schimmele

13 Health and Living Arrangement Transitions among China's
    Oldest-old 215
    Zachary Zimmer

14 Intergenerational Support and Self-rated Health of the Elderly in
    Rural China: An Investigation in Chaohu, Anhui Province 235
    Lu Song, Shuzhuo Li, Wenjuan Zhang and Marcus W. Feldman

15 The Effects of Adult Children's Caregiving on the Health Status of
    Their Elderly Parents: Protection or Selection? 251
    Zhen Zhang

16 The Challenge to Healthy Longevity: Inequality in Health Care and
    Mortality in China 269
    Zhongwei Zhao

Part IV Subjective Wellbeing and Disability

17 Successful Ageing of the Oldest-Old in China 293
    Peng Du

18 Impairments and Disability in the Chinese and American
    Oldest-Old Population 305
    William P. Moran, Sihan Lv and G. John Chen

19 Tooth Loss Among the Elderly in China 315
    Yun Zhou and Zhenzhen Zheng

20 Psychological Resources for Well-Being Among Octogenarians,
    Nonagenarians, and Centenarians: Differential Effects
    of Age and Selective Mortality 329
    Jacqui Smith, Denis Gerstorf and Qiang Li

21	**An Exploration of the Subjective Well-Being of the Chinese Oldest-Old** .... 347 Deming Li, Tianyong Chen and Zhenyun Wu
22	**Social Support and Self-Reported Quality of Life: China's Oldest Old** .... 357 Min Zhou and Zhenchao Qian
23	**Mortality Predictability of Self-Rated Health Among the Chinese Oldest Old: A Time-Varying Covariate Analysis** .... 377 Qiang Li and Yuzhi Liu
24	**Gender Differences in the Effects of Self-rated Health Status on Mortality Among the Oldest Old in China** .... 397 Jiajian Chen and Zheng Wu
25	**Epilogue: Future Agenda** .... 419 Zeng Yi

Index 429

Contributors

Dennis A. Ahlburg
Leeds School of Business, University of Colorado 8315, Koelbel UCB 419 Boulder, CO 80309, USA
dennis.ahlburg@colorado.edu

Heather Booth
Australian Demographic and Social Research Institute, The Australian National University, Canberra, Australia
heather.booth@anu.edu.au

G. John Chen
Division of General Internal Medicine, Department of Medicine,
Medical University of South Carolina, Charleston, SC, USA

Jiajian Chen
East-West Center, Research Program, Population and Health Studies,
1601 East-West Road, Honolulu, HI 96848, USA
chenj@eastwestcenter.org

Tianyong Chen
Key Laboratory of Mental Health, Institute of Psychology, Chinese Academy of Sciences Beijing, 100101, China

Peng Du
Professor, Director, Gerontology Institute, Renmin University of China, Beijing 100872, China
dupeng415@yahoo.com.cn

Matthew E. Dupre
Center for the Study of Aging and Human Development, Duke University, Durham, NC 27710, USA
med11@geri.duke.edu

Marcus W. Feldman
Morrison Institute for Population and Resource Studies, Stanford University, Stanford, CA 94305, USA
marc@charles.stanford.edu

Denis Gerstorf
Department of Human Development and Family Studies, The Pennsylvania State University, 114M Henderson Building, University Park, PA 16802, USA
dxg36@psu.edu

Danan Gu
Center for the Study of Aging and Human Development, Medical School of Duke University, Durham, NC 27710, USA
gudanan@duke.edu

Eric Jensen
Department of Economics and Thomas Jefferson Program in Public
Policy, College of William and Mary
Williamsburg, VA 23187, USA
eric_jensen@wm.edu

Deming Li
Key Laboratory of Mental Health, Institute of Psychology, Chinese Academy of Sciences, Beijing 100101, China
lidm@psych.ac.cn

Jianmin Li
Institute of Population and Development, Nankai University, Tianjin 300071, China
lijianm0075@sina.com

Qiang Li
Max-Plank Institute for Demographic Research, Konrad-Zuse-Str. 1,
18057 Rostock, Germany
li@demogr.mpg.de

Shuzhuo Li
Institute for Population and Development Studies, School of Public Policy and Administration, Xi'an Jiaotong University, 28 Xianning Road, Xi'an, Shaanxi, 710049, China
shzhli@mail.xjtu.edu.cn

Ruyan Liao
Industrial Relations Center, Carlson School of Management,
University of Minnesota, Minneaplois, MN 55455, USA

Yuzhi Liu
Center for Healthy Aging and Family Studies, Institute of Population Research, Peking University, Yiheyuan Road 5, Haidian Qu, Beijing, 100872, PR China
yuzhil@pku.edu.cn

Sihan Lv
School of Public Health, University of Sichun, Chengdu, Sichun, China

Hosik Min
Center on the Family, University of Hawaii Manoa, 2515 Campus Rd, Miller 103, Honolulu, HI, 96822

William P. Moran
Professor of Medicine, Director, Division of General Internal Medicine and Geriatrics, Medical University of South Carolina, Rutledge Tower, 12 Floor, 135 Rutledge Avenue, PO Box 250591,
Charleston, SC 29425-0591, USA
moranw@musc.edu.

Dudley L. Poston Jr.
Department of Sociology, Texas A&M University, College Station,
TX 77843, USA
dudleyposton@yahoo.com

Zhenchao Qian
Department of Sociology, Ohio State University, 300 Bricker Hall,
190 N. Oval Mall, Columbus, OH 43210, USA
qian.26@sociology.osu.edu

Christoph M. Schimmele
Department of Sociology, University of Victoria, 3800 Finnerty Road,
Victoria, BC V8W 3P5 Canada

Jacqui Smith
Department of Psychology and Institute for Social Research, University of Michigan, 426 Thompson Street, Ann Arbor MI 48106-1248, USA
smitjacq@isr.umich.edu

Lu Song
School of Management, Xi'an Jiaotong University; Institute for
Population and Development Studies, School of Public Policy and Administration, Xi'an Jiaotong University, 28 Xianning Road, Xi'an, Shaanxi, 710049, China
songlu@stu.xjtu.edu.cn

Rongjun Sun Ph.D.
Department of Sociology, Cleveland State University, 2121 Euclid Ave., Cleveland, OH 44115, USA
r.sun32@csuohio.edu

Denese Ashbaugh Vlosky
Office of Social Service Research & Development, School of Social Work, The Louisiana State
University, Baton Rouge, Louisiana 70803, USA
denese@lsu.edu

Zheng Wu
Department of Sociology, University of Victoria, 3800 Finnerty Road, Victoria, BC V8W 3P5 Canada,
zhengwu@uvic.ca

Zhenyun Wu
Key Laboratory of Mental Health, Institute of Psychology, Chinese Academy of Sciences, Beijing 100101, China

Zeng Yi
Center for Study of Aging and Human Development, Medical School of Duke University, Durham, NC 27710, USA
Center for Healthy Aging and Family Studies/China Center for Economic Research, Peking University, Beijing, China
zengyi68@gmail.com

Wenjuan Zhang
Center for Healthy Aging and Family Studies/China Center for Economic Research, Peking University, Beijing, 100871, China

Zhen Zhang
Max Planck Institute for Demographic Research Konrad-Zuse-Str. 1, 18057 Rostock, Germany
zhang@demogr.mpg.de

Zhong Zhao
Institute for the Study of Labor (IZA), Schaumburg-Lippe-Str. 5–9 D-53113 Bonn, Germany
zhao@iza.org

Zhongwei Zhao
The Australian Demographic and Social Research Institute, The Australian National University, Canberra, Australia
zhongwei.zhao@anu.edu.au

Zhenzhen Zheng
Institute of Population and Labor Economics, Chinese Academy of Social Sciences, Beijing 100732, China
zhengzz@cass.org.cn

Min Zhou
Department of Sociology, Ohio State University, 300 Bricker Hall, 190 N. Oval Mall, Columbus, OH 43210, USA
zhou.144@sociology.osu.edu

Yun Zhou
Institute of Population Research, Peking University, Beijing, China
zhouyun@pku.edu.cn

Zachary Zimmer
Department of Sociology, University of Utah, 260 S. Central Campus Drive, Room 214, Salt Lake City, UT 84112, USA
zachary.zimmer@ipia.utah.edu

Chapter 1
Introduction: Aging and Aged Dependency in China

Dudley L. Poston Jr. and Zeng Yi

Abstract This introductory chapter reviews China's demographic history so to provide a perspective for the empirical analyses of healthy aging and longevity in the chapters of the book that follow. In the chapter we also present trend data and population projection data from 1950 to 2050 for China's elderly population. We show that the aged dependency burdens on the producing populations of China have increased from the past (1950) to the present (2000) and will become even heavier in the decades ahead (to 2050). China is projected to be older than the U.S. in 2050; moreover, its transition to becoming an elderly population has occurred more quickly than in the U.S. These dynamics have important implications for the society, and are discussed in this chapter.

Keywords Aged dependency, Aged dependency burden, China, Dynamic equilibrium, Elderly population, Family planning, Fertility decline, Compression of morbidity, Oldest old population, One-child policy, Parent support ratio, Population aging, Population growth, Population projection, Prevalence of morbidity, The United States, Total dependency ratio, Youth dependency ratio

1.1 Introduction

The People's Republic of China is the most populous country in the world. In the twentieth century China experienced dramatic decreases in fertility and increases in life expectancy. These changes have produced, and will continue to produce, large proportions of elderly people, although China has only recently become a demographically old country. This book deals with the demographic, socioeconomic, and psychological dimensions of healthy aging and healthy longevity in China. This introductory chapter reviews China's demographic history so as to provide a perspective for the empirical analyses of healthy aging and longevity in the chapters that follow.

D.L. Poston Jr.
Department of Sociology, Texas A&M University, College Station, TX 77843, USA
e-mail: dudleyposton@yahoo.com

In this chapter we also present trend data and population projection data from 1950 to 2050 for China's elderly population. We show that the aged dependency burdens on the producing populations of China have increased from the past (1950) to the present (2000) and will become even heavier in the decades ahead (to 2050). China is projected to be older than the U.S. in 2050; moreover, its transition to becoming an elderly population has occurred more quickly than in the U.S.

Of special interest for this book is the size of China's older and oldest old populations. In this chapter we follow the practice of the U.S. Bureau of the Census (Velkoff and Lawson 1998) and refer to the older population as persons aged 60 and over, and the oldest old as those aged 80 and over.

In the world in 2000, there were 591 million older persons, and over 68 million oldest old.[1] Of the world's older population in the year 2000, over 21 percent of them (or 128 million) live in China, compared to 7.6 percent (or almost 45 million) living in the U.S. (see Table 1.1).

If the older population of China in the year of 2000 were a single country it would be the eighth largest country in the world, outnumbered only by the non-elderly population of China (1.1 billion), and the populations of India (almost 1.1 billion), the United States (285 million), Indonesia (212 million), Brazil (172

Table 1.1 Total population, older population, and oldest old population: world, China, and the U.S., 2000, and 2010–2050

Year	Total	Older	Oldest old
World			
2000	5,995,544,836	591,389,484	68,259,980
2010	6,830,906,857	755,327,646	103,181,481
2020	7,561,076,957	1,018,949,740	136,919,697
2030	8,213,573,346	1,355,545,346	190,254,664
2040	8,809,366,772	1,663,858,895	284,553,277
2050	9,297,023,938	1,981,995,384	399,466,279
China			
2000	1,268,985,201	128,215,415	11,069,279
2010	1,358,722,700	168,804,989	17,654,658
2020	1,422,937,380	240,217,728	24,018,400
2030	1,432,807,130	341,693,798	35,136,698
2040	1,410,644,753	395,615,825	57,409,084
2050	1,347,624,386	424,395,138	92,505,472
U.S.			
2000	272,639,608	44,947,333	8,930,406
2010	298,026,141	55,623,834	11,227,361
2020	323,051,793	73,769,020	12,400,055
2030	347,209,212	87,874,783	18,009,972
2040	370,289,996	93,088,015	26,216,372
2050	394,240,529	99,459,187	30,200,741

---

[1] Unless otherwise noted all data on population size and age composition for 2000 and projected years are from the United Nations (2003).

million), Russia (146 million), and Pakistan (143 million) (United Nations 2003). Of the 69 million oldest old in the world, over 16 percent (or about 11.4 million) live in China.

Until very recently China has tended to focus more on matters of population control than population aging (Poston and Duan 2000). In the early 1970s China came to grips with the burgeoning size of its population and established a nationwide fertility control program that stressed *later* marriages, *longer* intervals between children, and *fewer* children. However, the large numbers of children born during China's "baby boom" in the early 1960s caused China's leaders in the mid- to late 1970s to become increasingly worried about demographic momentum and the concomitant growth potential of this extraordinarily large cohort. Thus in 1979 they launched the One-Child Campaign, requesting most families to have no more than one child, especially those living in urban areas.

The birth control policies adopted since 1970, along with increasing levels of socioeconomic development, resulted in a drastic decline in China's fertility rate, from levels greater than six children per woman in the early 1950s to under two in the late 1990s, to 1.6 in the year 2005 (Population Reference Bureau 2006). The fertility decline experienced in China has produced, and will continue to produce, an unprecedented increase in the proportion of the elderly population.

It is important to recall that the relatively large numbers just noted for the older and oldest old population of China in 2000 are numbers that were generated during demographic regimes in which fertility and mortality rates declined. A consequence of these transitions, especially unanticipated in China, is the proportionately larger older and oldest old populations projected for the decades of this new century.

Table 1.1 shows population projection data for the total populations, the older populations, and the oldest old populations of the World, China, and the United States for the decennial years of 2010 through 2050. These projections are the so-called "medium" projections of the United Nations (2003). They assume for the U.S. and China that total fertility rates will increase/decrease slowly from their present levels and will stabilize at 1.85 in 2045–2050. Mortality is projected to decline only modestly between 2000 and 2050. And international migration for the two countries is "set on the basis of past international migration estimates and an assessment of the policy stance of the countries with regard to future international migration flows" (United Nations 2003: Vol. I, 23).

In 2020 there are projected to be over 1 billion older persons in the world; almost one-quarter of them (about 240 million) will be in China, and over 7 percent in the U.S. By 2020, there will likely be almost 137 million oldest old people in the world, with7 more than 19 percent of them living in China, and over 9 percent in the U.S.

By the midway point of this new century (in 2050), there are projected to be nearly two billion older persons in the world out of a total population of 9.3 billion. Of these almost two billion older persons, 424 million of them (nearly 22 percent) will be residing in China, with 99 million (more than 5 percent) in the U.S.

This projected number of 418 million older persons in China in 2050 is a remarkably large number. The number of older persons alive in the world in 2000 (591 million) is only 167 million more than the total number of older persons projected to be living in China in 2050.

In the world in 2050, there are projected to be over 399 million oldest old people, about 26 percent of them living in China. The 93 oldest old projected to be living in China in 2050 is 1.4 times larger than the total number of 68 million oldest old living in the entire world in 2000.

The U.N. projected numbers of elderly and the oldest old in China in the next five decades (cited above) are rather conservative for two main reasons. First, the mortality assumption used in the "medium" projection assumes that the Chinese life expectancy slowly increases in 2050 up to a level 50 years later which is even somewhat lower than that in Japan today (United Nations 2003: xxii). This is rather conservative, given that further increases in life expectancy are likely to occur in the future. Second, the U.N. projection employs the classic model life table approach which assumes that the age pattern of changes in death rates in the future is the same as that observed in the past, namely, that death rates decline faster at younger ages than at older ages. This approach has led to implausible values (almost zero) of projected death rates at some young ages when mortality levels are in fact very low (Buttner 1999: 8; United Nations, Population Division 1998: 7–8; Lee and Carter 1992: 666). If one adopts a more realistic approach which assumes that changes in death rates at each age are proportional to the age-specific death rates, the projected number of oldest old in China in 2050 would be 114 million under the medium mortality assumption (Zeng and Vaupel 1989; Zeng and George, 2002).

This more realistic approach implies a faster decline of mortality at the advanced ages than at the young ages when the mortality level is low, and does not produce almost zero death rates at young ages. In most developed countries, the rate of reduction of death rates at the older ages has been accelerated and substantially higher than those at younger ages, especially since the 1970s (Kannisto et al. 1994; Vaupel and 1994: 303; Vaupel et al. 1998: 855). This pattern is likely to occur in China in the next few decades. Therefore, the number of oldest old in China around the middle of this century could well be significantly larger than 100 million.

We have presented a series of demographic accounts, actual and projected, of the total, older and oldest old populations of the world and China. China has emerged in the past century as the largest country in the world with the largest numbers of older and oldest old people. And the absolute and relative numbers of older and oldest old will increase even more so in the decades of the next century.

But China was not always a country with a large population and with a large number of elderly. In the next part of this chapter, we look at China's demographic past. We compare the growth of China with that of Europe for the past 600 years, and with that of the U.S. for the past 200 years. We then consider the major factor responsible for the very large current and projected numbers of elderly in China, namely, the fertility transition. We also consider the very heavy dependency burdens on the producing populations of China, and we show that the burdens will get even heavier in the decades ahead.

1.2 Population Growth over Time in Chine

Figure 1.1 charts the growth of the populations of China, Europe, and the U.S. from 1400 to the present, and beyond to 2025. (Population data are provided for the United States, starting in 1800.)

In 1400 the populations of China and Europe numbered 75 million and 60 million respectively. In 1600 their population figures were 150 and 100 million respectively. The population of China increased more rapidly than Europe during China's last dynasty, the Qing (1644–1911); but by the beginning of the last century Europe had grown to 390 million, and by 1950 to 515 million, largely a result of its demographic transition (from high to low levels of mortality and fertility) which began in the 1700s. China in the meantime had slackened its rate of growth so that by 1950, a year after Mao Zedong and the Chinese Communist Party established the People's Republic of China, China's population was only 35 million larger than that of Europe. In the meanwhile the U.S. grew from 5 million in 1800 to 158 million in 1950 (Durand 1967; Sun 1988; Banister 1992a; Lee and Wang 1999).

From 1950 to 2000 Europe grew by 213 million persons, while during the same period China grew by 693 million, for a net population gain in the 50 years that was 3.3 times larger than Europe's. In the same period the U.S. increased from 158 million to 285 million, about one-half of the increase of Europe.

Most of China's tremendous increases in the last half of the last century occurred between 1950 and 1980 and were due to reductions in mortality and to high fertility before the 1970s. The country introduced extensive fertility control programs in the 1970s causing the birth rate to plummet to below replacement levels in the 1990s.

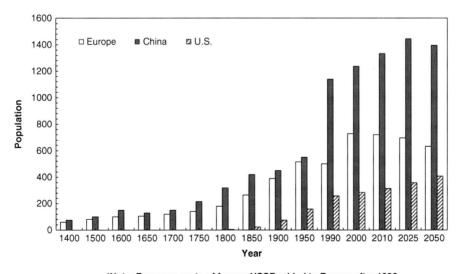

*Note: European parts of former USSR added to Europe after 1990

Fig. 1.1 Population size (in millions) of Europe*, China, and the U.S., 1400–2050

This very rapid fertility reduction occurring in a very large population is the main cause of the unprecedented high proportion of elderly people projected for China for the decades of the twenty-first century.

We focus now in more detail on China's demographic history and consider its population growth and change since the time of Christ. Of course, Chinese civilization began much earlier than the time of Christ, with the Xia Dynasty, the first dynasty of China, lasting from about the twenty-first century BC to the sixteenth century BC. But we do not have demographic records of the Chinese population in the centuries before Christ, other than an estimate of about 13 million at the start of the Xia Dynasty (Sun 1988: 9); a figure that is difficult to verify.

During the Han Dynasty, China took a population count in 2 AD, and it showed a population size of just under 60 million people (Banister 1992a).

Chinese population growth and declines over the 20 centuries since the time of Christ have almost always been associated with dynastic growth and decay. Typically, the beginning of a new dynasty was followed by a period of peace and order, cultural development and population growth. As population density increased, it often exceeded the availability of food and the struggle for existence was intensified. Then there would come a period of pestilence and famine and a consequent reduction in the size of the population.

Two thousand years of Chinese records and archives show that for all the centuries prior to the seventeenth century, China's population size increased and decreased at around 50–60 million. Indeed at the start of the Ming Dynasty (in 1368) the size of China's population was not much larger than it was at the time of Christ. For all the dynasties up until China's last dynasty, the Qing, China's population swayed roughly with the rise and fall of a dynasty (most dynasties reigned for about 200–300 years). The population grew during the initial years of the dynasty, but rarely exceeded 80 million. Population size would then fall so that one-third or sometimes one-half of the original population was decimated. Mortality then was too high to allow much of an increase in population.

To illustrate, from 1400 to 1500, the size of the Chinese population did not change appreciably, growing only by 25 million. It grew by another 50 million from 1500 to 1600 (see Fig. 1.1).

But since the mid-1700s after the establishment of the Qing Dynasty, there were ever slight reductions in mortality so that the population kept growing beyond the old limit of about 80 million. Indeed the Qing was the first dynasty to bring about and to maintain a population size much above 100 million. By 1850, there were over 420 million people in the country, six to eight times the traditional level (of 60–80 million) that was the demographic norm 200 years or so previously (Sun 1988; Banister 1992a).

The Qing was the only dynasty to live up to the perpetual Chinese ideal of "numerous descendants." It is indeed ironic that by achieving this ideal, not only was the Qing overthrown, but China's dynastic system of almost 4,000 years was eradicated. Previously, declines in population resulted in the collapse of the dynasties.

When Mao Zedong and the Chinese Communists took over the country in 1949, the population numbered about 550 million, a figure 30 percent higher than 100 years earlier. This 550 million was about ten times larger than China's historical equilibrium population (say around 60 million).

Furthermore, as a socialist regime, the People's Republic of China was founded on the premise that the past years of chaos, civil war, and political instability were now over. This resulted in a fresh impetus for population growth and led to levels of natural increase of a very high scale. Between 1950 and 1980, China added about another 433 million to its population (see Fig. 1.2). The population size of close to 1.3 billion in 2000 is almost ten times the size of the mid-seventeenth century population of around 130 million.

Here is another comparison. In the 300 years from 1650 to 1950, the average annual increase of China's population was around 1.5–1.6 million; this is 80–100 times greater than the annual increases in China before 1650. Since 1949, the average annual increase in population climbed to around 14 million per year. In the 1960s—the high growth years—the average annual increase was as high as 22 million. Even with the reduction of the birth rate in the 1970s, the average annual population increases have been nearly 20 million (see Fig. 1.2).

In ancient China, an increase of 14–22 million people took 700–1,000 years. In the 1960s, 1970s, and 1980s a population increase of between 14 and 22 million occurred almost every year.

Population size in China is projected to start tapering off in the second decade of this new century at about 1.4 billion. The United Nations (2003) projects that China will reach its largest population size of about 1.45 billion in 2030 (see Fig. 1.2), and will then begin to decline in size, falling back to almost 1.4 billion in the year 2050.

Since the early 1970s, China has experienced a pronounced and rapid decline in its fertility, to a level reached in 1992 of 1.8, and to less than 1.6 in 2005. This transition is one of the most dramatic fertility reductions in the historical demographic record.

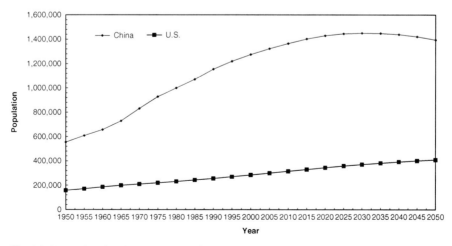

Fig. 1.2 Population size (in thousands) in China and the United States, 1950–2050

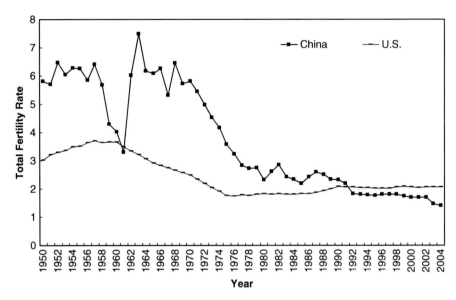

Fig. 1.3 Total fertility rates, China, and U.S.: 1950–2004

Figure 1.3 presents total fertility rates data for China from 1950 to 2004. China's TFR in 1950 was 5.8. It declined to 3.3 in 1961, then increased to 7.5 in 1963, fluctuated throughout the mid- and late 1960s, and has been rapidly falling ever since. It dropped below 3 in 1977, and fell by another half child by 1985; it dropped below 2 in 1992, and then to 1.7 in the late 1990s, and even lower in the early years of this new century.

The fertility decline from the late 1950s through the early 1960s is thought to have resulted from the "national hard times" (Chen 1984: 45) and famine experienced in China during and immediately after the Great Leap Forward. Coale (1984: 57) has written that in addition to "famine-induced subfecundity," the fertility decline was also due to the "disruption of normal married life."

During the 2 years from 1961 to 1963, the TFR increased markedly, from 3.3 to 7.5. According to Chen (1984: 45), this increase occurred in conjunction with the economic recovery in China. Coale (1984: 57) has added that the increase also "resulted from the restoration of normal married life, from an abnormally large number of marriages, and from the unusually small fraction of married women who were infertile." These years of the early to mid-1960s were the period of China's "baby boom." China experienced in the early 1960s the same kinds of growth that occurred in the U.S. after World War II. China's was of a shorter duration but of a significantly higher magnitude. China's TFR peaked at 7.5 in 1963. At the height of the U.S. "baby boom" in 1957, the TFR was 3.7.

Fertility began a sustained decline in 1968 through 1980. The Chinese demographer Chen (1984: 45) observed that since the 1970s, "family planning work has been widely carried out, and the total fertility rate has steeply declined year by year."

1 Introduction: Aging and Aged Dependency in China

Western demographers agree that this decline in the 1970s "was almost certainly the result of an increase in deliberate (fertility) control" (Coale 1984: 58; Mauldin 1982; Bongaarts and Greenhalgh 1985; Wolf 1986). There is also evidence showing significant variability in fertility among China's subregions today and in past decades. Those areas that have experienced the greatest reductions in fertility have had more "profound changes in socioeconomic structure" (Tien 1984: 385; also see, Birdsall and Jamison 1983; Poston and Gu 1987; Poston and Jia 1990; Poston 2000).

The somewhat lower TFRs in the early and mid-1980s have been viewed by some as due to a stringent implementation of China's population and family planning policies, particularly the "one child per couple" policy (Tsuya and Choe 1988; Tien 1989).

The total fertility rate then increased in the late 1980s to 2.6 in 1987. This increment was due in part to a relaxation of China's "one child per couple" policy. China was allowing more couples, particularly those in the rural areas, to have a second child. But the policies from the late 1980s to the 1990s have been sometimes implemented more stringently and sometimes less stringently, resulting in slight increases in the late 1980s, and then the decreases in the 1990s, leading to the current very low TFR.

These trends in fertility in the country as a whole are averages. There are vast differences in the fertility trends in urban compared to rural areas (Poston 1992). By the early 1970s, there was as much as a three child difference between the fertility rates of urban and rural women. Even in 1981, for instance, rural fertility was more than 100 percent greater than urban fertility.

Family planning programs have been well established in China since the beginning of the first family planning campaign in 1956. However, it was not until the establishment of the campaign in 1971 that fertility reduction became a national priority. Prior to that time, the role of the government in family planning had been at best intermittent.

In summary, the large population base and the high birth rates in the 1950s and 1960s accompanied with fast declines in mortality since 1950 have resulted in the huge number of elderly persons in China today and in the next few decades. It is the dramatic decline in fertility in the 1970s and 1980s that has produced in the 1990s and in the decades of this new century such a high percent of elderly people. Since 1992 China's TFR has been below the replacement level of 2.1. Birth cohorts are considerably smaller these days than a few decades ago. This much lower fertility, coupled with significant gains in longevity, both occurring in a country with a very large population base, have produced in China one of the most serious challenges ever, namely, the fastest absolute growth and percentage growth of the elderly ever witnessed in one country in human history.

1.3 Aged Dependency in China

A large number of elderly persons in a population is not problematic if there exists at the same time in the population a large number of producers. It is only when the ratio of elderly to producers becomes high that a host of economic, social, and related problems occur. In this section, we show empirically the degree currently of the dependency burden in China, and how much worse this burden will become in the years ahead.

We measure the total dependency in the populations of China, using the ratio of persons aged 0–14 and persons aged 65 and over to persons aged 15–64. The numerator consists of persons who typically are not employed, hence not serving actively as producers of goods, material resources, and sustenance. The denominator, persons aged 15–64, contains the productive members of the population, many of whom are in the labor force, all of whom, in varying ways, are producing foodstuffs, and related goods and services for the population; this ratio is multiplied by 100 and refers to the number of dependents in the population per 100 producers.

The total dependency ratio (TDR) may be subdivided into a youth dependency ratio (YDR), i.e., persons 0–14 divided by persons 15–64; and an aged dependency ratio (ADR), i.e., persons 65 + divided by persons 15–64.

Figure 1.4 presents the youth dependency ratios (YDR) for China and the U.S. for every 5 years from 1950 to 2050. Between 1950 and 2000, the YDRs have dropped for both countries; in 1950 the YDRs were 54 in China and 41 in the U.S. By 2000 they had declined to 36 in China and 33 in the U.S. Note also the increases in the YDRs for both countries during their baby boom years in the 1950s and 1960s. The YDRs of the two countries attained in 2000 are rather modest compared to those of other countries. To illustrate, the Gaza Strip had a YDR in 1995 of 114, the highest in the world, followed by Uganda (99), Ethiopia (97), and Libya (87).

The YDRs in China and the U.S. are not projected to change significantly between 2000 and 2050. The data in Fig. 1.4 indicate that in China the YDR will drop from 36 in 2000 to 26 in 2050. The United States is projected to have a higher YDR than that of China starting in the year of 2005, a situation that would not have been predicted, say back in 1965, when China's YDR was 72 and that of the U.S. was 50.

Figure 1.5 shows aged dependency ratios for China and the U.S. for every 5 years from 1950 to 2050. Unlike the situation with respect to the YDR in which both the

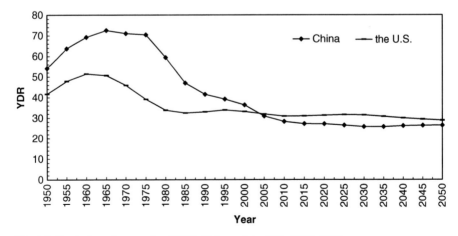

Fig. 1.4 Youth dependency ratios (YDR), China and the U.S., 1950–2050

1 Introduction: Aging and Aged Dependency in China

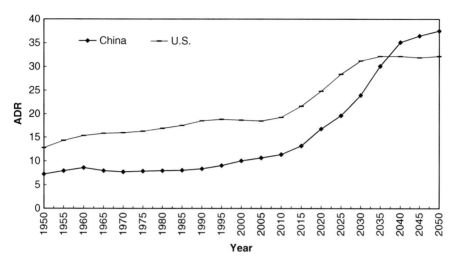

Fig. 1.5 Aged dependency ratios (ADRs), China and the U.S., 1950–2050

U.S. and China experienced major net decreases between 1950 and 2000, there have been modest increases in their ADRs. In China the ADR increased from 7 in 1950 to 10 in 2000. China's ADR in 2000 is a little higher than the world average, but not appreciably so. This is not an excessively high ADR.

To illustrate using different data that reflect the same phenomenon, in 2000 one-tenth of China's population was over age 60. By comparison, in 2000, ten countries, all in Europe, had more than 20 percent of their populations of the age 60+: Sweden had the highest percentage (22 percent), followed by Norway, Belgium, Italy, United Kingdom, Germany, Austria, Greece, Denmark, and Switzerland (United Nations 2003).

The aged dependency situation changes remarkably when we skip ahead 50 years to 2050. China will have become much older by 2050. China's ADR is projected to increase from 10 aged dependents per 100 producers in 2000 to 37 aged dependents per 100 producers in 2050 (Fig. 1.5).

By 2050 China is expected to have made the transition to a demographically very old country. In the 50 years since 2000, both China and the U.S. will have become demographically top-heavy. In 2050, 30 percent of China's population (over 418 million people) will be 60 years of age or older. The oldest countries in the world today, the European countries mentioned earlier, are nowhere near as old as China is projected to be in 2050.

Another way to consider changes in the age distribution of a population is with the parent support ratio (PSR), which takes the number of persons 80 years old and over, per 100 persons aged 50–64 (Wu and Wang 2004). The PSR is an indication of the relative burden of the oldest old population, i.e., the oldest old parents, on the population aged 50–64, i.e., the children of the oldest old parents. The PSR is an empirical ratio which is meant to reflect the degree of burden the oldest old have on their mid-age or younger elderly children. However, it does not take into account specifically and directly the actual burden. That is, the financial and emotional costs of caring for oldest old parents are not measured with the PSR. But the implication is that they will increase in a society with increases in the PSR.

Figure 1.6 presents parent support ratios (PSRs) for every 5 years from 1950 to 2050 for China and the United States. In 1950 both countries had very low PSRs. In China in 1950 there were less than 3 persons of age 80 and over per 100 persons aged 50–64. There was not much of an oldest old parent burden on the older children in either country in 1950.

By 2000 the parent support ratios had increased three-fold in both countries, for China from 2.5 in 1950 to 7.8 in 2000. The burden of oldest old parents on their children has increased in both countries, and the burden is almost three times as heavy in the U.S. as in China. In the U.S. in 2000 there were more than 21 elderly aged 80+ for every 100 persons aged 50–64, whereas in China the PSR was less than 8 persons aged 80+ per 100 aged 50–64.

Both countries are projected to experience even more dramatic increases by 2050. The PSR is expected to increase almost five-fold in China from 7.8 in 2000 to 35.5 in 2050. The PSR is projected to double in the U.S. from 21.4 in 2000 to 41.7 in 2050. The burden of oldest old parents on their children will be extremely high in 2050 in both countries. There are projected to be in the U.S. almost 42 persons aged 80+ for every 100 persons aged 50–64, whereas in China there are projected to be almost 36 persons aged 80 and over for every 100 persons aged 50–64. The PSR figure for China is projected to be slightly less than that for the U.S. But the change in PSRs between 2000 and 2050 for China will be much greater than for the U.S.

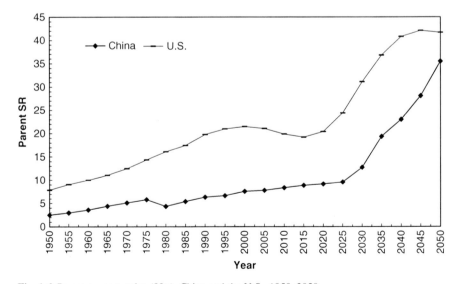

Fig. 1.6 Parent support ratios (80+), China and the U.S., 1950–2050

By 2050 China will have grown to be one of the oldest populations in the world, and a country with one of the heaviest aged dependency burdens of any population in the world. In the concluding section of this chapter we consider some of the implications of these findings.

1.4 Conclusion

The current and projected levels of the elderly population in China show that by the mid-point of the twenty-first century, China is projected to have become one of the oldest populations in the world with one of the heaviest old age dependency burdens. Two implications worthy of consideration are the transitions in the organization of eldercare by the family, and nuptiality.

Traditionally in China, the support of one's elder parents has been the responsibility of the sons. Often the parents lived with the oldest son, and either with or nearby the other sons. The eldest son and his brothers tended to be responsible for providing the parents with economic support. The sons would rely on one of their sisters, or sometimes on one or more of their wives to provide their parents with emotional support. These norms have been adjusted or modified in past decades, especially since the founding of the People's Republic in 1949, and particularly among urban residents. Nevertheless, the provision of economic and emotional support to one's parents has seldom been a major burden. As one might expect in a population with modestly high levels of fertility, there have usually been many more producers in the Chinese population than aged dependents. Traditionally these kinds of norms have not governed the U.S. family to the extent they have in China.

However, given in contemporary China the very low levels of fertility, as well as, since 1981, a highly unbalanced sex ratio at birth (Poston et al. 1997; Poston and Glover 2005; Poston and Morrison 2005), the provision of eldercare will be a major concern in the years of this new century. For one thing, as we have already noted, in the decades of this new century in China there are projected to be many more aged dependents per producers. In 2000, there were 10 aged dependents per 100 producers in China; but by 2050 in China there are projected to be 37 aged dependents per 100 producers; this is an astonishingly high number of old persons per 100 producing members in the population. The number of aged dependents per 100 producers in China in 2050 is projected to be 3.7 times larger than China's current number. This aged dependency ratio for 2050 will likely be one of the highest of any country in the world.

When we couple this very high ADR for the year 2050 for China with the abnormally high sex ratios at birth (SRBs) in China, the issue of eldercare provision in this new millennium becomes even more complex. According to data from the 2000 Census of China, the SRB in 2000 was near 120; it was 151 for second-order births, and 159 for third-order births. It has been estimated that based only on the births that have occurred in China between 1980 and 2001, there have already been born more than 23 million boys who will be unable to find Chinese brides when they reach marriageable age (Poston and Glover 2005; Poston and Morrison 2005). If these abnormally high SRBs continue into the decades of the twenty-first century (and there is no indication that they will not), this will mean that an enormously large number of Chinese males, perhaps near 100 million or more, will find it difficult, if not impossible, to meet females to marry. These single males will have the responsibility for providing both the economic and the emotional support for their parents.

In China in the decades of this new century, there will be a very large number of unmarried sons caring both economically and emotionally for their aged parents. There have already been born over 23 million such sons who will be unable to find brides to marry.

Unlike the case in past decades in China where there have usually been several married sons, along with their sisters, available to care for the elderly parents, the situation in the next 30–40 years will be different: there will be many, many more elderly—parents and aunts and uncles—requiring care, than has been the situation in the recent past. Moreover, many of the providers will be sons, perhaps only-born sons, without wives. The care of the elderly in the decades of this new century will not be without problems.

There are a couple of extenuating circumstances that could modify this rather negative prognosis and provide an alternative to male celibacy. The immigration to China of Chinese women from Hong Kong, Singapore, and Taiwan would of course enlarge the pool of wives; there is already some marriage migration of this form underway in China, but it would need to increase in magnitude (Davin 1998; Fan and Huang 1998). Polyandry is another possibility (see Cassidy and Lee 1989), and there is some evidence of its existence among some of China's minority populations (Zhang 1997; Johnson and Zhang 1991). Marriage market migration may be more likely than polyandry.

Is China in any position today to reduce its projected imbalance of elderly? The answer is yes. However, one of the better strategies for reducing the high percentage of elderly in future years is not a popular one among doctrinaire Chinese leaders (Poston and Duan 2000). It involves adjusting the country's so-called "one-child" fertility policy. The current "one-child" fertility policy requests about 63 percent of Chinese couples to have one child only.[2] Relaxing this policy and allowing between now and the year 2030 urban and rural fertility rates to increase would have a sizable impact (Zeng 2006, 2007); such a change would likely drop the aged dependency ratio and the percentage of the population aged 60+ by between 20 and 40 percent (see Banister (1992b) for another scenario). It would also likely reduce China's sex ratios at birth to more balanced levels. As Banister (1992b: 472) has noted, "China's population will age under any likely scenario, but the lower the fertility level, the more severe the aging."

---

[2] Based on 2000 census data and statistics of fertility policy regulations at the city and prefecture levels, Guo (2005) has shown that the one-child, 1.5-child, two-child, and three-child policy have been implemented among about 35.9, 52.9, 9.6, and 1.6 percent, respectively, of the population of China.

The analysis presented here of the demographic determinants in China portrays a situation that rings with irony. China solved its burgeoning fertility problem with an induced fertility transition, which was one of the most successful fertility reductions experienced by any country in the world. But it is the very success of this transition that has exacerbated the problem. Chinese fertility policies have limited the size of birth cohorts relative to their elders to a degree unprecedented in the less developed countries. And the very speed of the fertility transition has given China little time to evolve a non-familial old age support system to replace the traditional family. There is a growing awareness of such problems these days in China (Poston and Duan 2000). Although they were not foreseen at the inception of the fertility policies in the 1970s, debates about, and discussions of them are now underway.

It may well be that China will be forced to move to a country-wide "two-child & late-childbearing" policy in another decade or so, or later, in about 25 or 30 years, allowing all Chinese citizens to choose their family size and fertility timing (Zeng 2006, 2007). This would occur after the present "baby-boom" generation is beyond the childbearing years. Such a strategy would reduce significantly the percent of elderly shown in the projections presented earlier in this chapter, and the severity of the concomitant social problems.

However, fertility policy adjustments can only moderately reduce the percentages of the elderly as compared to keeping the current policy unchanged. But it cannot reduce the magnitude of the huge numbers of elderly and oldest old already born. Even if China moved to a two-child and late-childbearing policy in the next decade and allowed all couples to freely choose their family size and fertility timing in the next three decades, as is the case in other countries, the percentage of elderly in China in 2050 will still be around 24 percent and the magnitude of the huge numbers of elderly and oldest old persons will be at least as large as discussed earlier (Zeng 2006, 2007). This is simply because the elderly today and in the next 65 years have already been born. The un-reversible trend of declining mortality, especially at the older ages, will result in even substantially more elderly and oldest old persons. Thus, it is strategically important to gain a better understanding of healthy aging and healthy longevity so to be able to reduce the negative impacts of rapid aging. This is a main reason for publishing this book.

A fundamental question in aging studies is whether the mortality decline will be accompanied by a compression or an expansion of the period of disability and morbidity among the elderly population. The answer will determine the quality of life not only for the elderly but for all members of the society. The hypothesis of morbidity compression among the elderly was initially proposed by Fries (1980). The dynamic equilibrium hypothesis introduced by Manton (1982) assumes that the slowing down of the pace of progression of morbidity leads to an increase in the prevalence of light and moderate (but not severe) disability as mortality falls among the oldest old. The hypothesis of a pandemic of disability, however states that a decrease in the fatality rate at the oldest old ages leads to a significant increase in the prevalence of morbidity (Gruenberg 1977). In a careful evaluation of data from various national surveys, Freedman et al. (2002) have concluded that several measures of old-age disability and limitation show that improvements have occurred in recent decades in the U.S. And similar trends have been found in European populations (Bebbington 1988; Robine and Mormiche 1994; Van Oyen and Roelands 1994).

Is the period of disability among the elderly compressing or expanding with increasing life expectancy in developing countries in general, and in China in particular? What are the factors associated with these trends in the past? Is it possible in the future to realize morbidity compression or at least the dynamic equilibrium among elderly with a prolongation of the life span?

In general, there were no data resources that could be used to address these important questions before the initiation of the Chinese Longitudinal Healthy Longevity Survey (CLHLS); most of the chapters in this book use these new and important data. Motivated by the need to study factors associated with healthy aging and healthy longevity, this book aims to shed new light on the understanding of the population aging process; it seeks as well to introduce possible approaches to reduce the period of disability and morbidity while prolonging the life span, i.e., realizing the goals of healthy aging and healthy longevity.

The next chapter presents the CLHLS project's background and significance, objectives, study design, sample distribution, contents of data collected and research opportunities. These discussions are followed by chapters focusing on assessments and discussions from different perspectives and viewpoints of age reporting and data quality of the CLHLS. The remaining chapters are grouped into sections dealing with demographic, social, economic and psychological dimensions of healthy longevity in China. The last chapter (Epilogue) briefly summarizes the striking findings from CLHLS studies published elsewhere and outlines the future research agenda of CLHLS project.

Our goal in this volume is to contribute to the development of scientific knowledge related to morbidity compression, as well as the dynamics of the "slowing down" of the pace of progression of morbidity with mortality declines at the older ages. We hope that our book will provide knowledge that will be instrumental in part in the ultimate realization of the long-term dream of healthy longevity, that is, not only living longer, but living a healthier life.

References

Banister, J. (1992a), A brief history of China's population. In: D.L. Poston and D. Yaukey (eds.): *The population of modern China*. New York: Plenum, pp. 51–57
Banister, J. (1992b), Implications of the aging of China's population. In: D.L. Poston and D. Yaukey (eds.): *The population of modern China*. New York: Plenum, pp. 463–490
Bebbington, A.C. (1988), The expectation of life without disability in England and Wales. *Social Sciences and Medicine* 27, pp. 321–326
Birdsall, N. and D.T. Jamison (1983), Income and other factors influencing fertility in China. *Population and Development Review* 9, pp. 651–675
Bongaarts, J. and S. Greenhalgh (1985), An alternative to the one-child policy in China. *Population and Development Review* 11, pp. 585–618
Buttner, T. (1999), Approaches and experiences in projecting mortality patterns for the oldest old in low mortality countries. Working Paper No. 31. Statistical Commission and Economic

Commission for Europe. Conference of European Statisticians. Joint ECE-Eurostat Work Session on Demographic Projections.

Cassidy, M.L. and G.R. Lee (1989), The study of polyandry: A critique and synthesis. *Journal of Comparative Family Studies* 20, pp. 1–11

Chen, S. (1984), Fertility of women during the 42-year period from 1940 to 1981. In: China Population Information Center: *Analysis on China's national one-per-thousand sampling survey.* Beijing: China Population Information Center, pp. 32–58 (in Chinese)

Coale, A.J. (1984), *Rapid population change in China, 1952–1982*. Washington, DC: National Academy Press.

Davin, D. (1998), *Internal migration in contemporary China*. Basingtone, England: MacMillan Press.

Durand, J.D. (1967), The modern expansion of world population. *Proceedings of the American Philosophical Society* 3, pp. 137–140

Fan, C.C. and Y. Huang (1998), Waves of rural brides: Female marriage migration in China. *Annals of the Association of American Geographers* 88, pp. 227–251

Freedman, V.A., L.G. Martin, and R.F. Schoeni (2002), Recent trends in disability and functioning among older adults in the United States: A systematic review. *Journal of the American Medical Association* 288, pp. 3137–3146

Fries, J.F. (1980), Aging, natural death and the compression of morbidity. *New England Journal of Medicine* 303, pp. 130–135

Gruenberg, E.M. (1977), The failures of success. *Milbank Memorial Fund Quarterly* 55, pp. 3–24

Guo, Z. (2005), An analysis on fertility by regions implementing different policies in China based on the 2000 census data. In: *Reports of the research projects on 2000 census data analysis.* Beijing: The 2000 Census Office, National Bureau of Statistics of China (in Chinese)

Guo, Z., E. Zhang, B. Gu, and F. Wang (2003), Diversity of China's fertility policy by policy fertility. *Population Research* 27, pp. 1–10 (in Chinese)

Johnson, N.E. and K. Zhang (1991), Matriarchy, polyandry, and fertility amongst the Mosuos in China. *Journal of Biosocial Science* 23, pp. 499–505

Kannisto, V., J. Lauritsen, A.R. Thatcher, J.W. Vaupel (1994), Reductions in mortality at advanced ages: Several decades of evidence from 27 countries. *Population and Development Review* 20, pp. 793–810

Lee, R.D. and L. Carter (1992), Modeling and forecasting the time series of U.S. mortality. *Journal of the American Statistical Association* 76, pp. 674–675

Lee, K.Z. and F. Wang (1999), *One quarter of humanity: Malthusian mythology and Chinese realities*. Cambridge, MA: Harvard University Press.

Manton, K.G. (1982), Changing concepts of morbidity and mortality in the elderly population. *Milbank Memorial Fund Quarterly* 60, pp. 183–244

Mauldin, W.P. (1982), The determinants of fertility decline in developing countries: An overview of the available empirical evidence. *International Family Planning Perspectives* 8, pp. 119–127

Population Reference Bureau (2006), *World population data sheet, 2005*. Washington, DC: Population Reference Bureau.

Poston, D.L., Jr. (1992), Fertility trends in China. In: D.L. Poston, Jr. and D. Yaukey (eds.): *The population of modern China*. New York: Plenum, pp. 277–285

Poston, D.L. Jr. (2000), Social and economic development and the fertility transitions in mainland China and Taiwan. *Population and Development Review* 26 (supplement), pp. 40–60

Poston, D.L. Jr. and C.C. Duan (2000), The current and projected distribution of the elderly and eldercare in the People's Republic of China. *Journal of Family Issues* 21, pp. 714–732

Poston, D.L. Jr. and K.S. Glover (2005), Too many males: Marriage market implications of gender imbalances in China. *Genus* LXI (2), pp. 119–140

Poston, D.L. Jr. and B. Gu (1987), Socioeconomic development, family planning, and fertility in China. *Demography* 24, pp. 531–551

Poston, D.L. Jr. and Z. Jia (1990), Socioeconomic structure and fertility in China: A county level investigation. *Journal of Biosocial Science* 22, pp. 507–515

Poston, D.L. Jr. and P.A. Morrison (2005), China: Bachelor bomb. *International Herald Tribune* (September 14), p. 10

Poston, D.L. Jr. B. Gu, P. Liu, and T. McDaniel (1997), Son preference and the sex ratio at birth in China. *Social Biology* 44, pp. 55–76

Robine, J.M. and P. Mormiche (1994), Estimation de la valeur de l'esperance de vie sans incapacite en France en 1991 et elaboration de series chronologiques. *Solidarite Sante* 1 pp. 17–36 (in French)

Sun, J. (1988), The Chinese population: Its size and growth. In: China Financial and Economic Publishing House: *New China's population*. New York: Macmillan, pp. 9–14

Tien, H.Y. (1984), Induced fertility transition: Impact of population planning and socioeconomic change in the People's Republic of China. *Population Studies* 38, pp. 385–400

Tien, H.Y. (1989), Second thoughts on the second child: A talk with Peng Peiyun. *Population Today* 17 (April), pp. 6–9

Tsuya, N.O. and M.K. Choe (1988), Achievement of one-child fertility in rural areas of Jilin province, China. *International Family Planning Perspectives* 14, pp. 122–130

United Nations (2003), *World population prospects: The 2002 revision. Volume Two*. New York: United Nations.

United Nations, Population Division (1998), Extending population projections to age 100. Statement prepared by the Population Division, U.N., ACC Subcommittee on Demographic Estimates and Projections (ACC/SCDEP), Twentieth session, 23–25 June 1998. New York: United Nations.

Van Oyen, H. and M. Roelands (1994), New calculations: Estimates of health expectancy in Belgium. In: C.D. Mathers, J. McCallum, and J.M. Robine (eds.): *Advances in health expectancies*. Canberra: Australian Institute of Health and Welfare, pp. 213–223

Vaupel, J.W. and H. Lundstrom (1994), The future of mortality at older ages in developed countries. In: Wolfgang Lutz (eds.): *The future population of the world: What can we assume today?* London: Earthscan Publications, pp. 295–315

Vaupel, J.W., J.R. Carey, K. Christensen, T.E. Johnson, A.I. Yashin, N.V. Holm, I.A. Iachine, V. Kannisto, A.A. Khazaeli, P. Liedo, V.D. Longo, Y. Zeng, K.G. Manton, and J.W. Curtsinger (1998), Biodemographic trajectories of longevity. *Science* 280, pp. 855–860

Velkoff, V.A. and V.A. Lawson (1998), Gender and aging caregiving. *International Brief*, IB/98-3 (December). Washington, DC; U.S. Bureau of the Census.

Wolf, A.P. (1986), The preeminent role of government intervention in China's family revolution. *Population and Development Review* 12, pp. 101–116

Wu, C., and L. Wang (2004), Contribution of population control in creating opportunities for China arising from fertility decline should not be neglected. Presented at the International Symposium on the 2000 Population Census of China, Beijing, April.

Zeng, Y. (2006), Soft-landing of the two-child & late-childbearing policy in China. *China Social Sciences* 2, pp. 93–109

Zeng, Y. (2007), Options of fertility policy transition in China. *Population and Development Review*. Vol. 33 No. 2, pp. 215–246

Zeng, Y. and L. George (2002), Extremely rapid aging and the living arrangement of elderly persons: The case of China. In: *Living arrangements of older persons, population bulletin of the United Nations, Special Issue No. 42/43*. New York: United Nations

Zeng, Y. and J. Vaupel (1989), Impact of urbanization and delayed childbearing on population growth and aging in China. *Population and Development Review* 15, pp. 425–445

Zhang, T. (1997), Marriage and family patterns in Tibet. *China Population Today* 14, pp. 9–12

Part I
CLHLS and its Data Quality Assessment

Introduction *by Dudley L. Poston, Jr.*

The first section of this book deals specifically with the Chinese Longitudinal Healthy Longevity Survey (CLHLS). Most of the chapters focus on an assessment of its data quality. The one chapter that does not deal directly with issues of data quality is Zeng Yi's Chapter 2. In his chapter, Zeng introduces readers to the Chinese Longitudinal Healthy Longevity Survey (CLHLS), the data set used in almost all the analyses contained in the chapters in this book. His chapter includes the general goals and the specific objectives of the CLHLS, along with its organizational framework, study design, sample distribution and contents of the data collected. He also compares the CLHLS with other surveys that focus on elderly populations. The large population size, the focus on healthy longevity (rather than on a specific disease or disorder), the simultaneous consideration of various risk factors, and the use of analytical strategies based on demographic concepts, all make the CLHLS an innovative demographic data collection and research project.

The data collected in the CLHLS are extremely comprehensive and are able to illuminate a multitude of characteristics and conditions of the Chinese elderly respondents. Zeng writes that the questionnaire design was based on international standards and adapted to the Chinese cultural/social context and carefully tested by pilot studies/interviews. Questions were used that would shed light on risk factors for mortality and healthy longevity. Similarly, questions were not used that could not be reliably answered by the oldest-old because, perhaps of the respondents' low levels of education and/or poor hearing and vision. The data collected include items of family structure, living arrangements and proximity to children, self-rated health, self-evaluation on life satisfaction, chronic disease, medical care, social activities, diet, smoking and alcohol drinking, psychological characteristics, economic resources, caregiver and family support, nutrition and some health-related conditions in early life (childhood, adulthood, and around age 60). The CLHLS is surely the most complete survey of elderly ever undertaken in China, and rivals any such surveys ever conducted anywhere in the world.

D.L. Poston, Jr.
Department of Sociology, Texas A&M University, College Station, TX 77843, USA
e-mail: dudleyposton@yahoo.com

Chapters 3–6 present systematic and relatively detailed assessments of the age reporting and data quality of the CLHLS. In Chap. 3 Danan Gu undertakes a comprehensive review of the data quality of the third wave of the CLHLS (2002) in terms of proxy use, nonresponse rate, sample attrition, and reliability and validity of the major health measures. The CLHLS gathered extensive questionnaire data for 16,020 elderly aged 65+ in 2002. As noted in Zeng's discussion in Chap. 2, the survey was conducted in randomly selected counties and cities in 22 of China's provinces. The 2002 wave extended the age range of the sampled elderly to include the age range from 65 to 79 who were not included in the first two waves in 1998 and 2000. Among the 16,020 sample persons in 2002, 3,189 were centenarians, 3,747 were nonagenarians, 4,239 were octogenarians, and 4,845 were aged 65–79.

Gu writes that there are three major potential sources of error due to nonobservation (coverage error, nonresponse error, and sample error) and four major potential sources of errors due to observation or measurement (the interviewer, the respondent, the questionnaire, and the mode of interview). His chapter addresses these issues systematically and comprehensively in an attempt to assess the data quality of the CLHLS in 2002.

Gu first addresses the use of proxies. When they were used, about 90 percent of them were close relatives such as a spouse, children, and grandchildren. Multivariate analysis indicated that respondents with an older age, lower education, rural residence, lower cognitive functioning, and higher disability were more likely to use a proxy. It is not believed that proxy use resulted in significant bias or error. Gu then turns to item nonresponse. He notes that the CLHLS survey has encountered a small amount of nonresponse, both unit and item nonresponse. Two general approaches are recommended to compensate for these nonresponses, namely, weight adjustment and imputation. Gu addresses each of these approaches and suggests ways for their implementation.

He then turns to the very important issue of sample attrition. Of the 11,162 individuals interviewed in 2000, 56.3 percent were still alive at the 2002 wave, almost 30 percent died before the 2002 interview, and almost 14 percent were lost. Gu discusses possible reasons for this attrition and ways to take it into account in analyses of the data. Regarding the issue of inconsistent response, Gu notes that his analyses indicates this is not a major concern, and that it tends to increase slightly with age.

Gu next discusses issues of reliability of the major health measures. In the CLHLS, health has been conceptualized in a multidimensional manner, with a general emphasis on its physical and mental domains. Tests of the internal consistency for various measures (using Cronbach's alpha) are all satisfactory and most are more than satisfactory. Issues of homogeneity of the scales are also examined, with satisfactory results. Finally, Gu addresses several issues of validity, including content, construct, and criterion validity. He concludes his chapter with the observation that the evidence leads him to believe that the data quality of the 2002 wave of the CLHLS is generally good. We concur in this evaluation.

In Chap. 4, Zeng and Gu evaluate age reporting among the oldest-old, especially centenarians, in the CLHLS, using various indices of elderly age reporting and age distributions of centenarians in Sweden, Japan, England and Wales, Australia,

Canada, China, the USA, and Chile. Their analyses show that age reporting among the oldest-old interviewees (Han and six minority groups combined) in the CLHLS provinces is not quite as good as that in Sweden, Japan, and England and Wales, but is relatively close to that in Australia, more or less the same as that in Canada, better than that in the USA, and much better than that in Chile. As indicated by the higher density of centenarians, age exaggeration tends to exist among the six ethnic minority groups in the Han-dominated provinces. However, the authors cannot rule out and quantify the potential impacts of past mortality selection and better natural environmental conditions among these minority groups. They find that the age exaggeration of minorities in the CLHLS will not cause substantial biases in demographic and statistical analyses using the CLHLS data, since minorities comprise a rather small portion of the sample (6.8 percent in the baseline and 5.5 percent in the grand total sample of the 1998, 2000, and 2002 waves).

In Chap. 5, Booth and Zhao examine age reporting among the CLHLS respondents, using the first round of data collected in 1998. They state that the sample design tends to limit the use of traditional methods for assessing the accuracy of age reporting; innovative methods are thus adopted. They find that only the sample aged 100+ is fully representative of the population at that age. The age structure of centenarians is compared with populations with good age reporting, demonstrating some degree of age exaggeration. At ages 80+, their constructed estimates of age at childbearing show some age exaggeration, particularly in Guangxi and among several ethnic minorities. They conclude that some age exaggeration tends to increase with age. The findings of Booth and Zhao indicate the importance of continual examination of the quality of the age data.

In the final chapter in Section I, Gu and Dupre investigate a series of issues pertaining to mortality and morbidity in the 1998, 2000 and 2002 CLHLS. Among them are the accuracy of the mortality data recorded in the CLHLS, and the accuracy of chronic morbidity data reported in these first three waves. They also examine the accuracy of the mortality data for the non-Han nationalities.

In the first part of their chapter, the authors undertake an extensive evaluation of the CLHLS and census data and conclude that the mortality rates from the first three waves of the CLHLS are relatively reliable, with the exception of some recall errors by proxies in the respondents' date of death. Gu and Dupre correctly note that nearly 7 percent of the CLHLS baseline sample is comprised of non-Han people. Because ethnicity is a major sociodemographic factor in the study of aging, the accuracy of non-Han mortality data is very important to assess. If age reporting among the non-Han sample is shown to be inaccurate, it might be necessary to drop the non-Han people from the analysis. They thus undertake a detailed assessment of the accuracy of the non-Han mortality data, particularly with respect to age exaggeration and its impact on mortality estimates when age reports are not accurate. Their detailed statistical analyses show that even under the assumption of serious age exaggeration among the ethnic minorities, the impact on the estimation of overall mortality in the CLHLS is really quite negligible.

Gu and Dupre also show that possible age exaggeration among minority oldest-old in the CLHLS does not significantly affect the outcomes for estimates of indicators such as ADL, MMSE, self-reported health, as well as the estimates of other covariates in the multivariate statistical models using either cross-sectional or longitudinal CLHLS data.

The authors do, however, report that although overall mortality rates in the CLHLS during the period of 2000–2002 are reliable, observed mortality rates from the 1998 to 2000 CLHLS data were underestimated by 15–20 percent and by 5–20 percent for the regression-based rates before age 90; this, they note, is likely a potential consequence of nonrandom missing data. However, they believe that the impact of underestimating mortality is minor on the effects of predictors in statistical models. They suggest that researchers using the CLHLS data may wish to employ data from the interviews in 2000 and later or use pooled data when conducting survival analyses to produce more robust estimates. They note that if the distinction of first- and second-year mortality is needed for specific research designs, the results should be interpreted with caution because the first-year mortality rates tend to suffer from underestimation due to some recall bias by next-of-kin. Gu and Dupre evaluate several other issues of accuracy and reliability in the mortality and morbidity data in the CLHLS. Their analyses are extensive, and their recommendations for data users are worthy of attention.

In sum, all but one of the chapters in Section I are concerned in varying ways with the evaluation of CLHLS data. Zeng's Chap. 2 serves as a superb introduction to the CLHLS and is required reading for all researchers intending to use CLHLS data. Moreover, scholars wishing to better understand the growing empirical literature on healthy longevity based on the CLHLS also need to read Zeng's important and illuminating chapter. Following Zeng's chapter, Chaps. 3 through 6 present systematic and relatively detailed assessments of the age reporting and data quality of the CLHLS. Some of the chapters are more concerned with looking at the accuracy of the age data, and others evaluate the reporting of other issues pertaining to mortality and morbidity. The authors of all four chapters undertake detailed and very competent analyses, and provide expert and important recommendations for the community of scholars using the CLHLS data and reading the increasing literature based on the CLHLS data.

Chapter 2
Introduction to the Chinese Longitudinal Healthy Longevity Survey (CLHLS)

Zeng Yi

Abstract To provide readers with information on the data source and research opportunities inherent in the CLHLS data sets, we present in this chapter an introduction to the CLHLS. The chapter includes the general goals, the specific objectives of, the organizational framework, study design, sample distribution and contents of the data collected, and finally a comparison with other survey projects focusing on elderly populations.

Keywords Adult child sample, CLHLS, Data collection, Data source, Determinant, Centenarian, Elderly population, Extent of disability and suffering before dying, Family relation, Healthy longevity, Intergenerational relation, Next-of-kin, Oldest-old, Over-sampling, Refusal rate, Research opportunities, Sample distribution, Study design, Weight

2.1 Background and Significance

The number of oldest-old persons aged 80 and older in China is expected to climb from about 12 million in 2000 to 51, 76, and 114 million in 2030, 2040 and 2050, respectively, under the medium mortality assumption. With the medium fertility and medium mortality assumptions, the proportion of elderly aged 65 and over is expected to increase from 7 percent in 2000 to about 16 percent in 2030, and to more than 23 percent in 2050, while the oldest-old will constitute 14, 22 and 34.4 percent of the elderly population in 2000, 2030 and 2050, respectively (Zeng and George 2002). The main reason why the number of oldest-old will climb so quickly after the year 2030 is that China's "baby boomers," who were born in the 1950s and 1960s, will fall then into the category of the "oldest-old." The average annual increase rate of oldest-old persons between 2000 and 2050 is expected to be

---

Zeng Yi
Center for Study of Aging and Human Development
Medical School of Duke University, Durham, NC 27710, USA,
Center for Healthy Aging and Family Studies/China Center for Economic Research,
Peking University, Beijing, China
e-mail: zengyi68@gmail.com

around 4.4 percent in China, which is about 2.4, 1.9, 2.0, 1.8, and 1.7 times as high, respectively, as that in United Kingdom, USA, France, Germany, and Japan (United Nations 2005). Oldest-old persons are much more likely to need help in daily living as compared to the younger elderly. Data from the Chinese Longitudinal Healthy Longevity Survey (CLHLS), which is the prominent source of data of the analyses contained in this book show that the prevalence of disability in Activities of Daily Living (ADL), such as bathing, dressing and eating, increases dramatically with age from less than 5 percent at ages 65–69 to 20 percent at ages 80–84, and to 40 percent at 90–94 years of age. At ages 100–105, less than 40 percent are able to perform the basic activities of daily living (ADL) without help. The oldest-old consume services, medical care and receive transfers at a higher rate than younger elderly persons. Torrey (1992: 382) estimated that in the US, the cost of Medicare for the oldest-old is 77, 60, and 36 percent higher than that of elders aged 65–69, 70–74, and 75–79, respectively. The total cost of long-term care for the oldest-old is 14.4 times as high as that for younger elders aged 65–74.

The fact that the oldest-old sub-population is growing much faster than any other age group, and that they are also the most likely group to need help, indicates a significant need to investigate the demographic, socioeconomic, psychological and health status of the oldest-old. In the US, Canada, Europe, and some Asian and Latin American countries, efforts have been made to attract the attention of academics and policy makers to the concerns of the oldest-old (Suzman et al. 1992; Baltes and Mayer 1999; Vaupel et al. 1998). Some countries around the world have collected data from large samples of the old, with an over-sampling of the oldest-old. For example, 11 countries in the European Union have developed SHARE Surveys (Survey of Health and Retirement in Europe) and England had collected the ELSA (the English Longitudinal Survey of Aging), which are comparable to the American HRS (Health and Retirement Survey). In Japan the NUJLSOA (Nihon University Japanese Longitudinal Study of Aging) has been developed to be comparable to the US LSOA (Longitudinal Study of Aging). Longitudinal multi-wave studies are also available for the Taiwan area of China (SHLSE, Survey of Health and Living Status of the Elderly); Indonesia (The Indonesia Family Life Survey); and Mexico (MHAS, The Mexican Health and Aging Study). More studies are now being developed particularly in developing countries where less is known about the oldest-old.

Before the Longitudinal Healthy Longevity Survey study was launched in 1998 in China, little attention had been paid to ensure sufficient representation of the oldest-old in national surveys, and most studies on the elderly included few subjects aged 80 and over. Almost all published official statistics were truncated at ages 65 or 80. The surveys on the elderly thus had sub-sample sizes far too small for the proper evaluation of the oldest-old. For example, 20,083 elders aged 60 and above were interviewed under the 1992 Chinese national survey on support systems for the elderly; but among them, only 84 were aged 90+. These small sub-sample sizes made a meaningful analysis of the oldest-old sub-population impossible.

To fill in the data and knowledge gaps for scientific studies and policy analysis, the Chinese Longitudinal Healthy Longevity Survey (CLHLS) has been underway in China since 1998. Our main objective is to draw societal and governmental attention to scientific studies and practical program interventions for enhancing the well-being and life quality of the oldest-old and all other members of our society (Zeng et al. 2001).

To provide readers with information on the data source and research opportunities inherent in the CLHLS data sets, we present in this chapter an introduction to the CLHLS. The chapter includes the general goals, the specific objectives of, the organizational framework, study design, sample distribution and contents of the data collected, and finally a comparison with other survey projects focusing on elderly populations. Chaps. 3–6 present a systematic and relatively detailed assessment of the age reporting and data quality of the CLHLS.

2.2 Objectives and Organizational Framework

Our general goal is to shed new light on a better understanding of the determinants of healthy longevity of human beings. We are compiling extensive data on a much larger population of oldest-old aged 80–112 than has previously been studied, with a comparative group of younger elders aged 65–79. We propose to use demographic and statistical methods to analyze data culminating from the longitudinal surveys. We want to determine which factors, out of a large set of social, behavioral, biological, and environmental risk factors play an important role in healthy longevity. The large population size, the focus on healthy longevity (rather than on a specific disease or disorder), the simultaneous consideration of various risk factors, and the use of analytical strategies based on demographic concepts make this an innovative demographic data collection and research project.

Our Specific Objectives are as Follows:

- Collect intensive individual interview data including health, disability, demographic, family, socioeconomic, and behavioral risk-factors for mortality and healthy longevity.
- Follow-up the oldest-old and the comparative group of the younger elders, as well as some of the elders' adult children to ascertain changes in their health status, care needs and costs, and associated factors; and ascertain if they die and if so at what age, from what cause, the care that was needed and costs, and their health/disability status before death.
- Analyze the data collected to estimate the social, behavioral, environmental, and biological risk-factors that are the determinants of healthy longevity and mortality in the oldest-old.
- Compare the findings with results from other studies of large populations at advanced ages.

The organizational framework of the CLHLS is summarized in Table 2.1.

Table 2.1 The organizational framework

Sponsoring and supporting organizations	National Institute on Aging; United Nations Fund for Population Activities (UNFPA) and China National Foundation for Social Sciences joined NIA to co-sponsor the expanded survey in 2002; China National Natural Science Foundation and Hong Kong Research Grant Council joined NIA to co-sponsored the expanded survey since 2005; Peking University and Duke University have provided institutional support; Max Planck Institute for Demographic Research has provided support for international training
Principal investigator and steering committee	Zeng Yi, Principal Investigator, Duke University and Peking University. The steering committee of the Chinese research team of this project consists of (alphabetically listed): Guo Zhigang, Li Ling, Liu Yuzhi, Edward Tu, Xiao Zhenyu, Zeng Yi, Zhang Chunyuan
International longevity projects coordinator	James W. Vaupel, Coordinator of the Coordinated International Projects on Healthy Longevity in U.S., Europe, and China, Max Planck Institute for Demographic Research and Duke University
Data collection organizations	Peking University Center for Healthy Aging and Family Studies (CHAFS) and China Mainland Information Group

2.3 Study Design and Sample Distribution

The baseline survey and the follow-up surveys with replacement for deceased elders were conducted in a randomly selected half of the counties and cities in 22 of China's 31 provinces in 1998, 2000, 2002, and 2005. We will conduct the fifth follow-up wave in 2008. Han Chinese people, who generally report age accurately, are the overwhelming majority in the surveyed provinces. The surveyed provinces are Liaoning, Jilin, Heilongjiang, Hebei, Beijing, Tianjing, Shanxi, Shaanxi, Shanghai, Jiangsu, Zhejiang, Anhui, Fujian, Jiangxi, Shangdong, Henan, Hubei, Hunan, Guangdong, Guangxi, Sichuan, and Chongqing. The population in the survey areas constitutes about 85 percent of the total population in China.

In our 1998 baseline survey, we tried to interview all centenarians who voluntarily agreed to participate in the study in the sampled counties and cities; for each centenarian interviewee, one nearby octogenarian and one nearby nonagenarian of predefined age and sex were interviewed. In the 2002 and 2005 waves, three nearby elders aged 65–79 of predefined age and sex were interviewed in conjunction with every two centenarians. "Nearby" is loosely defined – it could be in the same village or on the same street, if available, or in the same town or in the same sampled county or city. The predefined age and sex are randomly determined, based on the randomly assigned code numbers of the centenarians, to have more or less randomly selected comparable numbers of males and females at each age from 65 to 99.[1]

Those interviewees who were still surviving in the follow-up waves were re-interviewed. In our 1998 baseline survey and 2000, 2002, and 2005 follow-up surveys, we tried to interview all centenarians who voluntarily agreed to participate in the study, in order to keep a large sub-sample of centenarians in each of the waves. Those elderly who were interviewed but subsequently died before the next wave were replaced by new interviewees of the same sex and age (or within the same 5-year age group).

We added a sub-sample of 4,478 elderly interviewees' adult children aged 35–65 in 2002. The adult children sub-sample covers the eight provinces of Guangdong, Jiangsu, Fujian, Zhejiang, Shandong, Shanghai, Beijing, and Guangxi. If an elderly interviewee had only one eligible child (i.e., aged 35–65 and living in the sampling areas), that child was interviewed. If an elderly interviewee had two eligible adult children, the elder child was interviewed if the elderly interviewee was born in the first 6 months, and the younger child was interviewed if the elderly interviewee was born in the second 6 months. If an elderly interviewee had three eligible adult children, the eldest, the middle, or the youngest child was interviewed if the elderly interviewee was born in the first 4 months, second 4 months or the third 4 months, and so on. Among the 4,478 adult children interviewed in 2002, 1,722 sons and 338 daughters co-resided with old parents and 1,410 sons and 1,008 daughters did not co-reside with old parents. Such sample distributions reveal a traditional Chinese social practice: most old parents live with a son; non co-residing sons usually live closer to old parents than do the non co-residing daughters.

To avoid the problem of small sub-sample sizes at the more advanced ages, we did not follow the procedure of proportional sampling design, but instead interviewed nearly all centenarians and over-sampled the oldest-old of more advanced ages, especially among males. Consequently, appropriate weights need to be used to compute the averages of the age groups below age 100, but no weights are needed when computing the average of the centenarians. The method for computing the age-sex and rural–urban specific weights, and the associated discussions, are presented in the Appendix to this chapter.

In sum, the Chinese Longitudinal Healthy Longevity Survey (CLHLS) interviewed 8,959 and 11,161 oldest-old aged 80–112 in 1998 and 2000, and 16,057

---

[1] We obtained the lists of names and addresses of the centenarians through the Chinese local aging committee network or the neighborhood/village residents committees, and randomly assigned a code to each of the centenarians. For those centenarians whose code ended in 0, 1, 2, 3,..., 9, we tried to find one nearby octogenarians aged 80, 81, 82, 83,..., 89, one nearby nonagenarians aged 90, 91, 92, 93,..., 99, one nearby elder aged 70, 71, 72, 73,..., 79, respectively. For those centenarians whose code ended in 5, 6, 7, 8, 9, we tried to find one nearby elder aged 65, 66, 67, 68, 69, respectively. The sex of the targeted elderly aged 65–99 is randomly determined, with equal numbers of targeted males and females. If the enumerator could not find (through the urban neighborhood or the village residents committees who have the households' registration records) the target with exactly the predefined age and sex, an alternative subject who is the same sex and in the same 5-year age group was also acceptable.

and 15,638 elderly aged 65–112 in 2002 and 2005, respectively. In the four waves, in total, 10,964, 14,384, 16,526, and 9,941 face-to-face interviews were conducted with centenarians, nonagenarians, octogenarians, and younger elders aged 65–79, respectively (see Table 2.2 for more detailed information). At each wave, the longitudinal survivors were re-interviewed, and the deceased interviewees were replaced by additional participants. Data on mortality and health status before dying for the 12,007 oldest-old (aged 80–112) who died between the waves of 1998, 2000, 2002, and 2005 and the 499 younger elders (aged 65–79) who died between 2002 and 2005 were collected in interviews with a close family member of the deceased. In our 2002 and 2005 surveys, we also interviewed (with follow-up) 4,478 elderly interviewees' adult children (aged 35–65) in the eight provinces.

2.4 Data Collected

An interview with some basic physical capacity tests was performed at the interviewee's home. The questionnaire design was based on international standards and was adapted to the Chinese cultural/social context and carefully tested by pilot studies and interviews. We emphasized questions that might shed light on risk factors for mortality and healthy longevity, and we sought to minimize questions that could not be reliably answered by the oldest-old, some of whom may lack education and may have poor hearing and vision. The data collected included family structure, living arrangements and proximity to children, self-rated health, self-evaluation on life satisfaction, chronic disease, medical care, social activities, diet, smoking and alcohol drinking, psychological characteristics, economic resources, caregiver and family support, nutrition and some health-related conditions in early life (childhood, adulthood, and around age 60). Activities of Daily Living (ADL) and cognitive function measured by the Mini-Mental State Examination (MMSE) were evaluated in all waves in 1998, 2000, 2002, and 2005. The capacity of physical performance was also evaluated in all waves by means of tests of standing up from a chair without using hands, picking up a book from the floor, and turning around 360°. As initially planned, Instrumental Activities of Daily Living (IADL) questions were not included in the 1998 baseline and 2000 follow-up surveys because the 1998 and 2000 waves interviewed the oldest-old only and the Chinese oldest-old are generally limited in IADL. We added IADL questions in our 2002 and 2005 surveys when we expanded our survey to cover both the oldest-old and the younger elderly aged 65–79.

The interview refusal rate among the Chinese oldest-old was very low: about 2 percent among those who were not too sick to participate with proxy assistance. This high rate likely is due to the fact that the Chinese oldest-old in general like to talk to outside people, plus they stay at home without a job or other duties. Many of them and their family members may also feel honored to participate in survey interviews concerning healthy longevity, as they may be proud of being a member of a long-lived group. Many of the disabled oldest-old agreed to participate in our

Table 2.2 Age and sex compositions of the samples of the 1998 baseline, and the 2000, 2002, and 2005 follow-up surveys

Age	Surviving interviewees			Newly added			Total			Deceased interviewees		
	Follow-up											
	M	F	T	M	F	T	M	F	T	M	F	T
1998 baseline survey												
80–89	–	–	–	1,787	1,741	3,528	1,787	1,741	3,528	–	–	–
90–99	–	–	–	1,299	1,714	3,013	1,299	1,714	3,013	–	–	–
100+	–	–	–	481	1,937	2,418	481	1,937	2,418	–	–	–
Total	–	–	–	3,567	5,392	8,959	3,567	5,392	8,959	–	–	–
2000 follow-up survey												
80–89	996	1,048	2,044	1,471	1,403	2,874	2,467	2,451	4,918	339	262	601
90–99	720	907	1,627	925	1,260	2,185	1,645	2,167	3,812	574	612	1,186
100+	262	891	1,153	256	1,022	1,278	518	1,913	2,431	348	1,213	1,561
Total	1,978	2,846	4,824	2,652	3,685	6,337	4,630	6,531	11,161	1,261	2,087	3,348
2002 follow-up survey												
35–65	–	–	–	3,132	1,346	4,478	3,132	1,346	4,478	–	–	–
65–79	–	–	–	2,456	2,438	4,894	2,456	2,438	4,894	–	–	–
80–89	1,454	1,411	2,865	673	672	1,345	2,127	2,083	4,210	481	367	848
90–99	948	1,236	2,184	590	858	1,448	1,538	2,094	3,632	543	677	1,220
100+	277	917	,194	442	1,685	2,127	719	2,602	3,321	292	930	1,222
Total	2,679	3,564	6,243	7,293	6,999	14,292	9,972	10,563	20,535	1,316	1,974	3,290
2005 follow-up survey												
35–65	2,183	820	3,003	–	–	–	2,183	820	3,003	30	5	35
65–79	1,849	1,805	3,654	959	935	1,894	2,808	2,740	5,548	271	228	499
80–89	1,147	1,231	2,378	741	748	1,489	1,888	1,979	3,867	721	627	1,348
90–99	522	824	1,346	943	1292	2,235	1,465	2,116	3,581	55	1,085	,940
100+	158	600	758	360	1462	1,822	518	2,062	2,580	50	1,636	2,086
Total	5,859	5,280	11,139	3,003	4,437	7,440	8,862	9,717	18,579	2,327	3,581	5,908

Table 2.2 (Continued)

Age	Surviving interviewees												Deceased interviewees		
	Follow-up			Newly added			Total								
	M	F	T	M	F	T	M	F	T				M	F	T
Total Sample size of the four waves (1998–2005)															
35–65	2,183	820	3,003	3,132	1,346	4,478	5,315	2,166	7,481				30	5	35
65–79	1,849	1,805	3,654	3,415	3,373	6,788	5,264	5,178	10,442				271	228	499
80–89	3,597	3,690	7,287	4,672	4,564	9,236	8,269	8,254	16,523				1,541	1,256	2,797
90–99	2,190	2,967	5,157	3,757	5,124	8,881	5,947	8,091	14,038				1,972	2,374	4,346
100+	697	2,408	3,105	1,539	6,106	7,645	2,236	8,514	10,750				1,090	3,779	4,869
Total	10,516	11,690	22,206	16,515	20,513	37,028	27,031	32,203	59,234				4,904	7,642	12,546

*M, Male; F, Female; T, Total (i.e. two-sexes combined); "–", not relevant.

healthy longevity study through proxy assistance by a close family member. Those who were too sick to participate with proxy assistance were not interviewed. Instead the interviewers answered the question "Why was the interview not conducted or not completed?" The answers to this question are used in data analysis to correct for selection bias. Refusal rates increase substantially among younger interviewees aged 65–79 (5.1 percent) and among adult children aged 35–65 (14.3 percent) because some of them did not want to devote their time to the interview.

One unique feature of the CLHLS study is that relatively comprehensive information on the extent of disability and suffering before dying was obtained for those interviewees who had died before the next wave by interviewing one of their close family members. Collected information before dying includes date/cause of death, chronic diseases, ADL, number of hospitalizations or incidents of being bedridden from the last interview to death, and whether the subject had been able to obtain adequate medical treatment when suffering from disease. If any of the ADL activities were disabled or partially disabled, then a question on the duration of the disability (or partial disability) would follow. The number of days spent bedridden before dying was also ascertained. Data on how many days before death the elder did not go outside and how many days before death the elder spent more time in bed than out of bed were collected. Information on socioeconomic and demographic characteristics, such as marital status, family structure, caregivers, financial situation, and living arrangement before death, were also collected.

2.5 Research Opportunities

2.5.1 Healthy Survival and Disability of the Elderly

Population aging accompanied with the fastest growth of the oldest-old is unavoidable. Hence the fundamental question is the following: how can the global community adequately face the challenges of aging and achieve the goals of healthy survival and declining disability, and not only survive, but also remain healthy up to advanced ages? Despite the significance of this question, little is known about why some people live into their 1980s, others into their 1990s, and a select few to age 100 or more; and why some people survive to advanced ages with good health while others suffer severe disability and diseases (Jeune 1995; Vaupel et al. 1998). We believe that research based on CLHLS data may contribute important new knowledge with which these basic questions may be addressed.

Improved knowledge on the determinants of healthy survival will help society to reduce the costs of taking care of the disabled elderly. Leon et al. (1998) have estimated that on average a one-month delay in nursing home placement could save as much as $1 billion annually in health services costs in the US Lessons learned from studying the oldest-old in Chinese society, where family and community-based care (rather than nursing home care) is the main institution for taking care of disabled elderly, may be a useful reference point for offsetting a devastating burden to the health care systems in the US and other Western countries. Improved knowledge about healthy longevity will also stimulate consumption and investment concerning healthy survival, and strengthen the human capital of the healthy elderly; all of this will certainly be useful in stimulating economic growth (Cutler and Richardson 1997; Morand 2002; Murray and Lopez 1996; Nordhaus 2002; WHO 2002). The information will not only be beneficial to Chinese societies but also to international businesses dealing with the huge markets of rapidly increasing elderly populations in Western countries and China.

2.5.2 Extent of Disability and Suffering Before Dying

Based on Medicare data, Lubitz and Prihoda (1983) showed that 28 percent of all Medicare costs were incurred by the 6 percent of enrollees who died within the next 12 months. If an individual experiences severe suffering for an extended period before dying, much pain and burden are brought on the individual, family, and society. Hence, it is important to study factors associated with both healthy survival and deceased elders' extent of disability and suffering before dying. In the CLHLS study, comprehensive data on health status, disability, and degree/length of suffering before dying were obtained about the deceased oldest-old interviewees by interviewing one of their close family members. As reviewed by George (2002), the use of surrogate or proxy responses from family members is appropriate in quality of dying research. However, George (2002) also found that in most previous quality of dying research, investigators appear to have selected a place for subject recruitment (e.g., hospice settings, palliative care units) and then simply enrolled the available patients, thereby introducing selection bias. Up to 2005, we collected relatively comprehensive data before dying from a sample of nearly 12,506 elders in the randomly selected half of the counties and cities forming the study area; as noted previously, this is an area constituting about 85 percent of the total population of China. This study has clear merit in this regard since it does not rely on a small sample from a single health care setting, as was done in most previous quality of dying research.

2.5.3 Intergenerational Family Relations and Healthy Longevity

Many studies consistently claim a strong association of family support with better health and lower elderly mortality (Anderson et al. 1999; Rogers 1996; Zunzunegui et al. 2001). Of all kin ties, the parent-child relationship remains the most important "stem" in the family support network (Wellman and Wortley 1990). A dyadic approach that collects and analyzes data from members of two generations has been proposed to study family relationships (Lye 1996; Thompson and Walker 1982). Examples of large-scale surveys that interview both parents and children are the US Longitudinal Study of Generations and the National Survey of Families and Households (NSFH). Using funds entirely from Chinese sources, we added a sub-sample of 4,478 elderly interviewees' adult children aged 35–65 in our 2002 survey. Follow-up surveys on these adult children and their elderly parents in 2005 provided unique data for studying intergenerational family relationships/transfers and their impacts on healthy longevity. This is particularly relevant in the Chinese cultural and social context, which tends to have a more valued family support system. As compared to other studies following the dyadic approach, our study has unique strengths: the mean age of old parents is 83.6 (SD=11.0) and the mean age of adult children is 50.3 (SD=8.6). About 60 percent of our paired-sample consists of oldest-old parent(s) aged 80–110 with a child who is also elderly or nearly elderly, and about 40 percent of the paired sample consists of old parents younger than age 80 with their relatively younger adult child. Our paired-sample is particularly useful for studying the association of healthy longevity with the family relationship between the oldest-old and their elderly children. To our knowledge, no study of this kind, with a large number of pairs of oldest-old parent(s) and their elderly children, has ever been conducted.

2.5.4 The Unique Features

Our continuing CLHLS project offers an unparalleled opportunity for studies of the determinants of healthy longevity. *First, research leverage will be gained by focusing on an extremely selected (and large) group of Chinese oldest-old.* The Chinese population is so huge that despite very high mortality in the past, the numbers of oldest-old persons are very large and continue to rapidly increase. The proportion of centenarians, nonagenarians, and octogenarians in China, however, is much lower than in developed countries. For example, there were about 5 centenarians per million in China in 1990s, compared with 25 per million in Japan and 50 per million in Western Europe (Jeune 1995). The Chinese oldest-old aged 80+ are an extremely select sub-population; they are the survivors of brutal mortality regimes of the past operating on birth cohorts of many millions. A focus on extreme cases is often a good way to gain research leverage at a reasonable expense. Research on the large but extraordinarily selected Chinese oldest-old should provide important leverage for better understanding healthy longevity in general.

Second, the age reporting of the Chinese oldest-old is reasonably good. Accurate age reporting is crucial in studies dealing with elderly people, especially the oldest-old. Often, older persons in developing countries cannot report their age accurately (Coale and Kisker 1986; Elo and Preston 1992; Mosley and Gray 1993). The age reporting of the Chinese oldest-old is reasonably reliable, based on the analysis by a wide variety of international and Chinese demographers such as Coale and Li (1991), Wang et al. (1998), and Zeng et al. (2001; 2002, using the CLHLS data), and as analyzed and discussed in Chaps. 4 and 5 of this book.

Third, this is the largest longitudinal study of the oldest-old and has a comprehensive approach. It includes oldest-old aged 80–110 with the largest sample size ever conducted, with younger elders aged 65–79 as a comparative group; moreover, data have been collected on the extent of disability and suffering before dying, oldest-old siblings-pairs, elderly interviewees' adult children, and information concerning the communities where the interviewees live. The CLHLS has the largest sample size of centenarians and nonagenarians compared to any other study in the world. It also has a larger sub-sample size of octogenarians than any other survey except the NLTCS and AHEAD in the US and is one of only ten studies that collected data on the extent of disability and suffering before dying; it is one of five studies that collected sibling-pairs data; and it is one of three studies that interviewed elderly subjects' adult children. The CLHLS and NECS are the only two studies that include all three of the above data collection components (extent of disability and suffering before dying, sibling pairs, and adult children) in one study. Further, the main sample of the NECS covers centenarians only, while the CLHLS covers persons in ages 80–110 in 1998 and 2000 and ages 65–110 in 2002 and 2005.

2.6 Data Availability and Contacts

2.6.1 Data Availability

The 1998 baseline, 2000, 2002, and 2005 follow-up healthy longevity survey data sets are being distributed by Peking University and Duke University. Researchers interested in using the data are expected to sign a Data Use Agreement. Raw data will then be provided freely.

2.6.2 Contacts

> Professor Liu Yuzhi, Executive Associate Director
> Peking University Center for Healthy Aging and Family Studies (CHAFS)
> Beijing, 100871, China; Tel: 0086-10-62756914; Fax: 0086-10-62756843
> E-mail: chafs@pku.edu.cn; Website : http://www.pku.edu.cn/academic/ageing
>
> Dr. Danan Gu, Research Scientist, Center for the Study of Aging and Human Development, Medical Center, Duke University;
> Durham, NC 27708 ; Phone: (919) 660-7532; Fax: (919) 668-0453
> E-mail: gudanan@duke.edu; Website : http://www.geri.duke.edu/china_study

Appendix: The Weights for Producing Correct Estimates of the Average of Entire Population of Old Persons, Based on the Healthy Longevity Sample Survey Data

The age (x), sex (s), and rural–urban residence (r) specific weight $w(x, s, r, t)$ in the survey year t is computed as

2 Introduction to the Chinese Longitudinal Healthy Longevity Survey (CLHLS)

$$w(x,s,r,t) = \frac{N(x,s,r,t)/\sum_x\sum_s\sum_r N(x,s,r,t)}{n(x,s,r,t)/\sum_x\sum_s\sum_r n(x,s,r,t)}$$

$$= [N(x,s,r,t)/n(x,s,r,t)] \left[\sum_x\sum_s\sum_r n(x,s,r,t)/\sum_x\sum_s\sum_r N(x,s,r,t)\right]$$

Where $N(x, s, r, t)$ is number of persons of age x, sex s, and residence r in year t, derived from the projected elderly population based on the last census 100 percent data tabulations for the 22 provinces where the CLHLS survey was conducted, and the estimated age–sex-specific survival probabilities between the census year and the survey year t. The $n(x, s, r, t)$ is number of persons of age x, sex s, and residence r, derived from the healthy longevity survey conducted in the year t. The weight is actually the multiplication of the ratio of $[N(x, s, r, t)/n(x, s, r, t)]$ and the overall sampling ratio in the survey year t. No weights are needed when we compute the average of the centenarians, since the survey attempted to interview all centenarians in the sampled areas.

The weight $w(x, s, r, t)$ is actually the ratio of age distribution of the entire elderly population in the survey year t to the age distribution of the sample in the year t. The weights for the over-sampled extremely old persons (e.g. 90+) are less than 1.0, and weights for under-sampled elders (e.g. age 65–69 to 80–85) are greater than 1.0.

The values of the weights vary (usually greater than 1.0 under age 88 and less than 1.0 above age 90), and it produces correct average proportions of certain attributes within age groups by using the weights. However, SPSS (or other software) would not produce correct p-values for testing the statistical significance of the differences of the proportions among different age groups, since the sub-sample size of the age groups are altered after weighting the individual cases. Therefore, the weights need to be adjusted to make sure that the sub-sample size within each age group after weighting is exactly the same as the true sub-sample size. Denote $C_j(s, r, t)$ as the adjusting factor for age group j (e.g. age group 90–95) with sex s and residence r; $T_j(s, r, t)$ as the total number of interviewed persons of the age group j with sex s and residence r. The following equations must be fulfilled: $C_j(s,r,t)\sum_x w(x,s,r,t)n(x,s,r,t) = T_j(s,r,t)$. Solving this equation, we obtain the adjusting factor, $C_j(s,r,t) = T_j(s,r,t)/\sum_x w(x,s,r,t)n(x,s,r,t)$.

The adjusted weights are $w'(x,s,r,t) = w(x,s,r,t)C_j(s,r,t)$. We should use the adjusted weights that produce both correct proportions, and correct sub-sample sizes and thus correct p-values for testing the statistical significance of the differences of the proportions among various age groups.

If one computes proportions of certain attributes of age groups with rural and urban combined, the adjusting factor is not rural–urban specific, but age group and sex specific: $C_j(s) = T_j(s)/\sum_x\sum_r w(x,s,r,t)n(x,s,r,t)$.

If one computes proportions of certain attributes of age groups with rural and urban combined and both sexes combined, the adjusting factor is neither rural–urban specific, nor sex-specific, but only age group specific: $C_j = T_j / \sum_x \sum_r \sum_s w(x,s,r,t)\, n(x,s,r,t)$.

Acknowledgments The author's work of this chapter is supported by the NIA/NIH grant (R01 AG023627-01) and China Natural Science Foundation grant (70440009).

References

Anderson, B.A., C.S. Kim, J.H. Romani, J.W. Traphagan, and J. Liu (1999), *Living arrangements and mortality risks of the urban elderly in Yunnan Province, China, 1995*. Research Reports No. 99-435. University of Michigan: Population Studies Center

Baltes, P.B. and K.U. Mayer (1999), *The Berlin aging study: aging from 70 to 100*. Cambridge: Cambridge University Press

Coale, A.J. and E.E. Kisker (1986), Mortality crossovers: reality or bad data? *Population Studies* 40, pp. 389–401

Coale, A.J. and S. Li (1991), The effect of age misreporting in China on the calculation of mortality rates at very high ages. *Demography* 28 (2), pp. 293–301

Cutler, D. and E. Richardson (1997), Measuring the health of the U.S. population. *Brookings Papers on Economic Activity: Microeconomics*, pp. 217–271

Elo, I.T. and S.H. Preston (1992), Effects of early-life conditions on adult mortality: A review. *Population Index* 58 (2), pp. 186–212

George, L.K. (2002), Research design in end-of-life research: state of science. *The Gerontologist* 42, pp. 86–98

Jeune, B. (1995), In search of the first centenarians. In: B. Jeune and J.W. Vaupel (eds): *Exceptional longevity: from prehistory to the present*. Odense: Odense University Press

Leon, J., C.K. Cheng, and P.J. Neumann (1998), Alzheimer's disease care: Costs and potential savings. *Health Affairs* 17, pp. 206–216

Lubitz, J. and R. Prihoda (1983), The use and costs of Medicare services in the last two years of life. *Health Care Financing Review* 5, pp. 117–131

Lye, D.N. (1996), Adult child-parent relationships. *Annual Review of Sociology* 22, pp. 79–102

Mosley, H.W. and R. Gray (1993), Childhood precursors of adult morbidity and mortality in developing countries: implications for health programs. In: J. Gribble and S. H. Preston (eds): *The epidemiological transition: Policy and planning implications for developing countries*. Washington, DC: National Academy Press. pp. 69–100

Morand, O.F. (2002), Economic growth, longevity, and the epidemiological transition. Working paper. University of Connecticut, Department of Economics. Available at http://www.econ.uconn.edu/working/2002-07.pdf

Murray, C.J.L. and A.D. Lopez (eds) (1996), *The global burden of disease*. Harvard School of Public Health, distributed by Harvard University Press, Cambridge, MA

Nordhaus, W.D. (2002), *The health of nations: The contributions of improved health to living standard*. Working Paper 8818, National Bureau of Economic Research. Available at http://www.nber.org/papers/w8818

Rogers, R.G. (1996), The effects of family composition, health, and social support linkages on mortality. *Journal of Health and Social Behavior* 37, pp. 326–338

Suzman, R.M., D.P. Willis, and K.G. Manton (1992), *The oldest old*. New York: Oxford University Press

Thompson, L. and A.G. Walker (1982), The dyad as the unit of analysis: conceptual and methodological issues. *Journal of Marriage and the Family* 44 (4), pp. 889–900

Torrey, B.B. (1992), Sharing increasing costs on declining income: the visible dilemma of the invisible aged. In: R.M. Suzman, D.P. Willis, and K.G. Manton (eds): *The oldest old*. New York: Oxford University Press, pp. 381–393

United Nations, Population Division (2005), *World population prospects: The 2004 revision*. New York.

Vaupel, J.W., J.R. Carey, K. Christensen, T.E. Johnson., A.I. Yashin, N.V. Holm, I.A. Iachine, V. Kannisto, A.A. Khazaeli, P. Liedo, V.D. Longo, Y. Zeng, K.G. Manton, and J.W. Curtsinger (1998), Biodemographic trajectories of longevity. S*cience* 280 (5365), pp. 855–860

Wang, Z., Y. Zeng, B. Jeune, and J.W. Vaupel (1998), Age validation of Han Chinese centenarians. *GENUS—An International Journal of Demography* 54, pp. 123–141

Wellman, B. and S. Wortley (1990), Different strokes from different folks: Community ties and social support. *American Journal of Sociology* 96, pp. 558–588

World Health Organization (WHO), (2002), *Healthy aging is vital for development*. Press Release WHO/24,9 April 2002

Zeng, Y. and L. George (2002), Extremely rapid aging and the living arrangement of elderly persons: The case of China. *Population bulletin of the United Nations, Special issue Nos. 42/43: Living arrangements of older persons*. New York: United Nations.

Zeng, Y., J.W. Vaupel, Z. Xiao, C. Zhang and Y. Liu (2001), The healthy longevity survey and the active life expectancy of the oldest old in China. *Population: An English Selection* 13 (1), pp. 95–116

Zeng, Y., J.W. Vaupel, Z. Xiao, C. Zhang, and Y. Liu (2002), Sociademographic and health profiles of oldest old in chiana. *Population and Devolopment Review* 28 (2), pp. 251–273.

Zunzunegui, M.V., F. Béland, and A. Otero (2001), Support from children, living arrangements, self-rated health and depressive symptoms of older people in Spain. *International Journal of Epidemiology* 30, pp. 1090–1099

Chapter 3
General Data Quality Assessment of the CLHLS

Danan Gu

Abstract This chapter provides a comprehensive review of data quality of the third wave of the Chinese Longitudinal Healthy Longevity Survey (CLHLS) in 2002 in terms of proxy use, nonresponse rate, sample attrition, and reliability and validity of major health measures. The results show that the data quality of the 2002 wave of the CLHLS is generally good. Some recommendations in use of the dataset are provided.

Keywords Accuracy of imputation, Bias, Convergent Cronbach's alpha coefficient, Data assessment, Discriminant validity, Don't know answer, Factual question, Full proxy response, Homogeneity, Imputation, Inconsistent responses, Internal consistency, Item nonresponse, Item-total, Item-total correlations, Knowledgeable proxy, Minimum reliability coefficient, Missing completely at random, Missing item, Missing value, Multiple imputation, Multiple item scale, Next of kin, Nonresponse, Nonresponse rate, Objective question, Proxy, Proxy reporter, Proxy response, Proxy use, Reliability, attrition, Significant other, Sources of error, Unit nonresponse, Validity

3.1 Introduction

This chapter provides a comprehensive review of the quality of the data from the third wave of the Chinese Longitudinal Healthy Longevity Survey (CLHLS) in 2002 in terms of proxy use, nonresponse rate, sample attrition, and reliability and validity of major health measures. A data quality assessment for the first wave in 1998 may be found elsewhere (Zeng et al. 2001).[1] The third wave of the CLHLS gathered

---

D. Gu
Center for the Study of Aging and Human Development, Medical School of Duke University, Durham, NC 27710, USA
e-mail: gudanan@duke.edu

[1] There is no systematical data assessment publication in English for the 2000 wave. According to the Chinese publication (Gu and Zeng 2004), the data quality of the CLHLS in 2000 is good.

extensive questionnaire data through interviewing 16,020 elderly aged 65+. The survey was conducted in randomly selected counties and cities in 22 of China's 31 provinces.[2] The 2002 wave extended its age range of the sampled elderly to include the age range from 65 to 79 who were not included in the first two waves in 1998 and 2000. Among the 16,020 sample persons in 2002, 3,189 were centenarians, 3,747 were nonagenarians, 4,239 were octogenarians, and 4,845 were aged 65–79.

The design of the CLHLS questionnaire is based on international standards and adapted to the Chinese cultural/social context and carefully tested by pilot studies/interviews. The CLHLS emphasizes questions that might shed light on risk factors for mortality and healthy longevity. An interview and a basic health examination at each wave were performed at the interviewee's home. Extensive data were collected including family structure, living arrangements and proximity to children, self-rated health, self-evaluation on life satisfaction, chronic disease, medical care, social activities, diet, smoking and alcohol drinking, psychological characteristics, economic resources, caregivers and family support, nutrition and other health-related conditions in early life (childhood, adulthood, and around age 60), activities of daily living (ADL) using the Katz ADL index (Katz et al. 1963), and cognitive function measured by the Mini-Mental State Examination (MMSE) (Folstein et al. 1975). Physical performance capacity was evaluated through tests of putting a hand to the back and neck, raising hands upright, standing up from sitting in a chair without using hands, picking up a book from the floor, and turning around 360 degrees. As initially planned, instrumental activities of daily living (IADL) questions were added in the 2002 survey.

According to Groves (1987), there are three major potential sources of errors due to nonobservation (coverage error, nonresponse error, and sample error) and four major potential sources of errors due to observation or measurement (the interviewer, the respondents, the questionnaire, and the mode of interview). This chapter follows this framework in its attempt to assess the data quality of the CLHLS in 2002. All data analyses to this end are conducted using STATA 8.0 and SPSS 12.0. Assessments of the accuracy of age reporting, mortality, and morbidity in the first three waves of the CLHLS are presented in Chaps. 4–6 of this volume.

3.2 Proxy Use

As frequently reported in most empirical studies, it is normal for a survey of the elderly to have more than 20 percent of respondents unable to complete the

---

[2] The 22 surveyed provinces are Liaoning, Jilin, Heilongjiang, Hebei, Beijing, Tianjing, Shanxi, Shaanxi, Shanghai, Jiangsu, Zhejiang, Anhui, Fujian, Jiangxi, Shangdong, Henan, Hubei, Hunan, Guangdong, Guangxi, Sichuan, and Chongqing. The population in the survey areas constitutes about 85 percent of the total population in China. Han Chinese people are the overwhelming majority in the 22 surveyed provinces. There were 631, 777, and 866 counties and cities in the 1998, 2000, and 2002 surveys, respectively. The increase in numbers of survey units in 2002 was mainly due to adding an elderly comparison group aged 65–79 who were not interviewed in 1998 and 2000; and partly due to an administrative boundary change in the later wave; or some selected counties/cities that had no centenarians in an earlier wave but had centenarians in a later wave.

questionnaire due to cognitive or linguistic impairments (Coroni-Huntley et al. 1986; DeHaan et al. 1993; Magaziner et al. 1988). Studies of the elderly that fail to use proxies often have a higher rate of nonresponse or missing data (Blazer et al. 1987). Therefore, proxies are frequently used as an alternative so to reduce elderly nonresponse, especially for the oldest-old because substantial proportions are usually incapable of providing accurate responses or even participating due to impaired hearing/vision, frail health or recall problems (Rodgers and Herzog 1992). Although it is not known for certain whether proxy information is similar to that provided by the subjects themselves (Pierre et al. 1998), there is a general consensus among investigators that proxy respondents should be used in research focusing on the oldest-old in order to avoid biasing the data in favor of healthy older persons (Rodgers and Herzog 1992).

Given that a proxy reporter is likely to be used, who then should be the proxy reporter? The existing literature in epidemiological studies suggests that validity varies considerably, depending on the relationship of the proxy to the respondent, the type of information sought, and the time period involved (Tang and McCorkle 2002). Caregivers may be more knowledgeable than personal friends and family members about physical health and functional symptoms of institutional respondents, although some studies have shown that caregivers tend to overrate the respondent's disability (Rothman et al. 1991); whereas family members may be more knowledgeable about personal, familial, and economic situations, and the like. Among family members, wives have been shown to be particularly reliable proxy reporters (Kolonel et al. 1977).

However, the use of proxy reporters rests to a large extent on the trade-off between nonresponse and inaccurate reporting. Errors due to unit and item nonresponse may be reduced by seeking information from proxies, but in such a case errors due to inaccurate responding may increase. To date, both of these assumptions remain unsubstantiated (Rodgers and Herzog 1992). However, it is widely understood that proxies can be used to report about factual issues, and produce fairly accurate information. Sometimes with a good questionnaire design, good quality data can be collected even on subjective questions (Basset and Magaziner 1988; Rodgers 1988; Rodgers and Herzog, 1992). The consistent finding across studies is that the accuracy of proxy ratings is high when the information sought is concrete and observable (Klinkenberg et al. 2003; Tang and McCorkle 2002).

In the CLHLS, questions such as self-rated health, life satisfaction, and MMSE tests on cognitive functioning are answered by the interviewees only. Other questions are answered by the interviewees themselves, as much as possible. For those who are not able to answer these questions, a close family member or another knowledgeable proxy (i.e., significant other) provides answers as indicated earlier. An indicator question is marked by the interviewer to signify whether the answer is provided by the interviewee or the proxy.

Table 3.1 shows the proportion of proxy use in the 2002 wave. Consistent with the first two waves, proxy use increases with age. Table 3.2 suggests that about 90 percent of proxies are close relatives such as a spouse, children, and grandchildren. Given the fact that proxies are used mainly to answer objective and factual

Table 3.1 Comparison of proportion of proxy in the CLHLS (%)

	Age group			
	65–79	80–89	90–99	100–105
1998 Wave				
Without proxy		61.50	36.57	16.43
Mix		38.02	62.23	81.18
Full proxy		0.48	1.19	2.39
2000 Wave				
Without proxy		62.05	37.41	15.92
Mix		37.39	61.07	80.64
Full proxy		0.57	1.52	3.43
2002 Wave				
Without proxy	88.40	64.35	36.91	19.04
Mix	11.39	35.43	61.57	76.41
Full proxy	0.21	0.21	1.52	4.55

The number of questions in the 2002 wave is slightly more than in the previous two waves, which might cause a relatively high percent of use of proxy. Full proxy means all questions except those that must be answered by the sampled person are answered by the proxy

questions in the CLHLS, the higher proportion of close relative proxies suggests that any potential bias is not substantial. Previous studies have shown that the level of agreement between respondents and proxies is influenced by a number of factors such as education, age, and living arrangement (Rothman et al. 1991; Tang and McCorkle 2002; Zsembik 1994). Our multivariable analysis indicates that respondents with an older age, lower education, rural residence, lower cognitive functioning, and higher disability are more likely to use a proxy. Our analysis further shows that the respondents with a proxy have a 20 percent more relative risk of death compared to those without a proxy (data not shown here).

The CLHLS did not obtain data comparing the agreement of responses between the proxy and the respondent. However, the small amount of full proxy responses in-

Table 3.2 Distribution of proxy subjects in the CLHLS (%)

	1998 wave	2000 wave	2002 wave[a]	2002 wave[b]
Spouse	5.55	5.18	5.02	32.00
Child or spouse of children	74.01	67.41	67.68	50.53
Grandchild or spouse of grandchild	12.37	16.56	15.92	8.00
Great grandchild or spouse of great grandchild	0.28	0.81	0.95	0.00
Sibling	0.24	0.18	0.20	0.84
Caregiver	2.76	4.34	5.27	2.11
Others	4.79	5.54	4.97	6.53

[a] Age 80–105
[b] Age 65–79

Based on two questions addressed to the interviewer "did anyone help the interviewee to answer any question?" and "who helped the interviewee to answer questions?". There is inconsistency between these two questions and actual proxy use in the questionnaire due to a misunderstanding of the questions by the interviewer (Zeng et al., 2001:112). The inconsistency rates are 13.2 and 6.2% for the 1998 and 2000 waves, and 3.9 and 2.6% for ages 80–105 and 65–79 in the 2002 wave, respectively

dicates the results may not be a big problem even if the bias between the respondent and proxy exists.[3] Researchers may also add an indicator variable for proxy use (i.e., whether the proxy is used or not for the sampled person) to adjust for such a bias if they think proxy answers could be problematic, as some other previous studies have done (e.g., Jenkins and Fultz 2005).

3.3 Nonresponse Rate and Incomplete Data

Nonresponse is an important indicator of data quality because it can bias survey estimates (Jay et al. 1993). Numerous studies indicate that nonresponse is greater for older adults than for younger adults (Herzog and Rodgers 1988), and nonresponse is a serious problem among older age groups and may be particularly high among those ages 85 and older (Herzog and Rodgers 1992). There are two types of nonresponse, namely, unit nonresponse and item nonresponse (Mohadjer et al. 1994).[4]

The unit nonresponse rate among the Chinese oldest-old was very low, about 4 percent, in the first three waves. This is because the Chinese oldest-old, in general, like to and have the time to talk to outside people, as they are at home without a job or other responsibilities. Many of the respondents and their family members may also feel honored to be interviewed about healthy longevity, as they may be proud of being a member of a long-lived group. Many of the seriously disabled oldest-old agreed to participate through proxy assistance provided by a close family member. Unit nonresponse rates tended to increase slightly among younger interviewees aged 65–79 (5.1 percent) because some of them apparently did not want to devote their time to the interview. One Japanese study of the elderly also finds a higher nonresponse rate in the lower age categories (Sugisawa et al. 1999). The amount of unit nonresponse error is difficult to measure, and thus efforts are often directed to minimize its occurrence (OMB 2001).

More recently, Lindner et al. (2001) recommend that steps should be taken to account for possible nonresponse error when a unit response rate is less than 85 percent. Although the CLHLS has a unit response rate higher than 85 percent, attention should be paid to item nonresponses, because a low unit nonresponse rate does not guarantee a low item nonresponse rate. Most data failures are due to a failure to obtain or record all-item information. A large amount of incomplete data for a particular item may indicate a problem with the translation of the item. Incomplete

---

[3] Proxies for the prevalence of chronic diseases and primary cause of death for the decedent persons are not reliable as indicated in Chap. 6 of this volume.

[4] The line between item and unit nonresponse is sometimes not clear. For example, if a completed questionnaire requires 90 percent of all possible items to be answered, it is possible that a number of partial interviews would be treated as unit nonresponses. On the other hand, if the required level of item responses is 80 percent for a completed questionnaire, the number of partial interviews treated as unit nonresponses would decrease and the unit response rate would increase (OMB 2001). In this study, a respondent who answered 60 percent or more of all possible items is coded as a valid unit response.

data might also indicate that respondents do not understand how to complete that part of the questionnaire. Data incompleteness can be classified into "Don't Know" (DK) and "Missing" categories, when the respondent refuses to answer or for other reasons.[5] DK usually occurs on questions related to historical information when the sampled person suffers recall problems, or when the proxy does not know about actual facts of the sampled person. Francis and Busch (1975) find that the oldest-old tend to give DK answers, and Herzog and Rodgers (1981) find that the oldest-old give DK answers more frequently on questions related to attitudes, feelings and expectations.

Table 3.3 shows that the average proportion of incompleteness of an item rated for each respondent in the CLHLS is less than 10 percent, much lower than some previous studies have reported (Wallace et al. 1992: 132). No difference is observed between the 2002 wave and the previous two waves. Table 3.4 summarizes the variables with incomplete answers of 2 percent or more.[6] Variables with the highest incomplete rate are "parents' ages at death." Although the incomplete rate declined in the 2002 wave compared to those in the 1998 and 2000 waves, it remains at

Table 3.3 Average percentage of item incompleteness of each respondent in the CLHLS (%)

Age	Males			Females		
	DK	Missing	Total	DK	Missing	Total
1998 Wave[a]						
80–89	4.64	0.48	5.12	5.45	0.63	6.07
90–99	4.58	0.65	5.22	7.23	0.75	7.98
100–105	6.03	0.75	6.78	8.53	1.03	9.56
2000 Wave						
80–89	2.26	1.51	3.78	2.91	1.97	4.87
90–99	2.61	2.06	4.67	3.54	2.82	6.37
100–105	3.22	2.50	5.72	4.42	3.52	7.94
2002 Wave						
65–79	2.09	0.95	3.04	2.32	1.33	3.65
80–89	2.74	1.58	4.32	4.01	2.23	6.24
90–99	3.80	2.01	5.81	4.84	2.69	7.52
100–105	4.60	1.95	6.55	5.95	2.74	8.69

Percentage of incomplete items (including don't know and refusal to answer) of each respondent is calculated based on the number of items that could be answered and the number of items answered by each respondent. Numerator in DK does not include "unable to answer" questions, which should be answered by the interviewee only

[a]The results for 1998 are different from Zeng et al., (2001) since the results of Zeng et al. (2001) did not include "don't know" in chronic diseases and did not include nonreported information about siblings or children

---

[5] Unlike most other studies, data incompleteness due to DK and missing in this paper is separately discussed from sample attrition.

[6] In the 1998 survey, there are 22 percent of respondents who did not know the name of the county in which they were born. There are 16 percent of respondents with missing lung flow data in the 1998 wave. These two variables are not listed in Table 3.4 since they were not asked in the 2000 and 2002 waves.

Table 3.4 Distribution of variables with more than 2 percent incomplete answers in the CLHLS (%)

	1998	2000	2002[a]	2002[b]
Eating style (D4 sets)	1.2–4.3	<2.0	<2.0	<1.0
Habit (i.e., smoking, drinking, exercise, physical laboring)	1.0–4.0	<2.0	<1.5	<1.5
Marriage history	>5.0[c]	>4.0[c]	>1.5[c]	>0.1[c]
Parents' age at death, and respondent's age at parents' death	30.0–40.0	27.0–35.0	25.0–30.0	7.8–10.5
Birth order and # of sibling	2.0–3.0	2.2	2.2	<0.5
Sibling information	>7.0[c]	>3.4[c]	>3.7[c]	>0.5[c]
Children's information	>3.0[c]	>2.7[c]	>2.2[c]	>1.0[c]
Blood pressure	3.0	3.2	0.9	<0.2
Height (Acromion-processus styloideus ulnae; right knee to the floor)	5.0	NA	0.0	0.0
Weight	8.0	1.7	0.0	0.0
Chronic diseases	7.0–10.0	5.0–8.0	3.6–7.5	2.4–5.0
Intergenerational transfers (upward)	NA	NA	6.5	7.5
Intergenerational transfers (downward)	NA	NA	6.3	7.0

[a] Age 80–105
[b] Age 65–79
[c] No upper boundary was provided here since the number of items that could be answered by each respondent is different, and the aggregated incomplete proportion is high for some items although the absolute number is not large due to the very small number of eligible respondents
NA, not applicable since there was no such question in 2000

more than 25 percent. Hence, extreme caution is recommended in dealing with such variables. If item nonresponses are missing completely at random, the estimates will not be biased (Allison 2002). The estimates might be biased, however, if item nonresponses are not completely at random. In such a case, tests should be made to detect any correlates. Prior studies have suggested that factors that might pertain to item nonresponse include age, sex, education, geographic region, and urban/rural residence (Jay et al. 1993). Our multivariable logistic results reveal that factors such as ethnicity, marital status, urban/rural residence, cognitive functioning, and self-reported health are all correlated to aggregated item nonresponses in each of the first three waves in the CLHLS. Those who are older, female, urban residents, of a minority ethnicity, not currently married, and in bad health are more like to have incomplete items, which is consistent with some previous studies in Western nations (Francis and Busch 1975; Herzog and Rodgers 1981).

Could item nonresponses that are conditional on a set of covariates introduce a bias in the estimation? Some studies argue that the effect of item nonresponses on outcomes does not depend on the difference between who gives the answers and who does not; rather, it depends on how the respondents who give answers differ from all those who are eligible to be interviewed (Norris and Goudy 1986; Kempen and van Sonderen 2002). In other words, if the response structure or pattern for those who answered the question is the same as the response structure or pattern for all sampled persons if they all could provide answers, then the estimates based on only those without nonresponse would be the same as the estimates based on the whole sample if all persons could answer the question.

Given that the CLHLS survey has encountered some level of nonresponse both in unit and item nonresponses, two general approaches could be applied to compensate for these nonresponses, namely, weight adjustment and imputation. Kalton and Kasprzyk (1986) note that weight adjustments are primarily used to compensate for unit nonresponses while imputation procedures are more likely to be used to compensate for missing items. In the CLHLS, a weight matching the post hoc distribution of age–sex–urban/rural residence in the sample with the distribution of the total population in the sampled 22 provinces is employed to reflect the unique sample design and compensate for unit nonresponses. This post hoc weight takes both the special design of the CLHLS and unit nonresponses of three basic demographic variables (i.e., age, sex, and urban/rural residence) into consideration (see Zeng et al. 2001 for detail). However, this weight has no relationship with other factors since their frequency distributions for the population are difficult to obtain and the weighting adjustment has the disadvantage of taking too many factors into consideration (Lepkowski et al. 1989). Researchers can create other weighting schemes if they have a reliable distribution for the total population in those 22 sampled provinces.

For compensating item nonresponses, Landerman et al. (1997) suggest using the mean if the incomplete rate of a particular variable is less than 2 percent; however, they argue that it is better to use regression or maximum likelihood methods to estimate nonresponse values when the incomplete rate is 2–5 percent, and to use multiple imputation to get estimates for nonresponse values when the incomplete rate exceeds 5 percent. With regression, maximum likelihood, or multiple imputation, biases in estimation can be lessened. Other strategies for dealing with this problem such as trimming bounds have also been suggested (Lee 2002). Other studies suggest treating the missing value as a special category if one is unable to ensure the accuracy of imputation (e.g., Hayward and Gorman 2004; Zimmer et al. 2002). The released CLHLS dataset does not provide imputed values for those variables with item nonresponses. If users of the CLHLS dataset want to impute the variable, they should follow the recommended approaches of Allison (2002), who provides a simple and very good theoretical background for how to handle item nonresponses. Most statistical packages such as SPSS, SAS, and STATA are capable of handling imputations or multiple imputations for item nonresponses. Our testing analyses show that the difference across different imputation approaches is not substantial, especially when the item nonresponse rate is less than 5 percent.

3.4 Sample Attrition

In longitudinal surveys, sample attrition (or data attrition, i.e., respondents lost in a follow-up survey) occurs when previous respondents migrated, refused to participate in the survey, became hospitalized, moved, or the address of a previous respondent was not sufficiently detailed.[7] Sample attrition is one of the most serious problems associated with longitudinal survey data. Similar to item nonresponse, sample attrition may distort the treatment/control comparison, depending on the type of attrition that takes place. If attrition is completely random with respect to all factors relevant to the outcome being measured, it leads to less precise estimates of program impacts (due to the reduction of the sample size), but does not lead to biased estimates (Mossel and Brown 1984). However, biased estimates might occur if sample attrition is correlated with some particular attributes, which may result in a lack of generalizability.

Out of a total of 11,162 interviewees in 2000, 6,291 (56.3 percent) were still alive at the 2002 wave, 3,335 (29.9 percent) died before the interview was held in 2002, and 1,536 (13.8 percent) were lost.[8] The proportion of attrition was higher between the 2000 and 2002 waves than between the 1998 and 2000 waves (9.6 percent). The true reason for the higher sample attrition in the period of 2000–2002 is not known. The frequency distribution indicates that the urban sample had a higher attrition in the 1998–2000 period than in the 2000–2002 period. We suspect this had something to do with more frequent resettlement of urban residents due to municipal construction and/or more frequent changes in re-delimiting the administrative boundary of counties and/or districts in the period, which would tend to cause more difficulty in locating previously sampled persons. Other possible reasons include unfavorable weather, transportation difficulties, and so forth. Compared with data attrition in surveys conducted in Western countries, the CLHLS has a similar proportion of data attrition. For instance, the proportion lost to follow-up in the 2-year interval in the second, third and fourth waves of the Longitudinal Study of Aging in the USA was 7.6, 12.1, and 16.0 percent, respectively (Mihelic and Crimmins, 1997). The proportion of respondents lost to a 2-year follow-up was 17.8 percent in a survey of Mexican elderly (Vellas et al. 1998).

Table 3.5 indicates that significant associations between sample attrition and variables in the model are observed except in self-reported health. Respondents who are female, physically and cognitively impaired, and with low social contacts are associated with higher attrition rates. This is consistent with prior findings in the literature (e.g., Powell et al. 1990; Sugisawa et al. 1999). However, unlike some previous research that finds older age is associated with a higher attrition (e.g., Slymen et al. 1996), we find that the age pattern is not significant from Wave Two to Wave Three.[9]

Urban respondents in the CLHLS are more likely to be lost to follow-up, partly because of changes that were made in administrative zones in urban areas as

---

[7] Those who died but followed up at the subsequent wave is not considered as a type of sample attrition in this study.

[8] Those who were lost to the follow-up also include some who actually died.

[9] There is a significant age pattern of sample attrition between Wave One and Wave Two: younger oldest-old are more likely to be lost to follow-up.

Table 3.5 Odds ratios of lost to follow-up by selected variables in the CLHLS

Variables	Lost to follow-up in 2000	Lost to follow-up in 2002
Females (males)	1.21*	1.20*
Age 90–99 (age 80–89)	0.68***	1.03
Age 100–105 (age 80–89)	0.51***	0.90
Rural (urban)	0.46***	0.70***
Minority ethnicity (Han)	0.46***	0.62**
1 + schooling (no schooling)	1.49***	1.23*
Currently married (not married)	1.04	1.07
Living alone (others)	1.19	1.17#
High proximity with children (low)	0.82*	0.80**
Poor ADL (good ADL)	1.14	1.32***
Poor MMSE (good MMSE)	1.09	1.20**
Self-reported poor health (good health)	0.97	1.00
Proxy (no proxy)	0.89	0.89#
Missing group 2 (missing group 1)	1.11	1.22**
Missing group 3 (missing group 1)	1.13	1.33***
2000 newly interviewed (1998 interviewed)	–	1.31***
N	8,805	10,844
-2LL	5298.1***	8506.6***

Age 80–105 only. Three missing groups are classified based on the missing rate of each respondent. Group1, <2%; Group 2, 2–5%; Group 3, >5%. #, $p<0.05$; *∗, $p<0.01$; ***, $p<0.001$.

indicated earlier, and partly because urban respondents have a higher mobility than their rural counterparts. Respondents of minority ethnicities are less frequently lost in follow-up surveys compared with Han respondents. Furthermore, respondents who have missing items of 5 percent or more have 13–33 percent more chance of being lost to follow-up compared with those respondents with missing values of less than 2 percent, after controlling for sociodemographic attributes and health conditions at previous waves. It is interesting to note that respondents using a proxy are less likely to be lost to follow-up in the following wave, possibly because they are less mobile, which makes them easier to locate.

As reported by Norris and Goudy (1986) and Kempen and van Sonderen (2002), the effects of sample attrition on outcomes depends on how reinterviewed respondents differ from all those who are eligible to do so. Furthermore, the strong linkage between sample attrition and its associates does not necessarily mean that the coefficients of predictors for outcomes of interest must be affected by sample attrition. Kempen and van Sonderen (2002) demonstrate that attrition might not always be a serious problem when associations between variables are the focus of a study, particularly when the proportion of dropouts is not too large, although a cross-sectional descriptive analysis at a later wave may be more affected by attrition. Therefore, it is unlikely that there will be significant problems in estimations in the CLHLS, with its relatively low sample attrition. All compensation approaches mentioned above for nonresponse items are fully applicable to deal with sample attrition wherever necessary.

Table 3.6 Inconsistent responses for selected items in the CLHLS

Inconsistent items	1998		2000		2002[a]		2002[b]	
	#	%	#	%	#	%	#	%
1. ADL fully dependent but can pick-up a book while standing	112	1.27	110	1.00	108	0.99	0	0.00
2. ADL fully independent but can't stand up from a chair	50	0.57	83	0.76	230	2.11	96	1.98
3. Can't stand up from a chair but does housework or fieldwork everyday	6	0.07	4	0.04	117	1.07	66	1.36
4. Reported bedsores but does housework or field work everyday	6	0.07	26	0.24	6	0.05	1	0.02
5. Had a proxy for answering some questions but interviewer didn't mark[c]	891	10.12	544	4.96	248	2.27	86	1.78

[a] Age 80–105
[b] Age 65–79
[c] This might be caused by interviewer's misunderstanding the question "Did anyone help the interviewee to answer any question?" They might have mistakenly understood it as referring only to those questions that must be answered by interviewee

3.5 Logical Error (Inconsistent Response)

Logical errors might occur across all questions due to inconsistent answers provided by interviewees, the carelessness of interviewers, and mistyping or miscoding of data entries. Tables 3.6 and 3.7 show that the inconsistency of responses given by interviewees or proxies is slightly higher in the 2002 wave compared with levels in the previous waves, although the inconsistency given by interviewers is lower in the 2002 wave. Inconsistent responses seem to increase slightly with age. The difference between genders is trivial.[10]

Table 3.7 Distribution of Inconsistent responses in the CLHLS

Ages	Males			Females		
	1998	2000	2002	1998	2000	2002
65–79			4.72 (2.75)			4.94 (2.95)
80–89	12.53 (2.41)	7.70 (2.35)	6.25 (3.90)	14.70 (2.93)	8.24 (2.49)	6.35 (3.60)
90–99	14.16 (3.93)	8.27 (2.92)	6.94 (4.42)	13.59 (3.44)	7.94 (3.78)	7.81 (5.59)
100+105	13.97 (5.46)	11.75 (6.39)	9.92 (7.24)	13.34 (4.71)	8.54 (4.61)	8.23 (6.65)

The figures in parentheses do not include interviewer's misunderstanding the question "Did anyone help the interviewee to answer any question?"

---

[10] A similar pattern was also observed in the National Long-Term Care Survey in the U.S. (Wallance 1992: 133).

3.6 Reliability of Major Health Measurements

In the CLHLS, health has been conceptualized in a multidimensional manner, with a general emphasis on physical and mental domains. The use of existing standardized instruments has the benefit of prior experience and information on measurement properties (Wallace and Herzog 1995), which is increasingly advocated as key outcome measures in health surveys (McHorney et al. 1994). Several translated Chinese versions of activities of daily living (ADL), instrumental activities of daily living (IADL), and the Mini-Mental State Examination (MMSE) have been developed and have been shown to be reliable and valid (e.g., Chou 2003; Zhang 1993; Zhang et al. 1998). However, these scales mainly focus on young adults or young elders. Their appropriateness for the oldest-old has not been determined. The CLHLS provides an opportunity to examine their reliabilities and validities among this rapidly growing subpopulation who need the most help but about whom we know very little. One unique feature of the CLHLS is that relatively comprehensive information on the extent of disability and suffering before dying was obtained by interviewing a close family member (next of kin). The 2002 wave gathered 3,340 questionnaires for the deceased aged 80 and over who died between the 2000 and the 2002 waves. The remaining parts of this chapter aim to provide a relatively detailed assessment on the reliability and validity for all major health domains in the CLHLS.

3.6.1 Internal Consistency of Multiple Items Scales

Internal-consistency reliability for selected measurements was estimated using Cronbach's alpha coefficient (Cronbach 1951). A minimum reliability coefficient of 0.70 has been recommended for group-level analyses, while reliability coefficients of 0.90 or greater have been suggested for individual-level analyses (Nunnally 1994; Stewart et al. 1992).

Table 3.8 shows that all Cronbach's alpha coefficients for the ADL scale (consisting of bathing, dressing, toileting, indoor transferring, continence, and eating) and MMSE in the 2002 wave are above the 0.70 criterion suggested for group comparisons, indicating good internal consistency. It is worth noting that the reliability of ADL before dying is higher than that for survivors, although questions of ADL before dying were all answered by next of kin. The IADL items are a combination of different sources derived from major surveys for elders around the world. The reliability for eight IADL items in the 2002 wave is also high, indicating the possibility of creating a scale. On the other hand, the data reported in Table 3.8 indicate that the reliability coefficients for negative and positive personality variables are lower than 0.70 if we exclude those who are too sick to be able to answer questions,[11]

---

[11] Questions related to personality and cognitive functioning must be answered by the interviewee themselves; no proxy is allowed in this regard. If the interviewee is too sick to answer a question, the interviewer marks "unable to answer" for that question.

Table 3.8 Reliability coefficients and validity for selected measures in the 2002 wave

Scales and measures[a]	Cronbach's alpha coefficient	N	Range of alphas if individual item deleted	Range of item-total correlations	% at floor[b]	% at ceiling[b]
Age 80–105						
Functioning of upper extremities (3)	0.833	10,912	0.753–0.794	0.668–0.710	5.3	80.0
Functioning of body mobility (2)	0.762	10,905	NA	0.617	7.8	44.8
Negative personality related variables (3)	0.891	10,953	0.815–0.892	0.753–0.822	–	–
Negative personality related variables (3)[c]	0.662	9,157	0.463–0.696	0.391–0.550	0.0	2.2
Positive personality related variables (4)	0.918	10,953	0.880–0.917	0.750–0.857	–	–
Positive personality related variables (4)[c]	0.453	8,838	0.344–0.416	0.237–0.307	0.5	10.0
ADL (6)	0.867	10,905	0.815–0.876	0.508–0.839	0.6	59.5
ADL for deceased persons between 2000 and 2002 (6)	0.939	3,188	0.916–0.930	0.624–0.902	15.9	22.9
IADL (8)	0.937	10,951	0.924–0.932	0.727–0.832	18.8	14.3
Mini-Mental State Examination (MMSE) (22)	0.984	10,945	0.983–0.984	0.771–0.901	10.9	19.7
Mini-Mental State Examination (MMSE) (22)[c]	0.888	6,971	0.877–0.887	0.286–0.659	0.1	30.6
Age 65–79						
Functioning of upper extremities (3)	0.892	4,843	0.839–0.858	0.775–0.798	1.0	92.5
Functioning of body mobility (2)	0.585	4,843	NA	0.418	0.6	86.4

Table 3.8 (Continued)

Scales and measures[a]	Cronbach's alpha coefficient	N	Range of alphas if individual item deleted	Range of item-total correlations	% at floor[b]	% at ceiling[b]
Negative personality related variables (3)	0.655	4,845	0.482–0.690	0.392–0.526	–	–
Negative personality related variables (3)[c]	0.625	4,758	0.438–0.678	0.351–0.500	0.2	15.2
Positive personality related variables (4)	0.630	4,845	0.530–0.596	0.414–0.490	–	–
Positive personality related variables (4)[c]	0.456	4,684	0.323–0.428	0.224–0.326	0.0	3.2
ADL (6)	0.817	4,843	0.745–0.854	0.355–0.826	0.1	93.1
IADL (8)	0.862	4,843	0.835–0.903	0.469–0.730	1.2	66.9
Mini-Mental State Examination (MMSE) (22)	0.952	4,844	0.948–0.951	0.574–0.777	0.6	54.6
Mini-Mental State Examination (MMSE) (22)[c]	0.788	4,461	0.764–0.789	0.120–0.568	0.0	59.3

[a]Some are not designed as scales in the questionnaire, but they are related variables to measure similar functioning. Our purpose here is to examine their reliability to see the possibility of generating scales later on
[b]Percentage of subjects with worst and best possible scores, respectively
[c]Excluding persons who were 'unable to answer' these questions. If persons are too sick to answer such questions that should be answered ONLY by interviewees, the answers for such questions are "unable to answer"
The figure in the parentheses indicates the number of items in corresponding scales or groups. NA, not applicable; –, not calculated since those close-end questions contain the "unable to answer" answer, which didn't provide any information regarding personality. Four newly added questions in the cognitive function section in the 2002 questionnaire are not included in generating the MMSE scale. Four newly added questions are mainly designed for the elderly aged 65–79 by the CLHLS research team. We suggest users drop these four variables in creating MMSE scores since they are not included in the MMSE scale proposed by Folstein et al. (1975).

implying that they might not be appropriate to use in scale generation.[12] It is also interesting to note that the reliability for some scales and variables among young elders is slightly lower than those for the oldest-old. The reasons for this are unclear. Smaller sample sizes among the younger elderly could be a possible cause. Further research is clearly warranted.

3.6.2 Homogeneity

The homogeneity of the ADL, IADL, and MMSE scales, and other potential scales is assessed by evaluating item-total correlations. Item-total correlations compute the correlation between an item and its own scale with the item of interest eliminated from the calculation of the score. Although some researchers argue that it is considered satisfactory if the item-total correlation reaches 0.40 or more for the purpose of comparison (Ware et al. 1980), other scholars suggest that the criterion for item-total correlations might be efficient if it exceeds 0.20, especially for categorical items that define the extremes of the scale range (Streiner and Norman 1995). Our results show that all item-total correlations (see Table 3.8) are over the minimum requirement suggested by Streiner and Norman (1995), and even over the 0.40 criterion, except those for personality if persons unable to answer questions are excluded. Note that for the ADL scale, if continence is deleted, both the reliability coefficients and the item-total correlations will be substantially improved, and the item-total correlation will pass 0.40 for young elders in the 2002 dataset, indicating the possibility of removing continence from ADL scales, as some recent studies have done (e.g., Jagger et al. 2001).

The percentage of respondents at the highest possible (ceiling) or lowest possible (floor) scores also should be noted. As for scales or variables related to functional limitations, a negative skew pattern is observed, indicating distributions with respondents scoring toward the positive end of the scales in the CLHLS. This is anticipated for a generally well elderly subpopulation.

3.7 Validity of Major Health Measurements

The validity of a measure in the health field has often been evaluated by its content, construct, and criterion validity (Gandek and Ware 1998). Content validity examines the extent to which a measure or questionnaire represents the universe of concepts or domains; that is, whether the measure offers an adequate sample of the content of a construct (Stewart et al. 1992). Construct validity is a process in

---

[12] In designing the personality scale, we did not follow existing scales because most scales are developed in Western countries, and might not be appropriate for use in China. We, therefore, selected some major items from various scales that we believed were appropriate for use with the Chinese elderly. Therefore, it is better to analyze these variables individually, which is confirmed by the results in Table 3.8.

which validity is evaluated in terms of the extent to which a measure correlates with variables in a manner consistent with theory (Stewart et al. 1992). Convergent and discriminant validity are at the foundation of construct validation. Convergent validity is supported when different methods of measuring the same construct provide similar results, whereas discriminant validity is supported when a measure of one underlying construct can be differentiated from another construct. In brief, high and consistent correlations were assumed between an item and its own scale, and significantly lower associations between that item and all other scales. If a scale is valid, items on which the scale is based should be related to each other (convergent validity) and not related to measures of different concepts (discriminant validity).

For establishing convergent and discriminant validity of the measures, relationships of selected scales and measures have been examined, and the results are presented in Table 3.9. ADL measures daily functioning in terms of eating, dressing, moving, toileting, continence, and bathing. IADL also measures daily functioning but with respect to more difficult tasks. If they are valid, they are expected to have a higher correlation between them and a higher correlation with functional capacity of extremities and body mobility than correlations with personality measures. On the other hand, if the personality measures are valid, positive and negative personality should have a higher correlation between each other than the correlations between them and other measures. Table 3.9 also presents the ranges of all possible correlation coefficients within scales (measurements) and across scales (measurements). It is apparent that all correlations between items within the same dimension or similar dimensions are much higher than correlations between items from different dimensions. Moreover, the correlations between the cognitive performance measures and the IADL index of cognitive functioning are positive but small, reflecting similar findings reported in the literature (Morris 1983; Wallace and Herzog 1995). It is clear that the results presented in Table 3.9 support a good convergent and discriminant validity for these measurements in the 2002 wave.

Another approach for testing construct validity of measures is factor analysis, which measures whether the same dimensional variables load on the same factor (Stewart et al. 1992). Our results support the good validity of these measures in the 2002 wave.[13]

3.8 Concluding Remarks

This chapter has examined the data quality of the 2002 wave of the CLHLS, mainly on proxy use, item incompleteness, sample attrition, and the dimensionalities of reliability and the validity of health condition measurements. Based on the results, we are generally pleased with the quality of the health indicators in the CLHLS. Analyses of health measures showed high reliability and validity on items that we were able to

---

[13] The results of factor analyses are not shown in this chapter due to limited space, but they are available upon request.

Table 3.9 Convergent and discriminant validity for selected measures in the 2002 wave[a]

Scales and measures[a]	FU	FB	NP	PP	ADL	MMSE	IADL
Age 80–105							
Functioning of upper extremities (3)	**0.68–0.72**	0.21–0.27	0.07–0.12	0.05–0.08	0.13–0.26	0.03–0.18	0.16–0.22
Functioning of body mobility (2)	0.21–0.27	**0.54**	0.12–0.16	0.09–0.13	0.19–0.40	0.08–0.26	0.37–0.48
Negative personality related variables (3)[b]	0.07–0.12	0.12–0.16	**0.31–0.57**	0.11–0.33	0.04–0.11	0.02–0.13	0.07–0.18
Positive personality related variables (4)[b]	0.05–0.08	0.09–0.13	0.11–0.33	**0.16–0.32**	0.02–0.12	0.00–0.15	0.03–0.18
ADL (6)	0.13–0.26	0.19–0.40	0.04–0.11	0.02–0.12	**0.23–0.63**	0.08–0.20	0.16–0.52
Mini-Mental State Examination (MMSE) (22)[b]	0.03–0.18	0.08–0.26	0.02–0.13	0.00–0.15	0.08–0.20	**0.09–0.78**	0.09–0.30
IADL (8)	0.16–0.22	0.37–0.48	0.07–0.18	0.03–0.18	0.16–0.52	0.09–0.30	**0.48–0.69**
Age 65–79							
Functioning of upper extremities (3)	**0.77–0.80**	0.12–0.14	0.02–0.06	0.00–0.07	0.02–0.19	0.00–0.09	0.13–0.17
Functioning of body mobility (2)	0.12–0.14	**0.38**	0.03–0.10	0.01–0.09	0.07–0.29	0.01–0.13	0.25–0.40
Negative personality related variables (3)[b]	0.02–0.06	0.03–0.10	**0.30–0.58**	0.11–0.33	0.00–0.10	0.00–0.10	0.02–0.17
Positive personality related variables (4)[b]	0.00–0.07	0.01–0.09	0.11–0.33	**0.13–0.26**	0.00–0.08	0.00–0.09	0.03–0.15
ADL (6)	0.02–0.19	0.07–0.29	0.00–0.10	0.00–0.08	**0.14–0.60**	0.00–0.18	0.10–0.44
Mini-Mental State Examination (MMSE) (22)[b]	0.00–0.09	0.01–0.13	0.00–0.10	0.00–0.09	0.00–0.18	**0.01–0.60**	0.01–0.18
IADL (8)	0.13–0.17	0.25–0.40	0.02–0.17	0.03–0.15	0.10–0.44	0.01–0.18	**0.30–0.68**

[a]Some are not designed as scales in the questionnaire, but they are related variables to measure similar functioning. Our purpose here is to examine their reliability to see the possibility of generating scales later on
[b]Excluding persons who were 'unable to answer' these questions. If persons are too sick to answer such questions that should be answered ONLY by interviewees, the answers for such questions are "unable to answer"
Correlation coefficients are Spearman Cofficients. Bold numbers are correlation coefficients between items within the same scale or group
FU, functioning of upper extremities; FB, functioning of body mobility; NP, negative personality measures; PP, positive personality. The main purpose of the tabe is to see the magnitude of correlation coeffients in terms of an absolute number. Therfore, negative coefficients have been represented as an absolute number. (4) For other notations see the note in Table 3.8

evaluate and exceeded widely used criteria. Therefore, we are confident that they are measuring meaningful underlying concepts, thereby permitting comparisons between groups. The results reported in this chapter also suggest that the Chinese translated version of the Katz ADL Index and the Chinese version of the MMSE are both reliable and valid for the oldest-old.[14] In sum, the evidence above has led us to believe that the data quality of the 2002 wave of the CLHLS is generally good.

We recommend that attention be given to the following issues and items. First, it is inappropriate and not recommended to generate a scale for personality measurements since their reliability is below the required cut off point. Second, higher proxy use is related to older age, lower education, rural residence, lower cognitive functioning, and higher disability. Therefore, it would be better to add an indicator variable (i.e., the presence or absence of a proxy) in the analysis when the aim of the proposed research is to examine the effects of these factors. Third, we find that item incompleteness and sample attrition are linked to age, gender, urban/rural residence, ethnicity, and health conditions. Although it is unlikely that these limitations will significantly affect results, sufficient attention must be paid to them in verifying and reporting the outcomes.

Acknowledgments The research reported in this chapter is supported by the National Institute on Aging grant (R01 AG023627-01) and the National Natural Science Foundation of China key project grant (No. 70533010).

References

Allison, P. (2002), *Missing data*. Thousand Oaks, CA: Sage
Basset, S.S. and J. Magaziner (1988), *The use of proxy responses on mental health measures for aged, community-dwelling women*. Paper presented at the 41st annual scientific meeting of the Gerontological Society of America, San Francisco
Blazer, D., D.C. Hughes, and L.K. George (1987), The epidemiology of depression in an elderly community population. *Gerontologist* 27, pp. 281–287
Chou, K.L. (2003), Correlates of everyday competence in Chinese older adults. *Aging Mental Health* 7 (4), pp. 308–315
Coroni-Huntley, J., D.B. Brock, A.M. Ostfeld, J.O. Taylor, and R.B. Wallace (1986), *Established population or epidemiological studies of the elderly: Resource data book*. Washington, DC: NIH Publication NO. 86-2443
Cronbach, L.J. (1951), Coefficient alpha and the internal structure of tests. *Psychometrika* 16, 297–334

---

[14] The generally good quality of CLHLS data is also due to the data quality control program used in the CLHLS. Before data entry, a three-stage check is employed: a local site check, provincial check, and final check at the Mainland Information Company in Beijing. Questionnaires are returned to participating sites for correction if local and provincial supervisors or supervisors at the Mainland Information Company find them with missing items or errors. Data entry is conducted at Peking University. In data entry, specific logic, range, and consistency checks between related items are added to the data entry program using EPI 6.0 software. Data double-entry is conducted at Peking University under professional supervision to minimize entry errors. A questionnaire is returned to the participating site for correction if a logic error is detected in the questionnaire not due to a coding or entry error.

DeHaan, R., N. Aaronson, M. Limburg, K. Langton-Hewer, and H. van Crevil (1993), Measuring quality of life in stroke. *Stroke* 24, pp. 320–326

Folstein, M.F., S.E. Folsein, and P.R. McHugh (1975), "Mini-Mental state": A practical method for grading the cognitive state of pattern for clinician. *Journal of Psychological Research* 12, pp. 189–198

Francis, J.D. and L. Busch (1975), What we know about "I don't know". *Public Opinion Quarterly* 39, pp. 207–218

Gandek, B. and J.E. Ware Jr. (1998), Methods for validating and norming translations of health status questionnaire: The IQOLA project approach. *Journal of Clinical Epidemiology* 51 (11), pp. 953–959

Groves, R.M. (1987), Research on survey data quality. *Public Opinion Quarterly* 51, *pp.* S156–S172

Gu, D. and Zeng, Y. (2004), Data assessment of the CLHLS 1998, 2000, and 2002 waves. In: Zeng, Y., Y. Liu, C. Zhang, and Z. Xiao (eds.): *Analyses of the determinants of healthy longevity*. Beijing, China: Peking University Press, pp. 3–22

Hayward, M.D. and B.K. Gorman (2004), The long arm of childhood: The influence of early-life social conditions on men's mortality. *Demography* 41 (1), pp. 87–107

Herzog, A.R. and W.L. Rodgers (1981), Age and satisfaction—data from several large surveys. *Research on Aging* 3 (2), pp. 142–165

Herzog, A.R. and W.L. Rodgers (1988), Age and response rates to interview sample survey. *Journal of Gerontology: Social Sciences* 43, pp. S200–S205

Herzog, A.R. and W.L. Rodgers (1992), The use of survey method in research on older Americans. In: R.B. Wallace and R.F. Woolson (eds.): *The epidemiologic study of the elderly*, New York: Oxford University Press, pp. 60–90

Jagger, C., A.J. Arthur, N.A. Spiers, and M. Clark (2001), Patterns of onset of disability in activities of daily living with age. *Journal of American Geriatrics Society* 49, pp. 404–409

Jay, G.M., J. Liang, X. Liu, and H. Sugisawa (1993), Patterns of nonresponse in a national survey of elderly Japanese. *Journal of Gerontology: Social Sciences* 48, pp. S143–S152

Jenkins, K.R. and N.H. Fultz (2005). Functional impairment as a risk factor for urinary incontinence among older Americans. *Neurourology and Urodynamics* 24, pp. 51–55

Kalton, G. and D. Kasprzyk (1986), The treatment of missing survey data. *Survey Methodology* 12, pp. 1–16

Katz S., A.B. Ford, R.W. Moskowitz, B.A. Jackson, and M.W. Jaffe (1963), Studies of illness in the aged. The index of ADL: A standardized measure of biological and psychosocial function. *Journal of the American Medical Association* 185 (12), pp. 914–919

Kempen, G.I.J.M. and E. van Sonderen (2002), Psychological attributes and changes in disability among low-functioning older persons: Does attrition affect the outcomes? *Journal of Clinical Epidemiology* 55, pp. 224–229

Klinkenberg, M., J.H. Smit, J.H. Deeg, D.L. Willems, B.D. Onwuteaka-Philipsen, G. van der Wal (2003), Proxy reporting in after-death interviews: The use of proxy respondents in retrospective assessment of chronic diseases and symptom burden in the terminal phase of life. *Palliative Medicine* 17, pp. 191–201

Kolonel, L.N., T. Hirohata, and A.M.Y. Nomura (1977), Adequacy of survey data collected from substitute respondents. *American Journal of Epidemiology* 106, pp. 476–484

Landerman, L.R., K.C. Land, and C.F. Pieper (1997), An empirical evaluation of the predictive mean matching method for imputing missing values. *Sociological Methods and Research* 26 (1), pp. 3–33.

Lee, D.S. (2002), *Trimming for bounds on treatment effects with missing outcomes*. Center for Labor Economics, University of California, Berkeley. Working paper no. 51

Lepkowski, J., G. Kalton, and D. Kasprzyk (1989), Weighting adjustments for partial nonresponse in the 1984 SIPP panel. *Proceedings of the Section on Survey Research Methods*. Alexandria, VA: American Statistical Association, pp. 296–301

Lindner, J.R., T.H. Murphy, and G.E. Briers (2001), Handling nonresponse in social science research. *Journal of Agricultural Education* 42 (4), pp. 43–53

Magaziner J., E.M. Simonsick, T.M. Kashner, and J.R. Hebel (1988), Patient proxy response comparability on measures of patient health and functional status. *Journal of Clinical Epidemiology* 41, pp. 1065–1074.

McHorney, C.A., J.E. Ware, J.F.R. Lu, and C.D. Sherbourne (1994), The MOS 36 item Short Form Health Survey (SF-36) III: Tests of data quality, scaling assumptions and reliability across diverse patient groups. *Medical Care* 32, pp. 40–62

Mihelic, A.H. and E.M. Crimmins (1997), Loss to follow-up in a sample of Americans 70 years of age and older: The LSOA 1984–1990. *The Journals of Gerontology* 52B, pp. S37–S48

Mohadjer, L., B. Bell, and J. Waksberg (1994), *Accounting for item nonresponse bias—National Health and Nutrition Examination Survey III*. National Center for Health Statistics, Hyattsville, MD

Morris, P.E. (1983), The validity of subjective reports on memory. In: J.E. Harris and P.E. Morris (eds.): *Everyday memory: Actions and absent-mindedness*. London: Academic Press, pp. 153–172

Mossel, P.A. and R.S. Brown (1984), *The effect of sample attrition on estimates of channeling's impact for an early sample*. Executive summary, U.S. DHHS, Washington, DC

Norris, F.H. and W.J. Goudy (1986), Characteristics of older nonrespondents over five waves of a panel study: Comments. *Journal of Gerontology* 41, pp. 806–807

Nunnally, J.C. (1994), *Psychological theory*. 3rd ed. New York: McGraw-Hill

Office of Management and Budget (OMB) (2001). *Measuring and reporting sources of errors in survey*. Statistical Policy working paper No. 31. Office of Management and Budget, Washington, DC

Pierre, U., S. Wood-Dauphinee, N. Korner-Bitensky, D. Gayton, and J. Hanley (1998), Proxy use of the Canadian SF-36 in rating health status of the disabled elderly. *Journal of Clinical Epidemiology* 51, pp. 983–990

Powell, D.A., E. Furchtgott, M. Henderson, L. Prescott, A. Mitchell, P. Hartis, J.D. Valentine, and W.L. Milligan (1990), Some determinants of attrition in prospective studies on aging. *Experimental Aging Research* 16, pp. 17–24

Rodgers, W.L. (1988), *Epidemiological survey of older adults: Response rate, data quality and the use of proxies*. Paper presented at the 41st annual scientific meeting of the Gerontological Society of American, San Francisco

Rodgers, W.L. and A.R. Herzog (1992), Collecting data about the oldest: Problems and procedures. In: R.M. Suzman, D.P. Willis, K.G. Manton (eds.): *The oldest old*. New York: Oxford University, pp. 135–156

Rothman, M.L., S.C. Hedrick, K.A. Bulcroft, D.H. Hickam, and L.Z. Rubenstein (1991), The validity of proxy-generated scores as measures of patient health status. *Medical Care* 29, pp. 115–124

Slymen, D.J., J.A. Drew, J.P. Elder, and S.J. Williams (1996), Determinants of non-compliance and attrition in the elderly. *International Journal of Epidemiology* 25, pp. 411–419

Stewart, A.L., R.D. Hays, and J.E. Ware (1992), Methods of constructing health measure. In: A.L. Stewart and J.E. Ware (eds.): *Measuring function and well-being—The medical outcome study approach*. Durham, NC: Duke University Press, pp. 67–85

Streiner, D.L. and G.R. Norman (1995), *Health measurement scales: A practical guide to their development and use*. 2nd ed. Oxford: Oxford University Press

Sugisawa, H., H. Kishino, Y. Sugihara, H. Okabayashi, and H. Shibata (1999), Comparison of characteristics between respondents and nonrespondents in a national survey of Japanese elderly using six year follow-up study. *Nippon Koshu Eisei Zasshi* 46 (7), pp. 551–562

Tang, S.T. and R. McCorkle (2002), Use of family proxies in quality of life research for cancer patients at the end of life: A literature review. *Cancer Investigation* 20 (7–8), pp. 1086–1104

Vellas, B.J., S.S. Wayne, P.J. Garry, and R.N. Baumgartner (1998), A two-year longitudinal of falls in 482 community-dwelling elderly adults. *The Journals of Gerontology* 53A (4), pp. M264–M274

Wallace, R.B. and A.R. Herzog (1995), Overview of the health measures in the health and retirement study. *The Journal of Human Resources* 30 (supplement), pp. S84–S107

Wallace, R.B., F.J. Kohout, and P.L. Colsher (1992), Observations on interview survey of the oldest old. In: R.M. Suzman, D.P. Willis, and K.G. Manton (eds.): *The oldest old.* New York: Oxford University, pp. 123–134

Ware, J.E., R.H. Brook, A. Davies-Avery, K. Williams, A.L. Stewart, W.H. Rogers, C.A. Donald, and S.A. Johnston (1980), *Model of health and methodology.* Santa Monica, CA: RAND Corporation

Zeng, Y., J.W. Vaupel, Z. Xiao, C. Zhang, and Y. Liu (2001), The healthy longevity survey and the active life expectancy of the oldest old in China. *Population* 13 (1), pp. 95–116

Zhang, M.Y. (1993). *Manual of evaluation of scales in psychiatrics.* Changsha, China: Hunan Science and Technology Press

Zhang, M., Z. Zhu, and P. Chen (1998), Community investigation of the activities of daily living (ADL) and medical conditions of the elderly in Shanghai. *Chinese Journal of Medical Science* 78 (2), pp. 124–127

Zimmer, Z., L. Martin, and M.C. Chang (2002), Changes in functional limitation and survival among older Taiwanese, 1933, 1996, and 1999. *Population Studies* 3, pp. 265–276

Zsembik, B.A. (1994), Ethnic and sociodemographic correlates of the use of proxy respondents: The National Survey of Hispanic Elderly People, 1988. *Research on Aging* 16 (4), pp. 401–414

Chapter 4
Reliability of Age Reporting Among the Chinese Oldest-Old in the CLHLS Datasets

Zeng Yi and Danan Gu

Abstract This chapter evaluates age reporting among the oldest-old, especially centenarians, in the Chinese Longitudinal Healthy Longevity Survey (CLHLS) based on comparisons of various indices of elderly age reporting and age distributions of centenarians in Sweden, Japan, England and Wales, Australia, Canada, China, the USA, and Chile. The analyses demonstrate that age reporting among the oldest-old interviewees (Han and six minority groups combined) in the 22 provinces in China where the CLHLS has been conducted is not as good as that in Sweden, Japan, and England and Wales, but is relatively close to that in Australia, more or less the same as that in Canada, better than that in the USA (all race groups combined), and much better than that in Chile. As indicated by the higher density of centenarians, age exaggeration exists in the six ethnic minority groups in the 22 Han-dominated provinces, although we cannot rule out and quantify the potential impacts of past mortality selection and better natural environmental conditions among these minority groups. We find that the age exaggeration of minorities in the CLHLS may not cause substantial biases in demographic and statistical analyses using the CLHLS data, since minorities consist of a rather small portion of the sample (6.8 percent at baseline and 5.5 percent in the grand total sample of the 1998, 2000, and 2002 waves).

Keywords Age exaggeration, Age heaping, Age misreporting, Age reporting, Animal year, Australia, Canada, Chile, China, Census data, Density of centenarian, Distributions of centenarians, England and Wales, Ethnic minority, Han Chinese, Han-dominated provinces, Japan, Kannisto–Thatcher Database, Late childbearing, Lunar calendar, Myer's index, Pre-designed, Ratio of centenarian, Ratio index, Sweden, The oldest-old, The USA, Western calendar, Whipple's index

---

Zeng Yi
Center for Study of Aging and Human Development,
Medical School of Duke University, Durham, NC 27710, USA,
Center for Healthy Aging and Family Studies/China Center for Economic Research, Peking University, Beijing, China
e-mail: zengyi68@gmail.com

4.1 Introduction

This chapter examines the quality of age reporting among the oldest-old in the Chinese Longitudinal Healthy Longevity Survey (CLHLS). Assessments of data quality pertaining to disability, cognitive reliability and validity, proxy use, non-response rate and data incompleteness, sample attrition, and logical consistency in the CLHLS may be found in Chap. 3 of this volume and in Zeng et al. (2001).

Age reporting is a crucial issue in the study of healthy longevity. Age exaggeration will cause an underestimation of mortality rates at higher ages (Coale and Li 1991). The literature has established that accurate age reporting is generally a feature of developed societies with a few exceptions (Coale and Kisker 1986, 1990; Ewbank 1981; Seltzer 1973; Thatcher 1981). On the other hand, although age reporting among elderly persons in most developing countries is poor mainly due to age exaggeration (Dechter and Preston 1991; Retherford and Mirza 1982; Rosenwaike and Preston 1984), previous studies have shown that the quality of age reporting appears to be relatively good in several developing countries where date of birth has longstanding astrological significance such as among the Han majority in China (e.g., Coale and Li 1991; Wang et al. 1998), Korea (Jowett and Li 1982) and some other countries or regions (Knodel and Chayovan 1991).

Specifically, Coale and Li (1991) have shown that Han Chinese (and some minorities who over the years have been culturally and residentially integrated with the Han) tend to use the Chinese lunar calendar (for older generations) and the Western calendar (for younger generations) plus animal year to remember their birthdays[1]. This is important for Chinese people because the precise date of birth is significant in making decisions on important life events such as matchmaking for marriage, date of marriage, and the date to start building a house, among other events. Therefore, Han Chinese, even if illiterate, can usually provide a reliable date of birth for themselves or for their close family members (Coale and Li 1991).

In the CLHLS data collection efforts we employed user-friendly forms for converting the reported birth dates of the Chinese lunar calendar into the Western calendar. The CLHLS asks for date of birth (rather than age directly) and computes the respondent's age after the survey by subtracting it from the date of the survey, because the Chinese system of calculating nominal age may make the response ambiguous.[2] Other information relevant to the date of birth such as genealogical record,

---

[1] Although the Chinese animal year cycle is 12, there are no preferences for reporting ages, which are a multiple of 12. Some people may prefer their children to be born in a particular animal year (preference varies with region and time period), which may cause the birth rate in such a year to be somewhat higher than that in other years. But once born, Chinese people remember their actual animal year precisely, according to Chinese cultural tradition. Combining the animal year and a Chinese calendar such as the Gan Zhi, which reports the year or the years since the establishment of the ruling period of an emperor (e.g. Guang Xu year) or the Republic of China (Ming Guo year), for old people, and the Western calendar for younger people generally helps Han Chinese people to accurately report their birth date.

[2] According to the Chinese nominal age system, a person is counted as one year old on the day of birth, and one year older with each Chinese new year's day so that the nominal age is exaggerated

ID card, and household registration booklet were also collected in the CLHLS to validate the sampled elder's age. The interviewers and supervisors also check the parents' age, sibling's age, and the children/grandchildren's age of the sampled person, and the age of the sampled person at marriage and at birth, and so forth, to further validate age reporting. An additional question was designed for each interviewer to provide his/her judgment on the validity of the sampled person's age in the interviewer section in the CLHLS questionnaire. If the sampled person reported her/his age to be over 105, the interviewer was instructed to obtain additional evidence or concurrence from the local residential committee and local aging committee. If any inaccuracy in the reported age or any other logical problem in the questionnaire was found, a re-interview or phone call regarding specific questions was conducted.[3]

We examined the quality of age reporting in the CLHLS (1998) through comparisons with Sweden (1970–79), Japan (1970–79), England and Wales (1970–79), Australia (1970–79), Canada (1960–69), the USA (1960–69), and Chile (1980–89), all with similar life expectancies at age 65[4] as compared to the 22 surveyed provinces in China in 1998[5]. The inter-country comparisons are based on the indices of age heaping, age-specific percentage distributions of centenarians, age ratios of centenarians proposed by Booth and Zhao (Chap. 5 in this volume), and the density of centenarians. It is commonly believed among international demographic experts in the aging fields that the data quality of age reporting among oldest-old people is

---

by one or two years as compared with the actual age. If one simply asks for age, someone may respond in nominal age and some others may provide actual age, which will result in false age records. During the logical checks on the completed questionnaires and during data processing, we discovered that some interviewers did not follow the instructions but simply put the nominal age in the blank cell of the questionnaire where they are supposed to convert the reported birth date into age. The interviewers obtained the nominal age either from the list of centenarians provided by the local aging committee, which sometimes consider the nominal age 100+ as qualified ages for issuing centenarians' subsides, or from conversations with the interviewees, although there is no question for directly asking age in the questionnaire. In these cases, the survey team corrected the nominal ages by the correct ages converted from the reported birth dates.

[3] For example, in 2002 the age reporting of 46 persons was inconsistent with what was recorded in the 2000 data set. The survey team re-visited them and corrected the errors case by case.

[4] Unlike Booth and Zhao (2007, Chapter 5 in this volume) who used Swedish data from 1943–52, data from England & Wales in 1950–55, and Japanese data in 1962–66 to compare to the 1998 CLHLS data in 1998 according to the closeness of Chinese life expectancy at birth around 1998, we selected the periods for inter-country comparisons based on the closeness to the Chinese life expectancy at age 65 ($e_{65} = 17$) around 1998. We used this approach because period life expectation at age 65 is more directly and closely related to age distributions at the oldest-old ages, but period life expectancy at birth largely depends on the mortality of infants, children, and young adults, which is not directly and closely relevant to the age distribution at the oldest-old ages. Therefore, an inter-country comparative analysis aimed at examining age reporting through constricting the age distributions among the oldest-old would be more robust if we use the closeness of life expectancy at age 65 instead of the life expectancy at birth as a criterion for selecting the periods of the data sources.

[5] The data for Sweden, E&W, Australia, Canada, and USA are from Human Mortality Database (http://www.mortality.org). The data for Chile are from the Kannisto and Thatcher Database on Old Age Mortality (http://www.demogr.mpg.de).

the best in Sweden and Japan, very good in England and Wales, and in Australia, acceptably good in Canada, not so good but acceptable for academic research in the USA and Chile. For example, the Kannisto–Thatcher Database on Old Age Mortality consists of 31 countries, including Sweden, Japan, England and Wales, Australia, Canada, the USA, and Chile, as well as the Singapore Chinese. The inclusion of these 31 countries was based on careful data quality evaluations by Kannisto and Thatcher: they determined that these data were sufficiently reliable and detailed for their demographic and comparative analysis on oldest-old mortality using uniform methods and measurements (Kannisto 1994). The Kannisto–Thatcher database classified these 31 countries into four relative categories, as follows: good quality, acceptable quality, conditional acceptable quality, and weak quality. Sweden, Japan, and England and Wales were in the "good quality" category. Australia was in the "acceptable quality" category. The data quality of Canada and US Whites was also considered to be generally acceptable; however, the oldest-old data of United States non-whites (and the US as a whole) and Chile were the least reliable, although they passed Kannisto–Thatcher's careful data evaluations and were therefore entered into the 31-country database of old age mortality (Kannisto 1994). Note that Chile was the only developing country (in addition to the data on the Singapore Chinese) which was entered into the Kannisto–Thatcher Database on Old Age Mortality.

4.2 Age Heaping

Given the specially designed CLHLS target-sampling procedure aimed at interviewing approximately equal numbers of males and females at each single age category from age 65 to age 99 (see Section 4.3 in Chap. 2 of this volume for more details), we could not use the age distribution of all interviewees in the CLHLS to examine age heaping. Therefore, data from the 2000, 1990, and 1982 censuses for the 22 CLHLS provinces were used to investigate age heaping as a general context for an analysis of CLHLS age reporting. Table 4.1 presents the percent distribution of sampled Han (93.2 percent at baseline and 94.5 percent in the three waves combined) and the six ethnic minority groups (6.8 percent at baseline and 5.5 percent in the three waves combined) in the CLHLS, as well as the Whipple's Index and Myer's Index, both of which are conventional measurements of age heaping based on census data. The census data indicate that there is no age heaping among the Han and the six minority groups in the 22 CLHLS provinces.

Since both the Whipple's Index and the Myer's Index do not focus on the very old ages, one may reasonably question its validity for identifying age heaping problems among the oldest-old (see Chap. 5 in this volume). Coale and Li (1991: 395) proposed an index to measure age heaping (digit preference) by deviation of the ratio of the number at each age to a two-stage moving average (the five-term average of a five-term average) from a perfect standard without any age heaping. Employing Coale and Li's method, we computed the average ratio at ages 85–105 for the 22 provinces in China in the 2000 census, Sweden, Japan, England and Wales, Australia, Canada, the United Sates, and Chile for selected periods in which each

4 Reliability of Age Reporting Among the Chinese Oldest-Old

Table 4.1 Ethnic composition and the age heaping indices

Ethnic group	Percent of the Sample in the 1998 CLHLS (%) ($N = 8,805$)*	Percent of the Sample in the 1998–2002 pooled CLHLS (%) ($N = 19,890$)*	Whipple's Index in census			Myer's Index in census		
			1982	1990	2000	1982	1990	2000
Han	92.75	94.03	101.5	100.5	101.1	1.48	2.85	2.04
Zhuang	4.41	3.48	100.1	102.1	104.3	2.79	2.25	2.88
Hui	1.31	1.01	101.4	102.4	105.7	1.81	2.71	2.69
Yao	0.57	0.42	101.1	101.1	102.8	3.58	2.28	2.50
Korea	0.11	0.07	103.2	104.3	104.1	1.96	2.33	1.96
Manchu	0.33	0.52	100.1	105.3	102.9	2.57	3.13	1.50
Mongolia	0.03	0.03	99.7	104.0	102.8	2.56	2.45	2.31

There are around 0.5% of respondents who are other ethnic minorities or whose ethnic identities are missing. They are excluded form the data in the second and third columns.
Based on the United Nations' criteria, Whipple's Index: <105 very good, 105–110 good, 110–125 so-so, >125 poor.
Myer's Index: <10 good, 10–20 so-so,
>20 poor. 1982 and 1990 data are cited from Zeng et al. (2001).

Table 4.2 Mean of the ratios (MR) of the number at each age to a two-stage moving average, ages 85–105

	Women		Men		Both Sexes	
	MR	%Diff Comp to Sweden	MR	%Diff Comp to Sweden	MR	%Diff Comp to Sweden
China, 22 provinces	0.793	5.4	0.753	3.9	0.782	5.1
Sweden, 1970s	0.753	0.0	0.725	0.0	0.744	0.0
Japan, 1970s	0.728	−3.3	0.689	−5.0	0.717	−3.6
England & Wales, 1970s	0.780	3.6	0.711	−1.9	0.768	3.3
Australia, 1970s	0.777	3.2	0.756	4.3	0.771	3.7
Canada, 1960s	0.789	4.8	0.783	8.0	0.786	5.7
USA, 1960s	0.823	9.3	0.809	11.6	0.818	10.0
Chile, 1980–90	0.860	14.2	0.849	17.1	0.857	15.2

The age-specific number of centenarians for 22 provinces in China in 2000 is estimated based on the age-specific percentage distribution in the CLHLS in 1998 and the total number of centenarians from the 2000 census (the 2000 census publication does not include age-specific numbers of centenarians).

country had a female life expectancy at age 65 which was close to that in China in 2000. We then calculated single-age-sex-specific and average odds for the ratio of each country compared to that of Sweden in the 1970s, considering the Swedish age distribution as perfect, that is, without any age heaping. The results shown in Table 4.2 indicate that the quality of age reporting at the oldest-old ages in terms of age heaping measured by the ratio index proposed by Coale and Li for the 22 provinces of China is not as good as that in Sweden, Japan, England and Wales, and Australia, but it is similar to that in Canada, it is better than that in the USA, and it is much better than that in Chile. So, we conclude that there is little age heaping among the oldest-old in the 22 provinces in China where the CLHLS was conducted.

4.3 Age Distribution and Age Ratios Among Centenarians

Age exaggeration may still exist even if there is no age-heaping because it is possible that people at very high ages may systematically tend to over-report their ages. It is, therefore, worthwhile to explore the quality of age reporting among the centenarians in the CLHLS in comparison with Sweden, Japan, England and Wales, Australia, Canada, USA, and Chile, using the single-year-age-specific percentage distributions of centenarians. We tried to interview all centenarians in the sampled cities and counties of the 22 CLHLS provinces. The age distribution of the interviewed centenarians in our survey should be therefore compatible with the national age distribution in the 22 provinces in China if there are no substantial age exaggerations in the CLHLS survey. Note that we purposely tried to have approximately equal numbers of male and female octogenarians and nonagenarians at each age from 80 to 99, who resided nearby the centenarians, and their age and sex were pre-designed based on the centenarians' code numbers which are randomly assigned. Thus, it makes no sense to compare the age distributions of octogenarians and nonagenarians interviewed in our survey to that of other countries. Rather, it is sufficient to compare the age distribution of the Chinese centenarians to that of centenarians in other countries for assessing the quality of age reporting of the oldest-old in our survey. There is no reason to suspect substantive age exaggerations among elders below the age of 100, if the age reporting of centenarians is acceptably good. This is because age exaggeration is much more likely among centenarians than among elders who are younger than age 100 and who live in the same area and share the same cultural traditions.

Single-year age-specific death rates are around (or higher than) 0.4 at ages 100 and over. Such extremely high mortality rates have dominated the shape of the age distribution of centenarians, which means that the effects of differentials in cohort size are minor. Therefore, the age distributions of centenarians of the European and Japanese populations, which have the highest data quality, look very much alike. Figure 4.1 is a comparison of the percentage age distributions of centenarians among Sweden, Japan, the USA, Chile and the CLHLS (including the Han and minorities in the 22 provinces). Table 4.3 presents detailed numerical results for the Sweden, Japan, England and Wales, Australia, Canada, the USA, Chile and the CLHLS. The comparisons shown in Fig. 4.1 clearly demonstrate that the age reporting among the Chinese centenarians in the 22 provinces is relatively close to, but not quite as good as, that in Sweden and Japan, but better than that in the USA (especially for males), and much better than that in Chile. Additional data in Table 4.3 show that the age reporting of the male centenarians in the CLHLS is relatively close to, but not as good as, that in England and Wales, more or less the same as that in Australia, and slightly better than that in Canada; the age reporting of the female centenarians in the CLHLS is not as good as that in England and Wales, and somewhat worse than that in Australia and Canada.

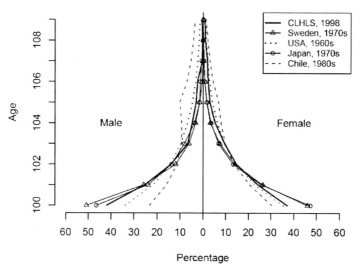

Fig. 4.1 A comparison of the age distributions of centenarians among Sweden, Japan, USA, Chile and CLHLS (including Han and minorities in the 22 CLHLS provinces)

Table 4.3 Comparison of percentage age distribution of centenarians, between CLHLS and selected countries

Female e_{65}	CLHLS	CLHLS(Han)	Sweden	Japan	England & Wales	Australia	Canada	USA	Chile
	1998	1998	1970s	1970s	1970s	1970s	1960s	1960s	1980s
	17.2	NA	17.4	16.6	16.4	16.9	16.7	16.7	17.6
Men									
100	41.9	42.7	50.8	46.4	50.3	41.0	37.9	33.8	23.3
101	25.4	25.2	25.8	23.9	26.0	24.7	24.1	21.7	15.9
102	14.0	13.9	11.9	13.8	12.9	15.5	15.2	14.3	11.4
103	6.7	6.6	6.2	8.2	6.4	9.5	9.1	9.8	9.4
104	4.8	4.6	3.3	3.8	2.7	5.0	5.6	6.8	9.4
105	3.1	3.1	1.5	1.6	1.0	2.2	3.6	4.7	10.2
106	2.5	2.4	0.4	1.2	0.4	1.1	2.1	3.3	7.1
107	0.4	0.4	0.2	0.7	0.1	0.7	1.3	2.4	5.4
108	0.4	0.2	0.0	0.2	0.1	0.3	0.8	1.8	4.3
109	0.6	0.7	0.0	0.2	0.0	0.1	0.4	1.4	3.6
Women									
100	37.2	37.7	46.0	47.3	44.5	43.4	42.8	35.4	30.4
101	24.3	24.3	26.5	25.9	25.7	25.7	25.6	22.8	20.6
102	14.8	14.8	14.1	13.3	14.4	14.8	14.6	14.7	13.7
103	9.5	9.5	7.4	6.8	7.9	8.0	7.9	9.5	9.8
104	5.3	5.0	3.2	3.5	4.1	4.0	4.2	6.3	7.9
105	3.0	3.0	1.6	1.7	2.0	2.1	2.5	4.1	6.8
106	2.2	2.1	0.7	0.8	0.9	1.0	1.4	2.8	4.4
107	1.9	1.9	0.3	0.4	0.4	0.5	0.6	2.0	3.0
108	1.0	1.0	0.1	0.2	0.2	0.3	0.3	1.4	2.1
109	0.8	0.8	0.1	0.0	0.1	0.2	0.1	1.1	1.4

Similar to the percentage age distribution of centenarians but with a stronger assumption of ignoring the cohort size effect, the age ratio of centenarians is defined as the number of persons at each age over the age of 100 divided by the number of persons at age 100 (see Chap. 5 in this volume). The results listed in Table 4.4 show that, if we consider Sweden as the best standard, the age reporting among centenarians measured by the age ratios of male centenarians in the CLHLS is not as good as that in Japan and England and Wales, more or less the same as that in Australia, slightly better than that in Canada, better than that in the USA, and much better than that in Chile; the age ratios of female centenarians in the CLHLS are not as good as those in Japan, England and Wales, and Australia, somewhat worse than that in Canada, better than that in the USA, and much better than that in Chile.

4.4 Density of the Oldest-Old and Centenarians

Another way of addressing the issue of age reporting among the oldest-old is to imagine that if the age exaggeration at very old ages is significant in a population, the reported proportion of the oldest-old persons among all elders and the total

Table 4.4 Comparison of ratio of number of centenarians at each age to age 100, the CLHLS and selected countries

Age	CLHLS 1998	CLHLS(Han) 1998	Sweden 1970–79	Japan 1970–79	England & Wales 1970–79	Australia 1970–79	Canada 1960–69	USA 1960–69	Chile 1980–90
Men									
100	10,000	10,000	10,000	10,000	10,000	10,000	10,000	10,000	10,000
101	6,050	5,907	5,076	5,155	5,169	6,022	6,360	6,430	6,800
102	3,350	3,264	2,348	2,964	2,566	3,770	4,004	4,239	4,900
103	1,600	1,554	1,212	1,774	1,268	2,307	2,394	2,888	4,033
104	1,150	1,088	644	814	538	1,212	1,472	2,018	4,033
105	750	725	303	355	205	530	942	1,393	4,367
106	600	570	76	250	75	280	565	975	3,067
107	100	104	38	146	25	172	331	725	2,333
108	100	52	0	42	11	84	205	527	1,833
109	150	155	0	42	5	20	97	405	1,533
Mean of the absolute deviation as compared to Sweden									
101–104	717.4(4)	633.3(3)	0.0	356.7(2)	118.2(1)	1007.9(5)	1237.6(6)	1573.9(7)	2621.6(8)
105–109	256.7(5)	237.9(4)	0.0	83.6(2)	25.3(1)	133.7(3)	344.7(6)	721.8(7)	2543.3(8)
101–109	461.4(4)	413.6(3)	0.0	205.0(2)	66.6(1)	522.2(5)	741.5(6)	1100.5(7)	2578.1(8)
Women									
100	10,000	10,000	10,000	10,000	10,000	10,000	10,000	10,000	10,000
101	6,541	6,452	5,762	5,464	5,774	5,918	5,978	6,427	6,795
102	3,978	3,922	3,066	2,801	3,231	3,413	3,413	4,140	4,519
103	2,549	2,530	1,605	1,439	1,779	1,840	1,835	2,677	3,216
104	1,429	1,332	690	745	916	919	979	1,775	2,618
105	798	808	353	362	442	484	586	1,167	2,233
106	602	554	161	179	205	230	330	790	1,453
107	518	494	64	92	95	108	150	557	983
108	266	269	16	36	39	77	69	402	684
109	210	210	16	5	12	35	17	307	449
Mean of the absolute deviation as compared to Sweden									
101–104	843.1(6)	778.2(5)	0.0	196.1(2)	143.8(1)	241.7(3)	270.4(4)	973.7(7)	1506.0(8)
105–109	357.0(6)	345.1(5)	0.0	17.1(1)	38.2(2)	65.0(3)	108.3(4)	522.5(7)	1038.3(8)
101–109	573.0(6)	537.6(5)	0.0	96.6(2)	85.1(1)	143.5(3)	180.3(4)	723.0(7)	1246.1(8)

The numbers in the parentheses are the rank of the mean of absolute deviation of the ratio of each age as compared to that of Sweden. The smaller the rank number is, the better the quality the age reporting is supposed to be.

population would be relatively large, compared with other populations with accurate age reporting. As shown by Coale and Kisker (1986), the proportions of those aged 95 or over among all elders aged 70 or over in the 23 countries studies by them with good data quality were all less than six per thousand. This proportion in the 28 countries with poor data because of age exaggeration by old persons extends from one percent to 10 percent (Coale and Kisker 1986: 398). The proportions of male and female Han Chinese aged 95 or over among those aged 70 or over in 1990 in all of China are 0.76 per thousand and 2.18 per thousand, proportions that are rather close to those of their Swedish counterparts (Wang et al. 1998).

Coale and Kisker (1986: 389–390) plotted values of e_{70} (life expectation at age 70) against values of l_{70}/l_5 (the conditional survival probability from age 5 to age 70), for the female populations in countries or regions with good data[6]. A very close relation between the e_{70} and l_{70}/l_5 values among countries or regions with good data is evidenced by a third-degree polynomial curve fitted by least squares (fig. 1 in Coale and Kisker 1986: 389). The plotting of the e_{70} against l_{70}/l_5 values for the female populations in countries with poor data[7] all lie far above the polynomial curve fitted to the data from the countries with good data (fig. 2 in Coale and Kisker (1986: 390). Wang et al. (1998) computed the ratio of the e_{70} against l_{70}/l_5 values for the Han Chinese female population in 1990. The Han Chinese ratio is almost exactly on the third-degree polynomial curve fitted to the data of populations with good age reporting.

Table 4.5 presents census data on the number of centenarians per million population, per million elderly aged 65 and above, and per million oldest-old aged 90 and above in the 22 provinces and other selected countries/periods with more or less the same female life expectancy at age 65 as in the 22 CLHLS provinces in China. There were 13.8, 191, 17,056 centenarians per one million population, per one million elderly aged 65+ and per one million oldest-old aged 90+ in the 22 provinces (including both Han and minorities) in China in 2000, as compared to 22.9, 155, 9,618 in Sweden in 1970–79, and 30.6, 218, and 12,360 in England and Wales in the 1970s, respectively. The census data in Table 4.5 demonstrate that the density of centenarians, which can be considered as one kind of quality of age reporting indicator, in the 22 provinces in China is not as good as that in Sweden and Japan, but is relatively close to England and Wales and Australia, more or less the same as that in Canada, better than that in the USA, and much better than that in Chile. However, the density of centenarians among the six minorities combined in the 22 provinces is substantially higher (worse) than that in Sweden, Japan, England and Wales, Australia and Canada, and moderately higher (worse) than that in the USA, but lower (better) than that in Chile.

---

[6] Countries or regions with good data include Sweden, Austria, Belgium, Czechoslovakia, Denmark, England, Finland, France, Germany, Hungary, Ireland, Italy, Japan, Luxembourg, Netherlands, Norway, New Zealand, Scotland, Switzerland and Taiwan, as indicated by Coale and Kisker (1986).

[7] The countries with poor data include Bolivia, Costa Rica, El Salvador, Guatemala, Honduras, Malaysia, Mexico, Panama, Peru, Philippines, Sri Lanka and Thailand, as indicated by Coale and Kisker (1986).

Table 4.5 Comparison of density of centenarians among selected ethnicities and countries

	Female e_{65}	Centenarians per million population among		
	Age 65	Age 0+	65+	90+
All Han Chinese in whole China	17.3	12.7	175	16,124
22 Provinces (Han & minorities)	17.2	13.5	184	16,441
Six minorities in the 22 provinces	18.4	33.7	524	32,672
Sweden, 1970–79	17.4	22.9	155	9,618
Japan, 1970–79	16.6	5.2	67	7,538
England & Wales, 1970–79	16.4	30.6	218	12,360
Australia, 1970–79	16.9	15.4	214	13,476
Canada, 1960–69	16.7	20.0	258	15,601
USA, 1960–1969	16.4	32.1	337	21,189
Chile, 1980–1990	17.6	34.9	662	33,464

The Chinese data are from the 2000 Census publications released by National Bureau of Statistics of China (2003a, b). The Chinese Female life expectancies at age 65 are estimated from published data by NBSC (2003a, b), adjusted for mortality rate underreporting at ages 94 and over.

The analysis presented above and in other studies including those by Coale and Li (1991), Wang et al. (1998) and Zeng et al. (2001) show that age reporting among the Han Chinese elderly, which is the population that constitutes the majority (about 92 percent) in China, is acceptably good. However, Coale and Li (1991) found that age reporting among some ethnic minorities, especially the Uyghur ethnicity in the Xinjiang autonomous region, was seriously biased with age exaggeration, leading to abnormal age patterns of mortality at old ages for China as a whole; once the Xingjiang data were excluded, the Chinese age pattern of mortality at old ages became normal as compared to other countries with accurate age reporting. We know of only one published study on age misreporting of China's ethnic minorities (Poston and Luo, 2004). Using Whipple's and Myers methods of evaluating the presence of age misreporting, it showed little evidence of age heaping among most of China's minorities in 2000. The Uyghur nationality was the major exception, showing a preference for digits 0 and 5, and an avoidance of digits 1, 3 and 9. Among most of the other minority groups, and among the Han, there was little if any evidence of age misreporting.

The summary indices in Table 4.5 show that the number of centenarians per million persons of all ages, per million elderly aged 65+, and per million oldest-old aged 90+ among the six minority groups in the 22 provinces, is much higher than that in Han Chinese and other populations with accurate age reporting (Sweden, Japan, England and Wales, Australia, and Canada), higher than that in the US, and somewhat lower than that in Chile. This fact leads us to seriously suspect that the oldest-old, especially centenarians, of the six minority groups in the CLHLS might well be exaggerating their ages.

On the other hand, we also suspect that the higher density of the centenarians among the six minority groups in the 22 CLHLS provinces may be partially due to differential mortality and natural environmental selection among these heterogeneous populations. Very old people of many of the ethnic minorities (except for Koreans[8]) in the 22 provinces have suffered poor living conditions and have had inadequate medical care in the past decades, and were more likely to be living in remote mountainous areas where the natural environment was more likely to have been well protected. Their misery in the past has been much worse than that of their counterpart cohorts among the Han Chinese, and among their counterpart cohorts in Sweden, Japan, England and Wales, Australia, Canada, and the USA, which could result in a mortality selection where minority persons who survived to very old ages were more robust in genetic and other biological characteristics, while those who were frail died before reaching old age. It is well known that much higher mortality rates at young ages in disadvantaged populations can produce relatively more robust old people as compared to those from advantaged populations (e.g. Coale and Kisker 1986; Horiuchi and Wilmoth 1998). Thus, the mortality rates among the minority oldest-old in the CLHLS might be lower because there are more robust survivors among them. The possible past mortality selection plus better natural environmental conditions in the minority areas in the 22 provinces might partially contribute to the higher density of centenarians among the six minority groups. It is, however, also highly possible that some minority oldest-old in the CLHLS survey areas indeed exaggerated their ages, which is our educated best guess. But, unfortunately, we do not have adequate data to quantitatively decompose the impacts of possible factors of age exaggerations, past mortality selection, and better natural environmental conditions on the higher density of centenarians among the six minority groups in the CLHLS areas.

4.5 Impacts of Possible Age Exaggerations Among the Six Minority Groups on the Analyses Using the CLHLS Data

Given the likelihood of age exaggeration among the six minority groups in the CLHLS data set and our inability to quantify the degree thereof, CLHLS data users may reasonably ask the following question: what will be the effects of the minority populations in the CLHLS dataset on statistical analysis, assuming that the impacts of past mortality selection and better natural environmental conditions are all negligible and that the higher density of centenarians among minorities is solely or largely caused by age exaggeration[9]? To address this important question, we conducted several additional analyses on the "association of late childbearing

---

[8] Koreans in China have an even higher socioeconomic status than Han Chinese and their age reporting could be as good as that of the Han, but again, we have no detailed data with which this can be verified.

[9] We appreciate very much that Booth and Zhao (Chapter 5 in this volume) thoughtfully raised this important question, which helped us to conduct careful comparative assessments to address the issue.

with health and healthy survival at the oldest-old ages," following exactly the same methodology, statistical modeling procedures and using the same dependent and independent variables as those used in Zeng and Vaupel (2004); however, in these new analyses we relied only on the CLHLS Han data, excluding the data for the ethnic minorities. We then compared the results of these additional analyses with the results of the original analyses in Zeng and Vaupel (2004), where they used the entire CLHLS data set including both the Han and minority data.

We listed the additional and the original estimates in parallel positions with additional estimates in parentheses in Tables 4.6, 4.7, 4.8, and 4.9[10]. It is clear that the two sets of parameter estimates including and excluding the six minority groups in the 22 Han-dominated provinces are fairly close to each other, indicating that the same qualitative conclusions concerning the association of late childbearing with health and healthy survival at oldest-old ages in China can be drawn based on the two sets of estimates. We also note that 66.7 percent, 28.7 percent, and 4.6 percent of the total number (216) of the estimates of the odds ratios and relative risks in the additional analysis excluding minorities indicate a slightly weaker, the same, and slightly stronger positive association of late childbearing with health and healthy survival at the oldest-old ages, respectively, compared to the original estimates including minorities.

The total number of estimates of the odds ratios and relative risks which are statistically significant in the additional analyses excluding the minority data is reduced by 13.5 percent (perhaps at least partially due to the 5.5–6.8 percent reduction of the sample size), but the direction and conclusion of the association of late childbearing with health and healthy survival at the oldest-old ages remains unchanged, compared to the analyses using both Han and minority data. This fact leads us to believe that the inclusion of the six minority groups in the CLHLS may not cause substantive bias in demographic and multivariate statistical analyses. This is mainly because the minority groups consist of a rather small portion of the samples: 6.8 percent of the total sample in the baseline survey and 5.5 percent of the grand total sample of the 1998, 2000, and 2002 waves combined[11].

Gu and Dupre (Chap. 6 in this volume) also show that possible age exaggeration among minority oldest-old in the CLHLS does not significantly affect the outcome for estimates of indicators such as ADL, MMSE, self-reported health, as well as the

---

[10] In Tables 4.6–4.9, the estimates which are not in the parenthesis are based on the whole CLHLS data set including both Han and the six minority groups cited from Zeng and Vaupel (2004); Estimates in the parenthesis are based on CLHLS data set excluding the six minority groups.

[11] In each follow-up wave of the CLHLS, the interviewees who died or were lost to follow-up were replaced by new interviewees with the same gender and age as those died or were lost to follow-up. We matched the gender and age only; there was no ethnicity requirement in the replacement process. The Han are the large majority in survey areas. Thus, it is usually much easier to find a replacement who is Han with the required gender and age, which is why new interviewees are more likely to be Han. This explains why the percent of minority groups in the CLHLS in later waves was somewhat lower than at baseline.

estimates of other covariates in multivariate statistical models using either cross-sectional or longitudinal CLHLS data.

4.6 Concluding Remarks

This chapter has evaluated patterns of age reporting among the oldest-old, especially centenarians, in the Chinese Longitudinal Healthy Longevity Survey (CLHLS) based on comparisons of the various indices of elderly age reporting and age distributions of centenarians among the countries of Sweden, Japan, England and Wales, Australia, Canada, China, the USA, and Chile. The analyses demonstrate that age reporting among the oldest-old interviewees (Han and the six minority groups combined) in the 22 CLHLS provinces is not as good as that in Sweden, Japan, and England and Wales, but is relatively close to that in Australia, more or less the same as that in Canada, better than that in the USA, and much better than that in Chile. As indicated by the high density of centenarians, however, age exaggeration does exist in the six ethnic minority groups, although we cannot rule out and quantify the potential impacts of past mortality selection plus better natural environmental conditions among these minority groups. Comparative analysis between including and excluding the six minority groupies suggest that age exaggeration of minorities in the CLHLS does not likely cause substantial biases in demographic and statistical analyses using the CLHLS data, since, as noted above, minorities comprise a rather small portion of the sample[12]. Of course, we must keep in mind that the minority groups in the CLHLS study areas might exaggerate their ages and some Han Chinese

Table 4.6 Effects (odds ratios) of late childbearing on health status of the oldest-old in China, based on multivariate logistic regression applied to data collected at 1998 baseline survey, adjusted for covariates of demographic characteristics, family support, social connections, and health practice

Focused covariates (category in parentheses is the reference group)	ADL disabled	MMSE impaired	Self-reported bad health	Depression symptoms
Women(0 birth after age 35)				
1 birth after age 35	0.890 (0.940)	0.949 (0.966)	0.923 (0.971)	0.854 (0.788*)
2 births after age 35	0.939 (0.984)	0.835# (0.836)	0.906 (0.960)	0.827 (0.763#)
3+ births after age 35	0.765* (0.857)	0.772* (0.768*)	1.031 (1.074)	0.626** (0.580**)
Men (0 birth after age 35)				
1 birth after age 35	0.954 (1.017)	1.111 (1.003)	1.002 (1.021)	0.846 (1.141)
2 births after age 35	0.879 (1.025)	0.937 (1.010)	0.858 (0.980)	0.757 (1.002)
3+ births after age 35	0.880 (0.977)	0.791 (0.873)	0.939 (0.981)	0.711 (0.805)

Differences between results including and excluding minorities for births after age 40 are minor, too. They are not listed in Tables 4.6–4.9, due to space limit but available upon request.
#$p < 0.10$; *$p < 0.05$; **$p < 0.01$; ***$p < 0.001$.

---

[12] It may be worthwhile to note that it is necessary to include the six minority groups in the CLHLS surveys to ensure adequate representation of minority groups in the study areas. US studies have a similar requirement.

Table 4.7 Effects of late childbearing (relative risks and odds ratios) on survival and healthy survival of the oldest-old in China between 1998 and 2000 based on multivariate Cox proportional hazards and ordinal logistic regression models

Models and focused covariates	Relative risk of Cox hazards models survival analysis		Odds ratios of ordinal logistic regression healthy survival analysis	
	Women	Men	Women	Men
Model I (0 birth after age 35)				
1 birth after age 35	0.941 (0.937)	0.910 (0.903)	0.914 (0.874#)	1.103 (0.979)
2 births after age 35	0.850* (0.893#)	1.077 (1.097)	0.758*** (0.764***)	0.930 (1.008)
3+ births after age 35	0.722*** (0.749***)	0.747*** (0.742***)	0.697*** (0.678***)	0.769** (0.738**)
Model II (0 birth after age 35)				
1 birth after age 35	0.941 (0.941)	0.919 (0.922)	0.977 (0.926)	1.209#(1.100)
2 births after age 35	0.858* (0.912)	1.100 (1.138)	0.847#(0.838#)	1.045 (1.183)
3+ births after age 35	0.715*** (0.754***)	0.800* (0.807*)	0.766* (0.753*)	0.895 (0.936)
Model III (0 birth after age 35)				
1 birth after age 35	0.949 (0.953)	0.919 (0.931)	0.967 (0.925)	1.157 (1.076)
2 births after age 35	0.861* (0.915)	1.116 (1.163)	0.828* (0.833#)	1.066 (1.208)
3+ births after age 35	0.720*** (0.763**)	0.821 #(0.841#)	0.747** (0.748**)	0.889 (0.965)
Model IV (0 birth after age 35)				
1 birth after age 35	0.977 (0.980)	0.947 (0.940)	1.003 (0.949)	1.169 (1.096)
2 births after age 35	0.902 (0.955)	1.181 (1.2136#)	0.854 (0.850)	1.111 (1.218)
3+ births after age 35	0.767** (0.797*)	0.889 (0.884)	0.791* (0.787*)	0.944 (1.014)

The category in parentheses in the column of "Models and Focused covariates" is the reference group in each case; Covariates in Model I are late childbearing plus demographic variables of age, gender, residence, education, and ethnicity. Model II is Model I plus covariates of family support and social connection. Model III is Model II plus covariates of health practices. Model IV is Model III plus covariates of health conditions; The number of degree of freedom for Model I to Model IV is 9, 17, 21, 25 respectively.

# $p < 0.10$; * $p < 0.05$; ** $p < 0.01$; *** $p < 0.001$.

Table 4.8 Ratio of survivorship (RS) of elders who had 1+, 2+, or 3+ births after age 35 to those who did not have such late births

$P_1(x)$	Ages 100–105 vs. 80–85				Ages 100–105 vs. 90–95				Ages 90–95 vs. 80–85			
	$P_1(80-85)$	$P_1(100-105)$	RS	p	$P_1(90-95)$	$P_1(100-105)$	RS	p	$P_1(80-85)$	$P_1(90-95)$	RS	p
Men												
1+ births	69.0(69.0)	71.0(70.1)	1.10(1.1)	0.449(0)	72.3(71.7)	71.0(70.1)	0.93(0.93)	0.596(0)	69.0(69.0)	72.3(71.7)	1.17(1.14)	0.106(0.203)
2+ births	50.1(49.5)	54.4(53.8)	1.19(1.2)	0.125(0)	49.9(49.3)	54.4(53.8)	1.20(1.20)	0.121(0)	50.1(49.5)	49.9(49.3)	0.99(0.99)	0.913(0.926)
3+ births	30.2(29.6)	35.6(34.3)	1.28(1.2)	0.036(0)	32.7(32.6)	35.6(34.3)	1.14(1.08)	0.281(0)	30.2(29.6)	32.7(32.6)	1.12(1.15)	0.233(0.164)
Women												
1+ births	49.8(48.8)	58.8(57.6)	1.44(1.4)	0.000(0)	51.3(50.3)	58.8(57.6)	1.35(1.34)	0.000(0)	49.8(48.8)	51.3(50.3)	1.06(1.06)	0.489(0.497)
2+ births	24.6(23.7)	34.2(32.7)	1.59(1.6)	0.000(0)	28.2(26.6)	34.2(32.7)	1.32(1.34)	0.000(0)	24.6(23.7)	28.2(26.6)	1.2(1.17)	0.059(0.134)
3+ births	8.9(8.0)	18.1(16.6)	2.27(2.3)	0.000(0)	12.6(11.1)	18.1(16.6)	1.53(1.59)	0.000(0)	8.9(8.0)	12.6(11.1)	1.48(1.44)	0.005(0.019)

$P_1(x)$: Percentage of elders who gave birth after age 35 among those aged x. p: Significance level of chi-square tests based on method of Mantel and Haenszel (1959) to test whether there is a statistically significant difference of survivorship between those with the fixed attribute and those without it.

Table 4.9 Ratio of healthy survivorship (RHS) of elders who had 1+, 2+, or 3+ birth after age 35 to those who did not have such late births

$P_1(x)$, $\pi(x+n)$

	Ages 100–105 vs. 80–85				Ages 100–105 vs. 90–95				Ages 90–95 vs. 80–85			
	P_1(80–85)	π(100–105)	RHS	p	P_1(90–95)	π(100–105)	RHS	p	P_1(80–85)	π(90–95)	RHS	p
Men												
1+ births	69.0(69.0)	71.4(70.3)	1.12(1.06)	0.506(0.717)	72.3(71.7)	71.4(70.3)	0.95(0.93)	0.778(0.700)	69.0(69.0)	72.3(71.2)	1.17(1.11)	0.151(0.358)
2+ births	50.1(49.5)	56.3(55.7)	1.28(1.28)	0.103(0.114)	49.9(49.3)	56.3(55.7)	1.29(1.29)	0.097(0.111)	50.1(49.5)	49.2(48.4)	0.96(0.96)	0.709(0.669)
3+ births	30.2(29.6)	39.3(37.5)	1.50(1.43)	0.010(0.029)	32.7(32.6)	39.3(37.5)	1.34(1.24)	0.069(0.196)	30.2(29.6)	31.5(31.4)	1.07(1.09)	0.556(0.445)
Women												
1+ births	49.8(48.8)	60.5(58.2)	1.54(1.46)	0.000(0.000)	51.3(50.3)	60.5(58.2)	1.45(1.37)	0.000(0.003)	49.8(48.8)	51.9(50.2)	1.09(1.06)	0.891(0.597)
2+ births	24.6(23.7)	36.7(34.1)	1.78(1.66)	0.000(0.000)	28.2(26.6)	36.7(34.1)	1.48(1.42)	0.000(0.002)	24.6(23.7)	29.0(26.7)	1.25(1.17)	0.045(0.188)
3+ births	8.9(8.0)	20.7(18.6)	2.68(2.61)	0.000(0.000)	12.6(11.1)	20.7(18.6)	1.81(1.82)	0.000(0.000)	8.9(8.0)	13.9(11.6)	1.67(1.51)	0.001(0.018)

$P_1(x)$: percentage of elders who gave birth after age 35 among those aged x. $\pi(x+n)$: percentage of elders who gave birth after age 35 among those age $x+n$ and ADL independent. p: Significance level of Chi-square tests based on method of Mantel and Haenszel (1959) to test whether there is a statistically significant difference of healthy survivorship between those with the fixed attribute and those without it.

oldest-old might also exaggerate (to a lesser extent) their ages. We thus need to exercise caution in our analyses and interpretations of findings.

Acknowledgments The research reported in this Chapter is supported by The National Institute on Aging grant (R01 AG023627-01) and National Natural Science Foundation of China key project grant (70533010). We thank very much the helpful comments provided by Dudley Poston, Heather Booth and Zhongwei Zhao.

References

Coale, A.J. and E. Kisker (1986), Mortality crossovers: Reality or bad data? *Population Studies* 40, pp. 389–401

Coale, A.J. and E.E. Kisker (1990), Defects in data on old-age mortality in the United States: new procedures for calculating mortality schedules and life tables at the highest ages, *Asian and Pacific Population Forum* 4, pp. 1–31.

Coale, A.J. and S. Li (1991), The effect of age misreporting in China on the calculation of mortality rates at very high ages. *Demography* 28 (2), pp. 93–301

Dechter, A. and S.H. Preston (1991), Age misreporting and its effects on adult mortality estimates in Latin America. *Population Bulletin of the United Nations* 31/32, pp. 1–16

Ewbank, D.C. (1981), *Age misreporting and age-selective underenumeration: Sources, patterns, and consequences for demographic analysis*. Washington, DC: National Academy Press

Horiuchi, S. and J.R. Wilmoth (1998), Deceleration in the age pattern of mortality at old ages. *Demography* 35 (4), 391–412

Jowett, A.J. and Y.Q. Li (1982), Age heaping: Contrasting pattern from China. *GeoJournal* 28 (4), pp. 427–442

Kannisto, V (1994), *Development of oldest-old mortality, 1950–1990: Evidence from 28 developed countries*. Odense: Odense University Press

Knodel, J. and N. Chayovan (1991), Age and birth date reporting in Thailand. *Asian and Pacific Population Forum* 5 (2–3), pp. 41–76

Mantel, N. and W. Haenszel (1959), Statistical aspects of the analysis of data from retrospective studies of diseases. *Journal of the National Cancer Institute* 22, pp. 719–748

Poston, D.L. Jr. and H. Luo (2004), Zhongguo 2000 nian shaoshu minzu de nian ling dui ji heshu zi pian hao (Age structure and composition of the Chinese minorities in 2000). *Zhongguo Shaoshu Minzu Renkou (Chinese Minority Populations)* 19 (3), pp. 9–15

Retherford, R.D. and G.M. Mirza (1982), Evidence of age exaggeration in demographic estimates for Pakistan *Population Studies* 36(2), pp. 257–270

Rosenwaike, I. and S.H. Preston (1984), Age overstatement and Puerto Rican longevity. *Human Biology* 56 (3), pp. 503–525

Seltzer, W. (1973), *Demographic data collection: A summary of experience*. New York: Population Council

Thatcher, A.R. (1981), Centenarians. *Population Trends* 25, pp. 11–14

Wang, Z., Y. Zeng, B. Jeune, and J.W. Vaupel (1998), Age validation of Han Chinese centenarians. *GENUS—An International Journal of Demography* 54, 123–141

Zeng, Y. and J.W. Vaupel (2004), Association of late childbearing with healthy longevity among the oldest-old in China. *Population Studies* 58 (1), pp. 7–53

Zeng, Y., J.W. Vaupel, Z. Xiao, C. Zhang, and Y. Liu (2001), The healthy longevity survey and the active life expectancy of the oldest old in China. *Population: An English selection* 13 (1), pp. 95–116

Chapter 5
Age Reporting in the CLHLS: A Re-assessment

Heather Booth and Zhongwei Zhao

Abstract Age reporting among respondents in the Chinese Longitudinal Healthy Longevity Survey is examined, using the first round of data collected in 1998. The sample design limits the use of traditional methods for assessing the accuracy of age reporting, and innovative methods are adopted. Only the sample aged 100+ is representative of the population at that age. The age structure of centenarians is compared with populations with good age reporting, demonstrating age exaggeration. At ages 80+, constructed estimates of age at childbearing show systematic effects consistent with age exaggeration, particularly in Guangxi and among ethnic minorities. Increasing age exaggeration with age is present in these data, which is at least partly the result of the age structure. These findings have implications for substantive analyses, and further examination of the quality of these data is needed.

Keywords Age exaggeration, Age heaping, Age misreporting, Age reporting, Age validation, Centenarian, China, Cluster sample, Data quality, Digit preference, England and Wales, Ethnic minorities, Guangxi, Han majority, Inaccuracy, Japan, Jiangsu, Large sample size, Longevity, Mean age at childbearing, Myers' Index, Non-response, Oldest-old, One Per Thousand Fertility Survey, Proportion of centenarians, Re-assessment, Regional variation, Sample design, Shanghai, Sweden, Whipple's Index, Yao, Zhuang

5.1 Introduction

The accuracy of age reporting is an important consideration for any demographic analysis, mainly because the existence of age misreporting often produces distorted results. Inaccuracies in age reporting are potentially a significant problem for studies of ageing and longevity on two counts. First, experience in many populations has

---

H. Booth
Australian Demographic and Social Research Institute, The Australian National University
Canberra, Australia
e-mail: heather.booth@anu.edu.au

shown that older people tend to misreport or exaggerate their age; and while the most serious problems occur in populations where literacy is low, more educated populations are not entirely free from such errors (Coale and Kisker 1986; Jeune 1995; Rosenwaike and Stone 2003). Second, the effect of misreporting at very old ages is often magnified by the shape of the age distribution.

The quality of age reporting among the majority Han population of China is generally good and is believed to be on a par with age reporting in many developed countries (Coale and Li 1991; Gu and Zeng 2004). This has been attributed to cultural factors (Wang et al. 1998). Among some ethnic minorities in China, however, age reporting at older ages has been found to be of poor quality largely because of age exaggeration (Coale and Li 1991). It is for this reason that nine provinces with sizeable minority populations were excluded from the Chinese Longitudinal Healthy Longevity Survey (CLHLS) (see Chap. 2 in this volume). The survey population is thus mostly of Han ethnicity, but 7 percent belong to minority ethnicities. This fact alone leaves open the possibility of some degree of age misreporting in the CLHLS data, and it is important to examine the quality of age reporting in the whole dataset.

The CLHLS is a rich source of data with a sufficiently large sample size to provide the statistical power necessary to undertake detailed analyses. The data have already been used to address many research questions regarding ageing and longevity, and the papers of this volume further augment this body of research. For such analyses, the quality of age reporting is of the utmost importance. Most forms of age misreporting, and in particular age exaggeration, will tend to lead to an older age distribution and the overestimation of longevity.

The purpose of this chapter is to undertake an objective examination of the quality of age reporting in the CLHLS. We first examine the evidence previously presented by others, and then re-assess the data more comprehensively using innovative methods and addressing reporting inaccuracies that have not previously been considered. We undertake this re-assessment in the spirit of Coale and Li (1991: 300) who stressed that all data "must be scrutinized critically, even when there are reasons to suppose that the data are accurate. Accuracy of most of the data does not mean that all of the data are accurate; as William Brass said, all data are guilty until proved innocent."

5.2 The CLHLS Data

Details of the CLHLS study design may be found in Chap. 2 of this volume. Only the 1998 (first wave) data are used for this study of the quality of age reporting.

5.2.1 Sample Design

The sample design is a cluster sample. For the first wave, approximately 50 percent of all counties and cities in 22 provinces were randomly selected. Among these selected clusters, 631 had centenarians, all of whom were included. For sampling purposes, the ages of the centenarians were obtained from the local ageing committees. These ages may have included some inaccuracies, particularly if based on the nominal age[1] of the person, and ages (in fact, dates of birth) were validated during the interview. Age validation resulted in a loss of 409 so-called centenarians to younger age groups, and the final number of validated centenarian respondents in the sample was 2,418.

For each centenarian, purposive sampling was used to randomly select one octogenarian and one nonagenarian from the population living nearby in such a way as to achieve approximately equal numbers of males and females at each single year of age. Thus weights are necessary in all analyses involving octogenarians and nonagenarians; these take account of (validated) age, sex and rural–urban residence (Zeng et al. 2001). After age validation, there were 3,528 octogenarian and 3,013 nonagenarian respondents in the survey.

Not all originally sampled individuals were interviewed. The 9,093 respondents[2] in 1998 represent a response rate of 88 percent. Non-responses were due to unavailability (too ill, deceased, or migrated) or refusal to participate. If those unavailable for interview are excluded, the response rate is 98 percent.

5.2.2 Reporting of Age

For all questions, every effort was taken to ensure the accuracy of responses: interviewers were extensively trained, and all training was standardized nationally. Detailed error checks and quality control mechanisms were incorporated into the interview procedure (Research Group of Healthy Longevity in China (RGHLC) 2000: 1–25; Xu 2001; Gu and Zeng 2004). In particular, respondent's age was subject to careful validation. The survey did not ask for age directly, but based this variable on date of birth. All reported dates of birth of respondents were validated by interviewers by reference to their household booklet and ID card,[3] Chinese calendar birth date and animal year, genealogical records if available, children's ages, siblings' ages, and so on (see Zeng and Gu, Chap. 4 of this volume for more detail).

5.2.3 Limitations for the Examination of Age Reporting

Conventional methods for the examination of the accuracy of age reporting rely heavily on the demographic stability of the true age distribution, often for the entire

---

[1] The nominal age is counted as exactly 1 year old at birth, increasing by 1 year each Chinese New Year Day. It is therefore up to 2 years greater than chronological age counted from zero at birth.

[2] This includes 134 sampled respondents whose validated age was <80.

[3] Household registration and ID cards were introduced between 1950 and 1990 and are largely based on self-reported age; they might thus be subject to age misreporting.

age range. The use of such methods for CLHLS data is severely limited by the fact that the age and sex distribution of the sample population is to a significant degree an artifact of the sample design. Only the sample population aged 100+ is proportionally representative by age of the Chinese population, that is, of the demographic processes shaping its structure. Thus, use of the age structure to examine age reporting is necessarily restricted to centenarians. Further, the examination is limited by the relatively small numbers at this age (481 males and 1,937 females) and by the restricted availability of valid and reliable distributions for comparison.

The sample design also compromises the applicability of digit preference detection methods including those by Coale and Li (1991) and Wang et al. (1998). The purposive selection of 80- to 99-year-old respondents was related to age because equal numbers of male and female octogenarians and nonagenarians at each single year of age were sought. Thus, neither randomness nor representativeness can be assumed at these ages, violating the basis of digit preference measures. It is also important to note that methods for the measurement of digit preference are unable to detect systematic reporting errors such as age exaggeration, which is the most likely and potentially the most serious source of error for studies of longevity.

Other sampling and related issues must also be taken into account when assessing data accuracy because of possible age-related bias. This is particularly important for centenarians because of the very high mortality rates at these ages. Non-responses may have introduced a bias toward a slightly younger age distribution,[4] equivalent to age under-reporting; however, this effect would enhance rather than detract from the findings reported here.

5.3 Previous Assessment of Accuracy of CLHLS Respondent's Age

The 1998 CLHLS data have previously been examined in relation to the accuracy of the reporting of respondent's age and judged to be "generally reliable" (RGHLC 2000: 6–17; Zeng et al. 2001). The main focus of this examination was a sex-specific comparison of the age distribution of CLHLS centenarians with Swedish centenarians in 1984–1993 (see Fig. 5.1). The differences observed were not considered sufficiently large to bring into question the quality of age reporting among younger centenarians (aged 100–105). However, it was concluded that at ages 106+, the quality of age reporting is questionable (RGHLC 2000: 6–17). Persons aged 106+ have been excluded from most existing analyses: they number 154 (1.7 percent of those aged 80+) comprising 131 females (2.4 percent) and 23 males (0.6 percent).

---

[4] This would occur if unavailability for interview increased with age, which seems likely for unavailability due to ill health and migration out of the study area to a relative's residence or an institution.

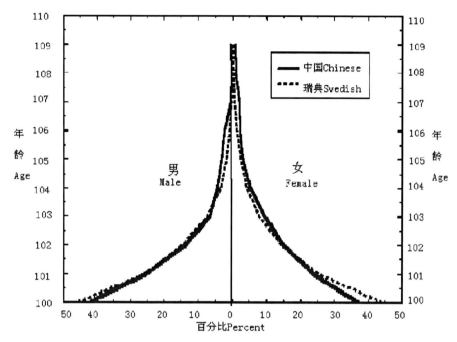

Fig. 5.1 Age distribution of centenarians, 1999 CLHLS and Sweden 1984–1993 (Reproduced with kind permission)

The choice of Sweden as a comparison population is well justified by its accuracy of age reporting, believed to be the best in the world. For centenarians, very high mortality rates dominate the shape of the age distribution, and cohort size effects are relatively small.[5] It would be expected that Chinese mortality rates would be no lower than the Swedish rates, and therefore that CLHLS centenarians would exhibit an age distribution as young as or younger than their Swedish counterparts. However, this is contra-indicated by the smaller CLHLS percentages aged 100 in Fig. 5.1, and the corresponding larger percentages aged 101+. Age misreporting in the CLHLS data is a probable explanation, and further examination is needed.

Other evidence, presented by Gu and Zeng (2004), in support of the reliability of age reporting in the CLHLS, consists of the similarity of the mortality pattern for Sweden in 1999 to the patterns derived from deaths among CLHLS respondents in 1998–2000 and 2000–2002. In fact, careful comparison of these data shows that mortality levels in the two populations are similar at ages less than 90, but at 90+ mortality is substantially higher in the CLHLS. This implies that the age distribution of centenarians will be younger for the CLHLS than for Sweden, which is again contrary to the smaller CLHLS percentages aged 100 in Fig. 5.1. Again, further examination is needed.

---

[5] These effects can be expected to be stable by age, and therefore not significantly influence the centenarian age distribution. In the 1890s (when these respondents were born) fertility rates were fairly stable and the age distribution of reproductive women was not subject to significant variation.

Additional supporting evidence cited by Zeng et al. (2001) and Gu and Zeng (2004) is the fact that age reporting in China's 1982 and 1990 Censuses was found to be reliable on the basis of Whipple's and Myers' indices of digit preference. However, these indices are based on all or most ages, so that accurate reporting at young ages would mask any misreporting at the oldest ages; hence no conclusions can be drawn from these indices about the oldest ages. Further, these indices measure digit preference, and do not address systematic biases such as age exaggeration or digit-neutral age misreporting.

An examination of age reporting in the 1982 Census by Coale and Li (1991) has been cited (Wang et al. 1998; Gu and Zeng 2004) as evidence of good age reporting in China with the exception of Xinjiang province[6]. However, the Coale and Li study concentrates on male mortality patterns at ages less than 100, and is thus unable to illuminate the evaluation of age reporting among centenarians or among females. Further, its reliance on the comparison of mortality patterns may not provide entirely solid evidence; this is due to the facts that age exaggeration produces underestimated mortality rates, and that an association between age exaggeration and relatively high mortality is likely to hinder the detection of age exaggeration. Similarly, if a proportion of the population reports their nominal age, the mortality curve will show no irregularity.

None of the above previously cited evidence is sufficiently convincing to establish without qualification the good quality of age reporting in the CLHLS. Indeed, some of the evidence would appear to indicate the presence of age misreporting, rather than its absence. The accuracy of age reporting thus remains open to question. It is therefore important to undertake further assessment of age reporting in the CLHLS.

5.4 A Re-assessment

This re-assessment of the accuracy of respondent's age in the CLHLS data adopts two approaches. The first focuses on centenarians, comparing the CLHLS with selected comparable populations. The second examines age reporting for the whole sample using unconventional methods that make use of age differences between respondents and their children. For both approaches, regional variation is also examined.

5.4.1 Age Reporting Among Centenarians

This examination of the accuracy of age reporting among reported centenarians involves a comparison with three populations: Sweden 1943–1952, England and Wales 1950–1955, and Japan 1962–1966. These countries were selected because

---

[6] Inaccurate age reporting among the Wei (or Weiwuer or Uyghur) ethnic minority was found responsible; the evidence included digit preference at ages 40–80 and age exaggeration at 110+. No evidence was provided to show that other provinces or ethnic minorities do not also exhibit age reporting inaccuracies.

of their known accuracy of age reporting (Kannisto 1994), and the specific periods were chosen to match on average China's 1998 life expectancy at birth of about 70 years and to provide sufficient numbers to reduce random fluctuation to an acceptable level. These populations span the age structure of the 22 provinces of China covered in the CLHLS, referred to here as China-22, as reported in the 2000 census. For China-22, the proportion aged 65+ is 7 percent; this is compared with 10, 11 and 6 percent for Sweden, England and Wales and Japan respectively. The proportion aged 90+ in China-22 is 8 per 10,000, compared with 11, 10 and 4 per 10,000 for Sweden, England and Wales and Japan respectively. The study population, that is the 22 provinces of China, is thus within the comparable population range.

The first question considered is: how many centenarians do we expect to find in a population? The exact number depends on both longevity and the overall population structure, but broadly similar numbers of centenarians per million are expected in populations with similar life expectancies at birth and similar structures. Given similar life expectancies and the above proportions aged 65+ and 90+ in the selected populations, it would be expected that China-22 in 2000 would have fewer centenarians per million than Sweden and England and Wales, but more than Japan in the selected years. Table 5.1, which shows the number of centenarians per million in these four populations, suggests that China-22 has relatively more centenarians than all three comparison populations: twice as many as Sweden and England and Wales, and seven times as many as Japan. Similarly, differentials of between 2 and 7 are found for the centenarian proportions among those aged 90+ and 65+ (Table 5.1). A likely explanation for such large centenarian proportions is age exaggeration, including reporting based on nominal age, among the very old in the Chinese census. The magnitude of these proportions is such that age exaggeration in the census must exist not only among the ethnic minority populations but also among the Han majority, as the size of the ethnic minority populations is too small to produce the differentials found.

Table 5.1 also shows centenarian proportions estimated from the 1998 CLHLS by using the 2000 Census to provide population data.[7] The much lower CLHLS

Table 5.1 Centenarians in the 1998 CLHLS and selected comparable populations

Population	Period	Centenarians per million population	Centenarians per million aged 90+	Centenarians per million aged 65+
Sweden	1943–1952	7	5,975	66
England & Wales	1950–1955	7	6,914	62
Japan	1962–1966	2	4,427	28
China (22 provinces)	2000	13.5	16,441	184
CLHLS	1998	4.5	5,475	62

Selected comparable populations cover periods when life expectancy averaged 70 years. CLHLS proportions are based on population data from the 2000 Census and assume that the selected clusters include half of the population of their respective provinces

---

[7] These estimates are based on the assumption that the selected clusters cover half of the population of their respective provinces.

proportions than in the 2000 Census for China-22 can be attributed to the better quality of age reporting in the CLHLS. The estimated CLHLS proportions fall within the range of the three comparison populations. It should be noted, however, that CLHLS centenarians per population aged 90+ and 65+ will be underestimated to the extent that the 2000 Census overestimates the size of these two age groups through age exaggeration. Further examination in greater detail is required.

A useful approach is to examine differences within the CLHLS data. Considerable variation in centenarian proportions is found among the provinces. This variation cannot be easily explained, except in terms of greater age exaggeration in certain provinces. The highest proportion of centenarians in the population[8] (14.9 per million) is found in Shanghai where mortality rates have been relatively low for at least 50 years and fertility began to decline earlier; this is not unexpected. However, equally as high (14.9) is Guangxi, a relatively undeveloped province in the south of China, where the centenarian proportion is expected to be comparatively low; it is suspected that age exaggeration is responsible. The third highest centenarian proportion is 9.8 in Jiangsu, which is one of the most developed provinces, located close to Shanghai in the east of China; Guangxi would not be expected to exceed this proportion. Age exaggeration and increasing age exaggeration with reported age are expected to be particularly marked in Guangxi because of its worldwide reputation for longevity. In contrast, age exaggeration is expected to be relatively limited in Jiangsu and Shanghai because of their higher socio-economic development for a longer period. These expectations are confirmed by the 1990 Census which recorded 64 centenarians per million aged 65+ in Shanghai and 62 in Jiangsu, while in Guangxi as many as 407 were recorded.

The age distribution of centenarians is also examined, using age ratios and the three comparable populations. It was first verified that there was negligible heaping on age 100 in the three comparable populations, as this would have resulted in age ratios that are too low. The possibility of heaping on age 100 in the CLHLS data was also considered. As digit preference measures could not be used, this was examined by calculating age ratios using ages 101 and 102 as the base instead of age 100. The results are generally consistent with those derived using 100 as the base (see Appendix 1, Tables 5.4 and 5.5), indicating minimal age heaping. The age ratios based on age 100, seen in Table 5.2, show that there are relatively more centenarians at each age above 100 in the CLHLS than in the commensurate mortality regimes. In other words, for each sex the CLHLS data exhibit an older centenarian age distribution than expected. Furthermore, the relative differentials between the CLHLS ratios and those for each of the three comparable populations are shown to increase with age. These ratios are consistent with increasing age exaggeration with reported age. At ages 101–104, the relative differentials are mostly greater for females than males, but at older ages they are more marked for males, as the ratio of ages 105+ to 100–104 also shows. Again, provincial comparisons show ratios for Guangxi to be particularly high: ratios for both sexes are 474 at age 102, and 105 at age 105.

---

[8] Again, these proportions are based on total population data from the 2000 Census and CLHLS centenarians.

Table 5.2 Age ratios for centenarians by sex, CLHLS and comparable populations

Age	Male				Female			
	Sweden 1943–1952	E&W 1950–1955	Japan 1962–1966	CLHLS 1998	Sweden 1943–1952	E&W 1950–1955	Japan 1962–1966	CLHLS 1998
100	1,000	1,000	1,000	1,000	1,000	1,000	1,000	1,000
101	382	518	532	605	506	534	467	654
102	221	273	286	335	265	292	242	398
103	88	142	143	160	123	152	142	255
104	59	65	91	115	62	79	64	143
105	15	21	0	75	31	41	39	80
106	0	3	0	60	12	20	27	60
107	0	0	0	10	6	10	21	52
108	0	0	0	10	0	3	15	27
109	0	0	0	15	0	1	9	21
110+	0	0	0	20	0	0	15	24
105+/100–104	8.4	12.4	0.0	85.8	25.2	36.3	66.5	107.5

Age ratios are expressed as the number of respondents aged 101, 102, etc. per 1,000 respondents aged 100. The ratio 105 +/100 − 104 is also per 1,000

Thus far, it has been demonstrated that there are more centenarians than expected in the CLHLS data, and that their age distribution is older than expected to an increasing extent with age. Both findings are much more marked in Guangxi. It has been suggested that the most likely explanation is increasing age exaggeration with reported age.

5.4.2 Age Reporting at Ages 80+

The suggestion of age exaggeration among centenarians raises the possibility that age misreporting also exists among younger respondents. As already noted, at ages less than 100, sample design limitations necessitate the use of unconventional methods for the assessment of age reporting accuracy. The approach used here is to examine the age difference between female respondents and their children, or the age at which the mothers bore their children,[9] by age of respondent. This approach is based on two underlying assumptions (note that no assumptions are made about reporting). The first assumption is that age at childbearing is constant across age: as female respondents bore their children during 1905–1970, before fertility restrictions were introduced and when fertility patterns were stable, this assumption is justified. The second assumption is that age at childbearing is unrelated to the survival of the mother. Accordingly, if the ages of mothers and their children were both reported accurately, mean age at childbearing would be constant across age. If some mothers over-reported their age, but their children's ages were better reported, the

---

[9] There was no direct question on age at childbearing. Male data were not examined; male reporting of offspring and their ages is usually of poorer quality than female reporting.

mother–child age gap would be artificially widened. Further, if mother's age exaggeration were more pronounced at older reported ages, the age gap would increase with age. If mothers accurately reported their own age, but under-reported their children's ages on average, a similarly widened age gap may appear.

The CLHLS obtained data on the ages of respondent's children, whether alive or dead. For deceased children, the age they would have been at the time of interview (had they survived) was reported. Mother's age at the birth of each child was calculated as the difference between the female respondent's age and the age of the child. The mean age at childbearing was calculated for all births (by averaging the average age for each individual mother), and for first and last births. These measures are shown by age of respondent[10] in Fig. 5.2. The trends so produced are highly sensitive to age misreporting in respondent's age, facilitating the detection of reporting error. Because any age exaggeration in an individual respondent's age will produce an equally exaggerated age at childbearing (by virtue of the method of calculation), a positive slope will result. The higher the proportion in any age group with exaggerated age, the steeper the slope. Thus, under the two assumptions, the trends in Fig. 5.2 are consistent with increasing proportions of respondents at age 90+ with exaggerated age.

It is also possible that an increasing age gap could be produced by increasing under-reporting of children's ages with respondent's age. However, the uniformity of the trends for different birth orders strongly suggests that they are influenced by the exaggeration of respondent's age, rather than by the under-reporting of

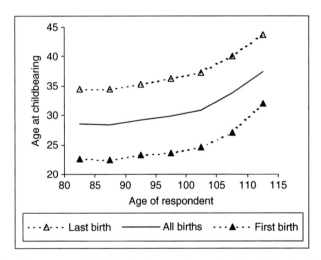

Fig. 5.2 Estimated mean age at childbearing by age of respondent by birth order. See Appendix 2 Table 5.6 for standard errors associated with these estimates

---

[10] As 94 percent of female respondents reported a least one child, conclusions drawn from these data about age reporting can be generalized to all female respondents.

children's ages, because only respondent's age is common to the calculation of all three measures.

Further evidence of age exaggeration is found in the regional variation in mean ages at childbearing. If age exaggeration is present, provinces with greater reported proportions of centenarians, such as Guangxi, would be expected to have higher estimated mean ages at childbearing. Table 5.3 is consistent with this expectation: the mean age at childbearing in Guangxi is 2.3 years greater than in Jiangsu and 3.8 years greater than in Shanghai. Similarly, the mean age at first birth in Guangxi is 1.4 and 2.0 years greater than in Jiangsu and Shanghai respectively, and the mean age at last birth is 2.6 and 4.8 years greater. That such differences are not genuine is supported by the fact that all respondents bore their children before the introduction of fertility restrictions when regional variation in fertility behavior was relatively low. This has in fact been shown by data in the 1982 One Per Thousand Fertility Survey, except for lower fertility in large cities such as Shanghai where fertility decline began earlier than elsewhere.

Figure 5.3 shows mean ages at childbearing by age of respondent for these three provinces and for the whole sample. The consistent patterns of increasing age at childbearing for Guangxi and Jiangsu indicate increasing age exaggeration with age, and the generally higher mean ages for Guangxi would appear to indicate a greater extent of age exaggeration at all ages in this province (given similar expected fertility behavior). For Shanghai, the patterns are neither consistent nor increasing, and there is no observable age exaggeration.

Why is age exaggeration more serious in Guangxi? A possible contributing factor is the high proportion of the Guangxi population who belong to the province's ethnic minorities, as most minorities report age less accurately than the Han majority. This possibility is examined in Fig. 5.3 with respect to two ethnic minorities, the Zhuang and Yao. The Zhuang are China's largest ethnic minority and comprise approximately one third of the population of Guangxi; the Yao are much smaller in number but reported a relatively large number of centenarians. Though minority numbers are relatively small, especially for the Yao, the three mean age at childbearing patterns are fairly consistent indicating age misreporting effects. For both minorities, mean ages at childbearing are generally higher than for Guangxi, suggesting a greater degree of age exaggeration at all ages especially among the Yao. Again, actual fertility behavior is not expected to differ appreciably between these populations.

Table 5.3 Estimated female mean age at childbearing and first and last birth, CLHLS 1998

Province	Number of respondents	Mean age at childbearing	Mean age at first birth	Mean age at last birth
Guangxi	708	31.46 (0.22)	24.53 (0.23)	38.11 (0.30)
Jiangsu	650	29.16 (0.20)	23.16 (0.19)	35.47 (0.28)
Shanghai	229	27.63 (0.32)	22.56 (0.31)	33.30 (0.50)
Total	4943	29.73 (0.08)	23.66 (0.08)	35.95 (0.11)

Based on respondents aged 80+. Standard errors in parentheses

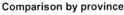
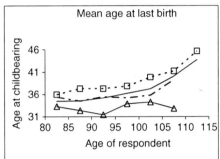
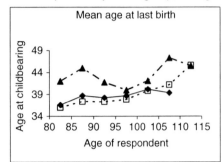
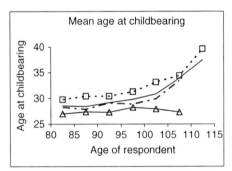
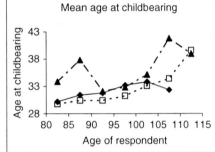
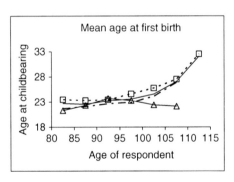
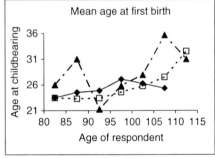

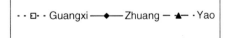

Fig. 5.3 Estimated mean age at childbearing by age of respondent by province and ethnic group. See Appendix 2 Tables 5.6 and 5.7 for standard errors associated with these estimates

5.5 Discussion

The examination in this chapter of the CLHLS centenarian age distribution appears to provide evidence consistent with increasing age exaggeration. This is somewhat different from the verdict of "generally reliable" at ages up to 105 (RGHLC 2000: 6–17). That verdict was based on a comparison of the CLHLS data with a single population, Sweden 1984–1993. However, this population does not provide a valid comparison: life expectancy is 77 years, the percentage aged 65+ is ten percentage points greater than for China-22 in 2000, and the proportion of centenarians in the population is 63 per million (see Table 5.1). The comparison presented in this chapter is with three populations that are considered comparable with the CLHLS data: they have equivalent life expectancies and their age structures (as measured by proportions aged 65+ and 90+) are similar to and span the Chinese age structure in 2000. The use of three populations broadens the comparison; and the inclusion of Japan provides an example of a population that has undergone a more recent and more rapid demographic transition, which is closer than the European populations to China's experience. Further, we base our analysis on age ratios because they are a more sensitive indicator than the percentage distribution. For these reasons, we consider our examination to be both more valid and more reliable than that presented in RGHLC (2000) and Zeng et al. (2001).

While increasing age exaggeration may be apparent in reported age, this does not necessarily indicate increasing age exaggeration when related to true age. This is because of the effect of the (true) age distribution. At the oldest ages, rapidly increasing mortality dominates the shape of the age distribution rendering cohort effects negligible and producing a concave distribution that tapers to zero. With such a distribution a constant rate (at true age) of age exaggeration will tend to have an increasingly large inflationary effect with age, giving an older observed age distribution. Wang et al. (1998) used this effect to support their argument that age reporting among the Han in the 1990 Census is generally good despite questionable numbers at ages 105+. If age exaggeration is present in the CLHLS data, the increasing differentials between the CLHLS age ratios in Table 5.2 and those for the three comparable populations can at least in part be attributed to this age distribution effect. However, this effect cannot exist without age exaggeration.

Neither does the older age distribution necessarily indicate that the data are affected only by age exaggeration. The concavity of the age distribution also means that the same constant rate of age under-reporting will have a smaller deflationary effect that decreases slightly with age. For equal under and over-reporting, the net effect is increasing inflation, and the degree of misreporting imbalance determines the degree of inflation (a modest imbalance toward under-reporting may also produce inflation). Thus the older centenarian age distribution in the CLHLS than in comparable populations could in theory be due to statistically symmetrical age misreporting (or slight net under-reporting). However, to produce the same effect on the age distribution, the extent of such age misreporting would have to be much greater than the extent of pure age exaggeration. Symmetrical but extensive age misreporting would have additional implications for the substantive analysis of these data, particularly if the direction of age misreporting were associated with other variables.

One irrefutable piece of evidence in support of age exaggeration is the fact that age under-reporting cannot have occurred across the upper limit of the age distribution. Thus the inflated age ratios at the oldest ages can only be the result of age exaggeration. It is highly likely that age exaggeration also extends to younger centenarians. Indeed, the smooth decline in the CLHLS ratios in Table 5.2 gives no indication of other patterns of age misreporting at younger ages. Thus, while it is impossible to determine from these ratios the exact extent and balance of age misreporting, it can be determined that age exaggeration is present, and that while this is not necessarily increasing with true age, it increasingly affects reported age at ages 101+.

The arguments related to the effect of the concavity of the (true) age distribution are also relevant to the evidence based on the age gap between female respondents and their children.[11] Thus, in part at least, the increasing trends in Fig. 5.2 could be produced by a constant rate of age exaggeration by true age. It might also be argued that because under-reporting of respondent's age will produce an under-estimated age at childbearing, this will also appear as a positive slope in Fig. 5.2.[12] However, the increasing slope would imply increasingly improved reporting with increasing age, to a theoretical limit of no under-reporting at the upper age limit, which is not only unlikely but could not produce the larger effects at older ages because of smaller numbers. (Again, any counterbalancing of misreporting would imply a much higher overall level of misreporting.) Further, under-reporting would imply higher (true) mean ages at childbearing; this issue is now addressed.

Comparison with directly reported ages at childbearing for the same cohort[13] of women in the 1982 One Per Thousand Fertility Survey shows that the CLHLS estimate of the mean age at first birth is too high. This could be the result of respondent's age exaggeration or child's age under-reporting or the omission of low-order births at younger childbearing ages. The same comparison shows the CLHLS estimated mean age at last birth to be too low, which could arise from respondent's age under-reporting or child's age exaggeration or the omission of higher order births at older childbearing ages. Clearly, if the reporting of respondent's age is responsible, age at first and last birth must exhibit a common direction of misreporting, which is not the case. If the reporting of first and last children's ages is responsible, the comparisons imply that the estimated childbearing period has been compressed at both ends. This is indistinguishable from the effect of omissions. Comparison with data in the 1982

---

[11] The age distribution effect does not apply to children's ages. Thus (increasing) under-reporting of children's ages is unlikely to explain the large increases in age at childbearing. The alternative explanation of over-reporting of children's ages to a decreasing degree with age is similarly flawed.

[12] If mother's and children's ages were equally misreported, there would be no effect in Fig. 5.2 because age at childbearing would be correct.

[13] The only cohort for which this comparison is possible is women aged 64–67 in the 1982 survey and 80–83 in the CLHLS.

One Per Thousand Fertility Survey suggests that around one child per woman has been omitted in the CLHLS. In addition, a high proportion of reported higher birth order children have missing age, which has the same effect as omissions of births per se. Thus omission is (at least partly) responsible for the relatively low age at last birth, and is likely to also be a factor in the high age at first birth.

There thus seems to be a contradiction between the explanations for the patterns observed. While omissions could explain the early age at last birth, found on average for CLHLS respondents, they cannot lead to the overestimated age at last birth found at very old ages (Fig. 5.2). Further, omissions are unlikely to produce the pattern observed in the mean age at childbearing. Thus, it is likely that the high level of omission for last births overwhelms the effect of respondent age exaggeration on age at last birth, in which case the late age at first birth found in the CLHLS can be at least partly attributed to age exaggeration.

A reviewer of this chapter questioned the validity of the above comparison on the grounds that the CLHLS did not go into the same level of detail in recording fertility histories as the 1982 survey, and moreover that the "memory capacity of women aged 80–83 [in the CLHLS] was certainly substantially weaker as compared to those aged 64–67 [in the 1982 survey]." This is precisely the point that we are making: the CLHLS data are not consistent with the much better quality 1982 data because of reporting inaccuracies, and that these inaccuracies relate in part to age reporting in the CLHLS. Moreover, the reviewer's observations and additional comment that "the reliability of the age at births reported by these extremely old women is certainly questionable" raise important further issues. First, given that the CLHLS age validation process was largely based on recall of personal events including marriage and childbearing (Xu 2001, and Chap. 4 of this volume), the weaker memory capacity and unreliable age at birth reporting of CLHLS respondents actually call into question the accuracy of validated age. Second, this dependence between age at childbearing and respondent's age invalidates the use of these data for the study of the relationship between longevity and age at childbearing, because reporting errors will result in their correlation. Third, the unreliability of reported fertility in the CLHLS (e.g., the omission of births) would have a considerable impact on the relationship between longevity and age at childbearing, again invalidating its study based on these data. As already noted, the CLHLS data likely omit around one child per woman.

The evidence supporting the increasing extent of age exaggeration in reported age is particularly strong for Guangxi. Centenarian proportions for Guangxi far exceed expectations, and their reported age distribution is exceptionally old. Further, the higher than average mean ages at childbearing suggest a greater extent of age exaggeration, which is in evidence from age 80. The significance of this higher level of age exaggeration in Guangxi for substantive analyses of the CLHLS data cannot be ignored. Indeed, preliminary analyses indicating that mortality rates of CLHLS centenarians between survey rounds are lower in Guangxi than in other provinces could be entirely an artifact of age exaggeration. Further, the variation in the degree of age exaggeration by province clearly has implications for the sample weights. The higher level of age exaggeration in Guangxi results in this province comprising

14 percent of CLHLS centenarians, and by virtue of the sample design 14 percent of the whole sample, when it comprises only 4 percent of the total population of China-22. The significance of age misreporting among particular ethnic minorities (e.g., Coale and Li 1991) should similarly be carefully considered.

In order to give some idea of the variability in the data, Tables 5.6 and 5.7 in Appendix 2 provide standard errors for the mean ages at childbearing shown in Figs. 5.2 and 5.3 respectively. It must be emphasized, however, that high variability (such as for the Zhuang and Yao ethnic minority populations of Guangxi) does not detract from the analysis and the validity of the conclusions drawn. Our concern is to demonstrate the quality of the CLHLS sample per se. The fact that the high mean ages for the Yao in Guangxi have relatively large standard errors does not detract from these observed means, nor from the fact that this group forms part of the sample and contributes to the inaccuracies in the data. It is noted, however, that in all but one cases standard errors are smallest for the age group 100–104, giving confidence in the patterns observed.

These conclusions about age reporting at ages 80+ are based on the assumption that age at childbearing is unrelated to the survival of the mother. This assumption is open to question: research findings show a positive relationship between late childbearing and longevity at the population level. However the exact nature of this relationship has yet to be determined. Several studies found no relationship between age at first birth and subsequent survival at ages 50+ (Smith, Mineau, and Bean 2002; Dribe 2004; Alter, Dribe, and van Poppe 2004), though Doblhammer (2000) found a significant positive association when comparing first births at age <20 (which accounted for only 7–12 percent of first births) with those at 20+. On the other hand, most of these studies found a significant positive relationship between age at last birth and survival at ages 50+, as have other studies (e.g. Perls, Alpert, and Fretts 1997; Müller et al. 2002).

Would relaxing this assumption lead to a different conclusion? A relationship between longevity and age at childbearing could in part explain the increasing mean ages at childbearing by age in Fig. 5.2. If there is a relationship in the CLHLS data between longevity and age at last birth, but not age at first birth (or a stronger relationship for age at last birth), its effect would be to widen the childbearing interval with increasing respondent's age. Such widening is in fact found for the whole sample (the interval increases from 11.8 years at age 80–84 to 13.0 at 105–109) and for Guangxi, but it is not found for Shanghai or Jiangsu where age reporting is more accurate, suggesting the absence of such a relationship. It is noted that the widening of the childbearing interval cannot be attributed to increasing omissions with age, as this would have the opposite effect. Thus the increasing means with age seen in Fig. 5.2 are attributed to age exaggeration and the effects of the concavity of the age distribution.

Using CLHLS data, Zeng and Vaupel (2004) found a statistically significant positive association between number of births after age 35 or 40 and the risk of survival as measured by deaths between the 1998 and 2000 waves of the survey. The number of births after age 35 or 40 is dependent on the respondent's age and on the age of each child because age at childbearing depends on these two ages. If, as suggested by this examination of data quality, age exaggeration exists among female respondents this finding could be spurious. The increasing level of significance of Zeng and Vaupel's finding with increasing number of children (2004: table 3) is also consistent with the effect of age exaggeration because each additional child must have been borne at an older age. The same is true of the increased significance at higher ages (2004: table 4a). Here, as in the above discussion, the effect of the age distribution comes into play to magnify the strength of the association at older ages.

In conclusion, careful examination of the CLHLS data has demonstrated the presence of age exaggeration in reported age as seen in the older than expected age distribution. The extent of exaggeration in the data appears to be low at the youngest ages, but increases with age especially after age 100. Much of the age exaggeration can be attributed to Guangxi and in particular Guangxi's ethnic minorities. However, the extent of exaggeration found in the data cannot be entirely accounted for by the minority populations. It must therefore be concluded that age misreporting or exaggeration at very old ages also exists among the Han majority, especially at ages 100+, even though age reporting is generally reliable in this population. While increasing age exaggeration may be due to the effect of the age distribution, it is nevertheless present in the data. Thus substantive analyses, especially those concerning longevity, must take age exaggeration into account. In particular, analyses using constructed variables based on respondent's age may well be open to spurious relationships with age. The recommendation that respondents aged 106+ be omitted from analyses only partially addresses this issue: though the largest biases will be removed by this means, biased results may still be obtained. Analyses of the effects of age exaggeration on substantive results based on the CLHLS data are beyond the scope of this chapter, but have subsequently been undertaken by others (Chap. 4 in this volume).

Finally, it must be emphasized that this chapter has sought to evaluate the quality of age reporting in the CLHLS through an objective examination of the data, including reporting problems that had not previously been investigated. The aim has been to identify whether age misreporting is present in the data, rather than to rank the CLHLS among various data sources according to the quality of age reporting. This initial analysis has demonstrated that the data are subject to the effects of age misreporting, and further evaluation of the quality of the data and the effect of age misreporting on analyses is still needed. Our ultimate interest is in the reliability of the conclusions about longevity based on CLHLS data.

Appendix 1: Examination of Evidence on Heaping at Age 100 in CLHLS Population

Age ratios for centenarians based on 101 and 102 give similar results for females as for ratios based on 100: the CLHLS distribution is older than any of the three comparable populations. Thus age heaping at age 100 does not affect the finding. For males, ratios based on 101 and 102 do not show an older age distribution at ages 102 and 103 suggesting that the extent of inflation (relative to 100) at ages 101, 102, and 103 does not increase; an older distribution is observed at ages 104 and above for all ratios.

Table 5.4 Age ratios based on age 101, CLHLS 1998

Age	Male				Female			
	Sweden 1943–1952	E&W 1950–1955	Japan 1962–1966	CLHLS 1998	Sweden 1943–1952	E&W 1950–1955	Japan 1962–1966	CLHLS 1998
101	1,000	1,000	1,000	1,000	1,000	1,000	1,000	1,000
102	577	527	537	554	524	547	519	608
103	231	274	268	264	244	284	305	390
104	154	125	171	190	122	148	136	218
105	38	41	0	124	61	78	84	122
106	0	7	0	99	24	37	58	92
107	0	0	0	17	12	18	45	79
108	0	0	0	17	0	6	32	41
109	0	0	0	25	0	1	19	32
110+	0	0	0	33	0	0	32	36

Age ratios are expressed as the number of respondents aged 102, 103, etc. per 1,000 respondents aged 101

Table 5.5 Age ratios based on age 102, CLHLS 1998

Age	Male				Female			
	Sweden 1943–1952	E&W 1950–1955	Japan 1962–1966	CLHLS 1998	Sweden 1943–1952	E&W 1950–1955	Japan 1962–1966	CLHLS 1998
102	1,000	1,000	1,000	1,000	1,000	1,000	1,000	1,000
103	400	519	500	478	465	520	588	641
104	267	237	318	343	233	271	263	359
105	67	78	0	224	116	142	163	201
106	0	13	0	179	47	68	113	151
107	0	0	0	30	23	33	88	130
108	0	0	0	30	0	11	63	67
109	0	0	0	45	0	2	38	53
110+	0	0	0	60	0	0	63	60

Age ratios are expressed as the number of respondents aged 103, 104, etc. per 1000 respondents aged 102

Appendix 2: Standard Errors for Figures 5.2 and 5.3

Table 5.6 Standard errors of mean ages at childbearing of total sample

	Respondent's age						
	80–84	85–89	90–94	95–99	100–104	105–109	110–114
Age at last birth							
Mean	34.48	34.52	35.22	36.25	37.26	40.08	43.75
SE	0.24	0.29	0.27	0.28	0.19	0.66	2.10
Age at childbearing							
Mean	28.55	28.41	29.15	29.82	30.85	33.89	37.52
SE	0.17	0.20	0.18	0.20	0.14	0.48	1.61
Age at first birth							
Mean	22.71	22.52	23.26	23.58	24.59	27.07	32.08
SE	0.17	0.19	0.18	0.20	0.15	0.52	2.26

Table 5.7 Standard errors of mean ages at childbearing by province and ethnic minority

Province/ ethnic group		Respondent's age						
		80–84	85–89	90–94	95–99	100–104	105–109	110–114
Age at last birth								
Shanghai	Mean	33.27	32.39	31.43	33.90	34.27	32.83	–
	SE	1.41	1.42	1.26	1.12	0.78	2.98	–
Jiangsu	Mean	35.35	34.48	35.49	35.20	35.81	39.29	–
	SE	0.73	0.83	0.83	0.78	0.41	2.16	–
Guangxi	Mean	35.97	37.33	37.29	37.84	39.85	41.22	45.67
	SE	0.66	0.80	0.70	0.80	0.52	1.59	2.46
–Zuang	Mean	36.60	38.71	38.28	38.71	40.20	39.38	–
	SE	1.07	1.12	1.12	1.49	0.81	2.81	–
–Yao	Mean	42.00	45.00	41.75	40.00	42.08	47.33	45.50
	SE	2.00	2.86	2.66	2.01	2.04	0.88	3.31
Age at childbearing								
Shanghai	Mean	27.00	27.37	27.28	28.24	27.95	27.33	–
	SE	0.88	0.92	0.90	0.76	0.46	1.89	–
Jiangsu	Mean	28.24	27.85	29.08	28.86	29.90	33.43	–
	SE	0.52	0.59	0.56	0.48	0.29	1.49	–
Guangxi	Mean	29.76	30.39	30.42	31.27	33.11	34.39	39.59
	SE	0.47	0.53	0.49	0.59	0.41	1.33	0.62
–Zuang	Mean	30.19	31.43	31.78	33.22	33.84	32.38	–
	SE	0.79	0.79	0.77	1.10	0.67	1.69	–
–Yao	Mean	33.88	37.84	32.11	32.86	35.15	41.81	38.98
	SE	2.62	3.45	2.77	2.50	1.58	3.12	0.76
Age at first birth								
Shanghai	Mean	21.30	22.33	23.76	23.27	22.43	22.16	–
	SE	0.59	0.86	0.95	0.86	0.47	1.91	–
Jiangsu	Mean	21.68	22.07	22.64	22.88	24.14	26.86	–
	SE	0.45	0.57	0.49	0.45	0.30	2.12	–
Guangxi	Mean	23.46	23.30	23.42	24.60	25.75	27.52	32.50
	SE	0.45	0.49	0.46	0.58	0.45	1.51	2.28
–Zuang	Mean	23.52	24.54	24.88	27.13	26.31	25.38	–
	SE	0.65	0.90	0.86	1.13	0.71	2.16	–
–Yao	Mean	26.00	31.00	21.25	25.88	28.00	35.67	31.00
	SE	4.00	5.05	2.78	2.50	2.49	5.92	2.38

Acknowledgments This chapter was originally presented at a seminar at the Max Planck Institute for Demographic Research (MPIDR), Rostock, Germany on 4 August 2004; we thank MPIDR for financial support to attend this seminar. Data for England and Wales, Sweden and Japan were obtained from the Human Mortality Database http://www.mortality.org. We thank Zeng Yi and two anonymous reviewers for helpful comments on an earlier draft.

References

Alter, G., M. Dribe, and F. van Poppe (2004), Childbearing history and post-reproductive mortality: A comparative analysis of three populations in nineteenth century Europe. Presented at the Annual Meeting of the Social Science History Association, Chicago (USA), 18–21 November 2004.

Coale, A. and E.E Kisker (1986), Mortality crossovers: Reality or bad data? *Population Studies* 40 (3), pp. 389–401.

Coale, A. and S. Li (1991), The effect of age misreporting in China on the calculation of mortality rates at very high ages. *Demography* 28 (2), pp. 293–301

Doblhammer, G. (2000), Reproductive history and mortality later in life: A comparative study of England and Wales and Austria. *Population Studies* 54 (2), pp. 169–176

Dribe, M. (2004), Long-term effects of childbearing on mortality: Evidence from pre-industrial Sweden. *Population Studies* 58 (3), pp. 297–310

Gu, D. and Y. Zeng, (2004), Data quality assessment of the CLHLS 1998, 2000, and 2002 waves. In: Y. Zeng, Y. Liu, C. Zhang, and Z. Xiao (eds.) *Analyses of the determinants of healthy longevity*. Beijing: Peking University Press, pp. 3–22 (in Chinese).

Jeune, B. (1995), In search of the first centenarians. In: B. Jeune and J. Vaupel (eds.), *Exceptional longevity: From prehistory to the present*, Odense: Odense University Press, pp. 11–24

Kannisto, V (1994), *Development of oldest-old mortality, 1950–1990: Evidence from 28 developed countries*. Odense: Odense University Press.

Müller, H.-G., J.-M. Chiou, J.R. Carey, and J.-L. Wang (2002), Fertility and life span: Late children enhance female longevity. *The Journals of Gerontology, Series A: Biological Sciences and Medical Sciences* 57 (5), pp. B202–B206

Perls, T.T., L. Alpert, and R.C. Fretts (1997), Middle-aged mothers live longer. *Nature* 389, pp. 133

Research Group of Healthy Longevity in China (RGHLC) (2000), *Data collections of the healthy longevity survey in China 1998*. Beijing: Peking University Press.

Rosenwaike, I. and L.F. Stone (2003), Verification of the ages of supercentenarians in the United States: Results of a matching study. *Demography* 40 (4), pp. 727–739

Smith, K.R., G.P. Mineau, L.L. Bean (2002), Fertility and post-reproductive longevity. *Social Biology* 49 (3–4), pp. 185–205

Wang, Z., Y. Zeng, B. Jeune, and J.W. Vaupel (1998), Age validation of Han Chinese centenarians. *Genus* LIV (1–2), pp. 123–141

Xu, Q. (2001), Evaluation of age-reporting among the elderly. *Marketing and Population Analysis* 3, pp. 1–12 (in Chinese)

Zeng, Y. and J.W. Vaupel (2004), Association of late childbearing with healthy longevity among the oldest-old in China. *Population Studies* 58 (1), pp. 37–53

Zeng, Y., J.W. Vaupel, Z. Xiao, C, Zhang, and Y. Liu (2001), The healthy longevity survey and the active life expectancy of the oldest old in China. *Population: An English Selection* 13 (1), pp. 95–116

Chapter 6
Assessment of Reliability of Mortality and Morbidity in the 1998–2002 CLHLS Waves

Danan Gu and Matthew E. Dupre

Abstract This chapter assesses the reliability of mortality and self-reported morbidity in the first three waves of the Chinese Longitudinal Healthy Longevity Survey (CLHLS). Results indicate that the observed rates of all-cause mortality reported in the CLHLS are underestimated by 15–20 percent between 1998 and 2000 and by 5–20 percent for ages 80–90 when based on hazard-model estimates; however, no such differences are found between the 2000 and 2002 waves. Our analyses further show that mortality rates over age 90 in the CLHLS are more reliable than those obtained from the census. The quality of self-reported morbidity and its population prevalence is generally quite good compared to other national data sets. However, the analyses suggest that information collected from next-of-kin should be interpreted with caution. We find that cause-specific mortality rates estimated from reports by the next-of-kin are substantially biased and that the prevalence of the decedents' morbidity reported by the next-of-kin is somewhat underestimated.

Keywords Accuracy of mortality data, Age exaggeration, Age reporting, All-cause mortality, Bias, Cause of death, Cause-specific death rates, Census data, China National Disease Surveillance Point System, Complex sampling design, Data assessment, Ethnic minorities, Extrapolation, First year death, First-year mortality rate, Han Chinese, Hazard model, Healthy longevity, Health and Retirement Study, Kannsito model, Lost to follow-up, Lunar calendar, Morbidity, Mortality, National Long-Term Care Survey, Next-of-kin, Reliability, Relative bias, Sample selection, Sample attrition, Second National Health Service Survey, Second year death, Second-year mortality rate, Self-reported chronic conditions, Simulation, Underestimation, Underestimation of mortality

---

D. Gu
Center for the Study of Aging and Human Development, Medical School of Duke University, Durham, NC 27710, USA
e-mail: gudanan@duke.edu

6.1 Introduction

The accuracy of mortality rates at old ages is critical for an understanding of human aging and longevity. Numerous studies have documented distinct patterns of mortality at older ages (Horiuchi and Wilmoth 1998; Kannisto 1994; Thatcher et al. 1998; Vaupel et al. 1998; Zeng and Vaupel 2003). For example, countries with high-quality data frequently show that age-specific mortality rates after age 80 tend to follow logistic, Kannisto, or quadratic patterns (Kannisto 1994; Thatcher et al. 1998; Vaupel et al. 1998). Departure from these patterns may signify age misreporting and produce inaccurate mortality estimates that will bias predicted outcomes and lead to erroneous conclusions.

Unlike respondent deaths in most longitudinal surveys in Western nations, which can be linked to a national death database with relatively accurate mortality reports, the survival status of respondents in longitudinal surveys in most developing countries are often collected through interviews. Consequently, mortality data from studies in developing countries may introduce inaccuracies due to age exaggeration and age-at-death misreporting by proxies, which are further confounded by sample attrition and selection. Such inaccuracies may well introduce biases in the estimation of associations between the study variables and mortality and lead researchers to draw incorrect conclusions. The first part of this chapter assesses the accuracy of mortality recorded in the first three waves (1998, 2000, and 2002) of the Chinese Longitudinal Healthy Longevity Survey (CLHLS).

The quality of morbidity data in surveys is important because it captures a major dimension of health and aging and is crucial for understanding healthy longevity. Conversely, inaccurate morbidity data will tend to produce biases in the associations between risk factors and disease (and other health indicators), as well as model misspecifications in the risk factors predicting subsequent mortality. However, age-specific patterns of morbidity at old ages are less well documented in the literature and, to date, there are no established criteria for comparing morbidity rates at the oldest-old ages across populations or nations. Indeed, the age patterning of chronic conditions among old persons tends to vary depending on the stage of the epidemiological transition, advances in medical technology, public awareness, and personal knowledge (see Klabunde et al. 2005; Myers et al. 2003; Robine and Michel 2004). The second aim of this chapter is to compare several external data sources to assess the accuracy of chronic morbidity reported in the first three waves of the CLHLS.

6.2 Mortality Reliability

6.2.1 Observed All-Cause Mortality

Han Chinese, as well as several ethnic minorities who are culturally and residentially integrated with the Han (e.g., Korean and Manchu), use the lunar calendar (animal year) to recall the date of their birth. This unique system for remembering an individual's age makes age exaggeration less common (see chap. 4 of this volume for more detail), ensuring that the mortality rates of Han and related Chinese ethnicities are generally reliable at advanced ages. As such, Coale and Li (1991) showed that mortality rates at the oldest-old ages were generally accurate in the third national census (in 1982) when excluding the Xinjiang Province, which has a heavy concentration of ethnic minorities. Some recent studies, however, show that the accuracy of mortality rates from censuses in China is improving (e.g., Banister and Hill 2004). Below we present comparative analyses of observed and estimated mortality using the CLHLS and data from the 2000 Chinese census in the 22 provinces where the CLHLS has been conducted. We also address the effects of potential age exaggeration by ethnic minorities on overall mortality risk in the CLHLS.

We calculate the person-years lived by each respondent for a specific period according to the date of birth, dates of the interviews for survivors, and the date at death for those who died between the survey waves. All analyses are weighted to account for the complex sampling design. We divide the weighted number of deaths at each age in a given period by the total weighted number of person-years lived during this period to obtain separate age-specific death rates for males and females.[1] Figure 6.1 compares the age-sex-specific death rates for the oldest-old in the CLHLS for the period of November 1999 thru October 2000 and for the period of January 2001 thru December 2001. For the 22 provinces in the 2000 census, the period is from November 1999 through October 2000.[2] Using the same estimation

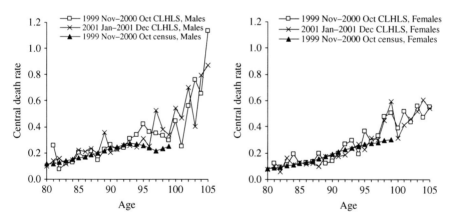

Fig. 6.1 Comparison of observed period-specific mortality rates in the CLHLS and the 2000 census (22 provinces)

---

[1] This chapter does not use $\frac{N_{x+n}^{T+n}}{N_x^T} = \exp\left(-\int_x^{x+1} \mu(t)dt\right)$ to estimate the central death rate which was employed by Zeng et al. (2004). However, the results in this chapter are quite similar to those of Zeng et al. (2004).

[2] Respondents who report being older than age 106 at baseline were excluded because of insufficient information to validate their age. Respondents younger than age 80 at baseline were excluded in this study for the 1998 and 2000 waves, while those younger than age 65 were excluded for the 2002 wave. The respondents are excluded because they are out of the targeted age range for the CLHLS sample.

method for the CLHLS and 2000 census, we find no major differences in mortality between the CLHLS and the 2000 census during the period from November 1999 to October 2000 (the CLHLS produces higher mortality rates than the 2000 census, especially after age 94).[3] However, we find that compared to the census, mortality in the CLHLS was underestimated by around 10–15 percent in the first year immediately after the interview for both the 1998 and 2000 waves (not shown), as has already been found by Goodkind (2004). This may be because the next-of-kin of deceased respondents misreported the timing of death to a later date (Goodkind 2004). If the first- and second-year deaths were not distinguished, as in most longitudinal studies, we would find no underestimation in mortality in the CLHLS in the period of 2000 and 2002; however, there would still be a 15–20 percent underestimation of mortality before age 90 for both males and females between 1998 and 2000 in the CLHLS (Fig. 6.2). Therefore, we conclude that the mortality rates from the first three waves of the CLHLS are relatively reliable, with the exception of some recall errors by proxies in the respondents' date of death.

6.2.2 Estimated All-Cause Mortality

Thatcher et al. (1998) examined the force of mortality (i.e., central death rate) at the oldest-old ages in thirteen developed countries across several recent decades. They used the observed population size and number of deaths for ages 80–98 and extrapolated for ages beyond 98. They found that the mortality patterns were modeled better with Logistic, Kannisto, and quadratic methods than with Gompertz, Weibulll, and Heligman and Pollar models for all the countries and across all three

---

[3] We summed the age-sex-specific number of deaths recorded for the period from November 1999 through October 2000 and the age-sex-specific number of persons counted in the 2000 census (i.e., October 31, 2000) for 22 provinces and applied the formula $d_x = d_x \frac{D_x}{0.5(P_x^{t-1}+P_x^t)}$ to calculate the age-sex-specific central death rate. Note that the National Bureau of Statistics of China (NBSC) and each provincial statistic bureau used $d_x = \frac{D_x}{0.5(P_x^t+P_{x+1}^t)}$ to calculate the central death rate for age x at time $t-1$ to time t (i.e., November 1999 through October 2000). This NBSC formula imposes negligible bias when the size of each birth cohort is similar and mortality is low. However, when mortality is high, especially at old ages, the formula suffers bias even though the size of each cohort is similar. In order to get P_x^{t-1}, the approach proposed by Vallin (1973) was employed to split all possible age-year death aggregates by cohort (i.e., the older cohort shares 2% more deaths for a given amount of death events at age x and year t). The Kannisto and Thatcher database on old-age mortality (80–120) used half versus half to split the death, although they admit their approach may also suffer minor biases. After splitting, we can easily estimate P_x^{t-1} and the central death rate. We also applied the same approach to estimate the central death rate in the 22 provinces from the 2000 census as we did for the CLHLS. We assume that each of person in P_x^t survives 0.5 person-year in age x, each of the deceased persons at age x survives 0.5 person-year in age x, and each person in P_{x+1}^t survives 0.5 person-year in age x before they reach the exact age $x + 1$. The central death rate is slightly higher for the person-year-based approach than for $d_x = \frac{D_x}{0.5(P_x^{t-1}+P_x^t)}$. In this chapter, we present only the person-year-based approach. Both the person-year-based approach and $d_x = \frac{D_x}{0.5(P_x^{t-1}+P_x^t)}$ are slightly smaller as compared to the NBSC published data for persons ages 90 and over.

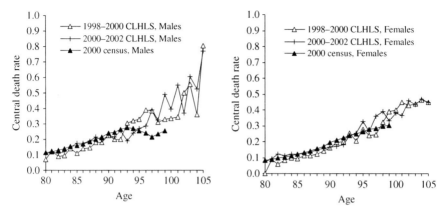

Fig. 6.2 Comparison of mortality rates between the CLHLS and 2000 census (22 Provinces) without distinguishing the first and second year mortality after the interview

decades. Zeng and Vaupel (2003) used the same methods to investigate mortality patterns for the oldest-old Han Chinese people in the 1990 census and found similar results. Zeng and Vaupel's findings also suggest that the mortality data for the oldest old Han Chinese are quite good in the 1990 census, although the observed rates of male mortality beyond age 94 are slightly underestimated.

We apply the same method to estimate mortality rates from ages 80–98 and extrapolate for ages 99–105 for the CLHLS 22 provinces in the 2000 census.[4] We present only the findings from the Kannisto model in this chapter. A comparison of mortality rates fitted from the Kannisto model with mortality rates obtained from the 2000 census indicates an underestimation for males after age 94 in the census data (not shown) and no difference for females. Therefore, we conclude that mortality rates for the oldest-old females in the 22 provinces in the 2000 Chinese census are reliable.

Because the CLHLS (or any survey in China) cannot produce reliable mortality rates for each age comparable to the census, we use regression models to estimate age patterns according to the survey data and then compare them with the established criterion (i.e., the Kannisto model, see Zeng and Vaupel (2003). Thus, mortality rates for two periods of the CLHLS (i.e., between the 1998 wave and the 2000 wave, and between the 2000 wave and the 2002 wave) are estimated using both semi-parametric Cox proportion hazard models and parametric exponential and Weibull hazard models.[5] We use an average of the estimates of these three

---

[4] In order to do this, the method described in footnote 3 is used to estimate the risk population and death for each age. We then follow Thatcher et al.'s (1998) method to estimate and extrapolate the age-specific mortality rates for ages 80–105. Zeng and Vaupel (2003) show that the Kannisto model is one of the best models in fitting the oldest-old mortality for Han Chinese.

[5] Weibull and exponential models produce similar results, but they produce higher rates than the Cox model. The mortality rates based on the Weibull or exponential models for the period of 2000–2002 are higher than those based on the Kannisto model. This suggests that different

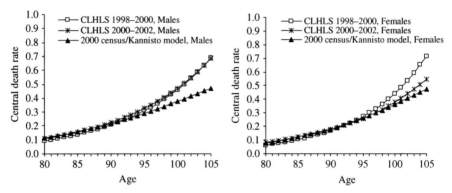

Fig. 6.3 Comparison of age-sex-specific mortality rates between the CLHLS based on hazard models and the 2000 census based on Kannisto models (22 provinces)

approaches to represent the model-based approach for the force of mortality in the CLHLS. Hazard models are conducted separately for males and females and sample weights are applied.

Figure 6.3 compares the central death rates for the two periods in the CLHLS using hazard models with those from the 22 provinces in the 2000 census using the Kannsito model. The relative bias of the model-based approach in the CLHLS is also compared with the census results based on the Kannisto model and is reported in Fig. 6.4. The results suggest that the age-sex-specific mortality rates are somewhat underestimated for the period of 1998–2000. There is an approximately 5–20 percent underestimation of mortality for both males and females before age 90 in the 1998–2000 interval. The average underestimations before age 90 are 12 percent for

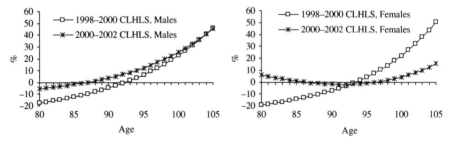

Fig. 6.4 Relative mortality bias between the CLHLS based on hazard models compared to census mortality fitted by Kannisto models (22 provinces)

---

analytical models will yield inconsistent findings. Other parametric models were also tested and the results were analogous to those produced by the Weibull and exponential models. According to Cleves et al. (2004: 197–200), parametric hazard models are superior to Cox hazard models in utilizing event data over the time intervals being studied. The gap in mortality rates (before age 90) between the estimated average rates from parametric models will be closer to those fitted by the Kannisto model.

males and 15 percent for females. However, there is no evidence of underestimation in the interval from the 2000 wave to the 2002 wave. Comparisons between the first three waves of pooled data (i.e., 1998, 2000, and 2002) and the results from the Kannisto model further show that there is an underestimation of mortality of roughly 8 percent before age 90 (results not shown).

6.2.3 The Effects of Potential Age Exaggeration on Mortality Rates

Approximately 7 percent of the CLHLS sample is categorized as being in a non-Han minority group. Because ethnicity is a major sociodemographic factor in the study of aging, the accuracy of non-Han mortality data will likely influence health and mortality-related patterns and their associations with other sociodemographic variables. If age reporting among the non-Han sample is not accurate and thus mortality estimates are not reliable, it may be necessary to drop these persons from the analysis; alternatively, researchers should retain the minority sample in the analyses if the data prove to be reliable. This section assesses the accuracy of mortality data among the non-Han minorities as they relate to age exaggeration and its impact on mortality estimates when age reports are not accurate.

As noted in Chap. 4 of this volume, the accuracy of age reporting for Han Chinese in the CLHLS is relatively reliable and roughly comparable to that of Canada and Australia. Chapter also shows that age reporting among non-Han Chinese is less accurate than among Han Chinese, but that the possible age exaggeration will not bias the association between health/survival and late childbearing in multivariate regression models. Gu and Zeng (2004) have demonstrated that age exaggeration among the oldest-old non-Han respondents in the CLHLS does not significantly affect estimates of disability, cognitive functioning, self-reported health, or their predictors either cross-sectionally or longitudinally. In this chapter, four simulations are performed using parametric (exponential) hazard models to further examine the effect of possible age exaggeration among ethnic minorities on mortality rates in the CLHLS. The simulations include: (1) each minority aged 85 and older over-reports their age by 2 years; (2) each minority aged 85–89 over-reports their age by 2 years, while each minority aged 90 and older over-reports their age by 5 years; (3) each minority aged 85–89 over-reports their age by 2 years, while each minority aged 90 and older over-reports their age by 8 years; and (4) each minority aged 85–89 over-reports their age by 2 years, each minority aged 90–94 over-reports their age by 8 years, and each minority aged 95 and older over-reports their age by 10 years.

Figure 6.5 demonstrates that even though we assume serious age exaggeration among ethnic minorities, its impact on the estimation of overall mortality in the CLHLS is negligible.[6] Zhang and Li (2004) find that there is no underestimation of

---

[6] The results in Fig. 6.5 are slightly different from Fig. 6.3 because Fig. 6.5 is based on the exponential hazard model while Fig. 6.3 is based on the averages from the exponential, Weibull, and Cox models. The simulation patterns based on the Weibull and Cox models are the same as those in Fig. 6.5. For illustrative purposes, we simply use the exponential model.

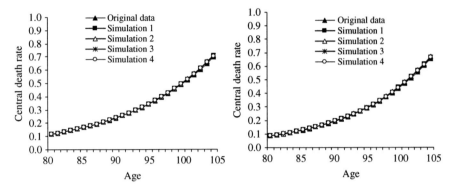

Fig. 6.5 Estimated impact of potential age exaggeration by ethnic minorities on overall mortality in the CLHLS, 1998–2002

mortality among minorities from 1998 to 2002 compared with Han Chinese using nonparametric Kaplan and Meier (KM) analyses.

6.2.4 Cause-Specific Mortality

Information on cause-specific death rates is important in determining the allocation of health investments and resources in public health policy and for identifying the priorities of national health services. The CLHLS includes information on the primary cause of death for the sampled decedent who died between survey waves (collected from his/her next-of-kin). Table 6.1 compares the cause-specific death rates in the CLHLS with those obtained from the China National Disease Surveillance Point System (DSP). Data from the DSP covers 145 sites in both urban and rural areas from 31 provinces, and they may be considered a proxy for the 22 CLHLS provinces. Although the DSP accounts for only one percent of the national population and undercounts the number of deaths, the DSP data provide sufficient confidence for the primary causes of death (Yang et al. 2005). Table 6.1 indicates that the CLHLS highly underestimated the cause-specific death rates of cancer, cerebrovascular disease, heart disease, and respiration and digestive diseases. Nearly half of the results in Table 6.1 underestimate mortality by more than 50 percent, particularly in rural areas, suggesting that the cause-specific data in the CLHLS are not reliable.[7] We suspect the primary reasons for this to be threefold. First, many next-of-kin did not know the cause of death because the majority of subjects died at home (over 90 percent), and many probably lacked medical tests and/or treatment prior to death. Second, many urban respondents—whose next-of-kin are more likely to provide the cause of death—were lost to follow up. Third, information on the cause of death was not collected from official government agencies in the CLHLS.

---

[7] Because we are unable to access the DSP data, we have no knowledge of the accuracy of the other cause-specific mortality rates in the CLHLS. We suspect it also would be underestimated.

Table 6.1 Comparisons of death rates (%) by main causes between the CLHLS (1998-2000) and China's national death rates (1998)

	Age group	Cancer	Cerebrovascular diseases (CVD)	Heart diseases	CVD and Heart dis.	Respiratory and digestive dis.
Urban						
CLHLS (pooled 1998–2002)						
Male	80–84	0.78	1.99	1.20	3.19	1.18
	85+	1.43	2.65	2.04	4.69	2.43
Female	80–84	0.89	1.53	1.07	2.60	0.81
	85+	0.61	1.80	1.34	3.14	1.26
China's Disease Surveillance Point System (1998)						
Male	80–84	1.93	3.28	2.56	5.84	3.30
	85+	1.90	3.75	4.16	7.91	5.40
Female	80–84	0.97	2.44	2.14	4.58	2.24
	85+	0.88	3.37	3.82	7.19	4.55
Rural						
CLHLS (pooled 1998–2002)						
Male	80–84	0.43	1.70	0.97	2.67	2.12
	85+	0.90	2.27	1.37	3.64	1.89
Female	80–84	0.43	1.36	0.79	2.15	0.95
	85+	0.33	1.61	1.03	2.64	1.73
China's Disease Surveillance Point System (1998)						
Male	80–84	1.16	3.36	2.27	5.63	5.51
	85+	0.91	4.53	3.31	7.84	7.86
Female	80–84	0.66	2.30	1.65	3.95	3.57
	85+	0.41	2.47	1.89	4.36	4.35

Age 85+ is the last age group available in the national death rates. China's Disease Surveillance Point System (DSP) was fully established in 1990 with 145 surveillance sites over the country in 31 provinces, covering approximately 10 million people. The DSP data are provided by Yong Li (2005). The CLHLS data are calculated by dividing the number of deaths in each age group by the number of person-years lived in those ages.

Over 30 percent of the next-of-kin reported the deceased respondents' cause of death as "old age/senility" or "natural death." There is a debate as to the acceptability of categorizing "old age" as the primary cause of death; however, the number of cases so reporting is increasing each year. In England and Wales in 2000, 2.3 percent of deaths were reported as "old age" (United Kingdom General Registry Office 2001). However, Hawley (2003) analyzed 4,300 cremations in the U.K. and found a significant underestimation of medical conditions among 300 deceased persons whose cause of death was recoded as "old age", which is likely due to undiagnosed conditions (Horner and Horner 1998).[8] Therefore, we speculate that the high proportion of "old age" deaths in the CLHLS is also likely resulted from undiagnosed conditions and should be used and interpreted with caution.

In summary, overall mortality rates in the CLHLS during the period of 2000–2002 are reliable. However, observed mortality rates from the 1998 to 2000 CLHLS data

---
[8] Seventy percent of deaths in the U.K. result in cremation (see Hawley 2003).

were underestimated by 15–20 percent and by 5–20 percent for the regression-based rates before age 90—a potential consequence of nonrandom missing data. Such underestimation of all-cause mortality likely induces some bias in calculations of life expectancy.[9] However, the impact of underestimating mortality is minor on the effects of predictors in statistical models.[10] Given the accuracy in age reporting among the oldest old in the CLHLS, and regardless of recall error in the date of event, one would expect little to no mortality underestimation in the CLHLS. The primary reason for such an underestimation of mortality is possibly because the sample lost to follow-up is not completely random. Chapter 3 shows that octogenarians interviewed in 1998 were two times and 50 percent more likely to be lost to follow-up by the 2000 wave compared to nonagenarians and centenarians, respectively. Most persons lost to follow-up were likely dead prior to the 2000 interview. From the period of 2000–2002, there is no age pattern associated with the sample lost to follow-up. We suggest that users of the CLHLS data utilize data from the interviews in 2000 and later or use pooled data when conducting survival analyses to produce more robust estimates. If the distinction of first- and second-year mortality is needed for specific research designs, the results should be interpreted with caution because the first-year mortality rates suffer from underestimation due to recall bias by next-of kin.

The underestimation of mortality is also found in some national longitudinal surveys in the United States, despite the linkage of datasets to the National Death Index. For example, Adams et al. (2005) observed an underestimation of mortality for elderly females ages 76–82 and ages 88 and older from two waves of the Assets and Health Dynamics among the Oldest Old Study (AHEAD).[11] Laditka and Wolf (2004) found differences in mortality between the 1982 interview of the National Long-Term Care Survey (NLTCS) and the data provided by the National Center on Health Statistics. Manton and Lamb (2005) also reported mortality differentials between the NLTCS data and other data sources. Because both the Health and Retirement Study (HRS/AHEAD) and the NLTCS have not released detailed information on the date of death, we are currently unable to examine the full extent of mortality underestimation in these two American surveys. The underestimation of mortality in these two surveys in the US might be due to sample

---

[9] The parametric hazard models produce slightly higher mortality rates than the Kannisto method, as indicated in the above note. Therefore, it is more appropriate to use parametric models when calculating life expectancy.

[10] We examined the impacts of mortality underestimation on the predictors in various hazard models. We divided the data between the 1998 and 2000 interviews into two intervals: the first year and the second year (after the 1998 initial interview). We found differences in the coefficients of the predictors between the first year and the second year data, as well as for the data between the 2000 and 2002 waves. However, when the data were not divided, as is the case for many studies, we found that the coefficients of the predictors were comparable to those based on the second year data. Furthermore, the results from the 1998–2000 dataset were similar to those of the 2000–2002 dataset.

[11] Authors did not provide data for males, therefore we do not know the degree of underestimation of mortality among males in the AHEAD study. To date, the mortality data from the Health and Retirement Study (HRS) and AHEAD have not been compared with other sources.

selection (i.e., sampled persons are healthier than the general population) or due to age exaggeration at the oldest-old ages, especially for African-Americans (Coale and Kisker 1990).

Given that there is little to no discussion in the literature on how to adjust for the underestimation of mortality in longitudinal data, we refrain from suggesting new or valid methods of handling this issue. In addition, we do not know whether the first-year underestimation of mortality is unique to the CLHLS or is a widespread phenomenon. Research on addressing the accuracy of mortality in longitudinal surveys should be explored further. Likewise, cause-specific mortality rates in the CLHLS suffer from substantial underestimation possibly because of inadequately informed next-of-kin and/or insufficient efforts in seeking official cause-of-death information. Therefore, we suggest that researchers avoid using cause-specific mortality from the CLHLS or interpret their results with caution when necessary.

6.3 Morbidity Reliability

Self-reported questionnaires offer an easy and less expensive alternative for collecting data on respondents' chronic conditions compared to data obtained through direct medical examinations. However, the accuracy and comprehensiveness of data obtained from self-report questionnaire depends to a significant degree on the respondents' health knowledge, ability to recall, and willingness to report (see Klabunde et al. 2005). The accuracy of self-reported chronic conditions compared to medical records, disease registries, or clinical and laboratory reports is inconclusive. For example, some studies suggest that self-reports of diabetes and heart diseases tend to be less accurate than medical records, while reports of arthritis are deemed reliable (Beckett et al. 2000). In contrast, others demonstrate that the accuracy of self-reported chronic conditions is high for diabetes (Midthjell et al. 1992) and moderately good for hypertension (Tormo 2000), whereas the accuracy of reporting other diseases such as arthritis is low (e.g., Kehoe et al. 1994). According to previous studies comparing self-reported conditions with medical records, disease registries, and clinical and laboratory experiments, underreporting of chronic diseases is common in developed nations (e.g., Gross et al. 1996; Hughes et al. 1993; Kehoe et al. 1994; Schrijvers et al. 1994). The underreporting of diseases in developing counties is largely the result of underdeveloped medical technologies and infrastructure, and the shortage of resources and skilled professional personnel in public health. Some studies also find that the accuracy of self-reported chronic conditions is associated with individual characteristics such as age or gender (e.g., Kehoe et al. 1994; Kriegsman et al. 1996). Others find that the accuracy of self-reported morbidity is related to the individual's knowledge, perceptions, awareness, and recognition of chronic disease symptoms (Apolone et al. 2002; Vergara et al. 2004; Satish et al. 1997). Still others indicate that morbidity rates are influenced by public health programs targeting specific populations (e.g., Zaslavsky and Buntin 2002).

To our knowledge, no existing studies have assessed and compared the accuracy of self-reported chronic conditions obtained from surveys and medical/clinical records in China. Therefore, we are unable to fully evaluate the reliability of self-reported conditions among the oldest old in the CLHLS. Fortunately, we can compare indirectly the prevalence rates of chronic disease in the CLHLS with other national sources in an attempt to begin to assess the accuracy of the self reports.[12]

Table 6.2 compares the self-reported prevalence rates for selected chronic diseases in the CLHLS in 2002 and the Second National Health Service Survey (SNHSS) in 1998. The findings strongly indicate that prevalence rates are much higher in the CLHLS than in the SNHSS, suggesting that the quality of self-reported morbidity in the CLHLS is high. Table 6.3 compares the CLHLS data with a second national survey on the elderly conducted by the China National Research Center on Aging (CNRCA) in 2000. The proportion of adults with one of more chronic diseases is quite similar across the two national surveys and, again, indicates relatively good data quality in the CLHLS.

The first two waves of the CLHLS (1998 and 2000) did not collect data for the younger elderly (ages 65–79). Hence we present comparisons of disease prevalence only among the oldest old from the first three waves of the CLHLS. The results in Table 6.4 indicate that the rates of disease are generally consistent across the waves, with some fluctuations (e.g., diabetes, Parkinson's disease). Despite the minor fluctuations, which are primarily associated with the small sample size and low prevalence, the validity of the prevalence rates appears to be strong in the CLHLS.

Prevalence rates for selected chronic diseases (reported by the respondent or proxy) for living adults were also compared to prevalence rates reported by the next-of-kin

Table 6.2 Comparisons of prevalence rates (%.) of chronic disease among the Chinese elderly aged 65 and older between the CLHLS in 2002 and the SNHSS in 1998

Diseases	SNHSS in 1998			CLHLS in 2002		
	Males	Females	Both sexes	Males	Females	Both sexes
Hypertension	62.9	57.9	75.2	157.6	193.1	176.4
Diabetes	11.4	10.5	15.0	24.8	38.6	32.1
Heart diseases	61.9	57.1	68.7	86.5	115.1	101.66
Stroke or CVD	34.0	31.3	36.2	67.7	52.2	59.5
Bronchitis, emphysema, pneumonia, asthma	194.2	178.9	84.6	166.1	107.4	126.3
Cataract or glaucoma	15.3	14.1	25.36	62.2	102.8	83.7
Gastric or duodenal diseases (digestive disease)	67.2	61.9	58.9	61.8	65.6	63.8
Dermatosis	8.3	7.7	4.0	19.6	18.8	19.2
Cancer (malignant neoplasm)	5.2	4.8	5.4	5.2	5.7	5.5
Psychosis	2.6	2.4	4.0	3.2	3.9	3.6

The Second National Health Service Survey is obtained from website at http://www.moh.gov.cn

---

[12] The prevalence rates of chronic conditions in two other national surveys used in this report are both based on self reports. There is a possibility that these two surveys and the CLHLS underreport the actual number of chronic conditions. The purpose of our comparisons is to examine the consistency between the CLHLS and other national sources, not to determine the true rates of chronic conditions in the population.

6 Assessment of Reliability of Mortality and Morbidity

Table 6.3 Comparisons of the percentages of elderly with chronic conditions between the CLHLS in 2002 and the CNRCA in 2000

	Urban			Rural			Urban and Rural		
	Males	Females	Total	Males	Females	Total	Males	Females	Total
CNRCA Survey 2000									
Age 65–79	67.07	71.97	69.46	51.93	57.98	54.72	59.26	65.14	62.05
Age 80+	61.72	63.66	62.88	52.80	57.17	55.23	56.50	60.14	58.59
Age 65+	66.60	70.88	68.74	52.03	57.85	54.79	58.98	64.41	61.62
CLHLS 2002[a]									
Age 65–79	64.71	67.59	66.19	53.83	58.31	56.14	57.69	61.57	59.69
Age 80+	69.19	68.15	68.55	58.64	62.49	61.06	62.24	64.36	63.56
Age 65+	65.19	67.68	66.50	54.37	59.01	56.84	58.19	62.03	60.22

Source: CNRCA (2003). *Data Analysis of the Sampling of Survey of the Aged Population in China*. Beijing: China Standard Press.
[a]Institutionalized respondents were excluded in the 2000 CNRCA survey.

Table 6.4 Comparisons of prevalence rates (%) for selected diseases among the Chinese oldest-old across three waves of the CLHLS by gender and urban/rural residence

Diseases	1998		2000		2002	
	Males	Females	Males	Females	Males	Females
Hypertension	15.61	18.46	15.55	19.09	15.55	18.54
Diabetes	0.99	1.13	1.64	1.65	1.68	1.53
Heart diseases	8.46	6.67	7.03	8.22	8.60	10.11
Stroke or CVD	4.04	2.98	4.76	3.35	5.85	5.05
Bronchitis, emphysema, pneumonia, asthma	16.52	12.07	13.99	10.57	16.52	12.33
TB	1.09	0.64	1.07	0.88	1.12	0.41
Cataract	13.76	15.21	8.12	10.41	9.30	12.15
Glaucoma	1.68	2.13	1.31	2.45	1.98	2.60
Prostate	8.30	–	4.47	–	5.27	–
Gastric or duodenal ulcer	4.25	3.84	3.71	3.86	4.75	5.15
Parkinson's diseases	1.12	0.41	0.63	0.27	0.73	0.27
Bedsore	0.75	0.59	0.78	0.87	0.58	0.49
Cancer (malignant neoplasm)	0.78	0.35	0.36	0.08	0.41	0.17
	Urban	Rural	Urban	Rural	Urban	Rural
Hypertension	16.86	17.67	18.17	17.58	22.71	14.77
Diabetes	2.41	0.45	2.72	1.13	3.48	0.66
Heart diseases	13.42	4.43	11.75	5.85	13.96	7.44
Stroke or CVD	5.37	2.42	5.28	3.19	7.49	4.33
Bronchitis, emphysema, pneumonia, asthma	13.68	13.72	13.83	10.87	13.56	14.12
TB	1.00	0.51	1.13	0.86	1.44	0.50
Cataract	22.06	11.16	14.90	6.97	15.56	9.00
Glaucoma	2.54	1.69	2.31	1.89	3.00	2.14
Prostate	6.18	1.73	3.50	0.91	3.57	1.21
Gastric or duodenal ulcer	4.20	3.89	4.75	3.34	5.13	5.18
Parkinson's diseases	0.77	0.62	0.60	0.30	0.67	0.36
Bedsore	0.86	0.55	0.65	0.93	0.69	0.52
Cancer (malignant neoplasm)	0.87	0.34	0.40	0.07	0.40	0.19

Table 6.5 Comparisons of prevalence rates (%) for selected diseases reported by decedents' next-of-kin in the first three waves of the CLHLS by gender and urban/rural residence

Diseases	1998 wave				Died between 1998 and 2000 waves			
	M	F	U	R	M	F	U	R
Hypertension	11.38	9.61	10.84	9.68	8.85	5.85	8.24	5.63
Diabetes	0.87	0.67	0.81	0.67	1.19	0.76	1.21	0.61
Heart diseases	8.62	6.18	8.30	5.82	8.54	7.23	8.41	6.98
Stroke or CVD	4.51	3.14	4.32	2.94	10.51	7.37	8.88	8.21
Bronchitis, emphysema, pneumonia, asthma	17.79	11.41	14.87	12.68	12.02	8.89	10.95	9.12
TB	0.71	1.14	1.21	0.73	0.71	0.71	0.58	0.86
Cataract or glaucoma	17.87	23.59	23.63	19.11	4.03	6.18	6.17	4.53
Prostate	7.83	–	3.80	2.27	3.24	–	1.61	0.92
Gastric or duodenal ulcer	3.00	2.43	3.11	2.14	2.53	1.95	2.07	2.27
Parkinson's diseases	0.79	1.09	0.81	1.16	0.47	0.24	0.35	0.31
Bedsore	1.19	1.33	1.27	1.29	1.66	1.24	1.38	1.41
	2000 wave				Died between 2000 and 2002 waves			
	M	F	U	R	M	F	U	R
Hypertension	12.83	11.09	12.00	11.42	9.60	8.41	10.07	7.70
Diabetes	1.50	1.00	1.27	1.12	2.55	1.84	2.47	1.74
Heart diseases	8.03	6.77	9.59	4.59	11.93	9.70	14.17	7.26
Stroke or CVD	5.78	4.18	5.91	3.66	11.55	9.25	11.16	9.31
Bronchitis, emphysema, pneumonia, asthma	13.05	10.35	12.12	10.55	15.68	12.44	13.81	13.84
TB	0.98	0.65	0.97	0.62	1.35	1.09	1.09	1.30
Cataract or glaucoma	11.55	16.27	16.77	11.67	5.63	8.76	8.20	6.89
Prostate	5.93	–	3.80	1.24	4.20	–	3.08	0.74
Gastric or duodenal ulcer	3.15	3.33	3.50	2.92	4.95	3.48	3.74	4.47
Parkinson's diseases	0.53	0.20	0.42	0.25	1.28	2.19	0.84	0.56
Bedsore	0.75	1.04	0.66	1.18	1.28	2.19	2.29	1.37
Arthritis	11.18	11.99	11.70	11.67	7.73	6.97	7.18	7.20
Dementia	3.90	5.62	5.31	4.35	2.93	3.88	3.38	3.54

when the respondent died (see Table 6.5). The results suggest some underestimation of disease prevalence for certain conditions reported by the next-of-kin. Therefore, researchers should be cautious when using these kinds of data that were collected for deceased persons. Prior research suggests more robust estimates are obtained by combining data from the last interview of the sampled person with data collected by the decedents' next-of-kin reported between waves (Marshall and Graham 1984).

In sum, the self-reported prevalence of chronic conditions is quite reliable in the CLHLS, compared to other national sources, whereas the morbidity of deceased subjects reported by the next-of-kin is not as reliable. According to previous research, however, the ADL disability information for deceased persons (reported by the next-of-kin) is shown to be reliable (Gu and Zeng 2004; Zeng et al. 2004). This discrepancy is understandable given that most of the kin resided and/or provided care for the respondent before his/her death. In addition, the underreporting of morbidity among the subsequently deceased is likely because many families did not (or could not) seek medical treatment for the respondent while the subject was alive, and consequently, family members may have been unaware of an existing disease.

6.4 Concluding Remarks

This chapter examined the reliability of mortality and morbidity data in the first three waves of the CLHLS. Compared with other data sources, the results demonstrate that mortality rates are reliably estimated from 2000 to 2002. However, before age 90 (from 1998 to 2000), we find some underestimation of mortality possibly due to a higher proportion of octogenarians lost to follow-up compared to nonagenarians and centenarians. The first-year mortality rates after the interview also tend to be underestimated. We suspect that this is related to recall bias in the date of death reported by the decedents' next-of-kin. Our results suggest that self-reported rates of chronic disease are reliable. It is believed that the good quality of the CLHLS data is largely attributable to the systematic operation and collection of survey data and the systematic age-verification procedures used for sampled persons. Because we are unaware of the implementation procedures of other surveys on aging in China, we are reluctant to infer the reliability of their mortality rates and the prevalence of self-reported chronic diseases.

We find that reports of morbidity by the next-of-kin and information on the cause of death reported by the next-of-kin are not quite as reliable. These findings suggest that the next-of-kin are not reliable proxies for gathering disease and cause-specific mortality data for the elderly in China. The appropriateness of using next-of-kin as proxies for information regarding end-of-life care research is still debated. Some studies indicate the appropriateness of such reports (see Tang and McCorkle 2002 for a review), while others do not (see Sprangers and Aaronson 1992 for a review). Further, the agreement between proxy-identified and respondent self-reported chronic conditions varies according to demographic and family characteristics, caregiving burden, and the health conditions of the sampled individuals and their proxies (Tang and McCorkle 2002). Given space limitations, we were unable to further explore the mechanisms associated with the estimation of proxy-rated chronic conditions. Additional research is clearly warranted to address this shortcoming. In addition, given that proxies are rarely used to determine cause of death in most end-of-life care studies, it is unclear whether such inaccuracies occur in the CLHLS. This is an issue requiring further investigation. Overall, our results suggest that greater interviewer training may be necessary to further improve CLHLS data collection by eliciting more accurate information from the respondents' next-of-kin.

Acknowledgments The research reported in this chapter is supported by The National Institute on Aging grant (R01 AG023627-01) and National Natural Science Foundation of China key project grant (70533010).

References

Adams, P., M.D. Hurd, D. McFadden, A. Merrill, and T. Ribeiro (2005), Healthy, wealth, and wise? tests for direct causal paths between health and socioeconomic status. *Journal of Econometric* 112, pp. 3–56

Apolone, G., A. Cattaneo, P. Colombo, C. La Vecchia, L. Cavazzuti, and F. Bamfi (2002), Knowledge and opinion on prostate and prevalence of self-reported BPH and prostate-related events: A cross-sectional survey in Italy. *European Journal of Cancer Prevention* 11 (5), pp. 473–479

Banister, J. and K. Hill (2004), Mortality in China 1964–2000. *Population Studies* 58, pp. 55–75

Beckett, M., M. Weinstein, N. Goldman, N., and Y.-H. Lin (2000), Do health interview survey yield reliable data on chronic illness among older respondents? *American Journal of Epidemiology* 151, pp. 315–323

China National Research Center on Aging (2003), *Data analysis of the sampling of survey of the aged population in China.* Beijing: China Standard Press

Cleves, M.A., W.W. Gould, and R.G. Gutierrez (2004), *An introduction survival analysis using STATA.* Revised edition, College Station, TX: STATA Corporation

Coale, A.J. and E.E. Kisker (1990), Defects in data on old-age mortality in the United States. *Asia and Pacific Population Forum* 4, pp. 1–31

Coale, A.J. and S. Li (1991), The effect of age misreporting in China on the calculation of mortality rates at very high ages. *Demography* 28 (2), pp. 293–301

Goodkind, D. (2004), *The mortality of China's oldest-old: Comparisons and questions from the Chinese Longitudinal Healthy Longevity Survey and the 2000 Census.* Presented at the workshop on Determinants of Healthy Longevity in China. Max Planck Institute for Demographic Research (MPIDR), Rostock, Germany, August 2–4, 2004

Gross, R., N. Bentur, A. Elhayany, M. Sherf, and L. Epstein (1996), The validity of self-reports on chronic disease: characteristics of underreporters and implications for the planning of services. *Public Health Review* 24 (2), pp. 167–182

Gu, D. and J. Lu (2004), *Age reporting of the minority oldest-old in the CLHLS.* Presented at the workshop on Determinants of Healthy Longevity in China. Max Planck Institute for Demographic Research (MPIDR), Rostock, Germany, August 2–4, 2004

Gu, D. and Zeng, Y. (2004), Sociodemographic effects on the onset and recovery of ADL disability among Chinese oldest-old. *Demographic Research* 11, pp. 1–42

Hawley, C.L. (2003), Is it ever enough to die of old age? *Age and Ageing* 32, pp. 484–486

Horiuchi, S. and J.R. Wilmoth (1998), Deceleration in the age pattern of mortality at older ages. *Demography* 35 (4), pp. 391–412

Horner, J.S. and J.W. Horner (1998), Do doctors read forms? A one-year audit of medical certificates submitted to a crematorium. *Journal of the Royal Society of Medicine* 91 (7), pp. 371–376

Hughes, S.L., P. Edelman, B. Naughton, R.H. Singer, P. Schuette, G. Liang, and R.W. Chang (1993), Estimates and determinants of valid self-reports of musculoskeletal disease in the elderly. *Journal of Aging and Health* 5, pp. 244–263

Kannisto, V. (1994), *Development of oldest-old mortality, 1950–1990: Evidence from 28 developed countries.* Odense, Denmark: Odense University Press

Kehoe, R., S-Y. Wu, M.C. Leske, and L.T. Chylack (1994), Comparing self-reported and physician-reported medical history. *American Journal of Epidemiology* 139, pp. 813–818

Klabunde, C.N., B.B. Reeve, L.C. Harlan, W.W. Davis, and A.L. Potosky (2005), Do patients consistently report comorbid conditions over time? *Medical Care* 43, pp. 391–400

Kriegsman, D.M., B.W. Penninx, J.T. van Eijk, A.J. Boeke, and D.J. Deeg (1996), Self-reports and general practitioner information on the presence of chronic diseases in community dwelling elderly: A study on the accuracy of patients' self-reports and on determinants of inaccuracy. *Journal of Clinical Epidemiology* 49 (12), pp. 1407–1417

Laditka, J.N. and D.A. Wolf (2004), *Duration data from the National Long-Term Care Survey: a foundation for a dynamic multiple-indicator model of ADL dependency.* Center for Policy Research, Working Paper No. 65. ISSN: 1525–3066. Maxwell School of Citizenship and Public Affairs, Syracuse University, New York

Li, Y. (2005), *Cause of death of the elderly in China.* School of Public Health. Maryland: Johns Hopkins University

Manton, K.G. and V.L. Lamb (2005), *U.S. mortality, life expectancy, and active life expectancy at advanced ages: trends and forecasts.* Paper presented at annual meeting of Population Association America, Philadelphia, March 30th–April 2nd, 2005

Marshall, J.R. and S. Graham (1984), Use of dual responses to increase validity of case–control studies. *Journal of Chronic Disease* 37, pp. 125–136

Midthjell, K., J. Holmen, A. Bjomdal, and P.G. Lund-Larsen (1992), Is questionnaire information valid in the study of a chronic disease such as diabetes? the Nerd-Trondelag Diabetes Study. *Journal of Epidemiology and Community Health* 46, pp. 537–542

Myers, G.C., V.L. Lamb, and E.M Agree (2003), Patterns of disability change associated with the epidemiological transition. In: J.M. Robine, C. Jagger, C.D. Mathers, E.M. Crimmins, and R.M. Suzman (eds.): *Determining health expectancies.* West Sussex, England: Wiley, pp. 59–74

Robine, J.M. and J.-P. Michel (2004), Looking forward to a general theory on population. *Journal of Gerontology: Medical Sciences* 59 (A), pp. 590–597

Satish, S., K.S. Markides, D. Zhang, and J.S. Goodwin (1997), Factors influencing unawareness of hypertension among older Mexican Americans. *Preventive Medicine* 26 (5), pp. 645–650

Schrijvers, C.T.M., K. Stronks, D.H. van de Mheen, J.W.W. Coebergh, and J.P. Mack-Enbach (1994), Validation of cancer prevalence data from a postal survey by comparison with cancer registry records. *American Journal of Epidemiology* 139, pp. 408–414

Sprangers, M.A.G. and N.K. Aaronson (1992), The role of health care providers and significant others in evaluating the quality of life of patients with chronic disease: A review. *Journal of Clinical Epidemiology* 45, pp. 743–760

Tang, S.T. and R. McCorkle (2002), Use of family proxies in quality of life research for cancer patients at the end of life: a literature review. *Cancer Investigation* 20 (7–8), pp. 1086–1104

Thatcher, A.R., V. Kannisto, and J.W. Vaupel (1998), *The force of mortality at ages 80 to 120.* Odense: Odense University Press. Online at http://www.demogr.mpg.de/Papers/Books/Monograph5/ForMort.htm

The United Kingdom General Registry Office (2001), Mortality statistics for England & Wales 1994 and 2000. http://www.gro.gov.uk/gro/content/

Tormo, M.-J., C. Navarro, M.D. Chirlaque, X. Barber, and the EPIC Group of Spain (2000), Validation of self diagnosis of high blood pressure in a sample of the Spanish EPIC cohort: overall agreement and predictive values. *Journal of Epidemiology and Community Health* 54, pp.221–226

Vallin, J. (1973), *La Mortalité par Génération en France.* depuis 1899. Paris: Presses Universitaires de France pp. 483 (Travaux et Documents) (in French)

Vaupel, J.W., J.R. Carey, K. Christensen, T.E. Johnson, A.I. Yashin, N.V. Holm, I.A. Iachine, V. Kannisto, A.A. Khazaeli, P. Liedo, V.D. Longo, Zeng, Y., K.G. Manton, and J.W. Curtsinger (1998), Biodemographic trajectories of longevity. *Science* 280 (5365), pp. 855–860

Vergara, C., A.M. Martin, F. Wang, and S. Horowitz (2004), Awareness about factors that affect the management of hypertension in Puerto Rican patients. *Connecticut Medicine* 68 (5), pp. 269–276

Zaslavsky, A.M. and M. J. Buntin (2002), Using survey measures to assess risk selection among Medicare managed care plans. *Inquiry* 39 (2), pp. 138–51

Zeng, Y., D. Gu, and K.C. Land (2004), A new method for correcting underestimation of disabled life expectancy and an application to the Chinese oldest-old. *Demography* 41 (2), pp. 335–361

Zeng, Y. and J.W. Vaupel (2003), Oldest-old mortality in China. *Demographic Research* 8, pp. 215–244

Zhang, Z. and Q. Li (2004), *Do possible age-exaggerations in minorities substantially impact the mortality risk of the oldest-old in the CLHLS?* Presented at the workshop on Determinants of Healthy Longevity in China. Max Planck Institute for Demographic Research (MPIDR), Rostock, Germany, August 2–4, 2004

Part II
The Effects of Demographic and Socioeconomic Factors

Introduction *by Danan Gu*

It has been well documented that health conditions and survival probabilities tend to decline with increases in age, and that women are disadvantaged in health but superior in survivorship (Martelin et al. 1998; Suzman et al. 1992; Zeng et al. 2002). The literature suggests that these differences in healthy longevity are likely due to socioeconomic status (SES), psychosocial, and behavioral factors, as well as to genetic components (Lantz et al. 1998; Stuck et al. 1999; Vaupel et al. 1998). With few exceptions, higher socioeconomic status (SES) is associated with better health status, lower levels of functional limitations and disability, and lower mortality risk across both nations and populations (Lynch 2006; von dem Knesebeck et al. 2000; Zhu and Xie 2007). The few exceptions are perhaps due to problems concerning the quality of the data, the sample size, measures of SES, and health outcomes (Martelin et al. 1998). However, despite important progress made in understanding the mechanisms underlying the association between SES and health outcomes in later life, a significant portion of previous research has seemed to ignore some important aspects of the impacts of SES on elderly health, such as childhood SES conditions. Focusing only on adult achieved statuses without any linkages to early life conditions or family background could well underestimate the influences of socioeconomic status on healthy longevity and miss important pathways through which SES affects health outcomes in later life (Preston et al. 1998). Indeed, exposure to life conditions at the various points across the life span may have either positive or negative effects toward the total stock of health capital, and such effects may well extend to older ages (Preston et al. 1998; Zeng et al. forthcoming). Alwin and Wray (2005) have recently articulated a life span perspective on social status and health and mortality; they contend that healthy longevity is a lifelong process and is influenced by multiple social contexts throughout the life span.

D. Gu
Center for the Study of Aging and Human Development, Medical School of Duke University, Durham, NC 27710, USA
e-mail: gudanan@duke.edu

Part II of this volume contains five chapters that use the survey data of the Chinese Longitudinal Healthy Longevity Survey (CLHLS) and explore further the associations between demographic and SES factors and healthy longevity at the very old ages. Poston and Min (Chap. 7) examine the effects of socio-demographic factors on mortality risk among 7,234 oldest-old men and women during the period of 1998–2000. They find that most socio-demographic variables are significant, even after controlling for various health related variables. They also look into the relative influence of each of these factors on mortality, and their analysis reveals that age is the most influential factor followed by some ADL factors and martial status. Given that the oldest-old normally need more social and medical care as compared to the young elderly, the findings of Chap. 7 are relevant and useful for developing better welfare policy recommendations concerning the elderly population.

In Chap. 8, Dennis Ahlburg and his colleagues use the 1998 CLHLS wave data and examine the impacts of childhood health/SES and adulthood SES on multiple health outcomes measured by Activities of Daily Living (ADL), chronic conditions, self-rated quality of life, self-rated health, interviewer rated health, and the number of times an individual suffered from serious illness during the past 2 years. They find that both childhood health/SES and adulthood SES play significant roles in determining several health outcomes at very advanced ages, in the presence of demographic, family support, and health practice factors. This indicates the robustness of the "long-arm" effects of early life conditions on healthy longevity, which is consistent with previous studies in the literature.

In Chap. 9, Jianmin Li uses data from China's 1982 census and the 1998 wave of the CLHLS, and compares the cohort survivorship ratio of the elderly Chinese in the period of 1982–1998 by different levels of educational attainment using fixed-attribute dynamics methods. The study finds that educational attainment is positively associated with survivorship among elderly Chinese, and the association is stronger for women than for men. According to Li, the protective effects of educational attainment on survivorship at late ages mainly occur through the enhancement of income-earning abilities in the labor market, increasing knowledge and the practice of health habits, and increased access to social welfare.

Zhong Zhao (Chap. 10) uses the 1998–2000–2002 CLHLS panel data and the CLHLS 2002 cross-sectional data, and analyzes associations between socioeconomic factors and health/mortality from a health capital perspective. Zhao uses the Grossman Model in which health is treated as capital that depends on investment and depreciation. Social support and life style are accommodated into the Grossman framework in Zhao's important study. Zhao argues that health status at the old ages is mainly dictated by the stock of health capital and the rate of depreciation. His study reveals that being in a white-collar occupation and having one's own bedroom significantly reduce mortality risk, while current economic resources and education play insignificant roles. According to Zhao, having a private bedroom likely captures the elder's permanent income and financial capacity. Moreover, current inputs into the health production function and contemporary changes in behavior and life style appear only to have incremental effects on the stock of health capital, which is determined by the individual's entire history. His analysis further shows that the accessibility of medical care services in childhood plays a significant role in determining health/survival at late ages, indicating that health capital is a kind of long-term investment. In line with most previous studies, his research finds that age, sex, urban/rural residence, marital status, and life style factors also play significant roles in reserving and depleting the stock of health capital at late ages.

Chapter 11 by Rongjun Sun and Yuzhi Liu investigates the effect of social engagement on mortality. In this analysis the authors use data from the first two waves of the CLHLS. Social engagement is measured by marital status and number of living children. Involvement in social activities was found to be significantly associated with a lower risk of mortality among the Chinese oldest-old in the 2-year period from 1998 to 2000, after accounting for health status, health behaviors, and other socio-demographic characteristics. The authors further showed that the effects of marital status and the number of living children do not depend on age; however, the beneficial effect of social activities on mortality gradually diminished with age, and was reversed at extremely old ages, when health status, health behaviors, and socio-demographic characteristics were controlled. The results tend to vary by place of residence and gender. According to Sun and Liu, retreating from social relations is not a passive but a positive shift for the very old–old, as narrowing social boundaries and focusing inwardly is a strategic adaptation that minimizes risks while maximizing benefits considering their limited resources and lower physical functioning.

We believe that each chapter in this section of the book makes meaningful contributions to the better understanding of the mechanisms of healthy longevity. The research reported in these chapters also show that the effects of demographic and socioeconomic factors on healthy longevity deserve further systematic examination. Perhaps multilevel modeling including macro data at the community level, and genetic data at the molecular level, would be an important next step (Hobcraft 2006).

References

Alwin, D.F. and L.A. Wray (2005), A life-span developmental perspective on social status and health. *Journal of Gerontology: Social Sciences* 60B, pp.7–14

Hobcraft, J. (2006), The ABC of demographic behavior: How the interplays of alleles, brains, and contexts over the life course should shape research aimed at understanding population processes. *Population Studies* 60 (2), pp. 153–187

Lantz, P.M., J.S. House, J.M. Lepkowski, D.R. Williams, R.P. Mero, and J. Chen (1998), Socioeconomic factors, health behaviors, and mortality: Results from a nationally representative prospective study of US adult. *Journal of the American Medical Association* 279 (21), pp. 1703–1708

Lynch, S.M. (2006), Explaining life course and cohort variation in the relationship between education and health: The role of income. *Journal of Health and Social Behavior* 47, pp. 324–338

Martelin, T., S. Koskinen, and T. Valkonen (1998), Sociodemographic mortality differences among the oldest old in Finland. *Journal of Gerontology: Social Sciences* 53B (2), pp. S83–S90

Preston, S.H., M.E. Hill, and G.L. Drevenstedt (1998), Childhood conditions that predict survival to advanced ages among African-Americans. *Social Science and Medicine* 47, pp. 1231–1246

Stuck, A.E., J.M. Walthert, T. Nikilaus, C.J. Bula, C. Hohmann, and J.C. Beck (1999), Risk factors for functional status decline in community-living elderly people: A systematic literature review. *Social Science and Medicine* 48, pp. 445–469

Suzman, R.M., D.P. Willis, and K.G. Manton (1992), *The oldest-old*. New York: Oxford University Press

Vaupel, J.W., J.R. Carey, K. Christensen, T.E. Johnson, A.I. Yashin, N.V. Holm, I.A. Iachine, V. Kannisto, A.A. Khazaeli, P. Liedo, V.D. Longo, Zeng, Y., K.G. Manton, and J.W. Curtsinger (1998), Biodemographic trajectories of longevity *Science* 280 (5365), pp. 855–860

von dem Knesebeck, O., G. Lüschen, W.C. Cockerham, and J. Siegrist (2000), Socioeconomic status and health among the aged in the United States and Germany: A comparative cross-sectional study. *Social Science and Medicine* 57, pp. 1643–1652

Zeng, Y., J.W. Vaupel, Z. Xiao, C. Zhang, and Y. Liu (2002), Sociodemographic and health profiles of oldest old in China. *Population and Development Review* 28, pp. 251–273

Zeng, Y., D. Gu, and K.C. Land (2007), The association of childhood socioeconomic conditions with healthy longevity at the oldest-old ages in China. *Demography* 44 (3), pp. 497–518

Zhu, H. and Y. Xie (2007), Socioeconomic differential in mortality among the oldest-old in China. *Research on Aging* 29, pp. 125–143

Chapter 7
The Effects of Sociodemographic Factors on the Hazard of Dying Among Chinese Oldest Old

Dudley L. Poston Jr. and Hosik Min

Abstract The oldest old population of China has grown in size and continues to increase. The growth is due to the combination of very low levels of fertility and longer life expectancy in China. However, there is not enough sociological and demographic literature about oldest old mortality in China. In this chapter we undertake Cox proportional hazard analyses and examine the effects of sociodemograhic factors on the hazard of dying among oldest old Chinese. Data on 7,234 oldest old Chinese who were interviewed in the Chinese Longitudinal Healthy Longevity Survey (CLHLS) were used in the analyses. The results indicate that sociodemographic variables such as age, sex, and marital status are strong predictors of the hazard of dying among the oldest old Chinese after controlling for other health-related variables. In terms of the relative impact among the covariates, the age variable is the most influential factor.

Keywords Chinese oldest-old mortality, Cox proportional hazard model, Hazard of dying, Hazard ratio, Influential factor, Kaplan–Meier Curve, Multicollinearity, Population aging, Semi-standardized hazard ratio, Sociodemographic factors, Strong predictors

7.1 Introduction

A common finding in the death and dying literature is that with increasing age, persons more and more so contemplate the fact of their mortality. Teenagers and persons in their 20s seldom think about their mortality, or for that matter, death in general. But when reaching the age of 60 and later, there is an increased tendency to consider one's impending mortality. The older one's age, the more frequent and intense this consideration and deliberation (Aiken 2001: 252–257; Quadagno 1999: 304–305).

D.L. Poston Jr.
Department of Sociology, Texas A&M University, College Station, TX 77843, USA
e-mail: dudleyposton@yahoo.com

These reflections about one's mortality are typically based on anecdotal evidence and newspaper stories about people who are dying or have died. There is little empirical evidence available about the actual hazard of dying among old people.

The Chinese Longitudinal Healthy Longevity Survey (CLHLS) provides a unique set of data to demographers enabling the development of empirical answers to questions about the older people's likelihood of dying, specifically the statistical hazard of death. The CLHLS is a multistage, stratified cluster survey of elderly respondents conducted in 631 randomly selected cities and counties in 22 provinces of China where the Han comprise the large majority. Although there is a rich literature on the hazard of dying, there is not much information at all about the hazard of dying among older people.

This chapter examines the hazard of dying among elderly Chinese males and females during the period of 1998–2000. Data for 7,234 male and female elderly Chinese who were interviewed in the CLHLS are used. These persons were of age 80 or older at the time of the first interview in 1998. By the time of the follow-up interview in 2000, 39 percent of them had died. After describing the hazard for different subgroups of the elderly Chinese, the effects on the hazard of dying of various sociodemographic factors are considered; these include factors such as age, sex, urban–rural residence at birth and at the time of the initial interview, nationality, marital status, education, daily activities, and self-rated health status. Before undertaking the hazard analyses, we estimate a survival curve to see the probabilities of surviving the hazard of death for each month of analysis time. We then show that many of the sociodemographic factors have important and significant effects on the hazard of death for oldest old Chinese.

7.2 Prior Studies

In this chapter we study the hazard of dying among elderly Chinese. The size of the population of elderly Chinese has increased enormously in the last few decades and will grow even more so in the future (Poston and Duan 2000; Zeng et al. 2002). Zeng and George (2000) note that the numbers of Chinese elderly will reach 330 million by the year 2050. Using United Nations projection data, Poston et al. (2004) report that of the almost two billion persons age 60 or greater projected to be living in the world in 2050, 418 million of them (nearly 22 percent) will be residing in China.

Irrespective of the set of projections that are used, they all indicate that the numbers of elderly in China have increased and will continue to do so. There are two main reasons: the dramatic fertility decline and the mortality reduction. For example, the total fertility rate in China was around 6.0 in the 1960s, and fell to under 2.0 in the 1990s (Poston 2000). This rapid fertility reduction has led to significant increases in the older population in China. Zeng (1989) projected that the elderly in China will comprise more than 20 percent of the population by 2050.

Hazard models are especially useful in the study of mortality mainly because time is involved in the nonoccurrence, and ultimately the occurrence, of one's death.

Hazard regression enables the researcher to examine the effects of covariates (independent variables) on the likelihood of experiencing death, while considering the factor of time. Hazard models have shown that variables such as age, sex, and race/ethnicity are influential predictors. There has not been an abundance of studies using hazard models in Chinese mortality. However, there are several analyses of Chinese mortality. Indeed Zeng et al. (2001) have reported that the dynamics of Chinese mortality are similar to those in other countries.

Among the variables mentioned above, age has been the most important factor in mortality. By improving medical care and standard of living, life expectancy lengthened significantly during the last century. Thus, most people now die by degenerative diseases, and not in their young ages (Quadagno 1999). Hence, when one looks at mortality by age, one usually finds a rectangular curve, which shows that mortality is high at the first year after birth, then drops down to almost zero for the next several decades of life, and then begins to increase again, usually after the age of 50 or 60 (Rowland 2003). For the oldest old, an additional year of age exactly increases the probability of death.

Generally, women have better survival probabilities than men (Liang et al. 2003; Rogers et al. 2000) for both biological and behavioral reasons. Women have genetic and hormonal advantages over men, and men often engage in more risk-taking behaviors compared to women. Women compared to men, however, are usually disadvantaged on socioeconomic factors that are related with mortality such as education and employment. These findings hold for the Chinese. Several scholars (Hao 1995; Xie 1996; see also Zeng et al. 2003) show that women live longer than men and that female mortality is lower than male mortality at most ages. In addition, Hao (1995), Yang (1986), and Zeng and Vaupel (2002) found that the higher mortality for men holds for the Chinese elderly.

Like women, minorities are also disadvantaged with regard to physical survival. Often they have lower levels of education and income. Hummer et al. (2000) found that minorities had higher mortality hazard rates than whites in the U.S. And Yusuf and Byrnes (1994) found the same results in China. The mortality of minorities in China was substantially higher than that for the Han. Xiong (1989) suggested that the lack of medical facilities and unhygienic conditions contributed to the higher mortality for minorities. He also reported that the mortality gap between minorities and the majority has been declining since the 1950s, but there is still a differential.

Marital status also influences mortality for several reasons. Trovato and Lauri (1989) argued that married couples are the most advantaged; they tend to have higher levels of social integration, greater reinforcement of healthy behaviors, and more economic resources. In contrast, divorced and widowed persons tend to experience more stress and emotional difficulties, leading to a relatively greater risk of dying. Campbell and Lee (1996) have reported that widowers have higher mortality than the currently married. Also, a mere change in marital status frequently results in increased stress and depression leading people to a higher hazard of death (Liang et al. 2003; Zick and Smith 1991).

Regarding residence, demographers have shown that urban dwellers have life expectancy advantages over rural dwellers (Fang 1993). This is largely due to the lack of health care service facilities, and less knowledge among rural residents about hygiene, diet, and health.

Several studies have reported other influential factors on Chinese mortality. For instance, socioeconomic status has also been shown to be very influential. More educated people have lower mortality (Xie 1996). Liu and Zhang 2004) show that among elderly Chinese, the higher the level of education, the better the health of the respondents, and thus, the better their survival probabilities.

7.3 Data and Methods

In this chapter we use data from the Chinese Longitudinal Healthy Longevity Survey (CLHLS). The CLHLS is a data-set dealing with the oldest old in China. Of the three surveys that have been conducted (in 1998, 2000, and 2002), we use the cohort data derived from the first two waves. Over 9,000 persons were interviewed in 1998; of these, 4,831 were reinterviewed in 2000. Although 3,368 of the respondents died between 1998 and 2000, information about the decedents (particularly their dates of death) is available. However, almost 900 respondents were lost and are thus not available for follow up in 2000. After taking into consideration other issues of missing data (because data on a few questions from the same respondents were missing or not available), we have available for analysis data on 7,234 persons, 39 percent of whom died in the 1998–2000 interval.

We undertake hazard analyses of the probability of dying between 1998 and 2000 for those 7,234 men and women who were interviewed in 1998. The dependent variable is a time-variant variable, namely, whether respondents who were interviewed in 1998 will pass away or not by the year 2000. The main advantage of hazard models is that they can analyze these time-dependent and -independent variables together (Allison 1984; Cleves et al. 2002). As previously mentioned, the dependent variable deals with whether respondents who were interviewed in 1998 will die or not by 2000. The analysis will take into account the number of months of time from the previous interview to the time of death or another interview in 2000.

The survival-time data consist of two variables; one is a dummy variable for each person indicating whether or not the event of death occurred during the observation period; the second is a variable measuring the number of months that have elapsed since the date of the interview in 1998 and the date of either the person's death, or the reinterview in 2000.

We first describe the survivor data for the 7,234 persons by graphing Kaplan–Meier (K–M) survival curves (Hamilton 1998; Kaplan and Meier 1958). Let n_t represent the number of persons who have not died and are not censored at the beginning of time period t, and d_t represent the number who die during time period t. The formula (below) is the Kaplan–Meier estimator of surviving beyond time t (i.e., not dying), and is the product of survival probabilities in t and the preceding periods:

$$S(t) = \prod_{j=t_0}^{t} \{(n_j - d_j) / n_j \} \qquad (7.1)$$

After describing the survival data for the 7,234 respondents, we next use Cox's partial-likelihood method to estimate a continuous time proportional hazards model of the hazard of death in the interval between 1998 and 2000. The Cox proportional hazards model assumes the following form when all the independent variables (covariates) are time-independent, that is, when their values do not change over time, as is the situation in this analysis:

$$\log h(t) = \log h_0(t) + b_1 x_1 + \ldots + b_k x_k \tag{7.2}$$

where $h_0(t)$ is an unspecified function of time t, and x_1 to x_k are sociodemographic covariates (independent variables), and b_1 to b_k are the parameters to be estimated. One feature of the Cox model that makes it so attractive is that the function of time does not have to be specified.

The dependent variable, $\log h(t)$, is the hazard rate, which is an unobserved value gauging the instantaneous probability that a respondent will die during the interval since the interview in 1998 (Allison 1984; Yamaguchi 1991).

The independent variables are the respondent's sex (male/female), nationality (Han/non-Han), unmarried (yes/no),[1] born in a rural area (yes/no), living in a rural area in 1998 (yes/no), educational attainment, age, limitations in daily activities such as bathing (yes/no), dressing (yes/no), continence (yes/no), feeding (yes/no), transferring (yes/no), and his/her self-rated health status (a 5-point scale scored as 1, Very bad; 2, Bad; 3, So-so; 4, Good; and 5, Very good).[2]

7.4 Results

We first describe the survival data for the entire population of elderly Chinese. In Figure 7.1 we have graphed a Kaplan–Meier survival curve of the probability of surviving death, $S(t)$, against the number of months between the date of the 1998 interview and the date of death. The K–M survival curve shows the probabilities of surviving the hazard of death for each month of analysis time. The K–M survival curve only describes the survival data. No covariates are considered in the calculation of the K–M curve. The curve steps down from a probability of 1.0 of surviving death during the month of the 1998 interview, to a probability near 0.75 by about the 30th month.

Now we turn next to estimating a Cox proportional hazard equation of the probability of dying among the oldest old Chinese. We present the descriptive statistics first, and then, the results from hazard analyses that model the transition of Chinese oldest old from being alive to dying during the 1998–2000 period. The results also

---

[1] Due to a small percentage of "separated" (1.27 percent) and "divorced" (0.66 percent) in the marital status variable, we combined these into "unmarried."

[2] We included several health-related variables in the model to see if sociodemographic factors hold even after controlling for these variables. We, however, excluded some health-related variables such as toileting and previous illness experiences due to high multicollinearity (see later discussion in the text).

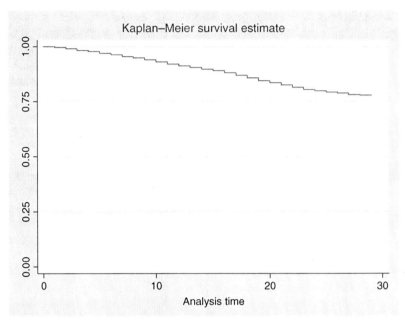

Fig. 7.1 Kaplan–Meier survival curve of the probability of surviving death by month: 7,234 oldest old persons, China, 1998–2000

present the semi-standardized hazard ratios to distinguish the relative impacts of the covariates on the hazard of dying.

Among the 7,234 Chinese oldest old in the sample, 39 percent of them were reported as died (see Table 7.1). The mean duration time at the risk of the hazard of dying was 21 months. However, one should keep in mind that this duration time also includes censoring for the Chinese oldest old who did not die during the survey period. Almost 60 percent of them were females. Over 90 percent of them were Han. Two thirds of them were living in rural areas (64 percent). Most of them were born in rural areas (86 percent). Over two thirds of them had no formal education, and were thus illiterate. One out of four had an elementary school education. A total of 8 percent of them had junior high and more education. The majority of the sample were unmarried (83 percent). Due to their old age, most of them were widowed. Regarding their age, 41 percent of the sample were in their 80s, 35 percent in their 90s, and the rest of them were centenarians.

For the most part, the Chinese oldest old had few limitations on their daily activities. For example, 29 percent of them had limitations on bathing, 15 percent of them had limitations on dressing, 7 percent of them had continence limitations, 10 percent had feeding limitations, and 14 percent had limitation on transferring. On average, the Chinese oldest old self-reported their health with a score of 3.6 (on a scale of 1.0–5.0), indicating that most of them were in fairly good health. Only 9 percent of them reported their health as bad or very bad, 43 percent as normal, and almost half of them (48 percent) as good or very good.

Table 7.1 Descriptive statistics for 7,234 oldest old persons, China, 1998–2000

Variable		Percentage	
Duration[a]	Months	21.1	Max: 29 min: 0
Died	Yes = 1	38.6	
Female	Yes = 1	58.7	
Han	Yes = 1	92.2	
Rural residence	Yes = 1	64.3	
Rural birth	Yes = 1	85.9	
Education	Years		
Illiterate		67.2	
Elementary		25.0	
Junior high +		7.8	
Unmarried	Yes = 1	83.4	
Age group			
80s		40.9	
90s		34.8	
100s		24.3	
Bathing limitations	Yes = 1	28.8	
Dressing limitations	Yes = 1	14.6	
Continence limitations	Yes = 1	7.3	
Feeding limitations	Yes = 1	9.7	
Transferring limitations	Yes = 1	13.8	
Self-rated Health[b]		3.6	Max: 5 min: 1

[a] Mean
[b] Mean, measured on 5-point scale

The Cox proportional hazard equation predicts the hazard of dying during the 1998–2000 interval, using the covariates of respondent's sex (male/female), nationality (Han/non-Han), unmarried (yes/no), born in a rural area (yes/no), living in a rural area in 1998 (yes/no), educational attainment, age, bathing (yes/no), dressing (yes/no), continence (yes/no), feeding (yes/no), transferring (yes/no), and self-rated health status.

Before conducting the hazard analysis, we ascertain whether there is significant multicollinearity present among the independent variables. An accepted way for detecting multicollinearity is to calculate the tolerance level of each independent variable. The tolerance level is calculated by regressing the independent variables on each of the other independent variables, one at a time, and subtracting the resulting R^2 value from 1. This value is known as the "tolerance," or the amount of independent variation, of the X variable (see Hamilton 1992: 133–134). So if the R^2 of the X variables regressed on another X variables is 0.75, this means that the X variable has a tolerance of 0.25, that is, that 25 percent of the variation in that particular X variable is independent of the other X variables in the model. Hence, the lower the tolerance, the more worried we should be about multicollinearity. In general, one ought to be worried if any of the tolerances is less than around 0.40. The statistical tolerances of the independent variables in this study are all above 0.44, with a mean tolerance of 0.75. Thus, there is no major problem regarding collinearity of the covariates.

Table 7.2 Cox proportional hazard model estimates of the effect of sociodemographic covariates, on the hazard of dying: 7,234 oldest old persons, China, 1998–2000

Variable	Hazard coefficients	Z-score	p-value	Hazard ratio	Semi-standardized hazard ratio
Female	−0.54	−5.95	*	0.58	0.76
Han	−0.12	−0.83		0.89	0.97
Rural residence	−0.07	−0.85		0.93	0.97
Rural birth	0.07	0.68		1.08	1.03
Education	−0.04	−3.12	**	0.96	0.88
Unmarried	0.35	3.16	**	1.41	1.14
Age group	0.61	9.24	*	1.84	1.62
Bathing limitations	0.20	1.76		1.23	1.10
Dressing limitations	0.37	2.01	***	1.45	1.14
Continence limitations	0.37	2.38	***	1.45	1.10
Feeding limitations	0.03	0.14		1.03	1.01
Transferring limitations	0.45	2.15	***	1.57	1.17
Self-rated health	−0.20	−4.09	*	0.82	0.85
Final log likelihood	−13064.086				
Likelihood ratio X^2	364.11		$p = 0.000$		

* $p < 0.000$; ** $p < 0.01$; *** $p < 0.05$

Table 7.2 reports the Cox proportional hazard estimates of the effects of the sociodemographic independent variables on the hazard of dying for the 7,234 oldest old Chinese respondents. The most important result in Table 7.2 is the very high and positive hazard coefficient for the age group variable. Among the oldest-old Chinese respondents in the sample, this variable has a coefficient of 0.61 (see Table 7.2). The older the respondent, the more likely he/she will experience the hazard of death, controlling for the effects of the other covariates in the equation. That is, this positive and significant hazard coefficient is net of the effects on the probability of dying of the other covariates of sex, nationality, rural residence, rural birth, education, marital status, daily activities, and self-rated health.

If we exponentiate the values of the hazard coefficients, that is, take their antilogs, we get hazard ratios (these are shown in the third column of data in Table 7.2). These exponentiated coefficients may be interpreted as follows: for each unit increase in the covariate, the hazard is multiplied by its exponentiated coefficient. Thus, if we compute $100(e^b - 1)$, we get the percentage change in the hazard with each one unit change in the explanatory variable.

The age group variable has a hazard ratio of 1.84. This means that moving from one age group to the next, say from an age in the 80s to an age in the 90s, increases the hazard of dying by 84 percent, or $[100 (1.84 - 1 = 0.84)] = 84$.

The results in the Cox model of the effects on dying of the other covariates confirm many of the hypotheses we set forth and discussed above. Females have a lower hazard of dying compared to males. The female variable has a hazard ratio of 0.58, meaning that females compared to males have a 42 percent lower hazard of dying. The more educated oldest old Chinese are 4 percent less likely to die compared to the less educated. Unmarried oldest Chinese persons have a hazard of dying that is 41 percent higher than that of married ones. Regarding health-related variables, people who have a limitation on dressing are 45 percent more likely to die compared to people who do not have such a limitation. People who have continence limitations are 45 percent more likely to die compared to those who do not. People who have limitations on transferring are 57 percent more likely to die compared to people do not have such limitations. Also, people who reported their health as good have an 18 percent lower hazard of dying than people who reported their health as bad.

Several of the hazard coefficients are not significantly different from zero, namely, Han, rural residence, rural birth, bathing, and feeding. For instance, among the oldest ages, minorities do not seem to have an advantage over the majority. That is, we do not find evidence here of the so-called mortality crossover. In many other populations, it has been found that prior to reaching their 80s, the majority group has the higher survival probabilities; during their 80s and thereafter, the minority group has the higher probabilities. For instance, in the U.S., there is a crossover between the survival probabilities of African Americans and Caucasian Americans at about age 87 or 88 (Nam 1995; Rogers et al. 2000: 66). However, we cannot state here that Chinese minorities experience a mortality crossover.

We ask now about the relative impacts of the covariates on the hazard of dying. One way of assessing such impacts on the hazard is to raise the hazard ratio of each covariate to the power of one standard deviation (Rabe-Hesketh and Everitt 2004: 223). We have produced such semi-standardized hazard ratios and present them in the last column of data in Table 7.2. Although there is a problem in the interpretation of the meaning of semi-standardized hazard ratios when the covariate is a dummy variable (cf. Long 1997), their values nevertheless indicate the relative effects of the covariates on the hazard of dying.

The semi-standardized hazard ratios indicate that the most influential covariate is that pertaining to age. When the age group variable is raised by one standard deviation, the hazard of dying increases by 62 percent. Variables such as transferring and dressing, which are related to daily activities, have the next most influential relative effects on the hazard of dying. The marital status variable is tied with the dressing limitation as having the next most influential relative effect on the hazard of dying.

7.5 Summary

In this chapter we have examined the effects of sociodemographic factors on the hazard of dying among oldest old Chinese during the period of 1998–2000. Data for 7,234 male and female elderly Chinese who were interviewed in the CLHLS were used. These persons were of age 80 or older at the time of the first interview in 1998. By the time of the follow-up interview in 2000, 39 percent had died. We first described the survival data by estimating a Kaplan–Meier survival curve to appraise the dynamics of longevity for the elderly Chinese.

A straightforward Cox proportional hazard model was then estimated using sociodemographic variables such as age, sex, urban–rural residence at birth and at time of the initial interview, nationality, marital status, education, and some health-related variables. Most of the sociodemographic variables were significant, even after controlling for the several health-related variables. The relative effects of the various sociodemographic factors on the hazard of dying were calculated and appraised. Age was the most influential among the sociodemographic factors, and then, health-related variables such as transferring and dressing and, then, martial status.

There are several implications from this research. First, it is worthwhile to study the mortality of the oldest old; this is justified in part because of the increasing importance of oldest old population in China. In addition, it is beneficial to study oldest old people who usually need more social and medical care because they have become more vulnerable with regard to death compared to the younger elderly. Second, as we already know, death is not only a biological process, but also a social outcome. The better the sociodemographic position, the lower the morality. Finding significant sociodemographic factors that influence mortality may help to develop better welfare policy. Finally, in future analysis, we need to compare the trends and patterns of the younger elderly with those of the oldest elderly to give us insights about the importance of sociodemographic factors on the hazard of death.

References

Aiken, L.R. (2001), *Dying, death, and bereavement*. Mahwah, NJ: Lawrence Erlbaum Associates

Allison, P.D. (1984), *Event history analysis: Regression for longitudinal event data*. Beverly Hills, CA: Sage

Campbell, C. and J.Z. Lee (1996), A death in the family: Household structure and mortality in rural Liaoning: Life-event and time-series analysis, 1792–1867. *History of the Family* 1, pp. 297–328

Cleves, M.A., W.W. Gould, and R.G. Gutierrez (2002), *An introduction to survival analysis using Stata*. College Station, TX: Stata Press Publication, Stata Cooperation

Fang, R. (1993), The geographical inequalities of mortality in China. *Social Science and Medicine* 36, pp. 1319–1323

Hamilton, L.C. (1992), *Regression with graphics: A second course in applied statistics*, Belmont, CA: Duxbury Press, Wadsworth, Inc. ISBN: 0534159001

Hamilton, L.C. (1998), *Statistics with Stata 5*. Pacific Grove, CA: Brooks/Cole

Hao, H. (1995), A study on the sex difference in mortality in China. *Chinese Journal of Population Science* 7, pp. 285–298

Hummer, R.A., R.G. Rogers, S.H Amir, D. Forbes, and W.P. Frisbie (2000), Adult mortality differentials among Hispanic subgroups and non-Hispanic whites. *Social Science Quarterly* 81, pp. 459–476

Kaplan, E.L. and P. Meier (1958), Nonparametric estimation from incomplete observations. *Journal of the American Statistical Association* 53, pp. 457–481

Liang, J., J.M. Bennett, H. Sugisawa, E. Kobayashi, and T. Fukuya (2003), Gender differences in old age morality: Roles of health behavior and baseline health status. *Journal of Clinical Epidemiology* 56, pp. 572–582

Liu, G. and Z. Zhang (2004), Sociodemographic differentials of the self-rated health of the oldest old Chinese. *Population Research and Policy Review* 23, pp. 117–133

Long, J.S. (1997), *Regression models for categorical and limited dependent variables*. Thousand Oaks, CA: Sage

Nam, C.B. (1995), Another look at mortality crossovers. *Social Biology* 42, pp. 133–142

Poston, D.L. Jr. (2000), Social and economic development and the fertility transitions in mainland China and Taiwan. *Population and Development Review* 26 (supplement), pp. 40–60

Poston, D.L. Jr. and C.C. Duan (2000), The current and projected distribution of the elderly and eldercare in the People's Republic of China. *Journal of Family Issues* 21, pp. 714–732

Poston, D.L. Jr., H. Luo, and H.K.M. Terrell (2004), The elderly populations and levels of aged dependency in China and the United States: Past, present and future. Paper presented at the Conference on Chinese Healthy Aging and Socioeconomics: International Perspectives, Duke University, August 20–21

Quadagno, J. (1999), *Aging and life course*. New York: McGraw-Hill

Rabe-Hesketh, S. and B. Everitt (2004), *A handbook of statistical analyses using Stata*. 3rd ed. Boca Raton, FL: Chapman and Hall/CRC

Rogers, R.G., R.A. Hummer, and C.B. Nam (2000), *Living and dying in the USA: Behavioral, health, and social differentials of adult mortality*. San Diego: Academic Press

Rowland, D.T. (2003), *Demographic methods and concepts*. New York: Oxford University Press

Trovato, F. and G. Lauri (1989), Marital status and mortality in Canada: 1951–1981. *Journal of Marriage and Family* 51, pp. 907–922

Xie, W. (1996), Mortality differential for various levels of education in China. *Chinese Journal of Population Science* 8, pp. 41–49

Xiong, Y. (1989), An analysis of the mortality of the ethnic minorities in China. *Chinese Journal of Population Science* 1, pp. 43–50

Yamaguchi, K. (1991), *Event history analysis*. Newbury Park, CA: Sage

Yang, S. (1986), Changes in elderly population and the elderly mortality of the Haimen County. *Population Research (Beijing)* 3 (2), pp. 35–38

Yusuf, F and M. Byrnes (1994), Ethnic mosaic of modern China: An analysis of fertility and mortality data for the twelve largest ethnic minorities. *Asia-Pacific Population Journal* 9 (2), pp. 25–46

Zeng, Y. (1989), Aging of the Chinese population and policy issues: Lessons from a rural–urban dynamic projection model. In: *1989 International Population Conference, New Delhi*. Liege, Belgium: International Union for the Scientific Study of Population 3, pp. 81–101

Zeng, Y. and L.K. George (2000), Family dynamics of 63 million (in 1990) to more than 330 million (in 2050) elders in China. *Demographic Research* 2, Article 5

Zeng, Y. and J.W. Vaupel (2002), Functional capacity and self-evaluation of health and life of oldest old in China. *Journal of Social Issues* 58, pp. 733–748

Zeng, Y., J.W. Vaupel, Z. Xiao, C. Zhang, and Y. Liu (2001), The healthy longevity survey and the active life expectancy of the oldest old in China. *Population: An English Selection* 13, pp. 95–116

Zeng, Y., J.W. Vaupel, Z. Xiao, C. Zhang, and Y. Liu (2002), Sociodemographic and health profiles of the oldest old in China. *Population and Development Review* 28, pp. 251–273

Zeng, Y., Y. Liu, and L.K. George (2003), Gender differences of the oldest old in China. *Research on Aging* 25, pp. 63–80

Zick, C. and K.R. Smith (1991), Marital transitions, poverty, and gender differences in mortality. *Journal of Marriage and Family* 53, pp. 327–336

Chapter 8
When I'm 104: The Determinants of Healthy Longevity Among the Oldest-Old in China

D.A. Ahlburg, E. Jensen and R. Liao

Abstract This chapter uses the first wave of the Chinese Longitudinal Healthy Longevity Survey to investigate the health status of the oldest-old in China. We found that the different measures of health collected in the survey were only moderately related. That is, there is not a single construct called "health." We found that work history was modestly related to some measures of health. We also found that childhood health and socioeconomic status were correlated with health even at advanced ages. To the best of our knowledge, this is the first study to examine this connection in developing countries and at such advanced ages.

Keywords Aging, Childhood socioeconomic status, Chronic conditions, Determinants, Functional limitations, Global health, Health status, Healthy longevity, Interviewer's assessment, Multidimensional concept, Oldest-old, Personality, Physical health measures, Quality of life, Regional differences, Self-rated health, Socioeconomic factors, Socioeconomic status, Subjective measures of health

8.1 Introduction

Forty years ago Lennon and McCartney speculated about what it would be like to be 64. Today it is appropriate to speculate about what it will be like to be 84 and not too far-fetched to contemplate one's state of well-being at 104. Such speculations are particularly relevant in China where the predicted rapid growth of the "oldest-old" (those aged 80 and older) has raised a number of social and economic concerns for individuals, families, and the state. The population of oldest-old is conservatively predicted to increase from about 12 million currently to 27 million in 2020 to about 100 million in 2050 (United Nations 2001). Concern over aging in China is based on the fact that the oldest-old consume a disproportionate share of medical care, social services, personal assistance, and government and private

---

D.A. Ahlburg
Leeds School of Business, University of Colorado, 8315, Koelbel UCB 419 Boulder, CO 80309, USA
e-mail: dennis.ahlburg@calorado.edu

transfers because the ability to lead an active daily life declines and disability rates increase dramatically with age (Zeng et al. 2002).

Relatively little is known about the health status of the elderly in developing nations and almost nothing about the numbers or health of the oldest-old. One exception is China where an international collaborative study of the oldest-old was initiated in 1997 (Zeng et al. 2001). The analysis in this chapter uses data from this collaboration to investigate different measures of health that reflect "healthy longevity" and the social, behavioral, and economic factors associated with these measures of health. A better understanding of these factors should allow better predictions of the likely impact of the rapid growth of the oldest-old population on private and public resources in China.

8.2 Data

The Chinese Longitudinal Healthy Longevity Survey (CLHLS) was conducted in 1998 in 631 randomly selected counties and cities in 22 provinces that are predominantly Han Chinese. This sampling strategy was chosen because age reporting, particularly at the older ages, among Han Chinese tends to be very accurate (Coale and Li 1991; Wang et al. 1997; also see some chapters in Part I of this volume). The survey collected extensive demographic, socioeconomic, health, and lifestyle data on those aged 80 and above. Because Wang et al. (1997: 94) found that age reporting among semi-super and super-centenarians is questionable, we restrict our study to those between 80 and 105 years of age. Ninety three percent of the sample is Han, 4.4 percent Zhuang, and 1.3 percent Hui. For a detailed description of the data and sampling procedures see Zeng et al. (2001, 2002).

8.3 Measure of Health

There is no universally agreed upon measure of health status because health is generally not directly observable. Health status is often thought of as a multidimensional concept which "reduce(s) to a single statistic or two, only with great difficulty" (Murray 2000: 512). Various measures of health have been used by researchers at both the micro-level and the macro-level of analysis. Generally speaking, health measures can be roughly categorized in three dimensions: subjective/objective; physical/psychological; and global/specific. However, these three dimensions are often intertwined (Miller 2001).

Subjective measures of health (self-rated health) represent subjective feelings about wellness or illness which are generally obtained by asking subjects to rate their own health conditions. Many researchers agree on the validity of self-rated health as a health measure because they have found that self-rated health is highly correlated with objective measures, such as clinical measures of morbidity, and it is a good predictor of mortality (Dwyer and Mitchell 1999; Farmer and Ferraro

1997; Geronimus et al. 2001; Idler and Benyamini 1997; Lynch 2003; Schoenfeld et al. 1994).

Many health studies include only physical health measures, like functional limitations, chronic conditions, physical fitness (Malina 2001), Body Mass Index (Murray 2000), or adulthood height (Murray 2000). A few also include measures of psychological health, such as depression (e.g., Karasek 1990; Lennon 1994). Among the physical heath measures, functional limitations and chronic conditions are most often-used. Functional limitations include three types of disabilities: disability in work, in mobility, and in personal activities (e.g., Geronimus 2001). Chronic conditions generally cover the most common health problems that threaten survival, function, and quality of life: heart disease, high blood pressure, lung disease like emphysema or lung cancer, breast cancer, any other type of cancer, diabetes, arthritis or rheumatism, osteoporosis (brittle bones), allergies or asthma, and ulcers, ulcerative colitis, or other digestive problems. Introducing psychological well-being into measures of health contributes to the literature on health measures by focusing researchers' attention on psychological health issues. However, it confounds the study of health effects because depression and other measures of psychological well-being often lead to physical ailments (Hayward et al. 2000).

Global health measures refer to a composite measure containing information on different aspects of a person or a group of persons' health status, for example, global self-rated measure of overall health status, a person's number of chronic conditions, or an index formed from several health measures (Mirowsky and Ross 2001; Ross and Wu 1995). In contrast, specific health measures are used to probe the effects on specific diseases, impairments and disabilities (Hayward et al. 2000).

In this study, we employed six measures of health: self-reported health (selfhealth); Activities of Daily Living (ADL); chronic conditions (diseasestatus); self-reported quality of life (selfqol); objective health status (intvhealth) and the number of times an individual has suffered from serious illness during the past 2 years (illness). The question "How do you rate your health at present?" was addressed to each subject to assess their self-reported health. The respondent chose one of the following answers: "very good," "good," "so-so," "bad," or "very bad." Since only a small percentage of people reported "very bad" (0.57 percent) and "bad" (8.43 percent), we combined these two groups and formed a four-scale self-reported health measure.

Measurement of ADL indicates an individual's functional capacity with respect to eating, dressing, getting in and out of a bed or chair, using the toilet, bathing, and continence (e.g., Katz et al. 1983; Zeng et al. 2002). Respondents were asked whether they have difficulties in (1) bathing, (2) dressing, (3) toilet, (4) transfer, (5) continence, and (6) feeding. Various ADLs are collinear to some degree because of co-morbidity which, in older populations, often occurs as a multiplicity of disease conditions rather than as a single form of co-morbidity (Kaplan et al. 1999). This collinearity recommends some form of combination of ADLs into more compact measures. We follow the suggestion of Zeng et al. (2002) who classified an individual as "active" if he or she needs no assistance in any ADL; as "mildly disabled" if he or she needs assistance in one or two ADLs; and as "severely disabled" if he or she needs assistance in at least three of these ADLs. They use ADLs as an indicator of functional capacity because they are "a reasonable proxy of health status, and a key element in attempts to measure quality of life" (Zeng et al. 2002: 264). ADLs are also closely related to care giving needs and health care use.

The measurement of chronic conditions is based on the respondent's answer to the question "Are you suffering from any of the following (chronic diseases)" and do you have a "disability in daily life (due to the chronic disease)." The types of chronic diseases listed in the survey include: (1) hypertension, (2) diabetes, (3) heart disease, (4) stroke, cerebrovascular disease, (5) bronchitis, pulmonary emphysema, asthma, pneumonia, (6) pulmonary tuberculosis, (7) cataract, (8) glaucoma, (9) cancer, (10) prostate tumor, (11) gastric or duodenal ulcer, (12) Parkinson's disease, (13) bedsore, and (14) others. The respondent is classified as "well" if none of the above chronic diseases caused a disability in daily life; if he or she suffered a disability from any chronic disease, the individual is classified as having a "mild chronic condition"; "severe chronic condition" refers to an individual who suffered a disability from two or more chronic diseases.

Self-reported quality of life is measured by asking the respondent "How do you rate your life at present" in terms of "very good," "good," "so-so," "bad," and "very bad," Only a small percentage of people reported "very bad" (0.21 percent) and "bad" (2.83 percent), so we combined these two groups and formed a four-scale measure of quality of life. The respondents were also asked whether and to what extent they suffered from a serious illness in the past 2 years. The survey question was "how many times have you suffered a serious illness which required hospitalization or caused you to be bedridden at home in the past 2 years?" The individual was classified as having "no serious illness" if he or she answered no serious illness; "some serious illness" if he or she reported some serious illness that did not cause the individual to be permanently bedridden; and the respondent was classified as "bedridden all year around" if that was the response given.

At the end of the survey, the interviewer was asked whether "the interviewee was: 'surprisingly healthy (almost no obvious ailments)'; or 'relatively healthy (only minor ailments)'; or 'moderately ill (moderate degrees of major ailments or illnesses)'; or 'very ill (major ailments or diseases, bedridden, etc.).'" Our "objective" measure of health status was thus based on the interviewer's assessment of the respondent's health status.

The majority of measures of health refer to prevalence or incidence but provide little information on severity. The count measures for ADLs and chronic diseases used here do capture severity to some extent. Because of co-morbidity, people with more severe symptoms tend to score higher on counts of ill-health conditions and ADLs (Dwyer and Mitchell 1999).

The distribution of the different health measures is shown in Table 8.1. Almost two-thirds of the respondents report no ADL constraints, and ninety percent report no serious illnesses, although over half report at least one or more chronic diseases. Although 57 percent report that their health is good or very good, 74 percent report that their quality of life is good or very good. Clearly, while good health is valued, it is not a necessary prerequisite to having a good quality of life; his may

Table 8.1 Distribution of health measures

Health measures	Percent
ADL Status	
Severe disability	18.24
Mild disability	18.93
Active	62.83
Disease Status	
Two or more illnesses	23.87
One illness	30.37
Well	45.76
Interviewer health rating	
Very ill	4.27
Moderately ill	11.19
Relatively ill	43.74
Surprisingly healthy	40.80
Self-reported health	
Bad	9.01
So-so	33.94
Good	44.29
Very good	12.76
Illnesses	
Bedridden	4.36
Some serious illness	6.42
No serious illness	89.22
Self-reported quality of life	
Bad	3.01
So-so	23.11
Good	56.10
Very good	17.75

speak to respondents' expectations about their health status. Interviewer's rating of respondent health appears to be heavily influenced by the respondent's disease status.

We investigated the correlation structure of the six measures of health used here to see if they could be combined to constitute a single measure of health status. The correlation structure and results of a factor analysis are shown in Table 8.2. In general, the correlations among the various measures are low. The correlations among the more objective measures are 0.2 to 0.3 and the correlation between self-rated health and quality of life is 0.43. Clearly, respondents value more than just good health. The correlations between quality of life and objective measures of health are very low. Thus, respondents' quality of life seems to be affected more by their subjective evaluation than by their objective evaluation of their health. Based on the correlations, it appears that the interviewers weigh both objective measures of health and the respondent's own evaluation in forming their view of the respondent's health. The first eigenvalue is only slightly greater than 1.0 and no other eigenvalue is greater than 1.0, so the factor-analytic evidence to suggest a single health measure

Table 8.2 Correlation matrix and factor analysis of the six health measures

	Correlation matrix						Factor loadings		
	ADL status	Diseases status	Self-reported health	Self q-o-l	Intrvr health rating	Illness	1	2	Uniqueness
ADL status	1.00						0.54	−0.25	0.64
Disease status	0.20	1.00					0.32	−0.13	0.88
Self-reported health	0.25	0.19	1.00				0.60	0.30	0.55
Self-reported quality of life	0.03	0.00	0.43	1.00			0.32	0.44	0.70
Interviewer health rating	0.54	0.26	0.42	0.18	1.00		0.70	−0.11	0.49
Illness	0.29	0.15	0.16	0.94	0.37	1.00	0.36	−0.17	0.84

Factor Analysis (Principal factors; two factors retained)

Factor	Eigenvalue	Difference	Proportion	Cumulative
1	1.48	1.07	1.08	1.08
2	0.41	0.42	0.30	1.38
3	−0.01	0.05	−0.01	1.37
4	−0.06	0.15	−0.04	1.33
5	−0.21	0.03	−0.15	1.18
6	0.24		−0.18	1.00

is weak.[1] The results suggest that the different measures of health collected in the survey capture different aspects of health status, and that they cannot be combined into a single measure of health.

8.4 Factors Associated with Healthy Longevity

Work history may affect morbidity and mortality through the type of work performed, the working conditions, or the fringe benefits associated with work, such as access to health care facilities or health insurance. Early studies of the effects of work on morbidity or mortality estimated occupational effects and assumed that any such differences were due to differences in the physical conditions encountered in different occupations such as pollution or risk of occupational injury. However, a more nuanced view is now taken. Occupations can differ on complexity (creativity, autonomy, and cognitive-skill demands), physical and environmental demands, social skill demands, and manipulative skill demands (Hayward and Gorman 2004; Karasek 1990; Marmot et al. 1997; Moore and Hayward 1990). Job complexity and job control emerge as having positive effects on health, though we note that it also is possible that occupation is a proxy for lifetime earnings (Zissimopoulus and Karoly 2003).

In this research undertaken here we investigate the effects of work through two variables in addition to a set of occupational indicators. Respondents were asked whether they had ever undertaken any physical work and, if so, when they started and when they stopped. A dummy variable was created for "ever did physical work," and a continuous variable for years of physical work was constructed. The occupational categories were: professional or government (7.4 percent), industrial (6.8 percent), commercial or service (9.1 percent), military or other (2.2 percent), housework (19.3 percent), and agriculture, forestry, or fisheries (55.2 percent, the excluded category in the regressions).

Some research suggests that adult morbidity is related to childhood life circumstances (in utero environment, nutrition, exposure to infectious diseases and environmental toxins, social and economic deprivation).[2] Initially, the effects were thought to be indirect and negative: childhood socioeconomic status (CSES) affects adult SES (ASES) which directly affects health (Kuh and Wadsworth 1993). Hayward and Gorman (2004) suggest that CSES can also affect education, and that CSES and education shape preferences for major lifestyle behaviors such as smoking, drinking, diet, and exercise which affect health. Recent research has suggested that there may also be direct effects of childhood health even after controlling for

---

[1] Only respondent and interviewer global health reports have factor loadings of 0.6 or above, generally taken as the cut-off level to be considered as loading on a factor.

[2] See, for example, Elo and Preston (1992), Fogel (1993), Kuh and Ben-Shlomo (1997), Hayward et al. (2000), Blackwell et al. (2001), and Hayward and Gorman (2004).

CSES and ASES.[3] Blackwell et al. (2001) found that the type of childhood illness differentially affects adult health. What matters most are infectious diseases. The importance of these findings is that health care policies targeted at children can have considerable long-term benefits for adult health.

We capture childhood health and socioeconomic conditions with a number of variables. Respondents were asked whether they were sick enough as a child to require care (childhood illness), and whether they went to bed hungry as a child (nutrition). These variables were coded one if the answer was "yes." Almost half of the sample reported being sick enough in childhood to require care, and 56 percent reported that they often went to bed hungry when a child. We also included a variable for the respondent's parity, because a number of studies in developed and developing countries have found that later children received a smaller share of household resources than earlier children; and they may suffer higher morbidity as a consequence. Childhood SES is proxied for by the father's occupation. Preston and Haines (1991) found that rates of infant mortality in the US in 1900–1910 were lowest in households in which the father was a farmer or a salesman and in which at least one parent was literate. We lack information on other potentially useful measures of CSES, such as parent's education and childhood household income. We do, however, have information from the respondents as to whether they received inadequate care when sick as a child. Fully 16.5 percent of the respondents had been sick enough as a child to require care but did not receive it. This variable should indicate a deprived childhood, either socially or economically, or residence in an area that lacked medical facilities. In an effort to control for the latter possibility we include a set of region of birth dummy variables and an urban/rural dummy variable.

Finally, regional differences in socio-economic factors and in the interpretation of the survey's health questions may occur between areas. There may also be urban/rural differences in the disease environment, so we include indicator variables for urban childhood residence and province of childhood residence in our regression analyses.

A number of other factors may affect longevity and are controlled for in our study. Health declines with age and may do so "quickly" for ADLs for the oldest-old and "slightly or moderately" for their self-rated health (Zeng et al. 2002). Zeng and Vaupel (2002) found that satisfaction with current life was almost unchanged between ages 80 and 94 years but declined slightly after that. Thus we add an age variable to our regression equations and also test for non-linear age effects by including age squared. Zeng et al. (2002) also find the health of oldest-old men to be better than that of women and speculate that this could reflect the fact that men work outdoors and therefore increase their capacity for maintaining the capabilities of daily living. It is also possible that the male advantage comes from higher education, pension, and income or the adverse selection of more frail males.

Education may affect health through its impact on risk taking, deferring gratification, and sense of control over one's environment (Elo and Preston 1992) or

---

[3] See, e.g., Blackwell et al. (2001); Kuh and Wadsworth (1993); Martyn et al. (1996).

through its impact on preferences for lifestyle behaviors and its impact on adult socioeconomic achievement (Hayward et al. 2000). We use a set of dummy variables to measure education. The categories are: zero (excluded category), few years (1–3 years), some (4–6 years), more (7–9 years), and well-educated (10 or more years).

We also control for a number of lifestyle variables that may be related to health, including having ever smoked, drank alcohol, or exercised. The survey asked respondents if their marriage had been "happy," "so-so," or "bad," If there is an effect of marriage on the morbidity of the oldest-old, we would expect to find lower morbidity and higher reports of quality of life among those who report being happily married.

Few studies of health investigate the impact of personality on health. An exception is Hayward et al. (2000) who suggested that significant differences between blacks and whites on satisfaction with friends and financial situation, and on a depression scale may contribute to racial differences in health. The CLHLS asked seven questions related to the respondent's personality. We factor analyzed the responses and found that they all load with a factor loading of 0.79 or more on a single factor. We combine these variables with equal weights to construct a personality index and expect that people with a more positive personality will report better health and quality of life.

Although longevity is moderately heritable in human populations (Ahlburg 1998; McGue et al. 1993; Mitchell et al. 2001), longevity is thought to contain only limited information on functional status, since some individuals can exhibit healthy functional survival but others disability-associated survival (Hadley 2000). However, Duggirala et al. (2002) have shown that at least one measure of biological aging in the Mennonite population has substantial genetic determinants. The CLHLS asked respondents if their parents were still alive and, if not, their age at death. Unfortunately, about one-third of the observations on parent's age at death are missing. Despite this limitation, we attempted to test for an association between parent's longevity and the respondent's health by including variables for mother's and father's age at death and dummy variables for missing values of these variables(Table 8.3).

8.5 Results

We find only modest support for the hypothesis that healthy longevity is related to work history, at least among the oldest-old. Whether a respondent has ever engaged in physical labor is positively associated with better ADL status (at the 0.01 level), self-reported health (at the 0.10 level), and interviewer rating of health (at the 0.06 level), but is not associated with the other three, arguably more objective, measures of health. It could be that physical work "hardens" the individual and protects him or her from health insults, or it could be that selection is operating. That is, only the hardiest survive physical labor and report relatively good health at advanced ages. When we added a measure of the number of years a respondent had done

Table 8.3 Variable definitions

Variable	Definition	Mean	Std. Dev.
adlstatus	A three-scale categorical variable indicating the individual's ADL ability: severe disable = 0; mild disable = 1; active = 2	1.44588	0.78226
diseasestatus	A three-scaled categorical variable indicating the individual's health status in terms of chronic diseases: severe chronic disease (two or more chronic diseases) = 0; mild chronic disease condition (1 chronic disease) = 1; relative few chronic diseases (no chronic disease) = 2.	1.21885	0.80529
selfhealth	A four-scale categorical variable indicating the individual's self-rated health: bad = 0; "so-so" = 1; good = 2; very good = 3.	1.60804	0.82082
selfqol	A categorical variable indicating the individual's self-rated quality of life: bad = 0; "so-so" = 1; good = 2; very good = 3.	1.88575	0.71907
intvhealth	A four-scale categorical variable indicating the interviewer's rating of the individual's health status: "very ill" = 0; "moderately ill" = 1; "relatively healthy" = 2; "surprisingly healthy" = 3.	2.21073	0.80393
illness	A categorical variable indicating the individual suffered from serious illness during the last 2 years before the interview: bedridden all the year around = 0; some serious illness = 1; no serious illness = 2.	1.84855	0.46458
physlabor	Equal 1 if the individual has done physical labor regularly.	0.78356	0.41184
physlabordur	Number of years of physical labor.	42.64654	27.28311
selfoccup1	Equal 1 if the individual's main occupation is professional, technical, or governmental, institutional or managerial personnel.	0.07439	0.26242
selfoccup2	Equal 1 if the individual's main occupation is agriculture, forestry, animal husbandry or fishery.	0.55196	0.49732
selfoccup3	Equal 1 if the individual's main occupation is industrial.	0.06758	0.25103
selfoccup4	Equal 1 if the individual's main occupation is commercial or service.	0.09052	0.28694
selfoccup5	Equal 1 if the individual's main occupation is military personnel or other occupation.	0.02283	0.14936
selfoccup6	Equal 1 if the individual's main occupation is housework.	0.19273	0.39447
pateroccup1	Equal 1 if father's main occupation is professional, technical, or governmental, institutional or managerial personnel.	0.05122	0.22046
pateroccup2	Equal 1 if father's main occupation is agriculture, forestry, animal husbandry or fishery.	0.71346	0.45217
pateroccup3	Equal 1 if father's main occupation is industrial.	0.02953	0.16929
pateroccup4	Equal 1 if father's main occupation is commercial or service.	0.12675	0.33271
pateroccup5	Equal 1 if father's main occupation is military personnel or other occupation.	0.06508	0.24668
pateroccup6	Equal 1 if father's main occupation is housework.	0.01397	0.11737

Table 8.3 (Continued)

Variable	Definition	Mean	Std. Dev.
sickaskid	Dummy variable equal 1 if the individual has been sick enough for care as a child.	0.49044	0.49994
nocareaskid	Dummy variable equal 1 if the individual did not receive adequate care when sick as a child.	0.16501	0.37121
hungaskid	Dummy variable equal 1 if the individual often went to bed hungry as a child.	0.56118	0.49627
age	Age	92.03044	7.40318
male	Dummy variable equal 1 if male.	0.40238	0.49041
noeduc	Dummy variable equal 1 if the individual received no schooling.	0.67055	0.47004
feweduc	Dummy variable equal 1 if the individual received 1–3 years' schooling.	0.14720	0.35432
someduc	Dummy variable equal 1 if the individual received 4–6 years' schooling.	0.09352	0.29118
moreduc	Dummy variable equal 1 if the individual received 7–10 years' schooling.	0.05013	0.21823
welleduc	Dummy variable equal 1 indicating the individual received more than 10 years' schooling.	0.03860	0.19265
eversmoke	Dummy variable equal 1 if the individual has smoked before or smoke presently.	0.32379	0.46795
everdrink	Dummy variable equal 1 if the individual drank before or drink presently.	0.34970	0.47690
everexercise	Dummy variable equal 1 if the individual exercised before or exercise presently.	0.34800	0.47636
birth1	Dummy variable equal 1 if the individual's birth order is one.	0.37640	0.48451
birth2	Dummy variable equal 1 if the individual's birth order is two.	0.25199	0.43418
birth3	Dummy variable equal 1 if the individual's birth order is three.	0.16222	0.36867
birth4	Dummy variable equal 1 if the individual's birth order is four or more.	0.10288	0.30382
birthurban	Dummy variable equal 1 if the individual's birthplace is urban area	0.15008	0.35717
personality	A composite variable showing the extent that the individual is optimistic	3.15366	0.58132

physical labor, the duration variable was positive and significant for most of the health measures and the variable for ever engaged in physical labor was now negative and significant for three of the health measures. When the duration variable was broken down into a series of dummy variables measuring work duration in decades, we found that the duration result was driven by the 13 percent of the sample who reported working 60 years or more. It is highly likely that the duration effect reflects causation from health to work, not from work to health. That is, only those who are healthy can work for very long durations. We attempted to create an instrument for work duration using the age at which the respondent first started physical work, but the instrument was not significant in any of the health regression equations. Either this is a poor instrument or the duration of work has no effect on health.

There are significant differences in several measures of health associated with different occupations. Professional and government workers report worse ADL status, and more diseases and illnesses compared to those who worked in agriculture, forestry, or fisheries. Industrial workers report worse ADL status and more diseases, commercial and service workers report fewer illnesses, military personnel report more diseases, and those who worked in the household report worse ADL status. It is interesting to note that these occupational and industrial differences in more objective measures of health do not translate into worse subjective reports of health or quality of life. As noted above, it is not clear whether these differences reflect occupational differences in working conditions, income, or occupation-related differences in access to health care. The relatively better health of primary-sector workers could reflect better nutrition due to better access to food.

Being sick as a child or going to bed hungry does not appear to affect adult health, at least among the oldest-old. However, not receiving adequate care when sick as a child is shown to be associated with poorer health at older ages and is statistically significant for all health measures except interviewer reported health. This finding could reflect either economic deprivation in childhood or a lack of local medical facilities. The regional and urban/rural dummies should control to some extent for differences in the availability of facilities. Birth parity is not related to health, except in the case of respondents who were fourth and higher order births. These individuals report better global health and quality of life. Further support for the importance of childhood SES comes from health differences related to the father's occupation. Children of farmers, foresters, and fishermen tend to have better ADL status and fewer diseases. The one exception is children who reported their father's occupation as "housework"; they tended to report better health.

A number of personal characteristics are associated with some measures of health status at advanced age. Like Zeng et al. (2002), we found that ADL status declines with age, but we did not find evidence of a non-linear decline. Nor did we find a significant effect of age on self-reported health. Two other significant age effects were found: a significant negative association with interviewer rating and a positive association with self-reported quality of life, although the latter association was negative and insignificant in a regression including only age and sex. No non-linear age effects were found for the other health measures. Unlike Zheng and Vaupel (2002), we did not find a decline in reported quality of life after age 94. In fact, we found those over 94 years to report a better quality of life. We did find those over 94 years to report fewer ADLs and more illnesses and interviewers to rate their health as poorer than younger respondents.

In general, males reported better health than females, but males reported lower quality of life than did females, although the latter association is positive and significant in a regression including only age and sex. The better educated seemed to have higher self-reports of health and quality of life than the less well-educated. Interviewers also rated the health of the more educated to be better. As noted above, the channels through which education is assumed to work, lifestyle choices, adult SES, and personality are all controlled for, so the impact of education is in addition to any

Table 8.4 Regression results

Variable	Adlstatus	Diseasestatus	Selfhealth	Selfqol	Intvhealth	Illness
physlabor	0.1144**	0.0154	0.0628	0.0085	0.0761*	−0.0435
selfoccup1	−0.2371***	−0.3083***	−0.0845	0.1244	−0.0999	−0.2313**
selfoccup3	−0.1607*	−0.2523***	−0.0850	−0.0095	−0.0869*	−0.1327
selfoccup4	−0.1032	−0.0499	0.0406	0.0107	−0.0026	−0.1534
selfoccup5	−0.1627*	−0.3082***	0.0235	−0.1215	0.0056	−0.1381
selfoccup6	−0.1137	0.0173	0.0641	0.1083*	−0.0180	−0.0589
pateroccup1	−0.1384	−0.1251	−0.0606	0.0167	0.0669	−0.0869
pateroccup3	0.0277	−0.0747	−0.0520	−0.0480	−0.0311	0.0114
pateroccup4	0.0032	−0.0744*	−0.0337	0.0153	0.0667	0.0545
pateroccup5	−0.1588*	−0.0203	−0.0799	0.0162	−0.0270	−0.2095***
pateroccup6	0.0344	−0.2306*	0.3199	0.1085	0.2698*	0.6988*
sickaskid	0.0002	−0.0296	−0.0084	0.1169*	−0.0761*	−0.0105
nocareaskid	−0.1529	−0.0921**	−0.1548***	−0.1605**	−0.0520	−0.1761**
hungaskid	−0.0059	0.0312	−0.0248	−0.0089	−0.0211	−0.0262
age	−0.0625***	0.0028	−0.0013	0.0073**	−0.0252***	−0.0041
male	0.2082***	0.2584***	0.0951***	−0.0903*	0.0923	0.0343
feweduc	0.0250	−0.0219	0.0680	0.0565	0.0975*	0.0109
someduc	−0.0467	−0.0370	0.1183***	0.0987**	0.1166	0.0325
moreduc	−0.0125	0.0329	0.1600***	0.1023	0.1555**	0.0084
welleduc	−0.1436	−0.2331*	−0.1432	0.1141	−0.1233	−0.2689**
eversmoke	0.0335	−0.0206	−0.0551	−0.0230	−0.0242	−0.0577
everdrink	0.0121	−0.0239	−0.0148	0.0418	0.0340	−0.0291
everexercise	0.1739***	−0.0112	0.1260***	0.1628***	0.2799***	0.1469*
birth1	−0.0833	0.0216	0.0717	0.0253	−0.0771	−0.0711
birth2	−0.0695	−0.0623	0.0573	0.0144	−0.0678	−0.0394
birth3	−0.0622	−0.0123	0.0793	0.0243	−0.0430	−0.0683
birth4	−0.0910	0.0254	0.1886***	0.1438**	−0.0128	−0.1388
personality	0.3227***	0.2107***	0.6549***	0.5593***	0.5033***	0.1533***
birthurban	−0.0610	−0.1148*	0.0109	0.0326	0.0167	−0.0421
_cut1	−5.9485***	0.4610	0.7030*	0.4291	−2.6861***	−1.8840***
_cut2	−5.1975***	1.2934***	2.0175***	1.8085***	−1.8276***	−1.2863**
_cut3			3.4776***	3.5279***	−0.3077	

*, **, and *** denote statistical significance at $p < 0.10$, 0.05, and 0.01 against two-tailed alternatives. Not reported in this table, but included in all regressions, was a set of indictor variables for province of birth.

effect through these channels. We found that reporting that one's first marriage was a "good" marriage was associated with better reported health.

Those who survive to very old ages seem to be little affected by life style choices with one very important exception. Exercise may impart benefits on more objective measures of health status (ADLs, illnesses) and on global reports of health and well-being. However, reverse causation is also a possibility because those who are healthy are more able to exercise. A positive personality appears to be good protection against the ravages of time. Irrespective of the specification of the model or the estimation technique, those with an optimistic personality not only reported a more positive outlook on their health and quality of life, but they also reported fewer ADLs, diseases, and illnesses. This could be because they set a higher threshold for what constitutes a "disease" or "impairment," or it could be that a more positive outlook somehow mediates other factors that can lead to poorer health.

We also tested a statement made by Zeng et al. (2002: 268) in their study of the same population we have studied. They concluded that "exceptionally long-lived people are likely to consider health to be good and view life as satisfactory, relatively independently of their capacity to perform daily activities." We added the ADL status variable to each regression as an explanatory variable (ignoring the fact that it was an endogenous variable). For all health measures except self-reported quality of life, those with better ADL status reported better health. So, in addition to finding support for at least part of the conclusion reached by Zeng et al, we find that their conclusion generalizes to other measures of health. This reinforces our conclusion that the health measures collected in the CLHLS capture different aspects of health.

Finally we checked the robustness of our findings by using different estimation approaches. We re-estimated the ADL and disease regressions by ordinary least squares, ordered probit (without grouping responses), and as count models (Poisson and negative binomial). Our results were indeed robust.

8.6 Conclusions

The CLHLS collected data on six measures of health. The correlations among the different measures were quite low and there was little statistical support for combining them into a smaller number of measures of "health." We agree with Murray's conclusion that health is a "multidimensional concept." The measures did not break down into a simple pattern of "more objective" measures such as ADLs, number of diseases, and number of illnesses, and "more subjective" self-reports of global health and quality of life. Although the effects of some variables tended to be similar for the more objective measures or the more subjective measures, this was not always the case. That is, the oldest-old in China seem to exhibit different forms of "health."

To the best of our knowledge, ours is the first study to investigate the impact of childhood health and socioeconomic status on adult health in a developing country and to study this relationship among the oldest-old. We found that childhood health and socioeconomic status had independent effects on adult health even at very advanced ages and even after controlling for adult socioeconomic status and lifestyle choices. The variable that was most important was whether a child received care for childhood illnesses. Those who received care reported better objective and subjective health at advanced ages. The importance of this finding is that the provision of health care services in childhood may well have very long-term returns in the form of improved adult health even at quite advanced ages.

Some other variables that were associated with at least some measures of better health were being happily married and exercising. We did not find that smoking and drinking had adverse health effects at advanced ages, perhaps because those most susceptible to these "vices" had already been removed from the population.

References

Ahlburg, D. (1998), Intergenerational transmission of health. *AEA Papers and Proceedings* 88 (2), pp. 265–269

Blackwell, D.L., M.D. Hayward, E.M. Crimmins (2001), Does childhood health affect chronic morbidity in later Life? *Social Science & Medicine* 52, pp. 1269–1284

Coale, A., S. Li. (1991), The effect of age misreporting in China on the calculation of mortality rates at very high ages. *Demography* 28 (2), pp. 293–301

Duggirala, R., M. Uttley, K. Williams, R. Arya, J. Blangero, and M.H. Crawford (2002), Genetic determination of biological age in the Mennonites of the Midwestern United States. *Genetic Epidemiology* 23, pp. 97–109

Dwyer, D. and O. Mitchell (1999), Health problems as determinants of retirement: are self-rated measures endogenous? *Journal of Health Economics* 18 (2), pp. 173–193

Elo, I.T., and S.H. Preston (1992), Effects of early life conditions on adult mortality: a review. *Population Index* 58 (2), pp. 186–212

Farmer, M.M. and K.F. Ferraro (1997), Distress and perceived health: mechanisms of health decline. *Journal of Health and Social Behavior* 39, pp. 298–311

Fogel. R. (1993), New sources and new techniques for the study of secular trends in nutritional status, health, mortality and the process of aging. *Historical Methods* 26, pp. 5–43

Geronimus, A.T., J. Bound, T.A. Waidmann, C.G. Colen, and D. Steffick (2001), Inequality in life expectancy, functional status, and active life expectancy across selected black and white populations in the United States. *Demography* 38 (2), pp. 227–251

Hadley, E. (2000), Genetic epidemiologic studies on age-specified traits. NIA Aging and Genetic Epidemiology Working Group. *American Journal of Epidemiology* 152, pp. 1003–1008

Hayward, M.D. and B. Gorman (2004), The long arm of childhood: the influence of early-life social conditions on men's mortality. *Demography* 41 (1), pp. 87–107

Hayward, M.D., E.M. Crimmins, T.P. Miles and Y. Yang (2000), The significance of socioeconomic status in explaining the racial gap in chronic health conditions. *American Sociological Review* 65, pp. 910–930

Idler, E. and Y. Benyamini (1997), Self-rated health and mortality: a review of twenty-seven community studies. *Journal of Health and Social Behavior* 39, pp. 21–37

Kaplan, G.A., M.N. Haan, and R.B. Wallace (1999), Understanding changing risk factor associations with increasing age in adults. *Annual Review of Public Health* 20, pp. 89–108

Karasek, R. (1990), Lower health risk with increased job control among white collar workers. *Journal of Organizational Behavior* 11 (3), pp. 171–185

Katz, S., L.G. Branch, M.H. Branson, J.A. Papsidero, J.C. Beck, and D.S. Greer (1983), Active life expectancy. *New England Journal of Medicine* 309 (20), pp. 1218–1224

Kuh, D. and B. Ben-Shlomo (1997), *A life course approach to chronic disease epidemiology*. New York: Oxford University Press

Kuh, D. and M. E. J. Wadsworth (1993), Physical health status at 36 years in a British national birth cohort. *Social Science and Medicine* 37, pp. 905–916

Lennon, M.C. (1994), Women, work, and well-being: the importance of work conditions. *Journal of Health and Social Behavior* 35 (3), pp. 235–247

Lynch, S. (2003), Cohort and life-course patterns in the relationship between education and health: a hierarchical approach. *Demography* 40 (2), pp. 309–331

Martyn, C.N., D.J.P. Barker, and C. Osmond (1996), Mothers' pelvic size, fetal growth, and death from stroke and coronary heart disease in men in the UK. *The Lancet* 348, pp. 1264–1268

Malina, R. (2001), Physical activity and fitness: pathways from childhood to adulthood. *American Journal of Human Biology* 13, pp. 162–172

Marmot, M.G., H. Bosma, H. Hemingway, E. Brunner, and S. Stansfeld (1997), Contribution of job control and other risk factors to social variations in coronary heart disease incidence. *The Lancet* 350, pp. 235–239

McGue, M, J.W. Vaupel, N. Holm, B. Harvard (1993), Longevity is moderately heritable in a sample of Danish twins born 1870–1880. *Journal of Gerontology* 48, pp. B237–B244

Miller, T. (2001), Increasing longevity and Medicare expenditures. *Demography* 38 (2), pp. 215–226

Mirowsky, J. and C.E. Ross (2000), Socioeconomic status and subjective life expectancy. *Social Psychology Quarterly* 63 (2), pp. 133–151

Mitchell, B.D., W. Hsueh, T.M. King, T. I. Pollin, J. Sorkin, R. Agarwala, A.A. Schaffer, and A.R. Shuldiner (2001), Heritability of life span in the old old Amish. *American Journal of Medical Genetics* 102, pp. 346–352

Moore, D.E. and M.D. Hayward (1990), Occupational careers and mortality of older men. *Demography* 27, pp. 31–53

Murray, J. (2000), Marital protection and marital selection: evidence from a historical-prospective sample of American men. *Demography* 37 (4), pp. 511–521

Preston, S.H. and M.R. Haines (1991), *Fatal years: child mortality in late nineteenth century America*. Princeton, NJ: Princeton University Press

Ross, C.E. and C. Wu (1995), The links between education and health. *American Sociological Review* 60, pp. 719–745

Schoenfeld, D.E., L.C. Malmrose, D.G. Blazer, D.T. Gold, and T.E. Seeman (1994), Self-rated health and mortality in the high-functioning elderly—A closer look at healthy individuals: MacArthur field study of successful aging. *Journal of Gerontology: Medical Sciences* 49, pp. M109–M115

United Nations (2001), *World population prospects. The 2000 revision. Volume I: Comprehensive Tables*. New York

Wang, Z., Y. Zeng, B. Jeune, and J.W. Vaupel (1997), A demographic and health profile of centenarians in China. *Longevity: To the Limits and Beyond*. Springer

Zeng, Y. and J.W. Vaupel (2002), Functional capacity and self-evaluation of health and life of the oldest old in China. *Journal of Social Issues* 58 (4), pp. 733–748

Zeng, Y., J.W. Vaupel, Z. Xiao, C. Zhang, and Y. Liu (2001), The Healthy Longevity Survey and the active life expectancy of the oldest old in China. *Population: An English Selection* 13 (1), pp. 95–116

Zeng, Y., J.W. Vaupel, Z. Xiao, C. Zhang, and Y. Liu (2002), Sociodemographic and health profiles of the oldest old in China. *Population and Development Review* 28 (2), pp. 251–273

Zissimopoulus, J. and L. Karoly (2003), Transition to self-employment at older age: the role of wealth, health, health measure, and other factors. Working paper WR-135. Santa Monica, CA: RAND.

Chapter 9
Association of Education with the Longevity of the Chinese Elderly

Jianmin Li

Abstract This study has clearly shown that educational attainment is positively and substantially associated with survivorship at old ages in China. The overall association is stronger among elderly women than among elderly men.

Keywords Census micro data, Childhood, Chinese elderly, Cohort, Education, Fixed-attributes dynamics, Longevity, Mortality, Ratio of survivorship, Socioeconomic status, Survivorship, Univariate analysis

9.1 Introduction

The public is interested in why some people are healthier and live longer as compared to other members of the same cohort. There is a growing interest in the relationship between health and socioeconomic status (SES), which is generally indicated by education, income, and occupation. A substantial body of literature demonstrates that lower SES individuals generally experience higher mortality compared to those with higher SES, and that SES is inversely related to health status (Brown et al. 2004; Daly et al. 2002; Feinstein 1993; Frank et al. 2003; Grundy and Holt 2001; Kim et al. 2004; Matthews et al. 2005; Rogot et al. 1992).

Some studies have focused on the relationship between SES and health status over the life course of individuals. For example, Davey Smith et al. (1997) found that socioeconomic factors over different states of the life course affect health and risk of premature death. Laaksonen et al. (2005) found that both childhood and adulthood economic difficulties had clear associations with health. In a study of a large sample of British women, Power et al. (2005) came to the conclusion that socioeconomic position in childhood was associated with adult mortality. Based on

---

J. Li
Institute of Population and Development, Nankai University, Tianjin 300071, China
e-mail: lijianm0075@sina.com

29 years of follow-up surveys in Alameda County, California, Frank et al. (2003) found that virtually all of the gradients associated with SES are inversely related to health outcomes, although there is no simple pattern of shape or evolution of shape over time. A study by Osler et al. (2005) showed that a father's occupational and social class is associated with adult mortality among members of the mother–father–offspring triad. But Lahelma et al. (2004) used cross-sectional survey data from the Helsinki Health Study in 2000 and 2001 to examine the pathways of the socioeconomic determinants of health. They found that among women, half of the variation in longstanding illness by education was mediated by occupational class and household income; inequalities by occupational class were largely explained by education; inequalities of income were explained by education and occupational class.

For a more comprehensive understanding of the influence of SES on the risks of morbidity and mortality at different stages in the life course, the following theoretical frameworks were developed and applied. Hertzman et al. (2001) used an interactive framework of society and the life course to explain self-rated health in early adulthood. Halfon and Hochstein (2002) developed a framework of life course health development (LCHD) to explain how health trajectories develop over an individual's lifetime. The framework shows that health is a consequence of multiple determinants operating in nested genetic, biological, behavioral, social, and economic contexts.

While other chapters in this book have employed the more comprehensive multivariate statistical modeling approach, this chapter investigates the long-term association of SES measured by educational attainment in childhood with the longevity of the Chinese elderly, following the simple approach of the Fixed-Attributes Dynamics approach (Zeng and Vaupel 2004).

9.2 Hypothesis

Generally speaking, there are five main kinds of factors influencing a person's life span: biological factors, individual's SES and life style/ behavior, the social and natural environment at the community level, scientific and technological development in general, as well as accidental events. However, it is impossible to include all of these factors in one study due to limits of data and space. Thus, in this chapter we will focus on the association of educational attainment in childhood with longevity. There are two reasons why we choose educational attainment as our measure of SES. First, the educational attainment of the elderly is a strong indicator of social and economic status and living conditions during childhood. Second, educational attainment is closely related to adulthood occupation and income, and thus can reflect SES and living conditions in later life. In this chapter, we investigate and test the hypothesis that the association of educational attainment in childhood is positively associated with longevity at old ages.

9.3 Data

The data used in this paper are from the 1998 baseline of the "Chinese Longitudinal Healthy Longevity Survey" (CLHLS) and the 1982 Census 1 percent micro data.[1] The 1998 CLHLS covered 22 provinces, with a valid sample size of 9,093 oldest-old interviewees.[2] At the time of the sample in 1998 the oldest-old were aged 80 and over, which means they were aged 64 and over in 1982.

9.4 Methodology

Zeng and Vaupel (2004) presented a basic method known as "Fixed-Attributes Dynamics (FAD)" to study the association between attributes fixed in early life (e.g. educational attainment, childhood SES, genetic composition, etc.) and survival at older ages. Using the FAD method, one can estimate the ratio of survivorship (RS) of those with the fixed attribute to those without the attribute. If the RS is greater than one, the attribute is positively associated with longevity, and vice versa. According to Zeng and Vaupel (2004), the FAD method is formulated as follows:

Let $N(x)$ denote the number of persons aged x; $p_1(x)$, the proportion of individuals who are x years old and have the fixed attribute; $s_1(x+n)$, the conditional survival probability from age x to $x+n$ for those who have the fixed attribute; $s_0(x+n)$, the conditional survival probability from age x to $x+n$ for those who do not have the fixed attribute; and $S(x+n)$, the conditional survival probability from age x to $x+n$ for all members of the cohort.

Because $N(x)p_1(x)s_1(x+n) = N(x)S(x+n)p_1(x+n)$, it follows that

$$s_1(x+n) = S(x+n)\left(\frac{p_1(x+n)}{p_1(x)}\right) \tag{9.1}$$

and

$$s_0(x+n) = S(x+n)\left(\frac{1 - p_1(x+n)}{1 - p_1(x)}\right) \tag{9.2}$$

Dividing (9.1) by (9.2) gives the Ratio of Survivorship (RS) for those with the fixed attribute to those without the attribute:

$$RS = \frac{s_1(x+n)}{s_0(x+n)} = \frac{(1 - p_1(x))p_1(x+n)}{p_1(x)(1 - p_1(x+n))} \tag{9.3}$$

---

[1] Educational attainment data collected from the CLHLS are years of schooling, while the data from the Census are the education levels. We converted the years of schooling data into various education levels: 0–2 years schooling is converted into the category "illiteracy and semi-illiteracy," 3–6 years into "primary school," 7–9 years into "middle school," 10–12 years into "high school," and 13 years and over into "college and above."

[2] Refer to the Research Group of Healthy Longevity in China (2000). *Data collections of the Health Longevity Survey in China 1998*, Chapter 2, Beijing: Peking University Press.

In the research reported in this chapter, the formula is:

$$RS_{\geq i} = \frac{(1 - p_{\geq i}(1982, x))p_{\geq i}(1998, x + 16)}{p_{\geq i}(1982, x)(1 - p_{\geq i}(1998, x + 16))} \quad (9.4)$$

where $n = 16$, i.e. the length of time period between 1982 and 1998 is 16 years; $\geq i$ denotes the educational attainment is at or above level i.

By the same logic for deriving formulas (9.1), (9.2), and (9.3), we can also estimate the Ratio of Survival of the elderly with different levels of educational attainment. The formula is as follows:

$$RS_{ij} = \frac{(1 - p_i(x))p_j(x + n)}{p_i(x)(1 - p_j(x + n))}(i \neq j)(i \neq j) \quad (9.5)$$

where i and j denote the level of educational attainment. $RS_{ij} > 1$ indicates that the likelihood of survival from age x to $x + n$ among those with educational level i is better than that among those with educational level j, and vice versa.

The FAD method assumes that the number of migrants is small or, alternatively, that migrants do not differ significantly from non-migrants with respect to the fixed attribute and survival. Note that migration in and out of the 22 surveyed provinces between 1982 and 1998 was negligible due to China's strict household registration control policy; in addition, in general the Chinese elderly do not like to move away from a familiar place. Thus, we can use elders aged 64 and over living in the 22 provinces during the 1982 census as the baseline of the cohorts of the sample aged 80 and over in 1998 in those same 22 provinces. This also implies that the analysis used in this chapter is not for the whole birth cohort born 80+ years before the survey year 1998. Rather, we investigate those cohort members who survived to at least 64 years old in 1982. By comparing the distributions of the educational attainment of the same cohort members in 1982 (aged 64+) and 1998 (aged 80+), we can investigate the relationship of educational attainment with the longevity of the Chinese elderly.

9.5 Results

The estimates of the $RS_{\geq \text{middle school}}$ and RS_{ij} are presented in Tables 9.1 and 9.2. We draw the following observations from these interesting results. First, all the estimates of $RS_{\geq \text{middle school}}$ for men and women at various age groups are much larger than 1.0. These estimates indicate that having a middle school or higher education substantially increases the likelihood of survival at old ages, as compared to those Chinese elderly whose education levels were below middle school.

Secondly, the values of $RS_{\geq \text{middle school}}$ for the female elderly in all age groups are higher than those for their male counterparts. This indicates that an education of middle school or higher may have greater effects on female longevity.

Table 9.1 Ratio of survival for *male* elderly by educational attainment

Survive from age group in 1982 to age group in 1998	RS$_{\geq\text{middle school}}$	RS$_{ij}$				
	Middle school and above/ below middle school	College/ high school	High school/ middle school	Middle school/ primary school	Primary school/ illiteracy	Illiteracy/ literacy
64–68 ~ 80–84	3.21	3.19	3.17	1.27	1.43	0.53
69–73 ~ 85–89	3.16	2.84	2.76	1.40	1.33	0.58
74–78 ~ 90–94	2.92	2.16	2.39	1.50	1.30	0.61
79–83 ~ 95–99	3.30	2.70	2.00	1.71	1.34	0.57
84–88 ~ 100–105	1.99	5.74	0.86	1.41	1.15	0.76

Source: 1998 CLHLS and the 1 percent census data of the 22 provinces in 1982

Thirdly, 38 out of 40 estimates of RS$_{ij}$ (College/High School; High School/ Middle School; Middle School/Primary School; Primary School/ Illiterate) for male and female elderly in various age groups are greater than 1.0. All of the estimates of RS$_{ij}$ (Illiterate/Literate) are much less than 1.0. These results show that in China the higher the education level, the larger the likelihood of survival at the oldest old ages.

9.6 Concluding Remarks

This study has shown that educational attainment is positively and substantially associated with survivorship at old ages in China. The overall association is stronger among the female elderly than among the male elderly; moreover, the higher the education level, the greater the likelihood of survival at old ages. In fact, education may affect elderly survival through other socio-economic variables. For example, education not only enhances income-earning abilities via occupation stratification, but also increases people's knowledge and practice of health habits, as well as their

Table 9.2 Ratio of Survival for *female* elderly by educational attainment

Survive from Age Group in 1982 to Age Group in 1998	RS$_{\geq\text{middle school}}$	RS$_{ij}$				
	Middle school and above/below middle school	College/ high school	High school/ middle school	Middle school/ primary school	Primary school/ illiteracy	Illiteracy/ literacy
64–68 ~ 80–84	7.22	2.69	2.07	1.94	2.21	0.31
69–73 ~ 85–89	8.99	1.32	1.37	3.59	2.15	0.30
74–78 ~ 90–94	4.65	1.10	1.00	1.87	2.54	0.35
79–83 ~ 95–99	6.07	1.33	1.56	3.68	1.27	0.48
84–88 ~ 100–105	3.00	0.43	4.44	0.70	2.40	0.40

Source: 1998 CLHLS and the 1 precent census data of the 22 provinces in 1982

access to medical services. All these factors contribute significantly to elderly health and survival, especially so in less developed societies such as China where the "institutional benefits" and "institutional protection" tend to differ between different occupational groups, which are largely determined by education levels.

The analysis presented in this chapter has two major limitations. First, since we use the FAD method, the cohort data represent a univariate analysis to ascertain the de facto association between educational attainment and survival while controlling for age and sex. The analysis does not determine whether the association of educational attainment with healthy survival is due to education itself, or whether it is caused or mediated by other factors. Second, the cohorts analyzed in this chapter are not based on information on complete birth cohorts due to the fact that those cohort members who died before age 64 are not included. Thus, it is not appropriate to draw conclusions about the association of educational attainment with survival at middle-ages.

References

Brown, A.F., S.L. Ettner, J. Piette, M. Weinberger, E. Gregg, M.F. Shapiro, A.J. Karter, M. Safford, B. Waitzfelder, P.A. Prata, and G.L. Beckles (2004), Socioeconomic position and health among persons with diabetes mellitus: A conceptual framework and review of the literature. *Epidemiologic Reviews* 26, pp. 63–77

Daly, M.C., G.J. Duncan, P. McDonough, and D.R. Williams (2002), Optimal indicators of socioeconomic status for health research. *American Journal of Public Health* 92 (7), pp. 1151–1157

Davey Smith, G., C. Hart, D. Blane, C. Gillis, V. Hawthorne (1997), Lifetime socioeconomic position and mortality: Prospective observational study. *British Medical Journal* 314 (7080), pp. 547–552

Feinstein J.S. (1993), The relationship between socioeconomic status and health: A review of the literature. *Milbank Quarterly* 71 (2), pp. 279–322

Frank, J.W., R. Cohen, I. Yen, J. Balfour, and M. Smith (2003), Socioeconomic gradients in health status over 29 years of follow-up after midlife: The Alameda County Study. *Social Science and Medicine* 57 (12), pp. 2305–2323

Grundy, E. and G. Holt (2001), The socioeconomic status of older adults: How should we measure it in studies of health inequalities? *Journal Epidemiology and Community Health* 55 (12), pp. 895–904

Halfon, N. and M. Hochstein (2002), Life course health development: An integrated framework for developing health, policy, and research. *The Milbank Quarterly* 80 (3), pp. 433–479

Hertzman, C., C. Power, S. Matthews, and O. Manor (2001), Using an interactive framework of society and lifecourse to explain self-rated health in early adulthood. *Social Science and Medicine* 53, pp. 1575–1585

Kim, S., M. Symons, and B.M. Popkin (2004), Contrasting socioeconomic profiles related to healthier lifestyles in China and the United States. *American Journal of Epidemiology* 159 (2), pp. 184–191

Laaksonen, M., O. Rahkonen, P. Martikainen, and E. Lahelma (2005), Socioeconomic position and self-rated health: The contribution of childhood socioeconomic circumstances, adultsocioeconomic status, and material resources. *American Journal of Public Health* 95 (8), pp. 1403–1409

Lahelma, E., P. Martikainen, M. Laaksonen, and A. Aittomaki (2004), Pathways between socioeconomic determinants of health. *Journal of Epidemiology and Community Health* 58 (4), pp. 327–332

Matthews, R.J., L.K. Smith, R.M. Hancock, C. Jagger, and N.A. Spiers (2005), Socioeconomic factors associated with the onset of disability in older age: A longitudinal study of people aged 75 years and over. *Social Science and Medicine* 61(7), pp. 1567–1575

Osler, M., A.N. Andersen, G.D. Batty, and B. Holstein (2005), Relation between early life socioeconomic position and all cause mortality in two generations: A longitudinal study of Danish men born in 1953 and their parents. *Journal of Epidemiology and Community Health* 59, pp. 38–41

Power, C., E. Hyppönen, and G. Davey Smith (2005), Socioeconomic position in childhood and early adult life and risk of mortality: A prospective study of the mothers of the 1958 British birth cohort. *American Journal of Public Health* 95 (8), pp. 1396–1402

Research Group of Healthy Longevity in China (2000). *Data collections of the Healthy Longevity Survey in China 1998*. Beijing: Peking University Press

Rogot, E., P.D. Sorlie, N.J. Johnson, and C. Schmitt (1992), *A mortality study of 1.3 million persons by demographic, social, and economic factors: 1979–1985 follow-up*. Publication No 92-3297. National Institutes of Health, PHS, DHHS. Bethesda, MD: NIH

Zeng, Y. and J.W. Vaupel (2004), Association of late childbearing with healthy longevity among the oldest-old in China. *Population Studies* 58 (1), pp. 37–53

Chapter 10
Analysis of Health and Longevity in the Oldest-Old Population—A Health Capital Approach

Zhong Zhao

Abstract Using the Chinese Longitudinal Healthy Longevity Survey, we study health in the oldest old population. Our study is based on the Grossman model, which suggests that health reflects the stock of health capital. Current inputs into the health production function, and contemporary changes of behavior and life style will only have an incremental effect on the stock of health capital, which is determined by health history. Our study supports this view. Furthermore, our results suggest that besides aging, there are other important factors contributing to poor health as measured by the Katz Index of Activities of Daily Living, and mortality, such as gender, urban residence, and marital status. Socioeconomic status, such as the financial resources and education level of the individual and of his/her spouse plays an insignificant role in the health of the oldest-old. There is an inverse relationship between health and risky behavior.

Keywords Decreasing degrees of independence, Environmental variables, Explanatory variables, Grossman Model, Health capital, Health endowments, Health outcome, Instrumental variables, Katz Index of ADL, Life style, Longevity, Mortality, Oldest-old, Probit model, Psychological factors, Socioeconomic status, Survival status

10.1 Introduction

Research on the relationship between socioeconomic status and health is an important and active research area in economics and the other social sciences. The gradient in health, the phenomenon that wealthier people are healthier, seems to have attracted a great deal of attention from researchers.

In economics, health is widely considered an important component of human capital. Since the seminal work of Grossman (1972), the Grossman Model has

---

Z. Zhao
Institute for the Study of Labor (IZA), Schaumburg-Lippe-Str. 5–9 D-53113 Bonn, Germany
e-mail: zhao@iza.org

become standard in studying health demands and determinants. Treating health as capital is the key insight of Grossman (1972). Health status—the stock of health—depends on two key factors: investing in health and depreciating health.

Applying the Grossman Model, economists have carried out numerous empirical studies of health in working adults and children (for example, see Case et al. 2002; Currie and Stabile 2003; Dustmann and Windmeijer 2000; Erbsland et al. 1995; Sickles and Yazbeck 1998; Wagstaff 1986, 1993). These studies have all added to our understanding of the socioeconomic health gradient.

In this chapter, we investigate the health of the oldest-old, i.e. people who are older than 80.[1] Though there are many studies of the health of the aging population (see Smith and Kington (1997) for a survey), few studies have focused on the oldest-old. Zeng et al. (2001) have observed that this is mainly due to the lack of data. Fortunately, the Chinese Longitudinal Healthy Longevity Survey (CLHLS), one of the largest surveys of the oldest-old population in the world (Koenig 2001), affords us a unique opportunity to explore issues related to the health of the oldest-old.

This chapter supplements previous studies of health in children, working adults, and the old. Understanding health in the aging population also has important policy implications. According to data from the most recent three Chinese censuses, there were 4.91, 5.57 and 6.96% people above the age of 65 in 1982, 1990 and 2000, respectively. The one-child policy contributed significantly to this trend[2] (see especially Chap. 1 in this book by Poston and Zeng). The aging issue has become an important topic in academia and in policy circles in China. A large share of resources has been spent on the aging population, and more will be needed.[3]

It is not surprising that we find that aging is a major contributor to poor health both in terms of dependence in activities of daily living (ADL) and in terms of mortality. However, other factors also play important roles in determining the health of the oldest-old. Among the many factors, we have found the following to be important: The effects of gender on ADL independence and mortality differ. Females tend to be more ADL dependent, but have a higher probability of survival. Oldest-old individuals living in urban areas are also more ADL dependent, but are less likely to die. The married oldest-old are more ADL independent and are more likely to survive. Socioeconomic status as defined by financial resources and the education level of the person and of his/her spouse all play an insignificant role in the health of the oldest old.

Our research also suggests that there exists reverse causality between health and risky behaviors, such as smoking and drinking. Without controlling for endogeneity, the coefficients for smoking and drinking have the wrong sign. After

---

[1] In this chapter, we define the population aged 65–79 as the old population, aged 80 and above as the oldest-old population. We refer to the old and oldest-old population collectively as the aging population.

[2] The one child policy is the family planning policy adopted by the Chinese government since the early 1980s. Loosely speaking, it means one couple can legally have only one child.

[3] In 1988, the oldest-old spent 25% of the Medicare budget in New York City, see Suzman et al. (1992: 6). This percentage is very likely growing along with the increase in life expectancy.

they are corrected for simultaneity bias by instrumental variable (IV) methods, the coefficients are consistent with theory and common wisdom.

We organize the remaining chapter as follows. Section 10.2 outlines the analytical framework based on the Grossman Model. Section 10.3 describes the dataset and descriptive statistics. Section 10.4 presents empirical results, and Section 10.5 presents the conclusion.

10.2 Theory

Economists have considered health as human capital for a long time. Mushkin (1962) believed that health and education were two important components of human capital. Becker (1964) regarded human capital as a consequence of long-term education, good health and good nutrition. Fuchs (1966) held a similar view.

Building on human capital theory, Grossman (1972) formulated a formal model to analyze health capital. The conceptual contribution of Grossman (1972) treats health as capital, where health status reflects the stock of health capital. Two key factors, the investment factor (investing in health) and the depreciation factor (depreciation of health capital) determine the stock of health, or health status. When the stock of health falls below a certain minimum level, people tend to die.[4]

However, unlike market goods or standard investments which can be bought from a store or stock market, it is not possible to purchase a unit of health. Conceptually, since the work of Grossman (1972), economists have modeled health as an output of household production (Becker 1965). In this model, according to Becker (1965: 495), households "combine time and market goods to produce more basic commodities that directly enter their utility functions" as stated by Becker. The main market goods used to produce health are health care services. This provides the theoretical foundation for the positive relationship between socioeconomic status and health because wealth and income determine budget constraints. People with higher socioeconomic status can afford to purchase more health services, which produces more health capital. China provides some basic health care services for its urban population; but even with universal health care the social gradient still exists (Decker and Remler 2004).

Education is complementary to health, and can improve knowledge of health practices among family members. Characteristics of other family members may also contribute to such knowledge. For example, more educated family members may be able to choose more qualified doctors, and may be more knowledgeable as to the harmful effects of risky behaviors. An educated individual can give advice to other family members. So, in theory, the education levels of all family members should have a positive effect on health.

---

[4] Vaupel (1998) notes living organisms and complicated equipment, e.g. cars, share a similarity in their trajectories of mortality at the advanced ages, which provides interesting evidence for health capital theory.

Life style, which is regarded as a key sociological aspect in studying health, can be accommodated in the Grossman framework. Life style, such as smoking, drinking, and going to the gym regularly affects health through two channels. One channel is household production technology which is captured in the health production process. Good/bad behavior will make the production of health more/less efficient. The other channel is the rate of depreciation. Good/bad behavior will decrease/increase the rate of depreciation, and thus will deplete the stock of health capital slower/faster. This channel is captured in the consumption of health.

It is important to note that one of the fundamental implications of the Grossman Model is that what really matters is health 'stock,' which is determined by the person's entire health history. Current inputs to the health production function and contemporary changes of behavior and life style will only have an incremental effect on the stock of health capital but are unlikely to change the stock of health significantly.

Though the Grossman Model was formulated based on working aged adults, its conceptual contributions are also relevant for old age groups. However, the model has different implications for older versus younger populations. Childhood is a period for accumulating health capital; in this period, investing in health stock dominates; there is minimal if any depreciation. For the aging population, especially the oldest-old population under study here, depreciation of health stock, rather than investment, dominates. The health status of the oldest-old is mainly dictated by the stock of health capital and the rate of depreciation. Health investment factors have a smaller impact. Current factors, such as current income or a recent change in risky behavior are unlikely to have a big influence, but historical factors, such as past life style, permanent income, and education level, should exhibit larger effects.[5]

10.3 Data

10.3.1 Data Set

The data set used in this chapter is the Chinese Longitudinal Healthy Longevity Survey (CLHLS). The CLHLS contains three waves of nationally representative panel data from 1998, 2000 and 2002. It covers 22 out of the 31 provinces in China.[6] In 1998 and 2000, for the most part, only people age 80 or above were included in the sample. In 2002, additional observations were added to the survey for individuals aged 65–79. In their 2001 paper, Zeng et al. describe the data set and survey design in detail and assess the quality of the data, especially age-reporting, concluding that

---

[5] Current income is the income for the current period including transitory income, and permanent income is income derived over the life-cycle.

[6] Xinjiang, Qinghai, Ningxia, Inner Mongolia, Tibet, Gansu, Guizhou, and Hainan are the nine provinces not in the sample. Excluding them from the survey is mainly due to the potential for inaccurate age-reporting, since all of these nine provinces have large proportions of minorities in their population; see Zeng et al. (2001).

"age-reporting in our 1998 survey is generally good" after comparing the CLHLS with Swedish data (also see the chapters in Part I of this book).

We use the 1998–2000–2002 panel data and 2002 cross-sectional data in our analysis. The 2002 cross-sectional data contain 16,064 observations. Among these 16,064 observations, 2,642 individuals were interviewed in 1998 and 2000, 3,674 were interviewed in 2000 only, and 9,748 were newly interviewed in 2002. Among the 9,748 new observations, 4,889 of the interviewees were 65–79 years old in 2002.

For the panel data, the 1998 data consist of 9,093 observations. Among them, 4,831 were re-interviewed in the 2000 survey, 3,368 were deceased before the 2000 interview and information on them was collected, and the rest of the observations (894) were lost to follow-up. In 2002, among the 4,831 interviewed in 2000, 2,642 survived and were re-interviewed, 1,604 were deceased and their information was collected, and 585 were lost to follow-up.

We restrict our analysis to those of the Han nationality to avoid possible inaccurate age-reporting of other ethnic groups. We also exclude the oldest-old living in nursing homes to circumvent potential systematic differences between the oldest-old living at home and the oldest-old living in nursing homes. There were 5.5% non-Han oldest-old and 4.6% nursing home residing oldest-old in the sample.

10.3.2 Health Outcome Variables

One of the major challenges in studying health is its measurement. One of the most popular and valid health measurements for the aging population is the Katz Index of Activities of Daily Living (ADL Index) (Katz et al. 1970).

The CLHLS survey administered a slightly modified Katz questionnaire (see table 10.1 in Katz et al. 1970), which contained questions on six categories of activities of daily living: bathing, dressing, toileting, transferring, continence and feeding.[7] Given the available information in the data set and the good reputation of the ADL Index, we construct and use the ADL Index as one measurement of health. The ADL Index ranges from Category A to Category G with decreasing degrees of independence in ADL functioning. These six categories do not exhaust all possible combinations. There is an "Other" category, which is more ADL dependent than Category A and Category B, and less dependent than Category G, but is incomparable with Categories C and D.[8]

---

[7] The only difference between table 10.1 in Katz et al. (1970) and the questionnaire in the CLHLS is item 2, dealing with feeding. The CLHLS uses "feeds self, with some help" instead of "feeds self, except for getting assistance in cutting meat or buttering bread," which is more consistent with Chinese eating habits.

[8] Katz et al. (1970) state there are usually less than 5% belonging to the "Other" Category. In our calculations based on 2002 cross-sectional data, we do not find any observation belonging to the category of "Other" in the CLHLS.

Table 10.1 Characteristics of Old and Oldest-Old in 2002

Label	Whole Sample		Age: 65–79 Sub sample		Age: 80–105 Sub Sample Male and Female		Male		Female	
	Mean	St. Er.	Mean	St. Er.	Mean	St. Er.	Mean	St. Er.	Mean	St. Er.
Age	85.98	11.56	71.90	4.28	92.44	7.34	90.29	6.77	93.86	7.36
Urban	0.46	0.50	0.45	0.50	0.46	0.50	0.47	0.50	0.46	0.50
Born in urban area	0.15	0.36	0.17	0.38	0.15	0.35	0.15	0.36	0.14	0.35
Female	0.57	0.50	0.50	0.50	0.60	0.49	0.00	0.00	1.00	0.00
Have own bedroom	0.87	0.33	0.90	0.30	0.86	0.35	0.90	0.30	0.83	0.37
Live alone	0.14	0.35	0.13	0.34	0.15	0.36	0.15	0.35	0.15	0.36
Self-reported quality of life: excellent	0.14	0.35	0.14	0.35	0.14	0.35	0.15	0.35	0.13	0.34
Self-reported quality of life: good	0.46	0.50	0.44	0.50	0.47	0.50	0.46	0.50	0.48	0.50
Self-reported quality of life: so so	0.33	0.47	0.35	0.48	0.31	0.46	0.33	0.47	0.30	0.46
Self-reported quality of life: poor	0.06	0.24	0.06	0.23	0.07	0.25	0.06	0.23	0.07	0.26
Self-reported quality of life: very poor	0.01	0.10	0.01	0.08	0.01	0.10	0.01	0.09	0.01	0.10
Self-reported health status: excellent	0.11	0.31	0.13	0.34	0.10	0.30	0.10	0.31	0.09	0.29
Self-reported health status: good	0.37	0.48	0.37	0.48	0.38	0.48	0.40	0.49	0.36	0.48
Self-reported health status: so so	0.35	0.48	0.35	0.48	0.35	0.48	0.34	0.47	0.36	0.48
Self-reported health status: poor	0.15	0.36	0.14	0.35	0.16	0.37	0.14	0.35	0.17	0.38
Self-reported health status: very poor	0.01	0.12	0.01	0.10	0.02	0.13	0.01	0.11	0.02	0.14
Self-reported change of health: better	0.11	0.31	0.13	0.34	0.10	0.30	0.10	0.30	0.10	0.30
Self-reported change of health: same	0.49	0.50	0.54	0.50	0.46	0.50	0.47	0.50	0.45	0.50
Self-reported change of health: worse	0.40	0.49	0.33	0.47	0.44	0.50	0.43	0.49	0.45	0.50
Looking on the bright side	0.77	0.42	0.78	0.41	0.77	0.42	0.81	0.39	0.74	0.44
Feel fearful or anxious	0.05	0.22	0.05	0.21	0.06	0.23	0.04	0.19	0.07	0.25
Feel lonely and isolated	0.08	0.28	0.06	0.24	0.10	0.29	0.08	0.28	0.11	0.31
Be happy as when you were younger	0.43	0.49	0.46	0.50	0.41	0.49	0.43	0.50	0.39	0.49
Feel useless with age	0.26	0.44	0.21	0.41	0.29	0.45	0.24	0.43	0.33	0.47
Eat meat/fish/egg often	0.69	0.46	0.71	0.46	0.69	0.46	0.71	0.45	0.67	0.47
Drink boiled water	0.96	0.19	0.96	0.20	0.96	0.19	0.97	0.18	0.96	0.19
Used tap water at age 60	0.28	0.45	0.39	0.49	0.23	0.42	0.26	0.44	0.21	0.41
Use tap water now	0.60	0.49	0.60	0.49	0.59	0.49	0.61	0.49	0.58	0.49
Smoke now	0.19	0.39	0.27	0.44	0.15	0.36	0.28	0.45	0.07	0.25
Smoked before	0.34	0.47	0.40	0.49	0.32	0.47	0.57	0.50	0.15	0.36
Drink now	0.20	0.40	0.24	0.43	0.19	0.39	0.29	0.45	0.13	0.33
Drank before	0.31	0.46	0.31	0.46	0.31	0.46	0.49	0.50	0.20	0.40
Exercise now	0.32	0.47	0.41	0.49	0.28	0.45	0.39	0.49	0.20	0.40
Exercised before	0.37	0.48	0.35	0.48	0.38	0.49	0.49	0.50	0.31	0.46
Engage in social activities	0.13	0.34	0.23	0.42	0.09	0.29	0.15	0.36	0.05	0.22

Table 10.1 (Continued)

Label	Whole Sample		Age: 65–79 Sub sample		Age: 80–105 Sub Sample Male and Female		Male		Female	
	Mean	St. Er.	Mean	St. Er.	Mean	St. Er.	Mean	St. Er.	Mean	St. Er.
Years of education	2.06	3.52	2.92	3.88	1.66	3.26	3.26	4.00	0.61	2.05
White collar employment	0.10	0.29	0.14	0.35	0.07	0.26	0.15	0.36	0.02	0.15
Have pension	0.21	0.41	0.31	0.46	0.16	0.37	0.31	0.46	0.06	0.24
Money is enough for expenses	0.81	0.39	0.82	0.38	0.81	0.39	0.83	0.38	0.80	0.40
Main financial resource is self	0.29	0.46	0.54	0.50	0.18	0.38	0.33	0.47	0.08	0.27
Family income divided by 1000	3.44	3.62	3.62	3.70	3.36	3.57	3.61	3.75	3.19	3.45
Present marital status	0.31	0.46	0.62	0.49	0.18	0.38	0.35	0.48	0.06	0.24
Years of education, spouse	2.09	3.45	2.79	3.76	1.77	3.24	1.00	2.64	2.28	3.50
White collar, spouse	0.07	0.26	0.12	0.32	0.05	0.22	0.03	0.17	0.06	0.24
Taken care by close relatives	0.93	0.25	0.95	0.21	0.93	0.26	0.93	0.26	0.92	0.27
Had enough medical care at age 60	0.86	0.34	0.90	0.30	0.84	0.36	0.88	0.33	0.82	0.38
Had enough medical care at childhood	0.41	0.49	0.41	0.49	0.42	0.49	0.45	0.50	0.40	0.49
Medical cost paid by government	0.12	0.33	0.19	0.39	0.09	0.29	0.17	0.38	0.04	0.19
Often went to bed hungry as child	0.65	0.48	0.65	0.48	0.65	0.48	0.63	0.48	0.66	0.47
Mother lived longer than 80	0.26	0.44	0.32	0.47	0.24	0.43	0.26	0.44	0.23	0.42
Father lived longer than 80	0.16	0.36	0.19	0.39	0.14	0.35	0.17	0.37	0.13	0.33
1+ siblings live longer than 80	0.37	0.48	0.19	0.39	0.45	0.50	0.49	0.50	0.43	0.50

Source: Author's calculation from 2002 cross-sectional data of CLHLS.
These descriptive statistics are calculated without considering the sampling weight.
In the CLHLS, the oldest-old population is over-sampled.

According to the author's calculations (not reported here) using the 2002 cross-sectional data for the old population, 97% of their indices are in the "A" Category. This percentage decreases to 76% for oldest-old population. The ADL Index is not significantly different across gender for the old population. But for the oldest-old population, females are more dependent in daily living activities than are males. Eighty one percent of the male oldest-old are in the "A" category, versus only 70% for females. Nonetheless, it is remarkable that a majority (76%) of the oldest-old are still totally ADL independent.

The old population becomes more dependent in activities of daily living as they age. At age 65, males and females have almost identical degrees of independence; however, females become more dependent than males with age (Fig. 10.1).

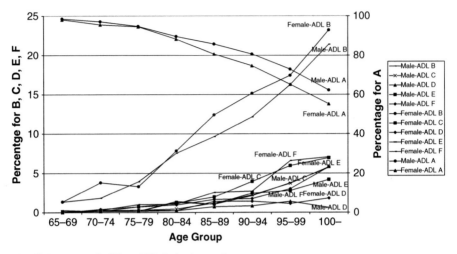

Fig. 10.1 Age trend of Katz ADL Index by gender.
Note: For Female and Male ADL A please refer to secondary y axis

The ADL Index is also a good indicator of mortality. At each age group, the group that survives has a better health status measured by the ADL Index.[9] We will use 2002 cross-sectional data to study ADL as a measure of health.

Another health outcome variable used in this chapter is mortality. To measure mortality, we use 1998–2000–2002 panel data. Our focus is the 2000 wave, and we only consider mortality between the 2000 and 2002 waves for those who were first interviewed in 1998. For the 2002 wave, we created a survival status for those still alive in 2000. Information in the 1998 wave allows us to control for historical factors.

10.3.3 Explanatory Variables

Table 10.1 compares characteristics of the old and oldest-old populations. The average age of the old and oldest old populations is 72 and 92 respectively.[10] The distributions of self-reported quality of life and self-reported health are very similar for these two groups, but the oldest-old report more loneliness, less happiness and more uselessness than the old population.

Compared with the old population, the oldest-old are also less likely to smoke, drink, exercise, and to participate in current social activities. In their earlier years, their drinking and exercise rates were similar to those in the old population. The oldest-olds are less educated, are less likely to have had a white collar job during their careers, have a lower pension coverage rate, are less self-sufficient financially, and have less family income.

---

[9] The results are not reported here, but are available from the author on request.

[10] The descriptive statistics are calculated without considering the sampling weights.

However, the oldest-old and the old share comparable characteristics on whether they live in an urban area, drink boiled water, use tap water, have enough money to cover expenses, have had enough medical services in childhood, and whether as children they often went to bed hungry. Table 10.1 summarizes the explanatory variables by gender for the oldest-old population. The average age for females was 94, and the average age for males was 90. Females were less educated, less likely to have had a white collar job, and have a lower pension coverage rate. The proportion of females, who smoke now, smoked in the past, drink now, drank in the past, exercise now, and exercised in the past are all lower compared with the male population.

Table 10.2 summarizes characteristics of the oldest-old in 2000 by their survival status in 2002. The deceased are 4.5 years older than those who survived, on

Table 10.2 Characteristics of the Survived and the Deceased Oldest-Old in 2000

Label	All		Survived in 2002		Deceased in 2002	
	Mean	St. Er.	Mean	St. Er.	Mean	St. Er.
Katz ADL Index in 2000	5.48	1.23	5.71	0.89	5.04	1.62
Katz ADL Index in 1998	5.74	0.85	5.83	0.63	5.56	1.11
Age	92.15	7.15	90.48	6.73	94.95	6.96
Urban residence	0.59	0.49	0.62	0.49	0.55	0.50
Born in urban area	0.15	0.36	0.17	0.38	0.12	0.32
Female	0.58	0.49	0.56	0.50	0.60	0.49
Live alone	0.13	0.33	0.14	0.35	0.11	0.31
Self-reported quality of life: excellent	0.21	0.41	0.23	0.42	0.18	0.38
Self-reported quality of life: good	0.45	0.50	0.43	0.50	0.49	0.50
Self-reported quality of life: so so	0.29	0.45	0.30	0.46	0.28	0.45
Self-reported quality of life: poor	0.04	0.20	0.04	0.19	0.05	0.21
Self-reported quality of life: very poor	0.01	0.08	0.01	0.08	0.01	0.08
Self-reported health status: excellent	0.15	0.36	0.16	0.37	0.13	0.33
Self-reported health status: good	0.39	0.49	0.42	0.49	0.33	0.47
Self-reported health status: so so	0.33	0.47	0.31	0.46	0.37	0.48
Self-reported health status: poor	0.12	0.32	0.09	0.29	0.16	0.37
Self-reported health status: very poor	0.01	0.11	0.01	0.10	0.02	0.13
Looking on the bright side	0.78	0.41	0.81	0.39	0.73	0.44
Feel fearful or anxious	0.04	0.20	0.03	0.18	0.05	0.23
Feel lonely and isolated	0.07	0.26	0.06	0.24	0.08	0.28
Be happy as when you were younger	0.39	0.49	0.43	0.49	0.33	0.47
Feel useless with age	0.20	0.40	0.18	0.39	0.24	0.43
Eat meat/fish/egg often	0.55	0.50	0.57	0.50	0.51	0.50
Drink boiled water	0.96	0.21	0.96	0.20	0.95	0.22
Used tap water at age 60	0.20	0.40	0.23	0.42	0.16	0.36
Use tap water now	0.50	0.50	0.53	0.50	0.46	0.50
Smoke now	0.16	0.37	0.18	0.38	0.13	0.34
Smoked before	0.33	0.47	0.33	0.47	0.34	0.48
Drink now	0.21	0.40	0.21	0.41	0.20	0.40
Drank before	0.32	0.47	0.35	0.48	0.27	0.44
Exercise now	0.35	0.48	0.41	0.49	0.25	0.43
Exercised before	0.32	0.47	0.35	0.48	0.27	0.44
Years of education	1.84	3.41	2.11	3.64	1.38	2.92

Table 10.2 (Continued)

Label	All		Survived in 2002		Deceased in 2002	
	Mean	St. Er.	Mean	St. Er.	Mean	St. Er.
White collar	0.09	0.29	0.12	0.32	0.05	0.22
Money is enough for expenses	0.84	0.36	0.84	0.36	0.85	0.36
Main financial resource is self	0.20	0.40	0.24	0.43	0.14	0.35
Family income divided by 1000	3.58	4.23	3.69	3.86	3.39	4.80
Present marital status	0.19	0.39	0.23	0.42	0.10	0.31
Years of education, spouse	1.95	3.35	2.20	3.63	1.54	2.78
White collar, spouse	0.08	0.26	0.10	0.30	0.04	0.19
Taken care by close relatives	0.92	0.28	0.91	0.29	0.93	0.26
Had enough medical care at age 60	0.95	0.22	0.95	0.22	0.95	0.23
Had enough medical care at childhood	0.66	0.47	0.66	0.47	0.66	0.47
Medical cost paid by government	0.11	0.32	0.14	0.35	0.07	0.25
Often went to bed hungry as child	0.55	0.50	0.55	0.50	0.56	0.50
Mother lived longer than 80	0.33	0.47	0.32	0.47	0.35	0.48
Father lived longer than 80	0.19	0.39	0.19	0.39	0.19	0.39
1+siblings live longer than 80	0.39	0.49	0.39	0.49	0.38	0.49

Source: Author's calculation from 2000 and 2002 panel data of the CLHLS.
These descriptive statistics are calculated without considering the sampling weight.
In the CLHLS, the oldest-old population is over-sampled.

average. The survived oldest-old self-report a higher quality of life and a better health status. Life style and behaviors, for example, smoking, drinking and exercising, are different for the two groups.

10.4 Results[11]

In this section, we analyze the relationship between socioeconomic status, life style, risky behaviors, and health. Health is measured by the ADL Index and by mortality. We use an ordered probit model to study the ADL Index and a binary probit model to study mortality. Besides socioeconomic status, we also control for psychological factors, life style variables, and environmental variables.

We address genetic differences in health endowments by including dummy variables to represent if the oldest-old has/had a mother, father or a sibling living longer than 80 years. We use an instrumental variable (IV) approach to control for simultaneity bias that arises from potential reverse causality between health and choice of life style.[12]

---

[11] We only present in this chapter the one set of results with the largest number of control variables. Results from other specifications are available from the author in a longer version of this manuscript.

[12] Moffitt (2005) provided a non-technical discussion on the IV approach targeted to demographers.

10.4.1 The Katz ADL Index

The Katz ADL Index divides people into six-categories from most independent to least independent in daily living. We use the ADL Index as a measurement of health, and estimate an ordered probit model to investigate the determinants of ADL independence of the oldest-old.[13]

Table 10.3 presents estimates from different model specifications based on the 2002 cross-sectional data. Aging is one of the main factors significantly related to health deterioration. Being female is negatively correlated with ADL independence, but the gender effect disappears when we introduce the additional covariates.

Current family income plays an insignificant role. Whether the oldest-old have a pension or not, have income enough to meet expenditures, and whether they provide their main financial resources, do not appear to be important factors.

Nonetheless, the oldest-old have better health if they have their own bedroom. One interpretation of this finding is that having your own bedroom captures the permanent income and the financial capacity of the whole family. From a health capital theory perspective, permanent income instead of current or transitory income affects the stock of health capital. Family resources, not personal resources, constrain household production. The education level of the oldest-old and of her/his spouse have insignificant albeit positive effects on health.

The effects of current medical care services and of medical care services at age 60 are insignificant, but there is a significant and positive effect of medical care services at childhood that persists into late-life. The oldest-old, who are married or who are taken care of by close relatives, such as by a spouse, their children, or in-laws, tend to have better health.

Among feelings of loneliness, happiness and uselessness, the feeling of uselessness is a good predictor of poor health. Whereas self-reported quality of life is insignificant, it becomes significant when we run the model separately for males and females. For males, a higher quality of life relates with better health, but the relationship for females is reversed. We do not have a good explanation for this finding.

At first blush, the life style and environmental variables seem to have the wrong signs. The coefficients for having tap-water now and having tap water at age 60 are both significantly negative. Although the coefficients for smoking and drinking in the past are negative, the coefficients for smoking and drinking in the present are both significantly positive.

In terms of the coefficients for tap-water, it is worth noting that the coefficients for living in an urban area are also significantly negative. This is consistent with the findings of Xu and Gu (2001), which show that the urban oldest-old are more ADL dependent than their rural counterparts. So the coefficients for tap-water may be biased by the confounding effect that oldest-old in urban areas are also more likely to have tap water.

---

[13] We use a scale from 1 to 6 to represent the oldest-old from least independent to most independent. A positive coefficient means a positive relationship between the explanatory variable and health status.

Table 10.3 Ordered Probit Model for Katz ADL Index

Label	Ordered Probit						Instrumental Variable Method					
	All		Male		Female		All		Male		Female	
	Coef.	P>|z|	Coef.	P>|z|	Coef.	P>|z|	Coef.	P>|z|	Coef.	P>|z|	Coef.	P>|z|
	(1)		(2)		(3)		(4)		(5)		(6)	
Age	−0.044	0.000	−0.041	0.000	−0.046	0.000	−0.029	0.000	−0.022	0.000	−0.031	0.000
Urban residence	−0.183	0.002	−0.339	0.001	−0.091	0.212	−0.102	0.013	−0.114	0.038	−0.077	0.196
Born in urban area	0.076	0.319	0.105	0.374	0.025	0.805	−0.008	0.886	0.015	0.823	−0.081	0.334
Female	−0.056	0.412					−0.033	0.496				
Have own bedroom	0.142	0.032	0.406	0.001	0.044	0.579	0.114	0.025	0.236	0.002	0.059	0.382
Self-reported quality of life: excellent	0.002	0.996	1.172	0.021	−1.007	0.056	0.112	0.630	1.170	0.001	−0.543	0.086
Self-reported quality of life: good	−0.070	0.831	0.866	0.081	−0.969	0.063	0.055	0.811	1.046	0.002	−0.543	0.081
Self-reported quality of life: so so	−0.094	0.771	0.889	0.073	−1.031	0.048	0.022	0.924	1.038	0.002	−0.620	0.045
Self-reported quality of life: poor	−0.081	0.809	0.562	0.274	−0.777	0.146	−0.016	0.946	0.737	0.033	−0.439	0.173
Self-reported change of health: better	0.145	0.080	0.062	0.658	0.205	0.054	0.096	0.104	0.017	0.820	0.158	0.073
Self-reported change of health: same	0.292	0.000	0.168	0.051	0.380	0.000	0.227	0.000	0.076	0.115	0.337	0.000
Looking on the bright side	−0.054	0.383	0.066	0.549	−0.122	0.117	−0.040	0.381	0.049	0.441	−0.103	0.107
Feel fearful or anxious	−0.098	0.400	−0.248	0.304	−0.073	0.592	−0.023	0.798	−0.116	0.415	0.035	0.767
Feel lonely and isolated	0.053	0.600	0.057	0.742	0.003	0.979	0.006	0.936	0.045	0.660	−0.066	0.552
Be happy as when you were younger	0.003	0.946	0.040	0.643	−0.038	0.553	−0.012	0.744	0.003	0.944	−0.027	0.604
Feel useless with age	−0.268	0.000	−0.215	0.026	−0.314	0.000	−0.209	0.000	−0.150	0.008	−0.267	0.000
Eat meat/fish/egg often	−0.034	0.540	0.137	0.151	−0.107	0.133	−0.025	0.527	0.041	0.441	−0.060	0.296
Drink boiled water	−0.225	0.111	−0.203	0.437	−0.301	0.079	−0.135	0.137	−0.071	0.584	−0.210	0.094
Used tap water at age 60	−0.165	0.015	−0.124	0.254	−0.199	0.023	−0.082	0.097	−0.067	0.274	−0.083	0.262
Use tap water now	−0.121	0.036	−0.140	0.171	−0.117	0.103	−0.068	0.092	−0.047	0.379	−0.076	0.190
Smoke now	0.161	0.060	0.189	0.089	0.049	0.727	−0.063	0.506	−0.049	0.639	−0.047	0.780
Smoked before	−0.148	0.027	−0.150	0.106	−0.123	0.221						
Drink now	0.249	0.002	0.371	0.001	0.139	0.242	−0.020	0.777	0.115	0.175	−0.177	0.108
Drank before	−0.134	0.051	−0.052	0.586	−0.186	0.070						
Exercise now	0.487	0.000	0.472	0.000	0.484	0.000	0.208	0.001	0.189	0.019	0.226	0.022
Exercised before	−0.134	0.028	−0.136	0.152	−0.117	0.155						
Engage in social activities	0.342	0.000	0.389	0.003	0.293	0.058	0.155	0.008	0.151	0.014	0.165	0.125

10 Analysis of Health and Longevity in the Oldest-Old Population

Table 10.3 (Continued)

Label	Ordered Probit						Instrumental Variable Method					
	All		Male		Female		All		Male		Female	
	Coef.	P>|z|	Coef.	P>|z|	Coef.	P>|z|	Coef.	P>|z|	Coef.	P>|z|	Coef.	P>|z|
	(1)		(2)		(3)		(4)		(5)		(6)	
Years of education	0.000	0.968	−0.007	0.613	0.007	0.746	−0.003	0.666	−0.006	0.381	0.010	0.526
White collar employment	−0.133	0.239	−0.031	0.816	−0.215	0.401	−0.053	0.494	0.026	0.721	−0.228	0.266
Have pension	−0.052	0.716	−0.285	0.140	0.430	0.099	0.032	0.733	−0.118	0.229	0.384	0.043
Money is enough for expenses	−0.030	0.672	−0.293	0.031	0.092	0.290	−0.047	0.366	−0.114	0.105	−0.017	0.820
Main financial resource is self	0.100	0.447	0.338	0.074	−0.222	0.277	0.027	0.732	0.152	0.081	−0.218	0.130
Family income	0.000	0.541	0.000	0.905	0.000	0.366	0.000	0.089	0.000	0.908	0.000	0.063
Present marital status	0.191	0.009	0.197	0.030	0.317	0.026	0.050	0.275	0.066	0.165	0.125	0.176
Years of education, spouse	0.006	0.520	0.000	0.988	0.005	0.641	0.004	0.551	−0.002	0.881	0.003	0.763
White collar, spouse	−0.234	0.037	−0.361	0.136	−0.199	0.124	−0.212	0.013	−0.235	0.114	−0.198	0.070
Taken care by close relatives	0.310	0.008	0.260	0.164	0.337	0.029	0.325	0.000	0.232	0.052	0.395	0.004
Have enough medical care now	0.055	0.516	−0.050	0.752	0.097	0.349	0.091	0.149	−0.023	0.799	0.158	0.071
Had enough medical care at age 60	0.044	0.522	0.234	0.053	−0.078	0.369	0.007	0.890	0.057	0.421	−0.064	0.364
Had enough medical care at childhood	0.115	0.036	0.175	0.050	0.082	0.251	0.067	0.080	0.044	0.358	0.089	0.125
Medical cost paid by government	−0.032	0.777	−0.117	0.405	0.125	0.575	−0.011	0.887	−0.118	0.136	0.175	0.311
Often went to bed hungry as child	0.091	0.100	0.201	0.025	0.012	0.868	0.032	0.405	0.062	0.210	0.007	0.902
Mother lived longer than 80	0.061	0.265	0.127	0.160	−0.015	0.828	0.048	0.213	0.073	0.134	0.000	0.999
Father lived longer than 80	−0.053	0.417	−0.070	0.500	−0.057	0.513	−0.033	0.477	0.012	0.832	−0.068	0.342
Sibling live longer than 80	−0.004	0.941	−0.032	0.691	0.030	0.619	0.011	0.740	−0.021	0.632	0.041	0.410
Constant							7.827	0.000	6.161	0.000	8.691	0.000
Cut–off point	−5.993		−4.463		−7.270							
Cut–off point	−5.699		−4.126		−6.985							
Cut–off point	−5.600		−4.042		−6.875							
Cut–off point	−5.349		−3.800		−6.612							
Cut–off point	−4.668		−3.057		−5.945							
Pseudo R2	0.098		0.129		0.089		0.101		0.104		0.101	
Number of Observation	3721		1628		2093		3721		1628		2093	

Estimation based on 2002 cross–sectional data of CLHLS.
We use (6, 5, 4, 3, 2, 1) to represent (A, B, C, D, E, F) in the Index of ADL

The mysterious positive sign for current smoking and drinking status (i.e. the paradox that smoking is good for your health) may well be caused by reverse causality, i.e. the current status of smoking and drinking are consequences of poor health in the past. Another possible explanation is that smokers and drinkers in poor health die first, and only the robust smokers and drinkers remain in the sample. It is well known that OLS estimates for both cases will be biased. We address this issue by using the IV approach. After we apply the IV method to the analysis of the status of smoking and drinking, and control for the endogeneity of these two variables, their coefficients become negative although insignificant,[14] which is consistent with theory and common wisdom.

Another reverse causality issue that perplexes economists is how poor health leads to lower socioeconomic status. For example, poor health may cause missing school days in childhood, which consequently lowers human capital, and leads to lower income thereafter. For those who are lucky enough to join the oldest-old club, it is unlikely that their health during childhood and adulthood were poor enough to significantly and negatively affect their socioeconomic status.

10.4.2 Mortality

Another health outcome variable analyzed is mortality. Mortality was studied in the interval between the 2000 and 2002 waves for those who were first interviewed in 1998. Information in the 1998 wave is used to control for historical factors. We use a probit model to study mortality. A positive coefficient indicates a positive relationship between the explanatory variable and the probability of death.

Table 10.4 summarizes the models' results. Like its effect on ADL independence, aging is an important factor contributing to death. Gender has differing effects on mortality and on the ADL Index. Being female tends to have a positive effect on the probability of survival, but a negative effect on ADL independence (see Table 10.4).

Similar to findings on ADL independence, the coefficients pertaining to current family income—having enough income to cover expenditures, and whether the respondent is his/her main financial resource—are insignificant. Unfortunately, we do not have information on whether the oldest-old individual has their own bedroom in the 2000 survey, and therefore we cannot compare the effect of this variable on ADL independence or its effect on mortality.

The educational level of the oldest-old and of her/his spouse has an insignificant effect on mortality.[15] For oldest-old females, having a white collar job or having a husband with a white collar job significantly reduces mortality. For males, only

---

[14] We use past status of smoking, drinking and exercising as instrumental variables.

[15] The finding that education is not important for the ADL Index and mortality is probably due to the small variation of education levels in the CLHLS. Average years of education are only 3.3 and 0.6 for males and for females, respectively.

10 Analysis of Health and Longevity in the Oldest-Old Population

Table 10.4 Probit and Instrumental Variables Models for Deceased Status in 2002

Label	Probit						Instrumental Variable Method					
	All		Male		Female		All		Male		Female	
	Coef.	P>|z|	Coef.	P>|z|	Coef.	P>|z|	Coef.	P>|z|	Coef.	P>|z|	Coef.	P>|z|
	(1)		(2)		(3)		(4)		(5)		(6)	
Index of ADL=E (Dependent)	0.273	0.295	−0.348	0.409	0.632	0.078	0.102	0.320	−0.041	0.874	0.204	0.146
Index of ADL=D	−0.192	0.529	−0.145	0.775	−0.226	0.562	−0.042	0.778	−0.030	0.935	−0.091	0.581
Index of ADL=C	0.019	0.933	−0.033	0.926	0.094	0.754	0.056	0.617	0.426	0.586	0.052	0.671
Index of ADL=B	−0.249	0.054	−0.505	0.026	−0.175	0.279	−0.022	0.812	0.176	0.785	0.025	0.756
Index of ADL=A (Independent)	−0.565	0.000	−0.727	0.000	−0.509	0.000	−0.082	0.558	0.132	0.860	−0.026	0.811
Age	0.042	0.000	0.029	0.000	0.054	0.000	0.013	0.000	0.000	0.978	0.015	0.000
Urban residence	−0.089	0.226	−0.119	0.304	−0.028	0.776	−0.033	0.332	−0.065	0.333	−0.020	0.679
Born in urban area	0.046	0.638	0.129	0.395	0.053	0.692	0.028	0.470	0.030	0.701	0.046	0.409
Female	−0.279	0.001					−0.118	0.277				
Live alone	−0.076	0.453	−0.115	0.500	−0.040	0.752	−0.044	0.456	−0.161	0.400	−0.017	0.790
Self-reported quality of life: excellent	0.615	0.232	4.730	0.000	0.602	0.286	0.151	0.314	0.729	0.500	0.128	0.505
Self-reported quality of life: good	0.842	0.099	4.914	0.000	0.880	0.114	0.201	0.192	0.825	0.452	0.197	0.303
Self-reported quality of life: so so	0.656	0.197	4.712	0.000	0.683	0.217	0.151	0.328	0.772	0.484	0.160	0.389
Self-reported quality of life: poor	0.867	0.095	5.326	0.000	0.572	0.321	0.221	0.181	0.803	0.383	0.169	0.393
Looking on the bright side	−0.087	0.279	0.020	0.875	−0.150	0.156	−0.021	0.521	0.062	0.637	−0.026	0.530
Feel fearful or anxious	0.198	0.250	0.180	0.535	0.235	0.294	0.048	0.430	0.060	0.677	0.077	0.369
Feel lonely and isolated	0.033	0.806	0.086	0.698	0.032	0.853	0.025	0.594	0.006	0.967	0.036	0.594
Be happy as when you were younger	−0.074	0.298	−0.143	0.175	−0.028	0.781	0.004	0.930	0.022	0.887	0.010	0.803
Feel useless with age	0.016	0.845	0.118	0.340	−0.064	0.557	−0.024	0.549	−0.033	0.790	−0.088	0.118
Eat meat/fish/egg often	−0.166	0.013	−0.195	0.058	−0.195	0.032	−0.035	0.367	−0.047	0.608	−0.026	0.551
Drink boiled water	0.147	0.353	0.285	0.280	0.093	0.652	0.067	0.266	0.093	0.545	0.042	0.591
Used tap water at age 60	−0.174	0.082	−0.218	0.155	−0.166	0.226	−0.045	0.184	−0.077	0.410	−0.048	0.334
Use tap water now	0.037	0.628	0.017	0.887	0.031	0.766	0.013	0.656	−0.003	0.964	0.008	0.868
Smoke now	−0.074	0.412	−0.174	0.121	0.146	0.353	0.058	0.922	−1.015	0.512	0.186	0.730

Table 10.4 (Continued)

Label	Probit						Instrumental Variable Method					
	All		Male		Female		All		Male		Female	
	Coef.	P>|z|	Coef.	P>|z|	Coef.	P>|z|	Coef.	P>|z|	Coef.	P>|z|	Coef.	P>|z|
	(1)		(2)		(3)		(4)		(5)		(6)	
Smoked before	0.084	0.260	0.097	0.367	0.032	0.767	0.068	0.675	0.268	0.367	0.082	0.617
Drink now	0.023	0.795	0.155	0.195	−0.146	0.276						
Drank before	−0.094	0.202	−0.028	0.787	−0.164	0.133						
Exercise now	−0.131	0.066	−0.147	0.146	−0.080	0.437	−0.413	0.303	−0.538	0.675	−0.629	0.075
Years of education	−0.002	0.850	−0.004	0.776	0.021	0.488	0.001	0.838	−0.004	0.565	0.019	0.134
White collar employment	−0.258	0.069	−0.084	0.590	−1.049	0.023	−0.062	0.190	−0.042	0.621	−0.222	0.075
Money is enough for expenses	0.101	0.297	0.024	0.869	0.181	0.181	0.047	0.230	0.076	0.571	0.078	0.153
Main financial resource is self	0.203	0.059	0.210	0.140	0.221	0.223	0.077	0.076	0.039	0.607	0.088	0.176
Family income	0.000	0.354	0.000	0.802	0.000	0.234	0.000	0.398	0.000	0.633	0.000	0.547
Present marital status	−0.454	0.000	−0.555	0.000	−0.218	0.285	−0.158	0.000	−0.171	0.051	−0.086	0.251
Years of education, spouse	−0.004	0.788	−0.029	0.287	0.000	0.983	0.001	0.869	−0.004	0.764	−0.001	0.910
White collar, spouse	−0.589	0.000	−0.828	0.023	−0.534	0.007	−0.128	0.008	−0.162	0.195	−0.126	0.059
Taken care by close relatives	0.328	0.015	0.474	0.027	0.303	0.090	0.154	0.352	−0.382	0.580	0.196	0.324
Have enough medical care now	0.027	0.862	−0.015	0.950	0.028	0.898	0.029	0.627	−0.035	0.810	0.029	0.718
Medical cost paid by government	−0.209	0.138	−0.236	0.167	0.132	0.643	−0.011	0.868	−0.112	0.320	0.159	0.148
Often go to bed hungry as child	−0.171	0.008	−0.308	0.002	−0.040	0.654	−0.062	0.013	−0.089	0.059	−0.028	0.478
1+ sibling live longer than 80	0.017	0.794	0.156	0.110	−0.072	0.424	0.001	0.984	−0.035	0.782	−0.021	0.530
Constant	−4.498	0.000	−7.471	0.000	−5.991	0.000	−0.904	0.087	0.322	0.812	−1.312	0.007
Pseudo R2	0.145		0.163		0.158		0.019		0.009		0.008	
Number of Observations	2001		908		1093		1994		902		1092	

Estimation based on 98-00-22 longitudinal data of CLHLS.
Dependent variable=1 means the respondent died between the 2000 and 2002 waves.

the occupation of wives was significant. The married oldest-old are more ADL independent and also have a higher probability of survival. Current medical care services do little to extend the life of the oldest-old.[16]

Unlike the findings for ADL independence, the oldest-old cared for by close relatives have a higher risk of death. Since the oldest-old are more likely to be taken care of by loved ones, reverse causality may also bias this estimate. Again we use the IV procedure to correct for possible simultaneity bias. After correction, the coefficient becomes insignificant. Whereas uselessness is a good predictor for ADL dependence, it is not for mortality. Self-reported quality of life is apparently a good predictor for male mortality, but not for female mortality.

The coefficients for the life style variables are insignificant but again have the wrong signs due to reverse causality. Controlling for endogeneity corrects the signs so that they are in-line with common wisdom. For example, the sign of the coefficient for smoking changes from negative to positive.

While the oldest-old in rural areas are less ADL dependent, they face a higher probability of death. This finding might be explained by the greater manual labor engaged in by the rural oldest-old, which is a benefit to their long-term health but which makes them more vulnerable to health shocks because of the lack of basic health care in rural areas.

10.5 Conclusion

In this chapter we used 1998–2000–2002 panel data and 2002 cross-sectional data to study health among the oldest-old. Our measurements for health are the Katz ADL Index and mortality. We find that aging is a major contributor to poor health both in terms of ADL independence and in terms of mortality, but there are other important factors, as well.

Gender has different effects on ADL independence and on mortality. Females tend to be more dependent in daily living, but have a higher probability of survival. The urban oldest-old are also more dependent in daily living, but are less likely to die. The married oldest-old are more independent in daily living and are more likely to survive. Socioeconomic status, such as financial resources and the educational levels of the oldest-old and his/her spouse play an insignificant role in the health of the oldest old. There seems to exist a reverse causality between health and choice of life style. After controlling for simultaneity bias by IV methods, the coefficients for risk behaviors are consistent with theory and common wisdom.

Our analytical framework is based on the Grossman Model. The fundamental implication of the Grossman Model is that what really matters is "stock." Current

---

[16] The effects of medical care services at age 60 and in childhood are also insignificant (results are not reported here). These findings might be affected by the very large portion of this sample with missing values on these two variables.

inputs into the health production function and contemporary changes of behavior and life style will only have an incremental effect on the stock of health capital, which is determined by the individual's entire history, but will unlikely change the stock of health significantly. Our research supports this view. We find that current family income plays an insignificant role. Nonetheless, an oldest-old individual will have better health if he has his own bedroom. The existence of a private bedroom captures the permanent income and the financial capacity of the whole family. The effects of current medical care services and of medical care services at age 60 are insignificant, but the significant positive effect of medical care services at childhood persists into late-life.

References

Becker, G.S. (1964), *Human capital*. New York: Columbia University Press for the National Bureau of Economic Research

Becker, G.S. (1965), A theory of the allocation of time. *Economic Journal* 75 (299), pp. 493–517

Case, A., D. Lubotsky, and C. Paxson (2002), Economic status and health in childhood: The origins of the gradient. *American Economic Review* 92 (5), pp. 1308–1334

Currie, J. and M. Stabile (2003), Socioeconomic status and child health: why is the relationship stronger for older children? *American Economic Review* 93 (5), pp. 1813–1823

Decker, S.L. and D.K. Remler (2004), How much might universal health insurance reduce socioeconomic disparities in health? a comparison of the US and Canada. *NBER Working Papers 10715*, National Bureau of Economic Research, Inc

Dustmann, C. and F. Windmeijer (2000), Wages and the demand for health-A life cycle analysis. *The Institute for Fiscal Studies*, Working Paper Series No. W99/20

Erbsland, M., W. Ried and V. Ulrich (1995), Health, health care, and the environment, econometric evidence from German micro data. *Health Economics* 4 (3), pp. 169–182

Grossman, M. (1972), On the concept of health capital and the demand for health. *Journal of Political Economy* 80 (20), pp. 223–255

Katz, S., T.D. Downs, H.R. Cash and R.C. Grotz (1970), Progress in development of the index of ADL. *Gerontologist* 10 (1), pp. 20–30

Koenig, R. (2001), Sardinia's mysterious male Methuselahs. *Science* 16, pp. 2074–2076

Moffitt, R.A. (2005), Remarks on the analysis of causal relationships in population research. *Demography* 42 (1), pp. 91–108

Mushkin, S.J. (1962), Health as an investment. *Journal of Political Economy* 70(5), pp. 129–157

Sickles, R.C. and A. Yazbeck (1998), On the dynamics of demand for leisure and the production of health. *Journal of Business and Economic Statistics* 16 (2), pp. 187–197

Smith, J.P. and R.S. Kington (1997), Race, socioeconomic status, and health in late life. In: L.G. Martin and B.J. Soldo (eds.): *Racial and ethnic differences in the health of older Americans*. Washington, DC: National Academy Press, pp. 106–162

Suzman, R.M., K.G. Manton, and D.P. Willis (1992), Introducing the oldest-old. In: R.M. Suzman, D.P. Willis, and K.G. Manton (eds.): *The oldest old*. New York: Oxford University Press, pp. 3–15

Vaupel, J. (1998), Demographic analysis of aging and longevity. *American Economic Review* 88 (2), pp. 242–247

Wagstaff, A. (1986), The demand for health: some new empirical evidence. *Journal of Health Economics* 5 (3), pp. 195–233

Wagstaff, A. (1993), The demand for health: an empirical reformulation of the Grossman Model. *Health Economics* 2 (2), pp. 189–198

Xu, Q. and D. Gu (2001), Comparing characteristics of health and mortality between rural and urban oldest-old in China. *Chinese Journal of Population Science*, Special Issue, pp. 83–88 (in Chinese)

Zeng, Y., J.W. Vaupel, Z. Xiao, C. Zhang and Y. Liu (2001), The healthy longevity survey and the active life expectancy of the oldest old in China. *Population: An English Selection* 13 (1), pp. 95–116

Chapter 11
The More Engagement, the Better? A Study of Mortality of the Oldest Old in China

Rongjun Sun and Yuzhi Liu

Abstract This chapter investigates the role of social engagement in mortality among the oldest old in China. Adopting the convoy network model, we differentiate engagement with close social ties (spouse and children) from other social activities. Weibull hazard models were employed to analyze the mortality risk of those aged 80 or above within a 2-year period between 1998 and 2000. While the results of the whole sample show significant effects of marital status and number of children alive on mortality without the interaction effect, the beneficial effect of social activities on mortality gradually diminished with age and was reversed at very old ages, when health status, health behaviors, and socio-demographic characteristics were controlled. But the results vary by place of residence and gender. The findings seem to suggest that more social engagement may not necessarily be better for the well-being of the elderly at very old ages.

Keywords Beneficial effect, Convoy network model, Disengagement theory, Identity accumulation hypothesis, Interaction, Mortality, Normal aging, Oldest-old, Social engagement, Social network, Socioemotional selectivity theory, Survival analysis, Weibull hazard models

11.1 Introduction

While the value of engaging in physical activities, including customary or habitual activities, for all aspects of well-being is well established in the literature (Belloc 1973; Berkman et al. 1983; Bygren et al. 1996; Cerhan et al., 1998; DiPietro 2001; Glass et al. 1999; Lennartsson and Silverstein 2001; Morgan and Clarke 1997; Morgan et al. 1991; Rowe and Kahn 1998; Welin et al. 1992), the importance of social engagement is less conclusive. Past research has demonstrated

R. Sun
Department of Sociology, Cleveland State University, 2121 Euclid Ave.,
Cleveland, OH 44115, USA
e-mail: r.sun32@csuohio.edu

that social integration, and social engagement in general, have positive impacts on physical and mental health, and on the survival of the elderly (Berkman and Breslow 1983; Berkman et al. 2000; Sabin 1993; Zunzunegui et al. 2003). Relating to others is usually seen as essential to well-being, while the lack of social ties may be viewed as a risk factor (Rowe and Kahn 1998). People in social relationships take on various roles. From a symbolic interactionist perspective, Thoits (1983) argued that individuals attached to social positions take on roles and identities that are defined by social expectations. These role requirements tend to give individuals a sense of purpose and meaning. She proposed an "identity accumulation hypothesis," where the possession of more roles or identities is related to higher levels of psychological well-being.

Disengagement theory, however, states that normal aging involves an inevitable withdrawal from social ties to prevent disruption to the social system (Cumming and Henry 1961). Although somewhat controversial, some studies have shown the relevance of this theory for the oldest old population. Johnson and Barer (1992) reported in their study of people aged 85 or above in San Francisco that less than 10 percent of their sample had weekly contact with siblings, and less than 30 percent had weekly contact with other relatives. Intensive interviews revealed that these elderly disengaged socially and psychologically by narrowing their social boundaries, loosening normative constraints, shifting their time orientation from the future to the present, and increasing introspection. Reductions in social contacts at very old ages may be attributed to declining physical function and the loss of family members all of which make it difficult to remain socially active. However, as some have cautioned, such reductions in social contacts should not be viewed merely as passive or involuntary. Instead, they may be a result of a lifelong selection process by which people strategically minimize their risks and maximize their well-being. Individuals constantly adapt their goals conditional on constraints in capacity and resources (Baltes 1997; Baltes and Carstensen 1996; Carstensen 1992). This idea is described in socioemotional selectivity theory (Carstensen 1991, 1992), which takes a life-span perspective that "views reductions in social activity in old age as reflective of the culmination of selection processes that begin early in life and have substantial adaptive value" (Carstensen 1991: 195). According to this theory, in old age, people selectively invest increasingly limited resources in their most valuable and intimate relationships and disregard the less important ones. Selectivity is an adaptive strategy to yield the greatest gain in well-being. This theme is also echoed in continuity theory (Atchley 1989). Individuals are believed to respond to changes by intentionally structuring their choices as they age, and they accept the fact that some of the changes brought about by normal aging can not be completely offset. It is regarded as *maladaptive* if individuals insist on continuing existing roles or social relationships when there is a loss of competence for doing so.

Positing a purposeful selection process suggests that it is important not to lump all social relationships together and treat them as equal when analyzing their impact on elderly well-being or survival. Instead, it is crucial to differentiate social relationships by their closeness to the individual. The convoy model (Kahn and Antonucci 1980), which distinguishes the structure of an individual's social network, serves such a purpose. The three concentric circles in the model represent three tiers of convoy members differentiated by their closeness to the individual at the center. The innermost circle consists of the closest ties, such as spouse and other close family members. These relationships are highly valued by the individual and tend to be stable throughout the years. The other two outer circles, which consist of others such as relatives, friends, neighbors, and co-workers, however, are more unstable and subjected to role changes of the individual. Kahn and Antonucci (1980) also pointed out that there are likely to be gains and losses in any of the circles of the convoy over an individual's life course. There are some empirical findings about the change in the composition of convoys over the life course. Using data from the General Social Survey, Morgan (1988) documented a downward trend in social network participation by age among those aged 60 and above even after available resources, such as income, education and health were controlled. Older respondents maintained fewer roles than younger respondents. They named fewer people with whom they discussed important matters or were in frequent contact, and relied more on family members and long-time relationships. Similarly, Lang and Carstensen's study (1994) showed that among Berlin residents aged 70–104 years, there was a significant negative correlation between social network size and age. The very old tended to limit social contacts to very intimate and close ties. They found a dramatic reduction in contacts with more distant social partners. In a study of the elderly in the United States, Carstensen (1992) found that at older ages, people became increasingly selective by maintaining the most rewarding relationships while reducing the number of less important relationships.

Based on selectivity and continuity theories, and the convoy model, it can be argued that one important reason for individuals to keep certain social relationships is because these ties have been persistently rewarding or beneficial. And these relationships are usually those that occupy the inner-most circle of the convoy model, like spouse and children. By the same token, the reason for individuals, especially the very old, to start disengaging from certain social relationships is because these ties have become less rewarding or the benefits have declined. These ties are usually those that occur in the outer circles of the convoy model, like friends or social organizations.

Based on the above discussions, two hypotheses may be proposed: (1) The beneficial effects of interacting with marital partners and the number of children alive, if any, are constant, and do not decline with age. (2) The beneficial effects of engaging in social activities with others diminish with age.

In addition, there are two conditional factors to consider: place of residence and gender. It is well-known that there are significant gaps between urban and rural areas in China in virtually all aspects of social development and living standards (Zeng et al. 2002). While in urban areas many retired elderly receive pensions and are covered by health insurance, in rural areas such benefits are rare, where most of the elderly rely on their families, especially their adult children for support. Furthermore, there are other differences that shape urban and rural living environments, such as the availability of health care facilities. Therefore, it is reasonable to perform analyses separately for the urban and rural samples.

Gender may also confound the effect of social relations, especially that of marriage. Marriage has been found to have benefits for both men and women (Waite 1995), but it is still controversial whether there is any gender difference in the allocation of such benefits. In other words, who benefits more from marriage, the wife or the husband? Gove (1972) introduced his sex-role theory when studying the relationship between mental illness and marital status. He argued that the stressful nature of women's roles in the society made marriage disadvantageous for women while advantageous for men. When he examined the relationship between mortality and marital status, for most of the causes of death, including suicide, accidents, and diseases, he found a consistent pattern: Although death rates were lower for both married men and women compared with the single, widowed, or divorced, the reduction in death rates was substantially higher for married men than for married women. However, more recent studies found no substantial difference between husbands and wives in their well-being (Mookherjee 1997; Pienta et al. 2000; Simon 2002). For example, Simon's study (2002) found that the benefits of marriage for depression apply equally to women and men.

In the Chinese context, there seem to be competitive forces that make gender differences in marital benefits uncertain. On the one hand, traditional values emphasize wives' submissiveness to their husbands (Freedman 1970). Wives are expected to take care of the well-being of all family members, including in-laws and children. It can be very stressful to engage in multiple family roles (wife, mother, and daughter-in-law) around the household; thus husbands are expected to benefit more from marriage than are wives. On the other hand, since the family is supposed to be the main arena where women play out their roles, they may tend to attach greater importance to family relations, including marital relations, compared to men (Shek 1995). Thus, marital quality may have a larger positive impact on the well-being of wives, as has been found among married Chinese adults in Hong Kong (Shek 1995). Both of these arguments have relevance, making it necessary to test the above hypotheses by gender.

11.2 Data and Methods

11.2.1 Data and Measures

The research reported in this chapter utilizes data from the Chinese Longitudinal Healthy Longevity Survey (CLHLS) and investigates the survival of the original sample between 1998 and 2000. For various reasons, 894 elderly (9.8 percent) in the original sample were not re-interviewed in the follow-up. Most of them were living in urban areas, where housing construction has been rapidly expanding, and many residents relocated within the 2-year period. Zeng et al. (2002) showed that the age reporting of the oldest old in the CLHLS is generally reliable up to the age of 105. Consequently, this analysis only included the elderly between 80 and 105 years of age, further reducing the sample size to 7,938. Data on at least one variable were missing for 1,051 cases. To maintain the sample size, we employed a random computation method to fill in these missing values. We first let a computer generate a random number for each missing case based on a uniform probability distribution. We also had a frequency distribution of each variable based on all the known cases, which we then matched with the computer-generated random numbers, assigning a corresponding category of a variable to each missing case. We also generated values for missing variables with the mean or the modal category. The results using both approaches are largely the same.[1]

The dependent variable was the duration of survival in months from the baseline interview in 1998 to the first follow up in 2000. The values of all the independent variables were taken from the baseline interview in 1998.

Marital status is a dummy variable with not married coded 0, and married coded 1. Number of children alive is directly obtained from the survey.

Social activities with others includes playing cards/mah-jong, and attending religious activities. Each activity was indexed at three levels: 0 = "never"; 1 = "sometimes"; 2 = "almost everyday." The sum of the two scores (from 0 to 4) was used for the analysis.

Among the control variables, physical activities fell into three categories: regularly performing physical exercise (0 = No, 1 = Yes), solitary-active activities (doing housework and gardening), and solitary-sedentary activities (reading newspapers/books, and watching TV/listening to the radio). The last two categories are customary activities that are less physically demanding relative to physical exercise but have been shown to have distinctive effects on the well-being and survival of the elderly (Bygren et al. 1996; DiPietro 2001; Glass et al. 1999; Lennartsson and Silverstein 2001; Morgan and Clarke 1997; Morgan et al. 1991; Welin et al. 1992). Each activity was indexed at three levels: 0 = "never"; 1 = "sometimes"; 2 = "almost everyday." A score for each type of customary activity was created by summing the indexes of the two activities within each category, ranging from 0 to 4.

In addition, as reported in the literature, physical and cognitive status (Anstey et al. 2001; Liang et al. 2000; Parker et al. 1992), health behaviors (Berkman et al. 1983; Kaplan et al. 1987b), and socio-demographic characteristics (Kaplan 1992; Kaplan et al. 1987a) are also associated with mortality; therefore, they were controlled as well.

Physical health status was measured by activities of daily living (ADL), physical performance, the experience of serious illness, and self-rated health. There were six questions regarding ADLs in the survey, bathing, dressing, using the toilet, transferring, eating, and continence, each of which was indexed at three levels: 0 = "needed a lot of assistance"; 1 = "needed some assistance"; 2 = "needed no assistance." A summed ADL score ranges from 0 to 12. An objective measure of functional status consisted of three indicators (Zeng and Vaupel 2002): standing up from a chair, picking up a book from the floor, and steps needed to turn around 360 degrees. Each

---

[1] A separate analysis with all the subjects with missing values deleted showed similar results as well.

of the three tasks was coded into three levels: 0 = "unable to perform"; 1 = "able with some assistance"; 2 = "able without assistance."[2] A physical performance score was created by summing these three measures, ranging from 0 to 6. The survey asked the number of times the elder suffered from serious illnesses that required hospitalization or caused him/her to be bedridden in the past 2 years. Because about 90 percent of the sample reported none, 6 percent reported a frequency between 1 and 10, and 4 percent reported having been bedridden all year long, this variable was coded as a dummy variable: 0 = "no serious illness"; 1 = "at least one serious illness."[3] Self-rated health was indexed in five levels: 0 = "very bad"; 1 = "bad"; 2 = "so-so"; 3 = "good"; 4 = "very good."

Cognitive status was measured by the Chinese version of the Mini-Mental State Examination (Zeng and Vaupel 2002), which is composed of 24 questions falling into five categories: orientation, registration, attention and calculation, recall, and language. For example, within the category of orientation, questions included, "What time of day is it right now (morning, afternoon, evening)?" and "What is the animal year of this year?" The answer to each question was coded as 0 for "wrong," and 1 for "correct."[4] A cognitive index was constructed by summing the scores of all 24 questions which ranged from 0 to 24.

Previous research has shown that not only current smoking, but also past smoking, have a detrimental effect on health (Berkman et al. 1983). Thus, smoking was coded into three categories: never smoked; currently smokes; and smoked in the past. Berkman et al. (1983) also found that for most age and sex groups moderate drinkers had a lower mortality rate than abstainers, light drinkers, or heavy drinkers. Since the effect of alcohol consumption may not be linear, a series of dummy variables was created with non-drinker as the reference category. The other three categories were: on average 1 *liang* (50 g) per day; 2 *liang* per day; 3 *liang* or more.

Age was measured as actual age in years in 1998. Gender was a dummy variable, with female coded as 0, male coded as 1. Since 72 percent of the elderly in the sample reported having no schooling, a dummy variable was created for educational attainment (0 for no schooling and 1 for some schooling). Urban residence was measured by a dummy variable with rural coded 0, and urban coded 1. Living arrangements were coded into four categories: living alone, living with spouse only in the community, living with other relatives in the community, and living in a nursing home.

---

[2] The answer to the third question ranges from 2 steps to 72 steps, and 2,072 subjects were coded as "unable to perform." Based on its distribution, this variable was categorized into three levels: 1 = "unable to perform"; 2 = "more than 10 steps"; 3 = "within 10 steps."

[3] Though there were questions that asked the respondent about certain diseases, including hypertension, diabetes, cataract, and cancer, it is quite likely that many diseases were not diagnosed among the elderly in rural areas where access to medical care was more limited than in urban areas. Therefore, this information was not included.

[4] There was one exception. One of the questions asked the respondent to name as many kinds of food as possible in 1 min. To be consistent with the format of the answers in other questions, based on the frequency distribution, the answer was coded "0" for those who gave 3 or fewer names and "1" for those who gave more than 3 names.

11.2.2 Statistical Methods

We used a Weibull hazard model to examine the mortality of these elderly individuals within the 2-year interval. A statistical test suggests that the Weibull distribution was a valid fit.[5] To address the research questions above, a series of nested models was fitted sequentially (Lennartsson and Silverstein 2001; Liang et al. 2000). The first model examined the marginal impact of marital status, number of children alive, and engagement in other social activities, without controlling for any other factors. The second model added all the control variables to investigate whether the effects of the various social engagements remained independent. The last model added the interaction effects of age with marital status, number of children alive, and other social activities to test whether the favorable effects of marital status, number of children alive and social activities diminished with age. To avoid redundancy, we reported all the models above for the whole sample. The findings of the effects of the key independent variables (marital status, number of children, and engagement in social activities) by residence and gender are summarized in a separate table.

11.3 Results

Tables 11.1 and 11.2 provide descriptive statistics for the categorical and continuous variables. The average age of the respondent was 92.3. Females accounted for three-fifths of the sample. The majority of the elderly lived in rural areas (65 percent) and had little schooling. Forty-one percent died in the 2-year period. About 16 percent were married. The means and standard deviations of customary activities at the baseline are given in Table 11.2. The elderly in this sample, on average, had two to three children alive. A separate analysis showed that about 25 percent were involved in at least one kind of social activity at baseline.

Results of the Weibull hazard models are presented in Table 11.3. The first model shows that all three measures of social relationships or engagements were significantly associated with mortality without controlling for any other characteristics. The hazard of death for married elderly was 47 percent lower than for those not married. An additional surviving child was associated with a 4 percent reduction in the death hazard. Engaging in other social activities lowered the death hazard by 26 percent.

In Model 2, when all control variables were added, the effects of marital status and number of children alive were weakened but were still statistically significant, especially the effect of marital status. The effect of social activities became insignificant.

As expected, age was associated with a higher likelihood of death, and men were more likely to die than women. Those who engaged in physical exercise, engaged

---

[5] The fact that the plot of $\log(-\log(S))$ versus $\log(t)$ is roughly a straight line, where S refers to the probability of survival and t refers to survival duration, suggests that adopting a Weibull distribution is a reasonable choice (Lee 1992).

Table 11.1 Frequency distribution of categorical variables ($N = 7,938$)

Variable	Frequency (%)
Survival status	
Dead	40.90
Alive	59.10
Gender	
Male	39.96
Female	60.04
Education	
No formal schooling	72.16
Some schooling	27.84
Urbanicity	
Urban	35.47
Rural	64.53
Marital status	
Married	15.76
Not married	84.24
Living arrangements	
Living alone	9.89
Living with spouse only	7.21
Living with other relatives	78.16
Living in a nursing home	4.75
Serious illness	
None	89.77
One or more	10.23
Smoking	
Never smoked	67.76
Currently smokes	17.36
Smoked in the past	14.88
Drinking	
None	76.00
One *liang*	10.33
Two *liang*	5.90
Three *liang*	7.77
Physical exercise	
Yes	26.05
No	73.95

in solitary-active activities, and engaged in solitary-sedentary activities had a 22, 15, and 7 percent lower hazard rate than those who did not, respectively. Most of the health status measures, except the presence of serious illness, were highly significant.

Neither education nor urban residence were related to mortality. There was no significant difference in mortality among those living alone, living with spouse only, or living with other relatives. However, compared to living alone, living in a nursing home was associated with a 30 percent lower death hazard rate. Both current and past smoking had detrimental effects on survival. The effects of drinking alcohol were not significant.

Table 11.2 Means and standard deviations of continuous variables ($N = 7,938$)

Variables	Mean	SD
Duration of survival	20.91	7.93
Age	92.27	7.37
ADL	10.48	2.69
Physical performance	4.34	1.95
Self-rated health	2.61	.83
Cognitive status	16.63	7.37
Number of children Alive	2.37	1.99
Social activities	.36	.69
Solitary-active activities	.78	1.01
Solitary-sedentary activities	.97	1.18

To directly test the two hypotheses—whether the beneficial effect of close ties and social activities diminishes with age—the interaction effects between age and marital status, number of children alive, and social activities were added in Model 3. With the addition of these variables an interesting pattern emerged: both the main and interaction effects of social activities were significant. It is apparent that social activities were not significant in Model 2 because the effect was suppressed when the interaction between social activities and age was missing in the model. To further confirm that the effect of social activities was not confounded by health status, in a separate analysis (not shown), we added the interactions of age with all the variables in Model 3 measuring health status (self-rated health, ADL, physical performance, the presence of serious illness, and cognitive status); none of them were significant. The main and age-interaction effects of social activities remained statistically significant. This finding, showing no age-dependent pattern of physical health, further excluded the possibility of a spurious effect associated with physical health.

While the main effect of social activities was negatively related to the hazard of death, suggesting that social activities reduced the death hazard, there was a positive interaction effect with age, indicating that the beneficial effect of social activities gradually eroded as age increased. For example, while social activities reduced the death hazard by 19 percent at the age of 80, they were reduced by only 7 percent at the age of 90 given other equal conditions. Their effects reached zero at about age 95 and steadily increased to larger positive values thereafter. This indicates that after the age of 95, social activities are associated with a higher, rather than a lower, hazard of death.

In contrast, neither the main effects nor the interaction effects of marital status or number of children alive were significant in the last model, although they were significant in Model 2. We do not believe that these results can be interpreted as an effect of social activities. Rather, we hold that when there is no interaction effect, adding interactions to the model merely adds multi-collinearity and likely makes none of the main effects significant (Agresti and Finlay 1997). Thus, there were virtually no interactions between age and marital status or number of children alive.

Findings based on the same modeling strategy are summarized in Table 11.4 with respect to the effects of social relations by residence and gender. Four sub-samples

Table 11.3 Estimates of hazard coefficients and hazard ratios of the weibull model of mortality ($N = 7,938$)

Covariate	Model 1		Model 2		Model 3	
	B	Exp(B)	B	Exp(B)	B	Exp(B)
Married (ref. not married)	−0.641***	0.527	−0.291***	0.748	0.101	0.904
Number of children alive	−0.043***	0.958	−0.021*	0.979	−0.040	0.960
Social activities	−0.305***	0.737	−0.013	0.988	−1.335***	0.263
Age			0.039***	1.040	0.035***	1.035
Male (ref. female)			0.449***	1.567	0.447***	1.564
ADL			−0.049***	0.952	0.050***	0.951
Physical performance			−0.085***	0.918	−0.085***	0.919
Serious illness (ref. no illness)			−0.049	0.953	−0.048	0.953
Self-rated health			−0.055*	0.946	−0.056**	0.946
Cognitive status			−0.022***	0.978	−0.023***	0.977
Some schooling (ref. no schooling)			0.001	1.001	0.002	1.002
Urban (ref. rural)			0.004	1.004	0.004	1.004
Living with spouse only (ref. living alone)			−0.167	0.846	−0.173	0.841
Living with others (ref. living alone)			−0.094	0.910	−0.094	0.910
Living in a NH (ref. living alone)			−0.360**	0.697	−0.350***	0.705
Physical exercise (ref. no)			−0.253***	0.776	−0.248***	0.780
Solitary-active activities			−0.163***	0.850	−0.162***	0.851
Solitary-sedentary activities			−0.070**	0.933	−0.069**	0.933
Currently smokes (ref. never)			0.174**	1.190	0.175**	1.191
Smoked in the past (ref. never)			0.168**	1.182	0.166**	1.180
Drinking 1 *liang* (ref. no drinking)			−0.041	0.960	−0.044	0.957
Drinking 2 *liang* (ref. no drinking)			0.029	1.030	0.028	1.028
Drinking 3 *liang* or more (ref. no drinking)			−0.058	0.944	−0.068	0.935
Married* age					−0.002	0.998
Number of children alive* age					0.000	1.000
Social Activities* age					0.014***	1.014

* $p < 0.05$; ** $p < 0.01$; *** $p < 0.001$;

Table 11.4 Major hazard coefficients of the weibull model of mortality by residence and gender

	Whole sample	Urban-male	Urban-female	Rural-male	Rural-female
Model 1					
Married	–[a]	–	–	–	–
No. of children alive	–	–	–	NS	–
Social activities	–	–	–	–	–
Model 2[b]					
Married	–	–	NS	–	NS
No. of children alive	–	NS	NS	NS	NS
Social activities	NS	NS	NS	NS	NS
Model 3[b]					
Married	NS	NS	NS	NS	NS
No. of children alive	NS	NS	NS	NS	NS
Social activities	–	NS	–	–	–
Married*age	NS	NS	NS	NS	NS
No. of children alive*age	NS	NS	NS	NS	NS
Social activities*age	+	NS	NS	+	+
Sample size	7,938	1,179	1,637	1,993	3,129

[a] " – " or "+" represents a significant coefficient, either negative or positive, at $p < 0.05$, "NS" = non-significant.
[b] All other covariates are the same as in Table 11.3.

were analyzed: urban-males, urban-females, rural-males and rural-females. The results based on the whole sample are also provided in the same table as a reference. There was no substantial difference in Model 1 across all the sub-samples. Two major findings emerged in Model 2. First, while being married was associated with a lower hazard of death for men in both urban and rural areas, it had no effect on women. Second, number of children alive was insignificant in all the sub-samples. The patterns found in the full sample regarding the effect of social activities were only manifested in the two rural sub-samples, not in the two urban samples.

11.4 Discussion

Using a national sample of the elderly in China aged 80 or above, the research reported in this chapter investigated the role of social engagement in mortality at very old ages. More specifically, we adopted the convoy network model to differentiate social relationships by their closeness to the individual concerned, and tested whether the effects of these different relationships or engagements changed with age. The existence of close ties, measured by marital status and number of children alive, was found to be significantly associated with a lower level of mortality in the whole sample independent of health status, health behaviors, and other socio-demographic characteristics. But these effects do not appear to depend on age. However, an age-related, downward trend of the beneficial effect of social activities on survival was documented. At the very old ages, the favorable effect of social activities was reversed.

Our statistical analysis, based on a large sample of the oldest-old in China, resonates with the findings of a study of the elderly aged 85 and above in San Francisco (Johnson and Barer 1997) and is consistent with selectivity and continuity theories (Atchley 1989; Baltes 1997; Balters and Carstensen 1996). Johnson and Barer (1997) observed that the oldest old were active agents, continuously readapting to new challenges in later life. Physical disability was described as an increasing constraint. To cope with it, the very old withdrew selectively from social relationships, simplifying and narrowing the boundaries of their social worlds. The elderly in their study deliberately reduced their social roles and only retained those relationships that they saw as beneficial, stating, for example, "I don't have the same desire to keep mixing and going out," (1997: 154) and "I don't want any more friends, I want less" (1997: 155). They did not view retreating from social relations as a passive response to their physical status. Rather, they regarded such disengagement as a positive change. They felt that it was time to focus inwardly and did not feel lonely or isolated. Two of them said, "I have reached the peaceful age of quietude," (1997: 154) and "I enjoy my privacy" (1997: 155). A 102-year-old woman said, "People impinge on you at my age; I need more time to be with myself" (1997: 155). They attached significant meaning to mundane activities.

These results seem to cast doubt on the universal benefits of social engagement as it applies to very old ages. In her elaboration of the "identity accumulation hypothesis," Thoits (1983) argued that the more roles played by an individual, the less psychological distress, because "(i)n taking the role of the generalized other, the individual perceives that he/she has been placed by others into recognized and meaningful social categories, or social positions" (1983: 175), which gives meaning and direction to one's life. Culturally sanctioned social expectations associated with roles, however, may not be relevant to the oldest old population. Atchley (1989) argued that the importance of some social roles tends to decline in later adulthood, releasing the old from role expectations. Among the oldest old they studied, Johnson and Barer (1992) reported a trend of both loosening normative constraints and increasing introversion. The elders tended to ignore social expectations regarding their relationships with others. Social conventions became less important and they did not care very much about what others thought of them. With respect to introversion, they said such things as, "More things are beyond my control, so I just roll with the waves" (1992: 360). "I have periods of aloneness where I feel unrelated to what is going on" (1992: 361). "...I am comfortable being alone. I'm pulling away from everything more and more..." (1992: 361). This shift in orientation towards introversion, congruent with selectivity and continuity theories, is a response to the elder's changing conditions and may explain why *less* social engagement is valued. Therefore, less social engagement may be better for elderly well-being at the very old ages.

But this finding does not apply equally to all sub-samples when place of residence and gender are taken into account. First of all, number of children alive is not significant in any of the sub-samples. Number of children is just a rough indicator of close social ties. Internal qualities, for example, emotional support, closeness and feelings of social embeddedness, were not addressed. We suspect that the significance of this variable in the whole sample is due to the larger sample size, which reduced the standard error of its effect. Though we also looked at children's proximity to their parents (and siblings' proximity), it was not significant either. More comprehensive measures of social ties, especially those that capture the quality of relationships, are definitely needed.

Secondly, the beneficial effect of marriage appears only to apply to Chinese elderly men, and not to women. This may be related to traditional values and practices in China regarding gender roles within marriage and the family.

Thirdly, the interaction effect of age and social activities is only significant for the rural samples. But before we rule out such an interaction effect in the urban areas, we believe more comprehensive measures of social activities are needed for confirmation. Due to the limitation of the data, only two kinds of social activities were included: playing cards/mah-jong and attending religious activities. Other forms of social activities, such as visiting friends and participating in other organizations, were not included. In Lennartsson and Silverstein's (2001) study to assess the role of social activities in mortality among the elderly in Sweden, they also included visiting friends, going to movies, theatres, concerts, or museums, and participating in study groups. When Walter-Ginzburg et al. (2002) examined social engagement and mortality among the old-old in Israel, they included measures of group activities, such as talking with family or friends, playing cards or board games, attending movies, and writing letters. Instruments that are more complete and suitable for describing social activities of the Chinese elderly will certainly strengthen this type of research.

In light of the limitations of this chapter in measuring social relationships and activities, caution is warranted in generalizing the findings. Nevertheless, we would like to draw attention to the uniqueness of the oldest-old population. The investigation of this special population should be placed in a life-course perspective. When discussing clarifying the distinctiveness of old age, Settersten (2006) highlights two tasks faced by gerontologists: "(1) to identify core features of human growth and maturation, as well as fundamental needs that must be met, in old age; and (2) to elaborate the differential expression of these needs and how they are or might be better met in a range of particular contexts" (2006: 9). We hope that our research has taken a step in this direction.

Acknowledgments This chapter is an extension of the study published on *Journal of Aging and Health* 18 (1), pp. 37–55, 2006. Unlike the previous study, which explored both physical and social customary activities, in the current study, we focused on social activities; we are especially interested in testing the social convoy model, which is absent in the previous study. In addition, the patterns of gender and residential sub-samples were not analyzed in the previous study.

References

Agresti, A. and B. Finlay (1997), *Statistical methods for the social sciences*. 3rd ed. New Jersey: Prentice Hall

Anstey, K.J., M.A. Luszcz, and L.C. Giles (2001), Demographic, health, cognitive, and sensory variables as predictors of mortality in very old adults. *Psychology and Aging*, 16 (1), pp. 3–11

Atchley, R.C. (1989), A continuity theory of normal aging. *The Gerontologist* 29 (2), pp. 183–190
Balters, M.M. and L.L. Carstensen (1996), The process of successful aging. *Ageing and Society* 16 (4), pp. 397–422
Baltes, P.B. (1997), On the incomplete architecture of human ontogeny. *American Psychologist* 52 (4), pp. 366–380
Baltes, M.M. and L.L. Carstensen (1996), The process of successful aging. *Ageing and Society* 16(4), pp. 397–422
Belloc, N.B. (1973), Relationship of health practices and mortality. *Preventive Medicine* 2 (1), pp. 67–81
Berkman, L.F. and L. Breslow (1983), Social networks and mortality risk. In: L.F. Berkman and L. Breslow (eds.): *Health and ways of living*. New York: Oxford University Press, pp. 113–160
Berkman, L.F., L. Breslow, and D. Wingard (1983), Health practices and mortality risk. In: L.F. Berkman and L. Breslow (eds.): *Health and ways of living*. New York: Oxford University Press, pp. 61–112
Berkman, L.F., T. Glass, I. Brissete, and T.E. Seeman (2000), From social integration to health: Durkheim in the new millennium. *Social Science and Medicine* 51 (6), pp. 843–857
Bygren, L.O., B.B. Konlann, and S. Johansson (1996), Attendance at cultural events, reading books or periodicals, and making music or singing in a choir as determinants for survival: Swedish interview survey of living conditions. *British Medical Journal* 313 (21), pp. 1577–1580
Carstensen, L.L. (1991), Selectivity theory: Social activity in life-span context. *Annual Review of Gerontology and Geriatrics* 11, pp. 195–217
Carstensen, L.L. (1992), Social and emotional patterns in adulthood: Support for socioemotional selectivity theory. *Psychology and Aging* 7 (3), pp. 331–338
Cerhan, J.R., C. Chiu, R.B. Wallace, J.H. Lemke, C.F. Lynch, J.C. Torner, and L.M. Rubenstein (1998), Physical activity, physical function, and the risk of breast cancer in a prospective study among elderly women. *Journal of Gerontology* 53A (4), M251–M256
Cumming, E. and W.E. Henry (1961), *Growing old: The process of disengagement*. New York: Basic Books
DiPietro, L. (2001), Physical activity in aging: changes in patterns and their relationship to health and functions. *Journals of Gerontology* 56A (Special Issue II), pp. 13–22
Glass, T.A., C. Mendes de Leon, R.A. Marottoli, and L.F. Berkman (1999), Population based study of social and productive activities as predictors of survival among elderly Americans. *British Medical Journal* 319 (8), 478–483
Gove, W.R. (1972), The relationship between sex roles, mental illness and marital status. *Social Forces*, 51, 34–44
Gove, W.R. (1973). Sex, marital status and mortality. *American Journal of Sociology* 70, 45–67
Ho, S.C. (1991), Health and social predictors of mortality in an elderly Chinese cohort. *American Journal of Epidemiology* 133 (9), pp. 907–921
House, J.S., R.C. Kessler, A.R. Herzog, R.P. Mero, A.M. Kinney, and M.J. Breslow (1992), Social stratification, age, and health. In: K.W. Schaie, D. Blazer, and J.S. House (eds.): *Aging, health behaviors, and health outcomes*. Hillsdale, NJ: Lawrence Erlbaum Associates, pp. 1–32
Johnson, C.L. and B.M. Barer (1992), Patterns of engagement and disengagement among the oldest old. *Journal of Aging Studies* 6 (4), pp. 351–364
Johnson, C.L. and B.M. Barer (1997), *Life beyond 85 years*. New York: Springer
Kahn, R.K. and T.C. Antonucci (1980), Convoys over the life course: Attachment, roles, and social Support. In: P.B. Baltes and O.G. Brim (eds.): *Life-span development and behavior*. New York: Academic Press, pp. 253–286
Kaplan, G.A. (1992), Heath and aging in the Alameda County Study. In: K.W. Schaie, D. Blazer, and J.S. House (eds.): *Aging, health behaviors, and health outcomes*. Hillsdale, NJ: Lawrence Erlbaum Associates, pp. 69–88
Kaplan, G.A., M.N. Hann, S.L. Syme, M. Minkler, and M. Windeby (1987a), Socioeconomic status and health. In: R.W. Amler and H.B. Dull (eds.): *Closing the gap: The burden of unnecessary illness*. New York: Oxford University Press, pp. 125–129

Kaplan, G.A., T.E. Seeman, R.D. Cohen, L.P. Knudsen, and J. Guralnik (1987b), Mortality among the elderly in the Alameda County Study: Behavioral and demographic risk factors. *American Journal of Public Health* 77 (3), pp. 307–312

Lang, F.R. and L.L. Carstensen (1994), Close emotional relationships in later life: Further support for proactive aging in the social domain. *Psychology and Aging* 9 (2), pp. 315–324

Lee, E.T. (1992), *Statistical methods for survival data analysis*. New York: Wiley

Lennartsson, C. and M. Silverstein (2001), Does engagement with life enhance survival of elderly people in Sweden? The role of social and leisure activities. *Journal of Gerontology* 56B (6), pp. S335–S342

Liang, J., J.F. McCarthy, A. Jain, N. Krause, J.M. Bennett, and S. Gu (2000), Socioeconomic gradient in old age mortality in Wuhan, China. *Journal of Gerontology* 55B (4), pp. S222–S233

Liu, X., A.I. Hermalin, and Y.L. Chuang (1998), The effects of education on mortality among older Taiwanese and its pathways. *Journal of Gerontology* 53B (2), pp. S71–S82

Mookherjee, H.N. (1997), Marital status, gender, and perception of well-being. *The Journal of Social Psychology* 137, 95–105

Morgan, D.L. (1988), Age differences in social network participation. *Journal of Gerontology* 43 (4), pp. S129–S137

Morgan, K. and D. Clarke (1997), Customary physical activity and survival in later life: A study in Nottingham, UK. *Journal of Epidemiology and Community Health* 51 (5), pp. 490–493

Morgan, K., H. Dallosso, E.J. Bassey, S. Ebrahim, P.H. Fenten, and T.H.D. Arie (1991), Customary physical activity, psychological well-being and successful aging. *Ageing and Society* 11 (4), pp. 399–415

Parker, M.G., M. Thorslund, and M. Nordstrom (1992), Predictors of mortality for the oldest old. a 4-year follow-up for community-based elderly in Sweden. *Archives of Gerontology and Geriatrics* 14, pp. 227–237

Pienta, A.M., M.D. Hayward, and K.R. Jenkins (2000). Health consequences of marriage for the retirement years. *Journal of Family Issues* 21 (5), 559–586

Rowe, J.W. and R.L. Kahn (1998), *Successful aging*. New York: Pantheon Books

Sabin, E.P. (1993), Social relationships and mortality among the elderly. *The Journal of Applied Gerontology* 12 (1), pp. 44–60

Settersten, R.A. Jr. (2006), Aging and the life course. In: R.H. Binstock and L.K. George (eds.): *Handbook of aging and the social sciences*. 6th ed. New York: Academic Press, pp. 3–19

Shek, D.T. (1995), Gender differences in marital quality and well-being in Chinese married adults. *Sex Roles* 32, pp. 699–715

Simon, R.W. (2002). Revisiting the relationships among gender, marital status, and mental health. *American Journal of Sociology* 107 (4), 1065–1096

Suzman, R.M., K.G. Manton, and D.P. Willis (1992), Introducing the oldest old. In: R.M. Suzman, D.P. Willis, and K.G. Manton (eds.): *The oldest old*. New York: Oxford University Press, pp. 3–14

Thoits, P.A. (1983), Multiple identities and psychological well-being: A reformulation and test of the social isolation hypothesis. *American Sociological Review* 48 (2), pp. 174–187

United Nations (2001), *World population ageing: 1950–2050*. Department of Economic and Social Affairs, Population Division. New York: United Nations

Waite, L.J. (1995), Does marriage matter? *Demography* 32, 483–507

Welin, L., B. Larsson, K. Svardsudd, B. Tibblin, and G. Tibblin (1992), Social network and activities relation to mortality from cardiovascular diseases, cancer and other causes: A 12 year follow up of the study of men born in 1913 and 1923. *Journal of Epidemiology and Community Health* 46 (2), pp. 127–132

Yu, E.S.H., Y.M. Kean, D.J. Slymen, W.T. Liu, M. Zhang, and R. Katzman (1998), Self-perceived health and 5-year mortality risks among the elderly in Shanghai, China. *American Journal of Epidemiology* 147 (9), pp. 880–890

Zeng, Y. and J.W. Vaupel (2002), Functional capacity and self-evaluation of health and life of oldest old in China. *Journal of Social Issues* 58 (4), pp. 733–748

Zeng, Y., J.W. Vaupel, Z. Xiao, C. Zhang, and Y. Liu (2002), Sociodemographic and health profiles of the oldest old in China. *Population and Development Review* 28 (2), pp. 251–273

Zunzunegui, M., B.E. Alvarado, T.D. Ser, and A. Otero (2003), Social networks, social integration, and social engagement determine cognitive decline in community-dwelling Spanish older adults. *Journal of Gerontology* 58B (2), pp. S93–S100

Part III
Living Arrangements and Elderly Care

Introduction *by Zeng Yi*

Previous research has demonstrated that living arrangements have a substantial impact on elderly care (e.g., Soldo and Freedman 1994; Angel et al. 1992; Suzman et al. 1992; Jamshidi et al. 1992). Even in developed countries, the elderly tend to depend heavily on children for emotional and psychological support, and occasionally for financial resources. In the developing world, as in China where pension and social security systems are not widely available, especially in the rural areas, the elderly depend on children as their main source of support in their old age.

Family structure and living arrangements of elderly have changed considerably in the past few decades in China. As shown by census data, the percentages of men aged 65 and over who did not live with children were 28.4, 29.5, and 37.9 in 1982, 1990, and 2000, respectively; the corresponding percentages for women aged 65+ were 24.7, 24.5 and 30.2.[1] These data indicate clearly that after 1990, the living arrangements of Chinese elderly are changing quickly towards living apart from their adult children. Such trends were likely caused by the rapid fertility decline plus substantial changes in social attitudes and economic mobility related to co-residence between older parents and adult children (Zeng and Wang 2003).

Rapid socio-economic development and urbanization should further increase people's preferences for independent living. The previous severe housing shortages have been and will continue to be relieved through housing reforms based on the market economy, which will allow more young people to live away from their parents. Increasing migration and job mobility will separate more old parents from their adult children. In the cultural context of Chinese society, however, filiality (*xiao*) has been a cornerstone for thousands of years and is still highly valued. The

---

Zeng Yi
Center for Study of Aging and Human Development, Medical School of Duke University, Durham, NC 27710, USA,
Center for Healthy Aging and Family Studies/China Center for Economic Research,
Peking University, Beijing, China
e-mail: zengyi68@gmail.com

[1] These figures are derived from table 2 in Zeng and Wang (2003).

philosophical ideas of filiality include not only respect for the older generations, but also the responsibility of children to take care of their elderly parents. Such ethnic and cultural traditions have been playing and will continue to play crucial roles in the living arrangements and care for the elderly.

Given the Chinese cultural, demographic and socioeconomic contexts, the patterns and changes in living arrangements have been and will continue to be strongly influencing caregiving, long term care service, quality of life, and health-related policy-making (Bian et al., 1998; Chen and Silverstein 2000; Olsen 1993; Pei and Pillai 1999; Zimmer and Kwong 2003). This is an important aspect of healthy longevity. Part III of this volume contains chapters that deal with this issue.

The longitudinal findings presented in Chap. 12 by Wu and Schimmele demonstrate that oldest old people living in family co-residences have superior psychological dispositions than individuals living alone or in nursing homes. The advantages of family co-residence are independent of differences in socioeconomic status, health status (including functional limitations), and demographic characteristics. Wu and Schimmele's chapter also discusses the differential effects of different types of living arrangements on the psychological disposition of individuals aged 80–89 years and those aged 90 years and older.

Chapter 13 by Zimmer tests the hypothesis that elders' coresidence with adult children and/or other family members more likely occurs when the elders' needs are greatest, for instance, when their health deteriorates or their spouse dies. He also investigates gender variation due to differences in authority and emotional bonds between older women and men and their families. Using CLHLS data, cross-sectional and transitional multinomial models were estimated to link health status measures with living arrangement outcomes. Results show that changes in living arrangements occur frequently. Functional limitations are more strongly associated with living arrangements than are other health indicators. Health indicators are more strongly related for those not married. Gender interactions show that a health change most likely triggers a living arrangement response for women. The chapter also discusses the implications of changing living arrangements on elderly health for a rapidly aging China.

Using data from the 2001 survey of "Welfare of Elderly in Anhui Province, China" by the Institute for Population and Development Studies of Xi'an Jiaotong University, Lu, Li, Zhang and Feldman (Chap. 14) conducted a logistic regression analysis of the effects of intergenerational financial, instrumental and emotional support on the self-rated health of Chinese rural elderly. Their results show that financial support, both received and provided, as well as mutual emotional support between parents and children, has positive impacts on the self-rated health of the elderly. However, instrumental support, both received and provided, tends to exhibit no relationship with the self-rated health of the elderly. Socio-demographic characteristics, socioeconomic status, as well as health status of the elderly are also found to affect their self-rated health.

An earlier analysis found that the oldest-old aged 80 and over in China who received care from children have a greater likelihood of dying than those who do not have such care (Zhang 2002). Could this result be interpreted that the care provided by children might be harmful to their oldest-old parents, or that the children selectively provided care to the already-frail elderly parents? Zhen's Chap. 15 provides important answers to this controversial and puzzling question. The analysis used CLHLS data collected from paired old parents and adult children (i.e. dyadic data) and estimated a joint model to carry out the dyadic analysis such that the protective and selective effects of caregiving could be specified comprehensively. This innovative empirical analysis that is presented in Chap. 15 shows that protection effects may be seriously mis-estimated when selection effects are ignored. After having successfully identified the selection effects, care provision to the elderly is found to be beneficial far more so than claimed in previous studies.

Chapter 16 by Zhao discusses the challenging issues of the negative impacts of inequality in health care on healthy longevity in China. The recent socio-economic reforms begun in the late 1970s in China have brought about both positive and negative changes in social security and health care systems. Some of the changes have created difficulties for the further improvement of public health. This chapter argues that while mortality has continued to fall in the last 25 years and life expectancy has now reached 72 years, China still faces serious challenges in further improving healthy longevity, induced by some major changes in health care systems. The chapter examines increasing inequality in health care and mortality across different areas and population groups.

In sum, the chapters in Section III of this volume examine the impacts of living arrangements and their dynamic changes, intergenerational support, the protective and selective aspects of caregiving from children, and inequality in medical care services on the physical and psychological health of the elderly in China.

References

Angel, R.J., J.L. Angel, and C.L. Himes (1992), Minority group status, health transitions, and community living arrangements among the elderly. *Research on Aging* 14 (4), pp. 496–521

Bian, F., J.R. Logan, and Y. Bian (1998), Intergenerational relations in urban china: Proximity, contact, and help to parents. *Demography* 35, pp. 115–124

Chen, X. and M. Silverstein (2000), Intergenerational social support and the psychological well-being of older parents in China. *Research on Aging* 22, pp. 43–65

Jamshidi, R., A.J. Oppenheimer, D.S. Lee, F.H. Lepar, and T.J. Espenshade, (1992), Aging in America: Limits to life span and elderly care options. *Population Research and Policy Review* 11, pp. 169–190

Olsen, P. (1993), Caregiving and long-term health care in the People's Republic of China. *Journal of Aging and Social Policy* 5, pp. 91–110

Soldo, B.J. and V.A. Freedman (1994), Care of the elderly: division of labor among the family, market, and state. In: Linda G. Martin and Samuel H. Preston (eds.): *Demography of aging*. Washington DC: National Academy Press, pp. 146–194

Suzman, R.M., K.G. Manton, and D.P. Willis (1992), Introducing the oldest old. In: R.M. Suzman, D.P. Willis, and K.G. Manton (eds.): *The oldest old*. New York: Oxford University Press

Pei, X. and V.K. Pillai (1999), Old age support in China: The role of the state and the family. *International Journal of Aging and Human Development* 49, pp. 197–212

Zeng Y. and Z. Wang (2003), Dynamics of family and elderly living arrangements in China: New lessons learned from the 2000 Census. *The China Review* 3 (2), pp. 95–119

Zhang, Z. (2002), The impact of intergenerational support on mortality of the oldest old in China. *Population Research* (*Chinese*) 26 (5), pp. 55–62

Zimmer, Z. and J. Kwong (2003), Family size and support of older adults in urban and rural China. *Demography* 40, pp. 23–44

Chapter 12
Living Arrangements and Psychological Disposition of the Oldest Old Population in China

Zheng Wu and Christoph M. Schimmele

Abstract Using data from the Chinese Longitudinal Healthy Longevity Survey (1998 and 2000), this chapter investigates whether different living arrangements influence the psychological disposition among the oldest old population (individuals aged 80+). This chapter's longitudinal findings demonstrate that oldest old people living in family co-residences have superior psychological dispositions than individuals living alone or in nursing homes. The advantages of family co-residences are independent of differences in socioeconomic status, health status (including functional limitations), and demographic characteristics. The chapter's findings also confirm that living arrangements have differential effects on the psychological disposition of selected age groups, with different types of family co-residential households having uneven effects between individuals aged 80–89 years and those aged 90 years and older.

Keywords Co-residence, Disability paradox, Fixed-effects, Healthy aging, Living arrangements, Mental health, Mixed models procedure, Oldest old, OLS models, Population health, Positive psychological disposition, Psychological disposition, Psychological hardiness, Psychological well-being, Random effects model, Restricted maximum likelihood, Robust psychological disposition, Salutogenesis, Two-level hierarchical linear model

12.1 Introduction

The Chinese oldest old population (individuals aged 80+ years) tripled between 1950 and 2000 because of improvements in population health and a corresponding reduction in the all-cause mortality rate (United Nations 2002). This demographic trend will continue be a factor in China's population transition for several more

---

Z. Wu
Department of Sociology, University of Victoria, 3800 Finnerty Road, Victoria, BC V8W 3P5 Canada
e-mail: zhengwu@uvic.ca

decades, with the oldest old projected to number over 30 million persons by 2025, amounting to a 160 percent population increase from the year 2000. By comparison, China's working-age population (individuals aged 15–64 years) is expected to increase 15 percent during this period. Hence, China is expected to experience a sharp transition in the potential old-age support ratio, which presents an emerging population health concern (for more discussion, see the first chapter in this volume by Poston and Zeng). The "greying" of China's population structure suggests that deficits of informal eldercare could well present hardships for the oldest old population. In China, families are the fount of eldercare, and a shortage of national financial resources continues to limit the number of individuals enrolled in state-funded eldercare programs (Pei and Pillai 1999; Zimmer and Kwong 2003). We can gauge the potential future effects of informal eldercare deficits on the healthy aging process by examining the current relationship between informal eldercare and patterns of well-being in the oldest old population.

The Chinese Longitudinal Healthy Longevity Survey (CLHLS) offers a unique data resource for understanding health and well-being outcomes among the oldest old because no other national dataset contains as large a sample ($N = 9,073$) of this special sub-population. This chapter employs a population health approach and uses the CLHLS data to investigate whether living arrangements influence psychological disposition in advanced ages. The chapter compares the Chinese oldest old in different living arrangements, as an indirect indicator of social resources and integration, to determine whether these conditions function as indicators of self-reported psychological disposition. Social resources and integration are well-established variables on health outcomes, with plentiful supplies of social support and higher social integration tending to augment an individual's abilities to prevent, minimize, and cope with negative life events or chronic life-strains (Pearlin et al. 1981; Thoits 1995). According to the stress process model, patterned health and well-being inequalities emerge through patterned differentials in health risk/stress exposure and through social disparities in an individual's cache of coping resources. Given that familial support is the foundation of social resources and social interaction for most oldest old Chinese people, we hypothesize in this chapter that living arrangements will have a significant effect on psychological disposition among this sub-population.

12.2 Background

The concept *healthy aging* refers to decreasing one's chance of mortality and compressing the number and duration of morbid conditions encountered during advanced ages (Baltes and Baltes 1990). However, as biological changes increase the general risk of disability and morbidity among the oldest old, healthy aging also depends on efficacious adaptation to unavoidable health transitions, such as coping with functional limitations or chronic illnesses. The *disability paradox* describes individuals with high coping attributes and high life satisfaction ratings despite facing health problems and disabilities (Albrecht and Devlieger 2000). In this respect, Antonovsky's (1979, 1987) innovative conceptual approach, termed *salutogenesis*, is important for understanding the association between positive thinking and healthy aging. Salutogenesis refers to "health-causing" processes, i.e., attitudes and behaviors that produce overall well-being; it the opposite of pathogenesis, or that which causes suffering. A positive psychological disposition is a salutogenic resource because psychological hardiness represents the capacity to manage negative experiences, particularly the health challenges prevalent in later life (Rodin 1986). Salutogenesis is germane for healthy aging among the Chinese oldest old because there is a connection between psychological hardiness and healthy aging in this population, even for individuals with less than optimal functional status (Smith et al. 2004; Zeng and Vaupel 2002).

In China, the merging of post-1978 economic reforms, increasing health care needs, and a declining old-age support ratio pose a serious impediment to healthy aging (Kwong and Cai 1992; Tracy 1991; Zeng et al. 2002). Under present economic conditions, the Chinese government has insufficient financial resources to guarantee full eldercare for the oldest old population (Grogan 1995; Liu et al. 1999). Indeed, fast population growth in conjunction with sluggish economic development prompted Chinese economic planners to reorganize the state's role in economic production and distribution, including shifting a large burden of pension and health care provisions onto the open market, essentially retreating from Maoist welfare principles. After the collapse of social welfare, an increasing proportion of health care costs is being recovered through user fees. For example, in the rural sector, home to about two-thirds of China's population, user fees comprise 85 percent of health care costs. Almost one-third of individuals now living beneath the poverty line dropped to this disadvantaged economic status through financial losses accrued from dealing with a serious illness (Hsiao 1995). Owing to these rising health care costs, 92 percent of rural elderly individuals have inadequate financial resources and 66 percent have unmet health care needs (Liu et al. 1995).

Post-reform health care inequalities are concentrated among the most vulnerable sub-population groups, such as seniors (Anson and Sun 2002; Gao et al. 2001; Shi 1996). The oldest old are the most likely to encounter relatively intense hardship from the transition from socialism to more market-based welfare distribution because need-based demand for health care and other support services tends to peak during late life; the oldest old demand a disproportionate amount of these resources because they experience a disproportionate amount of health and functional needs (Zeng et al. 2002). For example, among the Chinese oldest old, there is a sharp decline in functional status between the ages of 80 and 105 years, with age 85 being a significant threshold. In rural areas, about 12 percent of males and 16 percent of females aged 80–89 have mild or severe activities of daily living (ADL) disabilities. In urban areas, about 19 percent of males and 22 percent of females aged 80–89 have ADL disabilities. Among rural nonagenarians, about 25 percent of males and 35 percent of females have mild or severe ADL disabilities. In urban areas, these figures are about 35 and 47 percent, respectively (Zeng and Vaupel 2002).

In China, eldercare has traditionally been a family responsibility (termed *filial piety*), with the state providing little or no support. An altruistic model best exemplifies the pattern of intergenerational support in China. The altruistic model predicts that younger generations provide more support to elderly generations in a socio-cultural system that binds individual security and well-being to the family unit, and that the neediest family members will receive the most support (Zimmer and Kwong 2003). There is an upward bias in social support flows between Chinese parents and their adult children (Bian et al. 1998). About 55 percent of elderly parents receive regular support with grocery shopping, meal preparation, home cleaning and repairs, and personal care. About half as many adult children receive similar support from their parents. Under official policy (e.g., the Marriage Law of 1950), adult children (or grandchildren) are legally obliged to support their parents in old age (Tracy 1991). In rural China, individuals commonly obtain and depend on old age support through a married son. A comparatively high number of urban residents receive state pensions and support services, but these individuals also need family support (Olsen 1993).

As noted above, an aging population involves a shift in the potential old-age support ratio. In 1950, there were 13.8 working age (15–64 years) individuals per individual over age 65. By 2000, this figure was 10 working age individuals per individual over 65, and is projected to sharply decline in the future, reaching 5.2 by 2025 and 2.7 by 2050 (United Nations 2002). This trend suggests that the number of parents that urban adult males will have to support will increase from 0.7 in 1990 to 1.4 in 2030 (Lin 1995). The corresponding figures for rural males are 0.7 and 1.2, respectively. The number of co-residential years (caring time) will also increase. Urban females 40 years old in 2030 will spend an average of 17 years caring for elderly parents. This contrasts with 11 years of co-residence for their 1990 counterparts. Individuals aged 35–45 years in 2030 in urban areas will care for parents 5 more years and rural residents of the same age 4 more years than their 1990 counterparts. Despite these projections, the state has been reluctant to implement eldercare programs, which means that family-based social support will continue to be an important well-being indicator in old age (Tracy 1991).

12.3 Research Concepts

In this chapter we focus on the effect of familial and other living arrangements on variation in psychological disposition among the oldest old. The concept *psychological disposition* is defined according to a CLHLS instrument designed as a construct of optimism, conscientiousness, personal control, happiness, neuroticism, loneliness, and self-esteem (Smith et al. 2004). The literature suggests that psychological disposition accounts for considerable individual variation in health risk exposure, reaction to negative events and circumstances, and illness experiences (Gecas 1989; Kobasa et al. 1982). A robust psychological disposition modifies health attitudes, health behaviors, and health outcomes (Schieman and Turner 1998). A positive psychological disposition is a salutogenic resource because it represents the capacity to minimize and cope with negative life events or chronic life-strains (Rodin 1986). A recent report on the Chinese oldest old confirms this assumption, observing that psychological disposition predicts mortality and influences responses to widowhood, chronic health disorders, and functional limitations (Smith et al. 2004). This report indicates that a robust psychological disposition is characterized by salutogenic resources that promote coping and minimize pathogenic traits.

The concept *living arrangements* is meant to represent the expected cache of eldercare (or social resources) available to oldest old persons, and social integration. Social resources include financial, instrumental, informational, and emotional supports. In China, co-residence with one's children or grandchildren is the "core of support" for most seniors (Bian et al. 1998). The large majority (ranging between 69 and 82 percent) of oldest old individuals resides in extended family households (Zeng and George 2001). This pattern implies that living alone tends to represent childlessness or distant proximity from family members, which could amount to eldercare deficits. Even for those living near their children, the day-to-day reliability and quality of social support are presumably better in co-residential living arrangements because co-residence involves more frequent and closer contact with family members (Bian et al. 1998). In addition, co-residential living arrangements offer individuals a strong perception of projected security. Projected security refers to having someone who will provide long-term care and assistance, which is important for psychological well-being because it reduces the perceived threat of future problems (Ross and Mirowsky 2002). Co-residential living arrangements also provide opportunities for regular social interactions, and oldest old seniors may derive a sense of happiness from interactions with family members and self-esteem from contributing to the household.

We use a population health approach in our examination of differences in psychological disposition among the oldest old. The *population health* concept refers to population-based empirical inquiries into patterned differences in health risk factors, salutogenic resources, and health outcomes. This research is antecedent to public health policies, which refer to all community-level and government-sponsored strategies for promoting health and well-being and decreasing health risk factors (Kindig and Stoddart 2003). In some cases, the population health approach involves analyzing empirical data that are representative of conventional populations, such as national or regional populations, but this approach also includes more specific population groups, such as the oldest old. Most definitions or applications of population health derive from non-biomedical approaches to understanding health risk factors, salutogenic resources, and health outcomes. In this respect, population health is a constellation of social, psychosocial, economic, and ecological health risk disparities. The population health perspective represents a demand-side solution to rising health care costs and shortages. In sum, the population health approach emphasizes creating and maintaining healthy populations, rather than treating individuals diagnosed with an illness.

12.4 Hypotheses

1. Oldest old persons living in co-residential households will have better psychological dispositions compared to those living alone because co-residence implies a higher access to social resources and higher social integration.
2. Moving from co-residence to living alone will have a negative effect on psychological disposition because this involves a reduction in social resources.
3. Having children living in close proximity will increase psychological disposition because this contributes to available social resources.
4. The effect of living arrangements on psychological disposition will differ between those aged 80–89 years and those aged 90 and older because the demand for elder care increases with age.

12.5 Data and Methods

The empirical analysis uses data from the Chinese Longitudinal Healthy Longevity Survey (CLHLS), including the 1998 (T1) baseline survey and 2000 (T2) follow-up survey (see Chap. 2 of this volume for detailed information on the sample design). The survey covers a random selection of half the counties and cities in 22 provinces ($n = 631$), using multi-stage cluster sampling design, representing about 85 percent of the Chinese population. The survey endeavored to interview all centenarians in the included areas. For each centenarian respondent, an octogenarian (aged 80–89) and a nonagenarian (aged 90–99) living in the same or neighboring area, with predetermined age and sex characteristics, were interviewed. The survey aimed for an equal gender distribution at each age within the octogenarian and nonagenarian categories. The survey over-sampled extremely old individuals and males because these oldest old individuals are few in numbers. Extensive data were collected on health status, health indicators, and socio-demographics. A doctor or nurse gave each respondent a basic medical examination. The overall response rate was 88 percent, but this figure is 98 percent when the deceased, recent migrants, and individuals too infirm to participate are excluded. After removing cases where the key variables are missing, our study sample contains 3,625 individuals from the 2000 follow-up questionnaire (longitudinal data).

The dependent variable is psychological disposition. To measure psychological well-being, each respondent was prompted with the following statement: "People have their own disposition. Here are some statements of people's description of their disposition. How similar are you to these people?" The respondents were not prompted with this question in the 2000 follow-up wave, but the same construct of psychological disposition was used in both waves. The statements composing the disposition construct are as follows: I always look on the bright side of things; I like to keep my belongings neat and clean; I often feel fearful or anxious; I often feel lonely and isolated; I can make my own decisions concerning my personal affairs; The older I get, the more useless I feel; I am as happy now as when I was younger. In the baseline wave, there were five possible responses to each statement, including very similar (1), similar (2), so-so (3), not similar (4), and not similar at all (5). The possible responses in the 2000 follow-up were always (1), often (2), so-so (3), bad (4), and very bad (5). An overall assessment of psychological disposition was derived from the total score of the responses to these statements (Cronbach's alpha = 0.63).

The main independent variable is living arrangements, which was measured with a five-level categorical variable as follows: with a spouse only; with children and/or grandchildren; with siblings, parents, and/or others; in a nursing home; and living alone (reference group). These categories are mutually exclusive, but respondents living with children or grandchildren may be co-residents with other family members, such as parents, siblings, spouses, in-laws, and others.

The analysis included a three-level categorical variable to identify respondents who experienced a change in living arrangements between T1 and T2: moving into co-residence, moving out of co-residence, and other or no changes in living arrangements. About 10 percent of the target population experienced a change in living arrangements. Another dummy variable identified respondents having a child living in close proximity. Close proximity is defined as having a child living in the same village/neighborhood, the same township/district, or the same county/city. About 81 percent of the target population have a child living in close proximity. Table 12.1 presents the descriptive statistics for change in living arrangements, having a child living in close proximity, and all the other explanatory variables.

Socioeconomic status was measured with four variables. Education was measured in years. The distribution of education is skewed, with most respondents having little or no formal education. We experimented with different strategies to measure education (e.g., using a dummy indicator for respondents having no formal education). The results from these analyses are not substantively different from those reported in the paper. A dichotomous variable was used to indicate if the respondent's main occupation at age 60 was in agriculture or fishery. Over half of the target population were employed in agricultural or fisheries work. We used the husband's occupation to measure women's occupational status because many oldest old women were never employed outside the home. The primary source of financial support was measured in three levels: own income, family members, and government assistance. About 24 percent depended on personal income, 69 percent on families, and 7 percent on government assistance. Having fresh fruit (an important socioeconomic status indicator in China after controlling for urban–rural differences) was measured on a four-level ordinal scale: rare/never, occasionally, regularly except in winter, and almost everyday. On average, the elderly in the target population have an occasional frequency of having fresh fruit.

Physical health status was measured with three variables. Self-reported health was measured with a five-level ordinal variable, ranging from excellent (5) to poor (1). The mean score for self-reported health was high (3.8). A dichotomous variable was introduced to indicate the presence of any chronic health conditions, such as hypertension, diabetes, heart diseases, stroke, cataracts, cancer, bedsore, etc. Over half of the target population (51 percent) reported a chronic health problem. An ordinal scale was used to measure activities of daily living, which cover bathing, dressing,

Table 12.1 Definitions and descriptive statistics for control variables used in the multivariate analyses of psychological disposition: China, 1998–2000

Variable	Variable definition and code	Mean or %	SD
Change in living arrangements between T1 and T2			
Moving into co-residence	Dummy indicator (1 = yes, 0 = no)	4.2%	–
Moving out of co-residence	Dummy indicator (1 = yes, 0 = no)	5.9%	–
Other/no change	Reference category	89.9%	–
Proximity (T1)[a]	Dummy indicator (1 = have at least one child living in close proximity)	80.7%	–
Socioeconomic status			
Education	Completed years of schooling	2.09	3.67
Occupation at age 60	Dummy indicator (1 = agriculture/fishery, 0 = otherwise	53.4%	–
Source of financial support (T1)			
Own income	Dummy indicator (1= yes, 0 = no)	23.9%	–
Government assistance	Dummy indicator (1 = yes, 0 = no)	7.2%	–
Family	Reference group	69.0%	–
Fruit (T1)	Frequency of having fresh fruit (1 = rare/never, ..., 4 = almost everyday)	2.06	1.13
Health status			
Health (T1)	Self-reported health status in 5 levels (1 = poor, ..., 5 = excellent)	3.75	0.79
Chronic condition (T1)	Dummy indicator (1 = presence of any chronic condition, 0 = otherwise)	50.7%	–
ADLs (T1)	Number of limitations of activities of daily living (max = 6)[a]	0.41	1.04
Demographic variables			
Female	Dummy indicator (1 = yes, 0 = no)	62.6%	–
Age (T1)	Age in years	88.96	7.18

Table 12.1 (Continued)

Variable	Variable definition and code	Mean or %	SD
Marital status (T1)			
Widowed	Dummy indicator (1 = yes, 0 = no)	68.9%	–
Separated/divorced/never married	Dummy indicator (1 = yes, 0 = no)	3.6%	–
Married[a]	Reference category	27.5%	–
Children (T1)	Number of living children	2.63	2.10
Siblings (T1)	Number of living siblings	0.61	1.02
Ethnicity	Dummy indicator (1 = ethnic minority, 0 = Han)	7.4%	–
Rural residence (T1)	Dummy indicator (1 = yes, 0 = no)	63.5%	–
N		3,625	

Weighted means or percentages, unweighted N.
[a] See text for detailed description.

toileting, transferring, continence, and feeding. The prevalence of activities of daily living limitations is generally low, with the average number under one per elderly person.

The study controlled for several demographic variables. A dichotomous variable was used to indicate female gender. Women comprise about 63 percent of the target population. Age was measured in years. The mean age is about 89 years. A three-level categorical variable was used to measure marital status. The separated and the divorced were combined with never married respondents because these events were uncommon for the oldest old. Approximately 69 percent of the target population are widows, 28 percent are married, and less than 4 percent are separated, divorced, or never married. A continuous variable was used to indicate the number of living children and another for the number of living siblings. On average, elderly persons in the target population have 2.6 surviving children and less than one surviving sibling. A dummy variable was used to indicate minority group membership and another to indicate rural residence. The study population is comprised of about 7 percent ethnic minorities (non-Han Chinese). The majority of the target population (64 percent) lives in rural areas.

The study used random effects models in the data analysis (Laird and Ware 1982) to adjust for the cluster effects in the CLHLS and obtain valid estimates of parameters and standard errors. This statistical method is preferable over OLS models, which would underestimate the standard errors. In this study, we propose the following simplified random effects model:

$$y_{ij} = \mu_N + \sum_{k=1}^{m} x_{ijk}\beta_k + \alpha_i + \varepsilon_{ij}, \ i = 1, 2, \ldots, c, j = 1, 2, \ldots, n_i \quad (12.1)$$

where y_{ij} is the observed value of the dependent variable for the jth respondent in the ith county/city; μ_N is the intercept (the overall mean of the response measure); x_{ijk} represents the kth explanatory variable (including the four dummy variables indicating four different living arrangements and covariates in Table 12.1); β_k is the corresponding unknown fixed-effects parameter; $\alpha_i \sim$ iid $N(0, \delta_N^2)$, and $\varepsilon_{ij} \sim$ iid $N(0, \delta_\varepsilon^2)$. The first two terms on the right-hand side of Eq. 12.1 comprise the fixed effects part of the model, and $(\alpha_i + \varepsilon_{ij})$ forms the random effects part of the model. The variance components δ_N^2 and δ_ε^2 measure the variation, respectively, of counties/cities and respondents in terms of the response measure. Equation 12.1 can also be seen as a two-level hierarchical linear model (Bryk and Raudenbush 1992) because the respondents are nested in each city or country. The parameters and variance components in the random effects models were all estimated using the Restricted Maximum Likelihood (REML) method available in the SAS mixed models procedure (Littell et al. 1996).

12.6 Results

Using first wave data of the CLHLS, Table 12.2 shows that age tends to determine the type of oldest old living arrangement. Across all age groups, living with children or grandchildren remains predominant, but this living arrangement becomes significantly more prevalent with advanced age. In 1998, 62 percent of individuals 79–84 years of age reported living with children or grandchildren. The prevalence was 71 percent for those 85–89, 76 percent for those 90–94, 84 percent for those 95–99, and 85 percent for those 100 and above. Similarly, the prevalence of living with a spouse only and living alone steadily decreases with advanced age, which likely accounts for the increases in living with children or grandchildren.

Table 12.3 presents the unstandardized regression coefficients from the random effects models of the relationship between living arrangements (T1) and psychological disposition (T2). Model 1, which adds T1 mental health and demographic controls, offers a preliminary confirmation of Hypothesis 1. As the baseline model indicates, oldest old individuals living in family co-residential households tend to

Table 12.2 Living of arrangements of oldest old Chinese by selected age groups, 1998

Living arrangements	Age				
	<85	85–89	90–94	95–99	100+
Spouse only	14.8	8.4	4.5	2.5	0.1
Children/great/grandchildren	62.0	70.9	75.7	83.7	85.1
Siblings/parents/others	2.1	2.4	1.9	2.1	3.3
Nursing home	7.7	5.2	5.3	3.2	3.9
Living alone	13.5	13.0	12.6	8.5	7.7
Total	100%	100%	100%	100%	100%

Weighted percentages. Chi-square = 192.3 ($df = 16$), $p < 0.001$, $N = 9,093$.

have superior psychological disposition compared to those living alone. Those living in nursing homes are not significantly healthier than those living alone.

Model 2 in Table 12.3 examines whether a change in living arrangements (between T1 and T2) affects psychological disposition among the oldest old. Considering that undesirable or uncontrollable change is potentially harmful to an individual's mental health (Pearlin 1989), we hypothesized that a change in living arrangements would decrease psychological disposition because it usually represents an undesirable or uncontrollable event for the oldest old. For example, moving into a co-residential living arrangement may be indicative of recent widowhood or the onset of severe functional limitations (loss of control), both of which have negative implications for psychological health (Smith et al. 2002). Moving out of a co-residence usually implies an overall loss of social support and greater chances for social isolation. Confirming Hypothesis 2, Model 2 indicates that both moving into and out of a co-residential household have negative implications for psychological disposition.

Model 3 (in Table 12.3) examines the effects of having a child living in close proximity on psychological well-being among the oldest old. Although previous research indicates that proximity is an important indicator of overall social support available to older Chinese parents (Bian et al. 1998), our findings show a non-significant relationship between proximity and psychological well-being, falsifying Hypothesis 3.

The final model (Model 4) in Table 12.3 includes all previous variables and incorporates socioeconomic and health status controls. Given that socioeconomic status and health status are indicative of the need for social support, the effects of living arrangements and other variables may be suppressed (or intensified) because specific types of individuals are likely to be "selected" into specific types of living arrangements. For example, those individuals with comparatively high functional impairments may be selected into a co-residential household because their physical status reduces their ability to function independently. Such a selection effect could potentially suppress the benefits of co-residential living arrangements, for these households would thus include a disproportionate number of individuals with functional impairments or other health problems, which are health risk factors.

For the most part, however, the findings in the final model are consistent with previous models. The net effects of living arrangement, change in living arrangement (moving out of a co-residence), and proximity are unchanged from the preceding models. But the negative effects of female gender and widowhood disappear in the final model, suggesting that the health-damaging effects of these statuses may be a function of socioeconomic or health disparities. Interestingly, our findings imply that siblings are an important source of social support because the health advantage related to number of living siblings disappears when socioeconomic status and health status are held equal.

Finally, Table 12.3 indicates that individuals from rural areas report lower levels of psychological well-being compared to those from urban areas. This finding is not unexpected considering the deep disparities in state pension coverage and medical services between the rural and urban sectors (Pei and Pillai 1999).

The final model in Table 12.3 is repeated in Table 12.4 by selected age groups. We began the analysis by disaggregating the oldest old into two categories, octogenarians

Table 12.3 Random effects models of psychological disposition (T2) on living arrangements (T1): China, 1998–2000

Independent variable	Model 1	Model 2	Model 3	Model 4
Living Arrangements (T1)				
Spouse only	0.628*	–	–	0.689*
Children/great/grandchildren	0.682***	–	–	0.719**
Siblings/parents/others	1.015*	–	–	1.114*
Nursing home	0.115	–	–	0.385
Living alone[a]				
Change in living arrangements (T1, T2)				
Moving into co-residence	–	−0.848**	–	−0.377
Moving out of co-residence	–	−0.855**	–	−0.973***
Other/no change[a]				
Proximity (1 = yes)	–	–	−0.029	−0.005
Psychological disposition (T1)	0.253***	0.252***	0.254***	0.187***
Socioeconomic status (T1)				
Education	–	–	–	0.090***
Occupation at 60 (agriculture/fishery = 1)	–	–	–	−0.225
Source of financial support				
Own income	–	–	–	0.640***
Government assistance	–	–	–	0.201
Family[a]				
Fruit	–	–	–	0.280***
Health status (T1)				
Self-reported health status	–	–	–	0.318***
Chronic condition (yes = 1)	–	–	–	−0.228 †
ADLs	–	–	–	−0.174**
Demographic variables (T1)				
Female (yes = 1)	−0.587***	−0.582***	−0.585***	−0.188
Age	−0.321	−0.291	−0.291	−0.228
Age square × 100	0.173	0.158	0.159	0.127
Marital status				
Widowed	−0.476*	−0.2**	−0.589***	−0.308
Separated/divorced/never married	−0.170	−0.276	−0.393	−0.077
Married[a]				
Number of living children	0.045	0.049 †	0.045	0.045
Number of living siblings	0.115 †	0.122*	0.117 †	0.095
Ethnicity (minority = 1)	−0.134	−0.126	−0.127	−0.108
Rural residence (yes = 1)	−1.234***	−1.185***	−1.205***	−0.826***
Intercept	35.254***	34.423***	34.379***	29.960**
Covariance parameter estimate				
$\hat{\sigma}_e^2 \hat{\sigma}_N^2$	2.725***	2.675***	2.711***	2.413***
$\hat{\sigma}_N^2 \hat{\sigma}_e^2$	11.599***	11.604***	11.649***	11.321***
−2 REML log likelihood	19638	19635	19653	19540

[a] Reference category.

*$p < .05$ ** $p < .01$ *** $p < .001$ (two-tailed test);.

† $p < .05$ (one-tailed test).

Table 12.4 Random effects models of psychological disposition (T2) on living of arrangements (T1) by selected age groups: China, 1988–2000

	Age	
Independent variable	80–89	90+
Living arrangements (T1)		
Spouse only	0.725 †	0.609
Children/great/grandchildren	0.657*	0.730 †
Siblings/parents/others	1.360*	0.583
Nursing home	0.616	−0.069
Living alone[a]		
Change in living arrangements (T1, T2)		
Moving into co-residence	−0.144	−0.707
Moving out of co-residence	−1.215***	−0.516
Other/no change[a]		
Proximity (1 = yes)	0.214	−0.355
Psychological disposition (T1)	0.147***	0.220***
Socioeconomic Status (T1)		
Education	0.124***	0.036
Occupation at 60 (agriculture/fishery = 1)	−0.157	−0.269
Source of financial support		
Own income	0.851***	0.194
Government assistance	0.286	0.139
Family[a]		
Fruit	0.261**	0.374***
Health status (T1)		
Self-reported health status	0.362***	0.270 †
Chronic condition (yes = 1)	−0.238	−0.304
ADLs	−0.307**	−0.077
Demographic variables (T1)		
Female (yes = 1)	−0.175	−0.371
Marital status		
Widowed	−0.214	−0.202
Separated/divorced/never married	−0.062	0.042
Married[a]		
Number of living children	0.063	0.014
Number of living siblings	0.111	0.089
Ethnicity (minority = 1)	0.504	−0.682
Rural residence (yes = 1)	−1.032***	−0.669**
Intercept	20.368***	19.407***
Covariance parameter estimates		
$\hat{\sigma}_N^2$	1.740***	3.0***
$\hat{\sigma}_e^2$	11.491***	11.147***
−2 REML log likelihood	11426	8165

[a] Reference category.
* $p < .05$ ** $p < .01$ *** $p < .001$ (two-tailed test).
† $p < .05$ (one-tailed test).

and nonagenarians/centenarians. Small sub-sample sizes prevented further breakdown of the age groups. As expected, the findings confirm that living arrangements (and other variables) have differential effects on psychological disposition between these age groups, confirming Hypothesis 4. For octogenarians, all family co-residential households have a beneficial effect on psychological well-being. However, for those aged 90+, only living with children or grandchildren has a beneficial effect, with all other living arrangements having non-significant effects.

Living with a spouse only (or with siblings, parents, or others) becomes non-significant for those 90+ probably because the quality of social support a household can supply is partially a function of the mean age of all household members. Those individuals who are 90+ and living with a spouse only (or with siblings, parents, or others) may not have sufficient social support, despite living in a co-residential household. Assuming that the members of these households are roughly of similar age, the available social support would be significantly depreciated by the advanced age of each household member. For example, a 95 year-old spouse or sibling could not be expected to supply the same level of activities of daily living assistance as an 80 year-old spouse or sibling could.

12.7 Discussion and Conclusion

We argued in this chapter that living arrangements represent differences in social support and, therefore, explain the distribution of different psychological dispositions among the oldest old. We began our empirical analysis by examining the gross effects of different living arrangements on disposition. The results confirmed the primary hypothesis, indicating that individuals living in family co-residential households have better dispositions than those living alone. These results are consistent with well-established patterns indicating that elderly individuals benefit from social support provided by adult children, particularly when the demand for assistance is greatest, such as from functional impairments that impede independent living (Chen and Silverstein 2000; Silverstein and Bengtson 1994). After introducing socioeconomic and health status controls (including functional limitations), both highly predictive of the individual demand for social support, there was no significant change in the relationship between living arrangement and psychological disposition.

Two noteworthy trends are evident in these results. First, the differences between our selected family co-residential living arrangements are non-significant. That is, in analyses not reported here, findings indicate that living with children or grandchildren does not have a greater effect than living with a spouse only. Similarly, living in a family household (with children, grandchildren, or a spouse only) is no better or worse for psychological disposition than living with siblings, parents, or others. Second, the benefits of family co-residential living arrangements persist regardless of socioeconomic and health disparities, or variation in the need for instrumental support. These findings indirectly specify that emotional support is an especially relevant factor in the psychological disposition of the oldest old. Moreover, assuming that the main difference between our selected family co-residential living arrangements is structured by differences in the provision of instrumental support (with younger, multi-generation households presumably offering the best supplies), this type of support has a weaker effect on psychological disposition than expected.

The inferred effect of family-based emotional support is not surprising. Many oldest old individuals have few, if any, surviving friends given their advanced age. Further, the steady onset of functional limitations reduces their mobility, an obvious constraint on an active social life. From this perspective, family members are an important, and perhaps exclusive, source of social interaction and companionship for many oldest old people. To the extent that emotional support reflects belonging to a caring social network, it may have more profound effects on psychological disposition than instrumental support; it is dependent on having close family members or friends, whereas instrumental support can be obtained from far less personal sources. In fact, findings from prior research demonstrate that emotional support is less replaceable than instrumental support for the oldest old (Walter-Ginzburg et al. 1999).

While we continue to believe that instrumental support is crucial, its effect may be suppressed because a decline in personal control (i.e., increased dependence on others) often parallels the demand for instrumental support (Chen and Silverstein 2000). The presumed greater supply of instrumental support within younger, multi-generation households is not meaningless. Although we controlled for the expected demand for social support (i.e., socioeconomic and health disparities), if younger, multi-generation households do indeed contain the most individuals with relatively high functional impairment, then differences in personal control between these individuals and those from other household types may well explain why there are no significant differences in how our selected family co-residential households influence psychological disposition. Following this logic, a task of future research should examine how types of social support vary across different living arrangements, and how variations in the demand for specific types of social support pattern the distribution of psychological health in the elderly population.

This study contributed to the literature by further explicating the relationship between living arrangements and psychological disposition among the oldest old. Family co-residence remains the "core of support" for elderly people (Bian et al. 1998). Current demographic projections and social policies suggest that an increasing number of elderly individuals will have to live alone (Zeng et al. 2002). This means that these individuals may experience unmet instrumental and emotional needs. Further, primary eldercare givers will face an increasingly heavy burden of domestic work. Since women are the main providers of eldercare, this burden may reduce Chinese *women's* mental health through prolonged role strain, or even suppress equality by restricting women's capacity to enter the labor market full-time. In these respects, government intervention is required to foster healthy aging and ensure that the burden of eldercare does not become a persistent source of gender inequality.

Acknowledgments The authors gratefully acknowledge financial support from a Social Sciences and Humanities Research Council (SSHRC) grant. Additional research support was provided by the Department of Sociology, the University of Victoria. We thank Zeng Yi for providing the data for the analysis and Danan Gu for technical assistance. Direct all correspondence to Zheng Wu, Department of Sociology, the University of Victoria, P. O. Box 3050, Victoria, British Columbia, V8W 3P5 Canada. E-mail: zhengwu@uvic.ca

References

Albrecht, G.L. and P.J. Devlieger (1999), The disability paradox: High quality of life against all odds. *Social Science and Medicine* 50, pp. 757–759
Anson, O. and S. Sun (2002), Gender and health in rural China: Evidence from HeBei Province. *Social Science and Medicine* 55, pp. 1039–1054
Antonovsky, A. (1979), *Health, stress and coping*. San Francisco: Jossey-Bass.
Antonovsky, A. (1987), The salutogenic perspective: Toward a new view of health and illness. *Advances* 4, pp. 47–55
Baltes, P.B. and M.M. Baltes (1990), Psychological perspectives on successful aging: the model of selective optimization with compensation. In: P.B. Baltes and M.M. Baltes (eds.): *Successful aging: Perspectives from the behavioral sciences*. Cambridge: Cambridge University Press, pp. 1–34
Bian, F., J.R. Logan, and Y. Bian (1998), Intergenerational relations in urban china: proximity, contact, and help to parents. *Demography* 35, pp. 115–124
Bryk, A.S. and S.W. Raudenbush (1992), *Hierarchical linear models*. Newbury Park: Sage.
Chen, C. (2001), Aging and life satisfaction. *Social Indicators Research* 54, pp. 57–79
Chen, X. and M. Silverstein (2000), Intergenerational social support and the psychological well-being of older parents in China. *Research on Aging* 22, pp. 43–65
Gao, J., S. Tang, R. Tolhurst, and K. Rao (2001), Changing access to health services in urban China: implications for equity. *Health Policy and Planning* 16, pp. 302–312
Gecas, V. (1989), The social psychology of self-efficacy. *Annual Review of Sociology* 15, pp. 291–316
Grogan, C.M. (1995), Urban economic reform and access to health care coverage in the People's Republic of China. *Social Science and Medicine* 41, pp. 1073–1084
Hsiao, W.C. (1995), The Chinese health care system: Lessons for other nations. *Social Science and Medicine* 41, 1047–1055
Kindig, D. and G. Stoddart (2003), What is population health? *American Journal of Public Health* 93, pp. 380–383
Kobasa, S.C., S.R. Maddi, and S. Kahn (1982), Hardiness and health: A prospective study. *Journal of Personality and Social Psychology* 42, pp. 168–177
Kwong, P. and G. Cai (1992), Ageing in China: trends, problems, and strategies. In: D.R. Phillips (eds.): *Ageing in East and South-East Asia*. London: Edward Arnold, pp. 105–127
Laird, N.M. and J.H. Ware (1982), Random-effects models for longitudinal data. *Biometrics* 38, pp. 963–74
Lin, J. (1995), Changing kinship structure and its implications for old-age support in urban and rural China. *Population Studies* 49, pp. 127–145
Littell, R.C., G.A. Milliken, W.W. Stroup, and R.D. Wolfinger (1996), *SAS system for mixed models*. Cary: SAS Institute.
Liu, Y., W.C. Hsiao, Q. Li, X. Liu, and M. Ren (1995), Transformation of China's rural health care financing. *Social Science and Medicine* 41, pp. 1085–1093
Liu, Y., W.C. Hsiao, W.C., and K. Eggleston (1999), Equity in health and health care: the Chinese experience. *Social Science and Medicine* 49, pp. 1349–1356
Olsen, P. (1993), Caregiving and long-term health care in the People's Republic of China. *Journal of Aging and Social Policy* 5, pp. 91–110
Pearlin, L.I. (1989), The sociological study of stress. *Journal of Health and Social Behavior* 30, pp. 241–256

Pearlin, L.I., M.A. Lieberman, E.G. Menaghan, and J.T. Mullan (1981), The stress process. *Journal of Health and Social Behavior* 22, pp. 337–356

Pei, X. and V.K. Pillai (1999), Old age support in China: The role of the state and the family. *International Journal of Aging and Human Development* 49, pp. 197–212

Rodin, J. (1986), Aging and health: the effects of sense of control. *Science*, 233, pp. 1271–1276.

Ross, C.E. and J. Mirowsky (2002), Family relationships, social support, and subjective life expectancy. *Journal of Health and Social Behavior* 43, pp. 469–489

Silverstein, M. and V.L. Bengtson (1994), Does intergenerational social support influence the psychological well-being of older parents? the contingencies of declining health and widowhood. *Social Science and Medicine* 38, pp. 943–957

Schieman, S. and H.A. Turner (1998), Age, disability and the sense of mastery. *Journal of Health and Social Behavior* 39, pp. 169–186

Shi, L. (1996), Access to care in post-economic reform rural China: Results from a 1994 cross-sectional survey. *Journal of Public Health Policy* 17, pp. 347–361

Smith, J., M. Borchelt, H. Maier and D. Jopp (2002), Health and well-being in the young old and oldest old. *Journal of Social Issues* 58, pp. 715–732

Smith, J., D. Gerstorf and Q. Li (2004), Psychological resources for healthy longevity. Paper presented at a workshop on Determinants of Healthy Longevity in China, August, Max Planck Institute for Demographic Research, Rostock, Germany

Thoits, P.A. (1995), Stress, coping, and social support processes: where are we? What next? *Journal of Health and Social Behavior* 36, pp. 53–79

Tracy, M.B. (1991), *Social policies for the elderly in the Third World*. New York: Greenwood Press.

United Nations (2002), *World population ageing 1950–2050*. New York: United Nations.

Walter-Ginzburg, A., T. Blumstein, A. Chetrit, J. Gindin and M. Baruch (1999), A longitudinal study of characteristics and predictors of perceived instrumental and emotional support among the oldest old in Israel. *International Journal of Aging and Human Development* 48, pp. 279–299

Zeng, Y. and L. George (2001), Extremely rapid ageing and the living arrangements of the elderly: The case of China. In: *Living arrangements of older persons: Critical issues and policy responses*. New York: United Nations Population Division, pp. 255–287

Zeng, Y. and J.W. Vaupel (2002), Functional capacity and self-evaluation of health and life of the oldest old in China. *Journal of Social Issues* 58, pp. 733–748

Zeng, Y., J.W. Vaupel, Z. Xiao, C. Zhang, and Y. Liu (2001), The healthy longevity survey and the active life expectancy of the oldest old in China. *Population: An English Selection* 13, pp. 95–116

Zeng, Y., J.W. Vaupel, Z. Xiao, C. Zhang, and Y. Liu (2002). Socioeconomic and health profiles of the oldest old in China. *Population and Development Review* 28, pp. 251–273

Zimmer, Z. and J. Kwong (2003), Family size and support of older adults in urban and rural China. *Demography* 40, pp. 23–44

Chapter 13
Health and Living Arrangement Transitions among China's Oldest-old

Zachary Zimmer

Abstract This chapter begins with the notion that families in China act altruistically toward the old in that they operate as a single corporate unit, aiming toward the comfortable survival of all members. Coresidence with elders, based on this perspective, more likely occurs when needs are greatest, for instance, when health deteriorates or spouse dies. There is also the possibility of gender variation due to differences in authority and emotional bonds between older women and men and their families. These notions are tested. Cross-sectional and transitional multinomial models link health status measures with living arrangement outcomes. Results show changes in living arrangements occur frequently. Functional limitations are more strongly associated with living arrangements than are other health indicators. Health indicators are more strongly related for those not married. Gender interactions show health change most likely triggers a living arrangement response for women. Implications for a rapidly aging China are discussed.

Keywords Activities of daily living, Age, Aging, Altruism, China, Chinese Healthy Longevity Survey, Cognition, Health disorders, Family, Filial piety, Functional limitations, Gender, Health, Living arrangements, Living alone, Living with children, Living with others, Living with spouse, Longitudinal analysis, Marital status, Medical health, Mini-mental state exam, Multinomial regression, Oldest-old, Panel data, Social change, Social support, Traditional family, Transitional model

13.1 Introduction

The oldest-old are at a high risk of experiencing chronic and mental health disorders and diminished ability to undertake functioning tasks necessary for self-maintenance

---

Z. Zimmer
Department of Sociology, University of Utah, 260 S. Central Campus Drive, Room 214, Salt Lake City, UT 84112, USA
e-mail: zachary.zimmer@ipia.utah.edu

(Zeng et al. 2002). As a consequence, they require high levels of physical, emotional and material support. While support can hypothetically come from a variety of sources, including personal resources, public programs, and exchanges with individuals that are part of a social network, for the oldest-old of China, the first two are likely to be limited in scope, and reliance is weighted toward network sources of support (Yuan 1990; Zimmer and Kwong 2003). Support for the oldest-old in China is in part a function of living with or near kin, since those who coreside with family are more likely to receive the help they need compared to those who do not (Logan et al. 1998; Pei and Pillai 1999; Yan and Chi 2001). In fact, household composition for the oldest-old can be considered an indicator of well-being, and may be the result of a life-long strategic process that positions them for late-life security (Albert and Cattell 1994). This chapter examines living arrangements and living arrangement transitions. A central focus compares the types of transitions typical of those with health problems versus those whose health is relatively good.

13.1.1 Theoretical Perspective and Study Hypotheses

In the opening chapter of his edited volume *China's Revolutions and Intergenerational Relations*, Whyte (2003) notes that historically the needs of the old in China were thought to come ahead of other family members, and as far as grown children were concerned, the dutiful repayment of debt was necessary for having been born and raised by parents. Whyte (2003: 8) observes that "(w)hatever other insecurities they faced, most parents could rest secure in the knowledge that their children would place parental needs ahead of their own." Filial piety and dutiful repayment to older members of a family were facilitated through a system of coresidence with grown children, usually a married son. The son would ensure the material needs of the older adult, while the daughter-in-law would take care of physical needs. The Communist Party took over political power in China in 1949 and advanced policies that attempted to erode family functions and stress the importance of loyalty to the state (Leung 2001). Despite attempts to de-emphasize the family, filial support did not weaken to any great degree, and there were overall minimal changes in normative structures and living arrangements (Treas and Chen 2000; Whyte 2003). Today in China, older adults are integrated into a system where they both receive and provide support within the family, and they have a relatively high chance of living with grown children or other family members (Liang et al. 1992; Logan et al. 1998).

Enduring filial piety and high levels of coresidence between an older adult and their children are often explained using an "altruistic" perspective (Becker 1974; Lee et al. 1994; Hermalin 2002; Zimmer and Kwong 2003). This perspective presumes that in societies that adhere to altruistic ideals the family acts to pool resources, allocates them efficiently, and in this way assures the survival of each member. This assumes that physically able family members are obliged to provide support for those unable to take care of their own self-maintenance. It also assumes that those most in need will receive the greatest volume of support, even if they have little to offer in return, since their survival is critical for the well-being of the entire family.

Recent empirical evidence from China tends to substantiate the altruistic nature of the family by showing that support is most likely provided to older adults in greatest need, that, in turn, older adult needs underlie intergenerational coresidence, and that tendencies to coreside and receive support from coresident family members have changed little despite rapid changes in other social and economic structures (Lee et al. 1994; Logan et al. 1998; Pei and Pillai 1999; Whyte 1997, 2003). In this chapter we examine health status as an indicator, and assume that the altruistic tendency of the family will become most evident in cases where an older adult exhibits poor physical and/or mental health: those with poor health will be more likely to live with others, to move in with others over time, and to remain living with others over time.

Although the altruistic perspective has been useful in explaining high rates of coresidence, it tends to ignore several dynamics that result in variations in associations between health and living arrangements. First, effects are likely to differ across marital status (Friedman et al. 2003; Knodel and Ofstedal 2002). In a practical sense, marital status determines the set of living arrangements available to an individual. For instance, married individuals are at very low risk of living alone. This creates the analytical problem of different risk sets across different marital statuses. In a theoretical sense, the spouse is likely to be the one to provide support first and foremost; but older adults who are widowed or unmarried may be particularly vulnerable and therefore in greatest need. Having a spouse, then, is likely to influence the nature of relationships with living arrangements, and health may be a less consequential determinant for those with a spouse, while those without a spouse in poor health may be particularly likely to be living with children or other family members.

Second, there may be distinct gender differences in the tendency to receive support and live with others. A common perspective for examining the consequences of gender in the aging experience is to consider life-course effects, stressing the links between earlier and later stages of life (Friedman et al. 2003). For instance, because of a lifetime spent in marginal economic and decision-making positions, elderly women living in patriarchal societies such as China are often thought to be in a vulnerable position, not possessing the power or the economic authority needed to demand support from the younger generation (Mason 1992). An alternative view, posited in a number of recent studies, is that older women may actually command greater emotional loyalty than their male counterparts due to the recognition that they have sacrificed more in raising the family, and to the subsequent emotional bonds that develop between mothers and children (Beales 2000; Kandiyoti 1988; Knodel and Ofstedal 2002; Yount 2005). In addition, older women gain respect of the family in other ways, such as through their ability to continue to be functionally useful by conducting household tasks and acting as caretakers of grandchildren, even after male usefulness as income earners has declined. Hence, in some ways, the status of women and their ability to garner respect and support from family members surpasses that of men in later years.

Based on the above, we propose several hypotheses:

(1) The oldest-old with health problems are more likely to be in coresident living situations than are those without health problems.

(2) The oldest-old with health problems are more likely to move into a coresident living situation or remain coresiding with others than are those without health problems.

Given differences in effects across marital status, we also hypothesize that:

(3) The above relationships should be stronger among those not married as opposed to those married.

With respect to gender, we make no assumptions regarding differences in effects, but we do test for these with the introduction of interactions between gender and health.

Recent empirical work in China, in other Confucian societies, and throughout Asia, has generally supported the first of the above stated hypotheses by showing that living arrangements of older adults are at least partly a function of health; however, these studies do not especially focus specifically on the oldest-old, who would tend to be most likely to have health problems (Brown et al. 2002; DaVanzo and Chan 1994; Frankenberg et al. 2002; Kim and Rhee 1997; Knodel and Ofstedal 2002; Logan and Bian 1999; Martin 1989). Studies show that other important determinants of living arrangements, besides sex and martial status, are size of family, socioeconomic status and age. Studies have largely been based on cross-sectional data. However, living arrangements of older adults are dynamic, changes can occur over short time intervals, and these cannot be fully captured using cross-sectional data. The second hypothesis, which requires panel data, has thus received little attention.

13.2 Data, Measures and Methods

The data we use for the analyses conducted in this chapter are from the 1998 Chinese Healthy Longevity Survey and the 2000 follow-up. (The two waves will be referred to as time of origin and time of follow-up.) Details of the study and the sample can be found in a number of publications (Zeng et al. 2001, 2002, 2003) and in Chap. 2 of this volume. Our research is restricted to those living in the community at origin and those with any living children both at time of the initial survey in 1998 and at follow-up in 2000. This limits the final sample size at origin to 7,398 respondents. By the time of the 2000 survey (from this point forward referred to as the time of follow-up), 0.4 percent of the individuals had moved into a nursing home, 18 percent had died, and 11 percent were lost to follow-up. All results (including Ns) pertain to the weighted sample.

13.2.1 Living Arrangement

The outcome for this study is living arrangement at follow-up. No single agreed upon living arrangement classification scheme exists. Yet, the classification decision is an important one, given that the categories used and the scope of the definition has implications for transitions. For instance, dichotomously measured, as, say, living alone versus with others, the structure of a household may appear to be stable, that is, those living with others may rarely make a move to living alone, and those living alone may rarely move in with others. But, separated into multiple subcategories, for instance, ones that account for the specific relationships of each individual in a household, the structure may be deemed to be much more flexible. For instance, an individual may continue to live with others, but the relationships among individuals living in the household may change.

There have been several schemes put into use in cross-cultural and cross-national research that account for marital status, household headship, and living situation vis-a-vis offspring (Cowgill 1986; Kinsella 1990; Knodel and Ofstedal 2002; Shanas et al. 1968). These various schemes have been drawn together in a recent discussion by DeVos (2003). The classification used here is based on this latter discussion. It is meant to highlight important divisions in Chinese households while being parsimonious, but not so limiting as to underestimate important changes that take place in household structure. Perhaps the most basic distinction in China is the division between living with and without children (Logan et al. 1998; Zeng et al. 2001). Due to the age of the current sample, living with children means, in all cases, living with adult children. Those who do not live with children may be living alone or with a spouse only, depending upon their marital status. In addition, there may be others living in the household, such as siblings or grandchildren.

The outcome variable in this research first determines whether an individual lives with at least one child. If so, then the person is categorized as living with children, regardless of whether there are others also in the household. Other possible categories for married individuals not living with children are living with spouse only or living with spouse and others. We simply call the latter category "living with others." Other possible categories for those not married are living alone or living with others. When married and unmarried are combined, the living arrangement measure has four categories: with children; with spouse; alone; with others. But those currently married are not considered to be at risk of living alone and those currently not married, that is, those currently divorced, separated or widowed, are not considered to be at risk of living with a spouse.

13.2.2 Health

The first health measure is whether or not an individual has a functional limitation. This is determined by questions about ability conducting the following Activities of Daily Living (ADLs): bathing, dressing, toileting, getting out of bed or a chair, continence, and feeding (Katz et al. 1963). Those who report that they cannot conduct at least one of these tasks, on their own, have a functional limitation.

The second is reporting of a medical health problem. This is constructed from questions that ask whether respondents suffer from any of the following: hypertension, diabetes, heart disease, stroke, respiratory disease, tuberculosis, cataract, glaucoma, cancer, gastrointestinal disease, Parkinson's disease, or bedsores. Those who report at least one of these is considered to have a medical health problem, which, here forward, is simply referred to as a health problem. It is understood that these problems are often asymptomatic and may require diagnosis from a medical professional, while contact with a medical professional may not be frequent in some parts of China, particularly in rural areas. Therefore, caution needs to be taken when interpreting these results.[1]

The third is a measure of cognitive health using nineteen items from the Mini-Mental State Examination (MMSE) (Folstein et al. 1975). The MMSE and subscales derived from it have been used extensively to measure cognitive functioning of older adults across cultures, including the Chinese, and it has been found to be a valid measure of cognition, with high internal and content validity, among Chinese populations (Chiu et al. 1994; Herzog and Wallace 1997; Katzman et al. 1988; Ofstedal et al. 1999). The nineteen items include five on orientation (for example, naming the current day and month), six on word recall (words are mentioned and respondents are asked to repeat them), five on calculation (individuals are asked to subtract three from twenty, then three from the resulting total, and so on), and three on language (repeating a sentence and naming simple items that are shown to the respondent, like a pen). A continuous measure is calculated by scoring one point for each mistake a respondent makes. Those who score 0 are cognitively intact, while those scoring 19 have severe dementia.[2]

13.2.3 Covariates

Multivariate models control for other variables that have proven to be important determinants of living arrangements among older adults in Asian societies (Albert and Cattell 1994; Brown et al. 2002; DaVanzo and Chan 1994; Frankenberg et al. 2002; Knodel and Ofstedal 2002; Logan et al., 1998; Martin 1989; Zimmer and Kim 2001). These include socioeconomic characteristics, availability of children,

---

[1] Various coding schemes for both functional limitation and health problems were tried, including continuous measures and categorical measures dividing severity. A dichotomous coding was deemed best since both measures tend to be quite skewed, with very few respondents reporting more than one or two limitations or problems, and testing revealed few significant differences between those with more than one functional limitation or health problem.

[2] Regarding missing responses, there were 43 missing responses for functional limitation (before limiting the sample for those living in the community with children at both time points), but only four respondents had more than one missing response across the six items. For the 39 with only one missing response, they were considered to have a functional limitation if they reported a difficulty with one of the remaining five items. The other four cases were excluded from the analysis. For the health problems, there are quite a few missing responses (up to 10 percent of the population per item), and these are treated as not reporting the problem. Thus, health problems may be underestimated and this presents another reason to interpret with caution. For the MMSE score, there was a category for "cannot answer the question" for each of the 19 items. It is assumed that this represents a cognitive impairment and is coded as an incorrect response; however it is also possible that some legitimate missings are miscoded.

and marital status. Socioeconomic characteristics are age, gender, education, occupation and rural residence. Tests show that a linear treatment of age is adequate, while the other four measures are dichotomously measured.

Availability of children includes number of living children and having a son. For multivariate purposes, number of children is coded as dummy variables that contrast having one child with having two, three, four, or five or more children. This coding decision was determined by preliminary tests that found the measure to be related to living arrangement in a non-linear way.

Our analysis divides the sample into those not married and married at follow-up, but marital status at origin, measured dichotomously, is controlled. Changes in marital status over time run in both directions. Of those who originated married and survived to follow-up, about 21 percent (311 cases) were not married at follow-up, likely due to the death of a spouse. Of those not married at time of origin and surviving to follow-up, about 2 percent (79 cases) reported being married at follow-up. Table 13.1 provides means, standard deviations, and a description of the coding scheme, for all of the above determinants, measured at time of origin.[3]

13.2.4 Analytical Strategy

The analysis begins with descriptive breakdowns of living arrangements at time of origin and follow-up. Crosstabulations show living arrangement distributions at

Table 13.1 Means and standard deviations of study measures, determined at time of origin, and coding used for analyses

Measure	Mean	Standard deviation	Coding
Age	83.807	3.611	Continuous from 80 to 105
Female	0.619	0.486	1 if female, 0 if male
Education	0.374	0.484	1 if any, 0 if none
Agricultural occupation	0.538	0.499	1 if yes, 0 if no
Rural residence	0.659	0.474	1 if rural, 0 if urban
Number of children	3.723	1.913	Dummy variable coding contrasting 1 child versus 2, 3, 4, and 5 or more
Has a son	0.901	0.298	1 if yes, 0 if no
Is married at origin	0.275	0.446	1 if yes, 0 if no
Has functional limitation	0.180	0.384	1 if yes, 0 if no
Has health problem	0.479	0.500	1 if yes, 0 if no
MMSE score	2.894	3.956	Continuous from 0 to 19

---

[3] Again, with respect to missing responses, there were some, though very few, missing responses across the other determinants, and they were handled in the ways outlined in this note. These decisions were all made after further analysis of the data on a number of levels: 8 cases missing on marital status are likely not married and are coded as such; 34 cases with missing education likely have no education and are coded as such; 4 cases with missing occupation are coded as being non-agricultural; and 7 cases with missing ages are assigned the mean age of all other respondents.

both times, distributions at follow-up by living arrangement at time of origin, and competing risks, that is, being interviewed at follow-up versus moving into a nursing home, dying, or being lost to follow-up by living arrangement at time of origin.

The multivariate analysis follows closely the methodology used by Brown et al. (2002). Multinomial logistic regression equations regress the probability of being in a particular living arrangement at the time of follow-up on health and other covariates. For reasons referenced earlier, the sample is divided into those not married and those married. Two models are examined. The first looks at living arrangements at follow-up according to determinants measured at origin. The second adds living arrangement at origin. This changes the interpretation from effects on living arrangements to effects on living arrangement transitions (Brown et al. 2002). We code living with children and living with others at origin as dummy variables, which contrasts these situations with living independently. When these two equal 0, coefficients comment on the probability that someone who lives independently at time of origin moves in with children or with others by follow-up. Similarly, coefficients can comment on the probability that someone remains in a particular living arrangement. Therefore, a transition is defined as the probability of being in any living arrangement state at follow-up given a particular state at origin.

By first considering the influence of determinants on cross-sectional living arrangements, and then by considering the influence on transitions, we are able to assess whether living arrangements respond to determinants. For instance, if in the first model an individual with poor health is found to be more likely to currently live with children, and in the second model an individual with poor health is found to be more likely to change to living with children, then it is relatively certain that the event of poor health is triggering a response leading to living with children. As such, this hierarchical technique is equivalent to an event history. What is not tested is how quickly living arrangements respond to the determinants. This is because the data do not include the timing of events and rather look at discrete points in time.

The three health indicators used are interrelated. For instance, the existence of health problems or a cognitive disorder may lead to a functional limitation. We therefore test models with indicators entered simultaneously and use each entered independently. In order to assure that collinearity does not obscure results, we present coefficients for these independent tests.

The final procedure introduces interaction terms for gender by each health measure and tests whether and how living arrangements and transitions of women and men are influenced by health problems.

13.3 Results

Table 13.2, presented in three panels, shows several living arrangement distributions. For descriptive purposes, those married and not married are not separated in this table. Panel A presents the distribution living in four categories of living arrangement at time of origin and follow-up. This panel shows relative stability

Table 13.2 Living arrangement distributions

	Living arrangement state →	Alone	With spouse	With children	With others	Total N
Panel A	Percent living in this state at...					
	origin	12.2	12.8	68.3	6.8	7,398
	follow-up	14.0	12.3	67.1	6.6	5,182
Panel B	Percent of survivors originating in this state who...					
	lived alone at follow-up	64.8	7.1	5.9	16.5	632
	lived with spouse at follow-up	3.3	68.5	3.4	7.4	683
	lived with children at follow-up	25.1	19.2	87.1	39.0	3,541
	lived with others at follow-up	6.8	5.2	3.6	37.1	360
	changed status by follow-up	35.2	31.5	12.9	62.9	
Panel C	Percent originating in this state who...					
	were interviewed at follow-up	70.5	72.7	69.2	72.8	5,182
	were in nursing home at follow-up	0.7	0.7	0.3	0.4	30
	died before follow-up	17.4	13.2	20.0	14.8	1,364
	were lost to follow-up	11.5	13.3	10.6	12.0	823

in the distribution. For instance, the percent living alone increased only slightly from origin to follow-up, from 12.2 to 14.0 percent. The percent living with spouse and with children decreased slightly, and the percent living with others increased slightly. The most common arrangement is living with children, and in both survey waves this describes about 2/3 of respondents.

Despite stability in cross-sectional distributions, Panel B shows that when looking at follow-up status according to status at origin, there were, indeed, a good number of changes. For instance, among those who originated alone, about 65 percent were still living alone at follow-up while the remaining 35 percent were living in one of the other three arrangements, the most popular of which was living with children. Looking across originating states, it is clear that the majority of "movers" that is, those living in a different state at follow-up than at origin, moved in with children. Further calculations show that among those who originated alone, with spouse only, or with others, and ended up in a different arrangement, about 2/3 ended up living with children. The most stable state was living with children, as about 87 percent of those who originated in this state still lived in it at follow-up. The least stable state was living with others. Only 37 percent remained in this state at follow-up, while fully 39 percent of those living with others lived with children at follow-up. This suggests that living with others is often a transitory state. Of course, it is possible that those living with children are also living with others, but living with children takes priority in the coding scheme.

Further calculations also show that about one in four were living in a different arrangement at follow-up compared to at origin. This can only be an underestimate of the total number of changes that took place. Individuals may have experienced changes in the inter-survey period that are not recorded using discrete time data. For instance, an individual may move from living with children to living alone and back to living with children, and the data would record that their living arrangement remained stable. Also, there are other more sensitive types of changes not included in this categorization. For instance, an individual may change from living with a son to living with a daughter, or living with a grandchild to living with a sibling. In both cases, the categorization results in stability. It is difficult to determine the precise number living with different individuals at origin and at follow-up, but closer inspection of all available information on coresident individuals (for instance, their age and sex), suggests that it is well over half.

Attrition is often a problem in longitudinal studies. In the current dataset, about 11 percent of the original sample was lost to follow-up, and there is a possibility that these loses bias the results shown here and in the tables to follow. For this reason, we also present Panel C, which examines competing risks. Here we show the percent of those living alone, with a spouse, with children and with others, at time of origin, that survived, moved to a nursing home, died, and were lost to follow-up. It is noteworthy that the percent lost to follow-up is consistent across originating living arrangements. Therefore, there is no initial indication that those lost to follow-up are from a particularly unusual group with respect to the originating living arrangements. About 70 percent of the originating sample responded to the follow-up, and there is also very little difference in this proportion across originating living arrangement states. However, those originating living with children were most likely to die, a result that is consistent with older adults in poorest health moving in with children for their final years of life. Moves into a nursing home were very rare.[4]

13.3.1 Determinants of Living Arrangements and Living Arrangement Transitions

Table 13.3 presents multivariate models predicting living arrangements at follow-up separately for those not married and those married. The contrast category in both cases is the following: for those not married the models contrast the probability of living with children and living with others relative to the probability of living alone;

---

[4] We further examined loss to follow-up by comparing this group to survivors with respect to originating health status. The lost to follow-up group was a little more likely to have a functional limitation at origin (20.7 versus 14.0 percent) and to have a health problem at origin (55.4 versus 46.1 percent) although their MMSE scores were a little lower (means of .584 versus .612). These differences are not extremely large. Still, if poor health leads to changes in living arrangements, the main association being tested here, then it is possible that attrition will lead to an underestimate of the number of changes. It is also possible, since their health is somewhat worse at origin, that there are some unaccounted for deaths in the loss to follow-up group. We also examined the loss to follow-up group across other variables that are considered in the current analysis. They turn out to be fairly similar to the surviving group across most of these. Although we can never know for certain what happened to those who were not reinterviewed, the sum of these preliminary analyses did not indicate that the loss to follow-up group are particularly unusual with respect to any of the key variables being considered in this research.

Table 13.3 Multinomial logistic regression results showing log odds ratios for living with children or with others relative to living either alone or with spouse only, by marital status at follow-up

	Not married at follow-up				Married at follow-up			
	Model 1		Model 2		Model 1		Model 2	
	Lives with at least one child[a]	Lives with others[a]	Lives with at least one child[a]	Lives with others[a]	Lives with at least one child[b]	Lives with others[b]	Lives with at least one child[b]	Lives with others[b]
Sociodemographics								
Age	0.049**	0.101**	0.060**	0.115**	0.051*	0.049	0.042	.011
Female	−0.191***	−0.804**	−0.099	−0.761**	0.058	0.157	−0.131	0.310
Education	0.128	−0.172	0.244***	−0.090	0.247***	0.404	−0.014	0.545
Agricultural occupation	0.135	−0.453**	0.014	−0.658**	−0.008	−0.062	−0.361*	0.003
Rural residency	−0.155	0.184	−0.050	0.316	0.082	−0.635*	0.160	−0.626***
Availability of children								
Two children[c]	−0.275	0.221	−0.172	0.533*	−0.071	−0.065	−0.288	−0.831
Three children[c]	−0.524**	−0.293	−0.503	0.115	−0.852**	−0.176	−0.677***	−0.811
Four children[c]	−0.193	0.285	−0.140	0.658*	−1.538***	−0.558	−1.648**	−1.028*
Five or more children[c]	−0.501**	−0.507*	−0.374	−0.010	−1.211**	−1.243*	−1.073*	−1.889***
Has a son	0.889**	−0.792**	0.151	−0.810**	0.096	0.024	0.048	1.011***
Married at origin	−0.560**	0.538*	−0.143	0.915**	−0.368	−0.762	0.601*	−0.264
Health[d]								
Has functional limitation	0.672**	0.702**	0.299***	0.457**	−0.462*	−0.860	−0.269	−0.945***
	(0.788**)	(0.789**)	(0.794**)	(0.809**)	(−0.479**)	(−0.740)	(−0.278)	(−0.819)
Has health problem	0.096	0.369**	−0.058	0.427**	−0.185	0.450***	−0.187	0.638*
	(0.187*)	(0.440**)	(−0.015)	(0.463**)	(−0.253***)	(0.381)	(−0.227)	(0.567*)

Table 13.3 (Continued)

	Not married at follow-up				Married at follow-up			
	Model 1		Model 2		Model 1		Model 2	
	Lives with at least one child[a]	Lives with others[a]	Lives with at least one child[a]	Lives with others[a]	Lives with at least one child[b]	Lives with others[b]	Lives with at least one child[b]	Lives with others[b]
MMSE score	0.041**	−0.002	0.016	−0.028	0.018	0.031	0.033	0.028
	(0.056**)	(0.018)	(0.021)	(−0.011)	(0.012)	(0.015)	(0.031)	(0.014)
Living arrangement at origin[e]								
With children	–	–	3.432**	1.469**	–	–	2.861**	1.590**
With others	–	–	1.574**	2.830**	–	–	0.960**	3.382**
Constant	−3.182	−8.274	−5.787	−10.997	−3.294	−5.040	−4.561	−4.014
LL	−2639.5		−1977.7		−1006.1		−773.4	
LR χ^2 (df)	275.9 (28)		1599.4 (32)		119.8 (28)		585.2 (32)	
N	3919		3919		1228		1228	

* $p < .05$ ** $p < .01$ *** $p < .10$
[a] Reference category is living alone
[b] Reference category is living with spouse only or living alone
[c] Comparison category is having one child
[d] Results in parentheses are from models entering health items separately
[e] Comparison category is living alone or with spouse only

for those married, the models contrast the probability of living with children and with others relative to living with spouse only. Results shown are log odds ratios. Models were run with all health measures simultaneously and then with one at a time. The health related coefficients for the latter are in parentheses. These control for all other covariates. In general, results using health measures separately do not diverge from results using them simultaneously, although the former tend to be a little stronger.

It is clear that for those *not married*, functional limitations increase the likelihood of living with children and with others relative to living alone, and functional limitations relate to a higher probability of making these types of transitions. Reporting a health problem increases the probability of living with others, but has a minimal influence on living with children. Higher MMSE scores increase the probability of living with children, but have little influence on living with others or on living arrangement transitions. Thus, functional limitations have stronger effects on living arrangements than do health problems or MMSE scores. In some regards, this makes a good deal of sense. These are limitations in performing tasks necessary for daily survival, like bathing, dressing and toileting, and therefore those with limitations are truly in need of daily and regular physical assistance, and among those not married, a spouse is not available to provide such help. An individual with a health problem, like heart disease, or someone with minor cognitive problems, may or may not be in need of regular physical assistance. The type of assistance needed for those with functional limitations requires close living proximity, while those with health problems may get on quite well living independently, at least until problems manifest into functional disabilities.

Moving on to those *married*, functional limitations do not increase the probability of living with children or with others and do not increase the probability of transitioning into these arrangements. On the contrary, there are some significant negative associations present. Specifically, among the married, those with functional limitations are more likely to be living with spouse only than with children or others. So, it appears that having a functional limitation when married actually discourages living with individuals other than a spouse. Having a health problem increases the probability of living with others, but does little to the probability of living with children. Associations with MMSE scores are not significant.

In sum, there are some very clear distinctions in how health relates to living arrangements. On the one hand, functional limitations appear as the most critical health determinant, and for those not married, the results related to limitations support hypotheses (1) and (2). On the other, where limitations increase the chances of living with children and others for those not married, it has the opposite effect for those married. We will see shortly that these divergent effects may be explained further by differences across genders.

Among the other covariates, there are some notable effects. Among those not married, women are less likely than men to live with children and others, and are less likely to transition into these arrangements, meaning that they are more likely to live alone. This is particularly the case when contrasting living alone versus with others. In contrast, there is no significant association between gender and living arrangements or living arrangement transitions among those married. Higher age increases the probability of living with children and others, and of making these types of transitions, although the effects are stronger and more likely to be significant among those not married. Increasing numbers of children does not relate to higher probabilities of living with them. On the contrary, although there are some inconsistencies, those with more children are generally less likely to be living with a child. Perhaps it is the case that more children increases contact and the chances of receiving assistance from children living outside the household, diminishing the need for coresidence as a way to facilitate support, as has been speculated by others (Bian et al. 1998; Logan and Bian 1999). Those with more children are also less likely to live with those other than spouse or children, and of transitioning to this arrangement. Here, more children might be lowering the obligation or need of assistance from other family members. Having a son is thought to be important for coresidence, but it increases the chances of living with a child, and decreases the chances of living with others, only among those not married.

Being married at origin is an important determinant for those not married at follow-up. This may indicate that a change in marital status, that is, from being married to being not married, likely the result of becoming widowed, relates to a particular living arrangement. Specifically, among those not married at follow-up, being married at origin decreases the chances of living with a child but increases substantially the chances of living with others and making a transition to living with others. Therefore, it is likely that after becoming widowed older adults move in with others, which itself is the most transitory state.

Finally, Model 2 includes living arrangements at origin and indicates that there are higher chances of remaining in the same situation over time than of changing living arrangement states. It is worth mentioning that the coefficients for living with children and transitions to living with children tend to be stronger among the not married group, suggesting that the tendency to remain with children is slightly greater for this group than for those married.

13.3.2 Interactions with Gender

At time of follow-up, there were 876 married men, 360 married women, 1,009 unmarried men, and 2,937 unmarried women in the data. Using these samples, we next introduced gender and health interactions across those married and unmarried. Results are provided in Table 13.4. The table shows log odds ratios for several variables: being female, the three health measures, and the interactions between being female and health where significant. All other covariates are controlled for but are not shown since these results remain consistent with those shown earlier.

When significant interaction effects are present, the association between a health measure and living arrangement outcome differ between men and women. The gender specific effects can be determined by summing the coefficients. For instance, for those not married, Model 1 shows that the main effect of having limitations on

Table 13.4 Significant interaction effects for multinomial logistic regression log odds ratios predicting living arrangements

	Not married at follow-up				Married at follow-up			
	Model 1[a]		Model 2[b]		Model 1[a]		Model 2[b]	
	Lives with at least one child[c]	Lives with others[c]	Lives with at least one child[c]	Lives with Others[c]	Lives with at least one child[d]	Lives with others[d]	Lives with at least one child[d]	Lives with Others[d]
Female	−0.392**	−1.384**	−0.177	−1.177**	−0.328	−2.481**	−0.922**	−2.140*
Has functional limitation	−0.204	−0.540	−0.499	−0.884***	−0.635**	−.946***	−0.690*	−0.721
Has health problem	0.055	0.172	−0.131	0.220	−0.280***	−0.199	−0.418*	−0.015
MMSE score	0.001	−0.143*	0.038	−0.082	−0.002	−0.017	.002	−0.053
Female X Has limitation	1.135**	1.575**	1.047**	1.736**	0.535	0.348	1.282**	−0.353
Female X Has problem	0.057	0.311	0.099	0.324	0.394	3.058**	0.918**	2.861**
Female X MMSE score	0.048	0.169*	−0.027	0.057	0.071	0.203*	0.101*	0.236*

* $p < .05$ ** $p < .01$ *** $p < .10$
[a] Other covariates include age, education, occupation, rural residence, number of children, having a son, and marital status at origin
[b] Other covariates include age, education, occupation, rural residence, number of children, having a son, marital status at origin, and living arrangement at origin
[c] Reference category is living alone
[d] Reference category is living with spouse only or living alone

living with at least one child at follow-up is −0.204. This is essentially the effect for men, and indicates that unmarried men with functional limitations are less likely to live with a child than are unmarried men without limitations, although the result is not statistically significant, and therefore we cannot be confidant that the negative effect is not due to sampling error. The interaction effect suggests that the effect of functional limitations is +1.135 greater for females than for males. The net log odds ratio for living with a child relative to alone for non-married for women with functional limitations is (−0.392 − 0.204 + 1.135) = +0.539, while the net log odds ratio for non-married men with functional limitations is simply −0.204. Therefore, the hypothesis about functional limitation and living arrangements is supported for women but not for men, considering Model 1 for those not married.

More generally, there is substantial gender variation in the influence of health on the probability of living in various situations. Health by and large is more important for women across the models and measures of health. Women in poor health are much more likely than healthy women to live with children and others regardless of marital status, and they are much more likely to transition into these situations. Men in poor health are either less likely to be living with children and others, or their health has less influence on their living situation. For the married group, this may clarify some of the insignificance found in earlier models that did not include interactions. For instance, results indicate, among other things, that married men in poor health have a higher tendency of living with their spouse only in comparison to married men in good health. For married women, those in poor health are more likely to live with children and others in addition to their spouse. In short, poor health does not seem to trigger a living arrangement response in the same way for men as it does for women.

13.4 Discussion

The research reported in this chapter examined living arrangements and living arrangement transitions among Chinese aged 80–105 and considered whether poor health triggers coresidence responses, which would support the altruistic hypothesis about the workings of the Chinese family. Given that marital status dictates the types of living arrangement options available and that a spouse likely provides the most immediate source of support, associations were examined separately for those married and not married. Given that respect and authority in old age may vary by gender, and this might lead to different associations between health and living arrangements among men and women, hypotheses were tested separately for men and women by including interaction effects.

Descriptive results suggest a degree of fluctuation in living arrangements of the oldest-old. Although there is a greater chance of remaining stable than changing over time, about 1/4 did change their living arrangement over just a 2-year period, and the proportion changing is higher for those originating alone, living with spouse only, and living with others. The most popular living arrangement is living with children, and those living with children were most likely to remain in this arrangement.

Multivariate results provided some support for the three hypotheses. Among those not married, those in poor health are more likely than those in good health to be living with children and others as opposed to living alone, and they are more likely to move in with or stay with children and others over a 2-year period. Although this result supports the notion of altruism, mixed results for those married suggest the need to be cautious in the interpretation. Although it was hypothesized that effects of health would be weaker among those married, those married were not only found to be less likely to have their health influence their living arrangements, associations were at times found to be in the opposite direction. Perhaps children and others are obliged to coreside only where no alternative is available. If a spouse is present, on the contrary, they may even live elsewhere, leaving the coresidence and caring tasks to the spouse.

The situation is further complicated by significant gender interactions. Women in poor health, either having functional limitations, health problems, or high MMSE scores, are more likely than women in good health to move in with and stay living with a child or others, even if they are married. That is, poor health appears to trigger living arrangement responses toward coresidence with family members among women. In contrast, health appears to have less influence among men. It is tempting to interpret this finding as husbands not being regarded as best equipped to provide needed support for ailing wives and so the most rational option when a women is in poor health is for both the husband and wife to live with children or others. In contrast, married men with health problems are expected to receive care from their spouse, so when spousal assistance is available, children and others may even stay away. Yet, several other possible explanations exist. According to some, kin-keeping roles that women adhere to throughout their lives, such as their caretaking of the children and general provision of family services, build emotional ties to their families, and lead to stronger bonds by the time they reach old age (Kandiyoti 1988; Yount 2005). As such, children and other family members may feel greater attachment and obligation to older women than they do to older men. This leads to women's health problems prompting responses of support from the family, where men's health problems are treated with greater indifference. The results here are consistent with this argument. It may also be the case that men are likely to obtain support when they are in poor health, but that this support comes from non-coresidents. In this instance, it may be important to examine pseudo-coresidence, or residing not with but very near a family member. Finally, it may also be necessary to examine the health of both husband and wife. Married women who are older than 80 are likely married to men who are older still. Since age and health are closely correlated, married women with health problems may be more likely than married men with health problems to have spouses whose health is also poor. Where both husband and wife suffer from health disorders, the need to coreside with children and others would be greatest.

The results of this research have some important implications for a demographically changing China. The number and proportion who are very old is beginning to increase slowly, but over the course of the next several decades extreme population aging will be taking place, and the oldest-old will be the fastest growing segment of the population (Zeng and George 2001). The main reason for this is a drastic reduction in fertility, from over 5 or 6 children per average family in the 1950s to less than 2 today, coupled with some increase in longevity, as well as the huge size of baby-boom cohorts born in the 1950s and the 1960s (Poston 1992). The upcoming generation of oldest-old will be survivors from a period where fertility rates were high, but their families will be small, thus inverting the typical population pyramid. The swiftness of the change has led to dire predictions regarding the possibility of unmet needs for those most vulnerable (Cheung 1988; Du and Guo 2000; Gui 2001). For one, it is questionable whether the current levels of support, and the current rates of coresidence, are sustainable. But, results such as those presented here provide some optimism that the needs of the oldest-old may continue to be met by the family in cases where there is no alternative. It does this by suggesting that those perceived to be in greatest need may be most likely to coreside and therefore receive regular support. However, there are several caveats. First, health was found to be more predictive among women, and it is uncertain how and whether very old men are receiving needed levels of support. Second, it is uncertain whether associations reported in this research will remain intact in the future. Changes in China are not limited to family size; other social and economic transformations, such as increased rural to urban migration and changes in the nature of the employment sector, will influence rates of coresidence. Third, other sources of support, including state programs, may decrease the need for coresidence, and rapid socioeconomic development currently taking place may change the distribution of personal resources that are available.

Acknowledgments An earlier version of this paper was published in *Research on Aging*, 27, pp. 526–555. The author wishes to thank Sage Publications for providing permission to publish this revised version.

References

Albert, S.M. and M.G. Cattell (1994), Family relationships of the elderly: Living arrangements. In: S.M. Albert and M.G. Cattell: *Old age in global perspective: Cross-cultural and cross-national views*. New York: G.K. Hall & Co., pp. 85–107

Beales, S. (2000), Why we should invest in older women and men: The experience of HelpAge International. *Gender and Development* 8, pp. 9–18

Becker, G.S. (1974), A theory of social interactions. *Journal of Political Economy* 82, pp. 1063–1093

Bian, F., J.R. Logan, and Y. Bian (1998), Intergenerational relations in urban China: Proximity, contact and help to parents. *Demography* 35, pp. 115–124

Brown, J.W., J. Liang, N. Krause, H. Akiyama, H. Sugisawa, and T. Fukaya (2002), Transitions in living arrangements among elders in Japan: Does health make a difference? *Journal of Gerontology: Social Sciences* 57, pp. 209–220

Cheung, F.C.H. (1988), Implications of the one-child family policy on the development of the welfare state in the People's Republic of China. *Journal of Sociology and Social Welfare* 15, pp. 5–25

Chiu, H.F.K., H.C. Lee, W.S. Chung, and P.K. Kwong (1994), Reliability and validity of the Cantonese version of the mini-mental state examination—a preliminary study. *Journal of Hong Kong College of Psychiatrists* 4, pp. 25–28

Cowgill, D.O. (1986), *Aging around the world*. Belmont, CA: Wadsworth

DaVanzo, J. and A. Chan (1994), Living arrangements of older Malaysians: Who co-resides with their adult children? *Demography* 31, pp. 95–113

DeVos, S. (2003), Research note: Revisiting the classification of household composition among elderly people. *Journal of Cross-Cultural Gerontology* 18, pp. 229–245

Du, P. and Z.G. Guo (2000), Population aging in China. In: D.R. Phillips (ed.): *Ageing in the Asia-pacific region: Issues, policies and future trends*. New York: Routledge, pp. 194–209

Folstein, M.F., S.E. Folstein, and P.R. McHugh (1975), Mini-mental state: A practical method for grading the cognitive state of patients for the clinician. *Journal of Psychiatric Research* 12, pp. 189–198

Frankenberg, E., A. Chan, and M.B. Ofstedal (2002), Stability and change in living arrangements in Indonesia, Singapore, and Taiwan, 1993–1999. *Population Studies* 56, pp. 201–213

Friedman, J., J. Knodel, B.T. Cuong, and T.S. Anh (2003), Gender dimensions of support for elderly in Vietnam. *Research on Aging* 25, pp. 587–630

Gui, S. (2001), Care of the elderly in one-child families in China: Issues and measures. In: I. Chi, N.L. Chappell, and J. Lubben (eds.): *Elderly Chinese in pacific rim countries: Social support and integration*. Hong Kong University Press, pp. 115–124

Hermalin, A.I. (2002), Theoretical perspectives, measurement issues, and related research. In: A.I. Hermalin (ed.): *The well-being of the elderly in Asia*. Ann Arbor, MI: The University of Michigan Press, pp. 101–141

Herzog, R.A. and R.B. Wallace (1997) Measures of cognitive functioning in the AHEAD study. *Journal of Gerontology: Psychology Sciences* 52, pp. 37–48

Kandiyoti, D. (1988), Bargaining with patriarchy. *Gender and Society* 2, pp. 274–290

Katz, S., A.B. Ford, R.W. Moskowitz, B.A. Jackson, and M.W. Jaffee (1963), Studies of illness in the aged: The index of ADL, a standardized measure of biological and psychosocial function. *Journal of the American Medical Association* 185, pp. 914–919

Katzman, R., M. Zhang, O.-Y. Qu, Z. Want, W.T. Liu, E. Yu, S.-C. Wong, D.P. Salmon, and I. Grant (1988), A Chinese version of the mini-mental state examination: Impact of illiteracy in a Shanghai dementia survey. *Journal of Clinical Epidemiology* 41, pp. 971–978

Kim, C.S. and K.O. Rhee (1997), Variations in preferred living arrangements among Korean elderly parents. *Journal of Cross-Cultural Gerontology* 12, pp. 189–202

Kinsella, K. (1990), *Living arrangements of the elderly and social policy: Cross-national perspective*. Washington, DC: Bureau of the Census, Center for International Research

Knodel, J. and M.B. Ofstedal (2002), Patterns and determinants of living arrangements In: A.I. Hermalin (eds.): *The well-being of the elderly in Asia: A four-country comparative study*. Ann Arbor, MI: University of Michigan Press, pp. 143–184

Lee, Y.J., W.L. Parish, and R.J. Willis (1994), Sons, daughters and intergenerational support in Taiwan. *American Journal of Sociology* 99, pp. 1010–1041

Leung, J.C.B. (2001), Family support and community-based services in China. In: I. Chi, N.L. Chappell, and J. Lubben (eds.): *Elderly Chinese in pacific rim countries: Social support and integration*. Hong Kong University Press, pp. 171–188

Liang, J., S. Gu, and N. Krause (1992), Social support among the aged in Wuhan, China. *Asia-Pacific Population Journal* 7, pp. 33–62

Logan, J.R. and F. Bian (1999), Family values and coresidence with married children in urban China. *Social Forces* 77, pp. 1253–1282

Logan, J.R., F. Bian, and Y. Bian (1998), Tradition and change in the urban Chinese family: The case of living arrangements. *Social Forces* 76, pp. 851–882

Martin, L.G. (1989), Living arrangements of the elderly in Fiji, Korea, Malaysia, and the Philippines. *Demography* 26, pp. 627–643

Mason, K.O. (1992), Family change and support of the elderly in Asia: What do we know. *Asia-Pacific Population Journal* 7, pp. 13–32

Ofstedal, M.B., Z. Zimmer, and H.-S. Lin (1999), A comparison of correlates of cognitive functioning in older persons in Taiwan and the United States. *Journal of Gerontology: Social Sciences* 54, pp. 291–301

Pei, X. and V.K. Pillai (1999), Old age in China: The role of the state and the family. *International Journal of Aging and Human Development* 49, pp. 197–212

Poston, D.L. (1992), Fertility trends in China. In: D.L.J. Poston and D. Yaukey (eds.): *The population of modern China*. New York: Plenum Press, pp. 277–286

Shanas, E., P. Townsend, D. Wedderburn, H. Friis, P. Milhoj, and J. Stehouwer (1968), *Old people in three industrial societies*. New York: Atherton Press

Treas, J. and J. Chen (2000), Living arrangements, income pooling and the life course in urban Chinese families. *Research on Aging* 22, pp. 238–262

Whyte, M.K. (1997), The fate of filial obligations in urban China. *The China Journal* 38, pp. 1–31

Whyte, M.K. (2003), China's revolutions and intergenerational relations. In: M.K. Whyte (ed.): *China's revolutions and intergenerational relations*. Ann Arbor: University of Michigan Press, pp. 3–32

Yan, S. and I. Chi (2001), Living arrangements and adult children's support for the elderly in the new urban areas of mainland China. In: I. Chi, N.L. Chappell, and J. Lubben (eds.): *Elderly Chinese in pacific rim countries: Social support and integration*. Hong Kong University Press, pp. 201–220

Yount, K.M. (2005), The patriarchal bargain and intergenerational coresidence in Egypt. *The Sociological Quarterly* 46, pp. 139–166

Yuan, F. (1990), Support for the elderly: The Chinese way. In: Y. Zeng, C. Zhang, and S. Peng (eds.): *Changing family structure and population aging in China: A comparative approach*. Beijing: Peking University Press, pp. 341–358

Zeng, Y. and L.K. George (2001), Extremely rapid ageing and the living arrangements of the elderly: The case of China. *Population Bulletin of the United Nations* 42/43, pp. 255–287

Zeng, Y., J.W. Vaupel, Z. Xiao, C. Zhang, and Y. Liu (2001), The healthy longevity survey and the active life expectancy of the oldest old in China. *Population: An English Selection* 13, pp. 95–116

Zeng, Y., J.W. Vaupel, Z. Xiao, C. Zhang, and Y. Liu (2002), Sociodemographic and health profiles of the oldest old in China. *Population and Development Review* 28, pp. 251–273

Zeng, Y., Y. Liu, and L.K. George (2003), Gender differentials of the oldest old in China. *Research on Aging* 25, pp. 65–80

Zimmer, Z. and S.K. Kim (2001), Living arrangements and sociodemographic conditions of older adults in Cambodia. *Journal of Cross-Cultural Gerontology* 16, pp. 353–381

Zimmer, Z. and J. Kwong (2003), Family size and support of older adults in urban and rural China: Current effects and future implications. *Demography* 40, pp. 23–44

Chapter 14
Intergenerational Support and Self-rated Health of the Elderly in Rural China: An Investigation in Chaohu, Anhui Province

Lu Song, Shuzhuo Li, Wenjuan Zhang and Marcus W. Feldman

Abstract Data from the survey "Welfare of Elderly in Anhui Province, China" conducted in October 2001 by the Institute for Population and Development Studies of Xi'an Jiaotong University are used in a logistic regression analysis to determine the effects of intergenerational financial, instrumental and emotional support on the self-rated health of Chinese rural elderly. The results show that financial support, both received and provided, as well as mutual emotional support between parents and children have positive impacts on self-rated health of the elderly. However, instrumental support, both received and provided, exhibits no relationship with the self-rated health of the elderly. Socio-demographic characteristics, socioeconomic status, as well as the health status of the elderly are also found to affect their self-rated health.

Keywords Anhui Province, Coresidence, Elderly, Emotional Support, External Resources, Financial Support, Instrumental Support, Intergenerational Support, Internal Resources, Perceived Health Status, Logistic Regression, Rural China, Self-Rated Health, Social Support, Socioeconomic Status, Subjective Health Status, Survey Of Welfare Of Elderly In Anhui Province

14.1 Introduction

Self-rated health (SRH) is a global assessment of personal health status and is determined in this study by answers to the following question: "How is your health? Very good, Good, Average, Poor, or Very Poor?" SRH differs from other more complicated health status indices because it assesses perceived health status and reflects an integrated evaluation of physical, mental, behavioral, and social well-being, some of which are difficult to measure (Deeg and Bath 2003; Idler and Benyamini 1997;

L. Song
School of Management, Xi'an Jiaotong University; Institute for Population and Development Studies, School of Public Policy and Administration, Xi'an Jiaotong University, 28 Xianning Road, Xi'an, Shaanxi, 710049, China
e-mail: songlu@stu.xjtu.edu.cn

Kaplan and Camacho 1983). In a relatively short time, SRH has become one of the most commonly used measures of health status (Idler and Benyamini 1997; Morgan and Kunkel 1998; Qi 1998). It is the index used to assess the health status of the population in several large-scale epidemiological surveys in the USA and in the Zutphen study in Netherlands (Idler and Angle 1990; McGee et al. 1999; Pijls et al. 1993).

According to reviews of SRH, respondents are likely to draw on different sources for their health self-assessments, including family history, severity of current illness, possible symptoms of diseases not yet diagnosed, trajectory of health status over time, as well as availability of external resources (such as social support) and internal resources (such as the perception of control), all of which may affect the SRH of the elderly (Helweg-Larsen et al. 2003; Idler and Benyamini 1997). Social support in particular has direct effects on SRH (Idler and Kasl 1991; Krause 1987; Liu et al. 1995). A number of studies investigating the relationship between social support and health have shown that, in general, people who have better social support enjoy better health status, while lack of social support is associated with increased mortality (Idler and Kasl 1991; Krause 1987; Liu et al. 1995) and poor health (Krause et al. 1998; Litwin 1998).

Although children, friends or neighbors, the community, and the government may all provide some social support to the elderly, the family is the critical social institution in providing support to its members in China and other nations in which the Confucian culture is deeply embedded (Frankenberg et al. 2002). Since the vast majority of Chinese rural elderly have no access to any insurance or formal support (say, from the government or the community), the family is the predominant source as for older individuals in rural areas. It is widely recognized that children play an important role in the health and well-being of the elderly, but there are few studies concerning familial intergenerational support on the health status of the elderly in rural China.

14.2 Model Specification

14.2.1 Framework

A theoretical framework that illustrates the effects of intergenerational support on the self-rated health of the elderly using existing literature and relevant observations about the elderly in rural China is shown in Fig. 14.1.

The majority of studies concerning intergenerational support focus on the support received by elderly parents as opposed to the support provided by elderly parents to their children (Ofstedal et al. 1999; Shi 1993). The latter may also contribute to the health status of elderly parents. Beckman (1981) found that elderly parents who provided support to their adult children experienced a greater sense of well-being than those who did not. A more recent study found that in contrast to elderly parents who only received support, those elderly parents who both provided and received

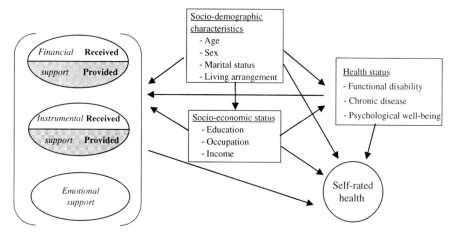

Fig. 14.1 Conceptual framework for relations between intergenerational support and self-rated health of elderly rural Chinese

support were more satisfied with life, while elderly parents who did not exchange any support with their children were less satisfied (Kim and Kim 2003). Liu et al. (1995) investigated the association between social support, both received and provided, and the health status of the elderly and found that emotional support played a crucial role in affecting an elder's health status. Accordingly, to reflect the functional contents of social relationships, social support was classified into emotional social support and instrumental social support. Under the present circumstances in rural China, many elders without pensions or medical insurance are dependent on the financial support of their adult children. This financial support is so crucial that in this analysis we separate it from the two other kinds of transfers between generations, namely instrumental support and emotional support; thus, there are three categories of intergenerational support.

Some studies suggest that elderly parents' needs are the basis of flows between the generations (Hermalin 1995; Lee et al. 1994). In analyzing the effects of intergenerational support on self-rated health, we controlled for the following variables: socio-demographic characteristics, socio-economic status, and health status. It is assumed that these factors not only directly affect the self-rated health of the elderly, but also serve as mediating factors through a correlation among the three control variables or between these and the three types of intergenerational support.

Socio-demographic variables are assumed to have direct effects on self-rated health. Age is generally considered as a substitutive variable indicating subjective health status (Chen 1998). The oldest old tend to rate their own health as better than the young old when physical health variables are controlled (Idler 1993). Even so, the association between age and self-rated health is not complete. In addition, elderly women tend to suffer more functional disabilities and to rate their own health worse than men (Idler 2003). Being married may lead to better objective and subjective health (Angel and Thoits 1987). Living arrangements of the elderly are assumed to be related to intergenerational support (Hermalin et al. 1996), but some characteristics of Chinese society and culture may entail that the health status of the elderly leads to their living arrangements, or vice versa. Research on Chinese cities reveals that coresidence usually caters more clearly to parents' than to children's needs (Logan et al. 1998).

Numerous studies document strong, positive associations between socioeconomic status (SES) and health status, while attempts to determine the precise mechanisms of the SES-health relationships have been inconclusive (Adler et al. 1993; Smith 1999). In general, education increases efficiency in transforming resources into health (Hayward et al. 2000; Smith 1999). For example, educated people may seek medical care immediately when problems occur, thus preventing further deterioration and facilitating recovery. The economic status of elderly parents can influence their living conditions and health status, as well as financial support from their children.

This study views health status as consisting of three aspects: functional disability, chronic disease, and psychological well-being. Functional status and chronic disease are assumed to directly affect the level of SRH (Liu et al. 1995; Sugisawa et al. 1994). Since SRH is an integrated measure of subjective health, the psychological well-being of the elderly can directly affect their SRH.

14.2.2 Hypotheses

Financial support is divided into monetary support and in kind support. The former consists of medical assistance and financial assistance that are essential and primary for the elderly. More than two-thirds of elderly Chinese are dependent on financial assistance from their children (Shi 1993; Xu and Yuan 1997; Yao 2001). As institutional support for cooperative medical care in rural areas has substantially diminished since the 1980s, about 700 million rural people must pay for their medical care themselves (World Bank 1997). Thus, children provide not only the costs associated with daily living for their elderly parents, but also the cost of medical care, so that the level of financial support received directly influences the medical treatment and health status of Chinese rural elderly. On the other hand, parents have the obligation to raise their children, help them get married, and provide them with houses, with the result that the elderly may be overburdened by the financial support they provide. With economic development, the flow of financial support is no longer predominantly unidirectional, but goes both to and from elderly Chinese parents (Guo and Chen 1998; Wang and Ma 2002). In light of the above discussion, the following hypothesis is suggested:

Hypothesis 1: *Financial support received will improve elders' self-rated health; financial support provided will lead to a deterioration of elders' self-rated health.*

Instrumental support consists of living assistance and daily activity assistance, including washing clothes, cleaning, grocery shopping, baby-sitting, and so on. Not only do elderly parents who are healthier and more mobile require less assistance, they are also more likely to provide assistance to their children. Although Liu et al. (1995) suggested that increased instrumental support from elders would deteriorate their health status, there was no empirical justification for this position. Many studies have supported the claim that the need for instrumental support is closely related to the health status of the elderly (Shi 1993; Xiong 1998). Silverstein and Bengston (1994) found that elderly parents preferred to be independent as long as possible, but support from children became important when they had disabilities in functional activities. Coresidence and instrumental support seemed to be detrimental for older couples. Another study found that the well-functioning elderly who received the most frequent instrumental support exhibited a greater risk to their future health than similarly healthy elders who receive less frequent instrumental support. This was interpreted to mean that social support may induce excessive dependency by subverting the self-reliance of older individuals (Seeman et al. 1996). Therefore we examine for the relationship between instrumental support and SRH of the elderly with the following hypothesis:

Hypothesis 2: *Instrumental support received will cause a deterioration of the elders' self-rated health; instrumental support provided will also cause a deterioration of the elders' self-rated health.*

Emotional support includes listening, talking, paying attention, comforting, consulting, and respecting, and is considered to be a crucial factor in the health status of the elderly (House and Kahn 1985; Sugisawa et al. 1994). While a negative association between emotional support and elders' health status has been observed in China (Wang 1990), data from the Wuhan area suggested that emotional support, both received and provided, was positively linked to the SRH of the elderly (Liu et al. 1995). We therefore assume that emotional support between generations is a process of mutual communication, so that such support becomes reciprocal, and hence may be of benefit to one's subjective health. The following hypothesis concerning the impact of emotional support on the SRH of the elderly is thus suggested:

Hypothesis 3: *Emotional support will improve the elders' self-rated health.*

14.3 Methods

14.3.1 Sample

The data used in the research reported in this chapter are derived from the survey "Welfare of Elderly Survey in Anhui Province," which was carried out in April 2001 by the Institute for Population and Development Studies of Xi'an Jiaotong University, in conjunction with the University of Southern California. The sample was identified using a stratified multistage method to select potential respondents in 12 randomly selected rural townships; six villages were randomly selected from each rural township. The respondents were from among all residents aged 60 and over; people 75 years of age and older were over-sampled. Of the 1,800 individuals identified as eligible respondents, 1,715 completed the survey, yielding a response rate of 95.3 percent. After excluding respondents without children and cases with missing data on relevant study variables, 1,604 respondents remained for analysis.

14.3.2 Measurement

Definitions, measures, and other descriptive statistics for the dependent variables, independent variables, and control variables are found in Table 14.1.

Table 14.1 Descriptive statistics ($n = 1604$)

Variables	Mean / Percentage	SD
Dependent variables		
Self-rated Health: (Poor)	26.5	
Good	73.5	
Independent variables		
Intergenerational support		
Financial support received	2.72	0.67
Financial support provided	0.91	1.14
Instrumental support received	4.18	8.53
Instrumental support provided	3.64	7.28
Emotional support	7.30	1.32
Control variables		
Health status		
Functional disabilities	0.95	1.78
Chronic disease	1.98	1.52
CES-D	15.47	3.94
Socio-economic status		
Education[a]: (Illiterate)	78.9	
Literate	21.1	
Occupation[a]: (Farming or housework)	95.3	
Non-agricultural work	4.7	
Income[a]: (No income)	55.2	
Having income	44.8	
Socio-demographic characteristics		
Age[a]: (69−)	45.8	
70–79	41.3	
80+	12.9	
Sex[a]: (Male)	46.1	
Female	53.9	
Marital status[a]: (Unmarried)	42.8	
Married	57.2	
Living arrangement[a]: (Living without children or grandchildren)	48.8	
Living with children or grandchildren	51.2	

[a] Per 100.

14.3.2.1 Dependent Variables

Self-rated health, the dependent variable, was originally assessed by a four-point-scale response to a question *"How is your health?"*: (1) very good, (2) good, (3) average, (4) not so good. In our data analysis, we condensed this classification into two scales: (1) good (including "very good," "good" and "average"), (2) poor ("not so good"). Self-rated health is coded 1 if "good," while 0 if "poor" is considered to be the benchmark in the data analysis.

14.3.2.2 Independent Variables

The independent variable, intergenerational support, is divided into financial support, instrumental support, and emotional support.

Financial support received is measured by the question, *"Did the child send you (or your spouse living with you now) money, food or gifts?,"* and is a measure of the total amount received from each child during the past 12 months. Response options are the following: 0="none," 1="less than 50," 2="50–99," 3="100–199," 4="200–499," 5="500–999," 6="1,000–2,999," 7="3,000–4,999," 8="5,000–9,999," and 9="More than 10,000." The median value of each interval is taken as the amount of financial support from all children. The sum of financial transfers received from all children is considered to be the financial support received by one elderly person, and we take the log of the sum as the financial support received. Financial support provided is measured by the question, *"Did you (or your spouse living with you now) send your child money, food or gifts?"* The scale for financial support provided is the same as that for financial support received.

Instrumental support received from children (including children's spouse and grand-children) to elderly parents during the past 12 months is assessed in two areas: (1) *household tasks,* such as cleaning the house and washing clothes, and (2) *personal care tasks,* such as bathing and dressing. One of the following four options is selected for each area: (1) Every day=7.5; (2) At least once per week=1.5; (3) Several times per month=0.5; and (4) Seldom=0. The scoring follows the method proposed by Bian et al. (1998). Summing the frequency of instrumental support from each child, the total score is considered as the instrumental support received by an elder. The scale for instrumental support provided is the same as that for instrumental support received.

Emotional support is assessed by three questions adapted from the intergenerational solidarity inventory (Mangen et al. 1988): (1) *"Overall, how close do you feel to (this child)?"* (2) *"Overall, how well do you and (this child) get along together?"* (3) *"How much do you feel that (this child) would be willing to listen when you intend to talk about your worries and troubles?"* The items are coded as follows: 1="Not at all close/not at all well/not at all"; 2="Somewhat close/somewhat well/somewhat"; 3="Very close/very well/very much." An additive scale is computed for each child, ranging from 3 to 9, with a higher score indicating a better parent–child relationship. The alpha reliability coefficient for these items is 0.86. Different from the financial support and instrumental support, the average level of these responses reflects the quality of intergenerational relations. We take the average score across all children for each elderly parent as the measure of emotional support between the elder and the children, avoiding multicollinearity between emotional support and number of children.

14.3.2.3 Control Variables

We assess health status using three scales: functional status, chronic disease, and psychological well-being. Functional status is measured as the sum of six items reflecting difficulty in performing personal activities of daily living (PADL). Personal activities of daily living include bathing, putting on and taking off clothes, walking around the room, using the toilet, and eating a meal (Katz et al. 1963). An elder is considered as functionally limited in a given activity if he or she has any degree of difficulty in performing that activity without help. Functional status is measured by the number of functional limitations. Chronic disease is indicated by the number of chronic illnesses reported by the respondents, ranging from 0 to 12 items. It is noted that, due to poor medical service conditions, rural Chinese elderly may not know whether they have a chronic disease until their activities of daily living are affected. Such self-reported illnesses may therefore be underestimated and not reliable. Psychological well-being is measured using nine questions based on the CES-D scale (Radloff 1977). Each of the nine items is coded *1= "rarely or none of the time," 2= "some of the time," and 3= "most of the time"*; they are then summed, resulting in a depression scale ranging from 9 to 27, with a higher score indicating greater depression. The alpha reliability coefficient for the nine items is 0.77.

For the SES variables, education is coded as 0 if illiterate, and 1 if literate; most of the rural population have little education. Occupation (before retirement) is categorized in two groups: farming or housework, representing lower SES; and non-agricultural work, representing higher SES. If the respondent and spouse received income from any work or pension during the past 12 months, this is classified as having income.

The socio-demographic variables include age, sex, marital status and living arrangement. Self-rated health may not depend entirely on the decline of functional abilities and physical health due to aging; for this reason the oldest old tend to assess health status comparatively better, due to selection by death (Zeng et al. 2004). Age is represented by three groups: (1) younger than 70 = "0," (2) 70–79="1," and (3) 80 or older="2." Marital status is coded as a dummy variable, with the unmarried (widowed, divorced, and never-married) contrasted with married. Living arrangement is categorized in two groups: living without children or grandchildren, and living with children (including grandchildren).

14.3.3 Statistical Methods

We employ logistic regression for the analysis. Controlling for socio-demographic characteristics, SES, and health status, we test for an overall statistical association between intergenerational support (including financial support, instrumental support and emotional support), both received and provided, and self-rated health.

14.4 Results

14.4.1 Description of Sample

Our data show that a large majority of elderly parents receive financial transfers from children. In contrast, less than half of the sample contribute financial assistance to their children, and the money contributed to children is far less than that received from children. In terms of transfers of instrumental support, 43.1 percent of elderly parents receive some household help or personal care from their children, and 35.3 percent provide similar types of assistance to their children. Parents contribute assistance to their children less often than they receive it from their children. Emotional support from children tends to be strong; the mean score is 7.30 (SD = 1.32) on a nine-point scale (see Table 14.1). Thus, support from children to elderly parents is the main flow of assistance between the generations. The main support from children to elderly parents is financial, while the main support from elderly parents to children is instrumental.

In terms of health status, on average, respondents report difficulty performing about one task. The average chronic disease score is 1.98 out of a possible 12, and the average depression score is 15.47 out of 27. For SES, most (78.9 percent) have no formal education, and a large majority, 95.3 percent, is currently or previously engaged in farming or housework. Only 44.8 percent have work-related or pension-related income. Almost 46 percent (45.8 percent) are young old, while 12.9 percent are oldest old. Slightly more than half of the sample (53.9 percent) is female, and 57.2 percent are currently married. Almost half (48.8 percent) live without their adult children or their grandchildren, and more than one-half live with both adult children and grandchildren

14.4.2 Results of Logistic Regression

We use binary-logistic regression to test for the effects of intergenerational support on SRH via a set of hierarchical models. Four blocks of variables are sequentially added to the model, representing, in order of entry, intergenerational support, health status, socio-economic status, and socio-demographic characteristics. Table 14.2 presents the results of the logistic analysis.

Model 1 tests for the relationship between intergenerational support, both received and provided, and the self-rated health. The more financial support received from their children, the higher is the probability that elderly parents are in "good" SRH (OR=1.151). The elderly who are provided with more financial support are more likely to be in "good" self-rated health (OR=1.325). Increased instrumental support

Table 14.2 Results of logistic regression of intergenerational support on self-rated health ($n = 1604$)

Independent variables	Model 1	Model 2	Model 3	Model 4
Intergenerational support				
Financial support received	1.151+	1.467***	1.394***	1.410***
Financial support provided	1.325***	1.144*	1.155*	1.180*
Instrumental support received	0.979**	0.996	0.996	0.994
Instrumental support provided	1.010	0.998	0.996	1.007
Emotional support	1.069*	1.252***	1.184***	1.129**
Health status				
Functional disabilities		0.729***	0.737***	0.721***
Chronic disease		0.723***	0.723***	0.741***
CES-D		0.958**	0.936***	0.913***
Socio-economic status				
Education (Literate)			1.253	1.566*
Occupation (Non-agricultural work)			2.192**	2.002**
Income (Having income)			0.958	0.861
Socio-demographic characteristics				
Age group				
70–79				1.103
80 or older				2.429**
Sex (Female)				1.843***
Marital status (Married)				1.175
Living arrangement (Living with children or grandchildren)				1.330*
−2LL	1796.48	1522.65	1509.25	1473.61
Df	5	8	11	16

The reference categories of the categorical variables are omitted, including illiterate, farming or housework, no income, younger than 70, male, unmarried, living without children or grandchildren
*** $p < 0.001$; ** $p < 0.01$; * $p < 0.05$; + $p < 0.1$

received reduces the probability of being in "good" self-rated health (OR=0.979); for instrumental support provided, the reverse appears to be true but is not significant. In addition, elders with more emotional support and closer relationships with their children are more likely to have "good" SRH (OR=1.069). On the whole, financial support, both received and provided, instrumental support provided, and emotional support have positive associations with self-rated health.

For Model 2, when the entire block of health status variables is employed, financial support and emotional support remain significant, but instrumental support no longer shows a significant direct effect on SRH. Health status variables are significant predictors of SRH. The elderly with more functional disabilities have a lower probability of being in "good" SRH (OR=0.729), and increased chronic disease is observed to decrease "good" SRH (OR=0.723). In addition, with a higher CES-D score, the elderly are less likely to assess themselves in "good" SRH (OR=0.958).

Model 3 includes the SES variables; the inclusion of these produces no change in the effects of intergenerational support. Non-agricultural work, which represents higher SES is shown to have a statistically significant positive effect on SRH (OR=2.192), while education and income are not significant.

In controlling for socio-demographic variables, model 4 shows that the variables related to intergenerational support, including financial support, both received and provided, and emotional support remain significantly related to SRH. The literate elderly are more likely to assess themselves as in "good" health compared to the illiterate elderly (OR=1.566). Older age leads to a greater probability of being in "good" SRH; that is, the oldest old are more likely to be satisfied with their health status than are the young old (OR=2.429). There is a gender difference, with the female elderly more likely to be in "good" SRH than the male elderly (OR=1.843). Those living with children or grandchildren have a greater probability of reporting "good" SRH than those living without children (including grandchildren) (OR=1.330).

Worth special mention are the effects of physical health and psychological well-being, which are employed as intermediate variables in the model. For example, the previously mentioned significant effects of instrumental support received are no longer significant when physical health and psychological well-being are taken into account, suggesting that instrumental support received is related to the health status of the elderly; that is, to some extent, instrumental support is determined by the physical health of the elderly.

14.5 Discussion

We have examined the proposition that in rural China intergenerational support, both received and provided, affects the SRH of the elderly, and the affects depend on the type of support.

Controlling for the needs of the elderly, we find that financial support, both received and provided, has a strong positive impact on SRH. Increased financial support received enhances the subjective health of the elderly. Since the majority of Chinese elders depend on financial assistance from their children or others (Du and Wu 1998), the standard of living and the medical care of the elderly are entirely determined by the financial support received. In contrast to Hypothesis 1, financial support provided is positively linked to the SRH of the elderly, suggesting that increased financial assistance from elderly parents has not caused a deterioration of their health status. This may be because elderly parents can generally afford to give financial assistance to their children without risk to their own health status. There is evidence that elderly who are economically independent are more likely to hold a high status in their family and in turn to receive instrumental assistance and affective support from their children, instead of financial support (Chen 2000). Overall, financial support from children to parents is the primary flow of financial transfer between the two generations, and financial support received has a stronger impact on the SRH of the elderly than does financial support provided.

Hypothesis 2 concerning instrumental support is partially confirmed. Needy parents generally get more support (Lee et al. 1994); that is, there is a negative association between the instrumental support received and the elder's subjective health. But there is not a negative relationship between instrumental support provided and SRH, as suggested by Hypothesis 2. It is assumed that as an important part of the reciprocal flow of resources between generations and in compensation for financial assistance received, instrumental support provided to adult children may enhance feelings of usefulness and the self-worth of elderly parents, thus improving their health status. Controlling for health status variables, instrumental support is no longer significant, which suggests that instrumental support is associated with the elder's SRH through physical health and psychological well-being.

Finally, consistent with Hypothesis 3, we find that emotional support between generations improves the subjective health of the elderly. It has been reported that whether a given type of social support has an impact on health status depends strongly on the need for this support (Liu et al. 1995). According to Maslow's Need Theory (Maslow 1970), the dominance of a higher need would not take place before a specific need is satisfied. Given that most Chinese rural elderly have insufficient safety and security needs, a lack of affective assistance representing psychological needs would not cause a deterioration of their health. Traditionally, in the Chinese culture, support for the aged involves not only material assistance but also affective support (Yao 1996), and the two aspects are expected to be reinforcing (Wang 1990). Our research finds that the direction of the effect of financial support received on self-rated health is consistent with that of emotional support, whereas the direction of effect of instrumental support received is reversed. These opposite effects produce an ambiguous result. Does the positive association occur because more emotional support improves health status, or because there is a reversed linkage between these two factors? The research reported in this chapter reveals that there is a strong positive relationship between emotional support and the subjective health of the elderly.

On the whole, we find that financial support, both received and provided, has positive impacts on the SRH of the elderly in rural China. Although instrumental support, both from and to the elderly parents, is not shown to be significantly linked to subjective health, we find that emotional support may improve subjective health. This suggests that the closer the parent–child relationship, the better the health status and psychological well-being of the elderly parents.

Unlike earlier studies that have found that older women tend to experience prolonged debilitating illnesses, functional impairments and worse self-rated health (Penning and Strain 1994; Rahman and Liu 2000; Silverstein et al. 1997), we find that the SRH of older women is better than that of older men. This may be because, with a similar level of physical health, older women are more satisfied with their health status than older men. It may also be due to the selection effect of death. We also find that living arrangement affects the SRH of the elderly. This suggests that when the elderly coreside with adult children, quality of family support is improved to meet the needs of the elderly parents.

We have not assessed the impact of net flows between generations on self-rated health of the elderly. With the availability of longitudinal data, the underlying causal linkages between intergenerational support and old-age, and self-rated health may be tested more effectively. Further research is clearly warranted as to the mechanism by which intergenerational flows between elderly parents and sons and daughters affect well-being and health status of the old-old. Such an analysis may contribute to addressing the important problems of sex preference and very high sex ratio at birth in China.

Acknowledgments The research reported in this chapter is jointly supported by a grant from the Fogarty International Center, NIH (R03.TW01060-01A), the 2nd period of the National 985 Project of the Ministry of Education and Treasury Department of China (07200701), and the Program for New Century Excellent Talents in University (NCET) of the Ministry of Education of China (NCET-04-0931).

References

Adler, N.E., W.T. Boyce, M.A. Chesney, S. Foldman, and L. Syme (1993), Socioeconomic inequalities in health. *Journal of the American Medical Association* 269 (24), pp. 3140–3145

Angel, R.J. and P. Thoits (1987), The impact of culture on the cognitive structure of illness. *Culture, Medicine, and Psychiatry* 11, pp. 465–494

Beckman, L.J. (1981), Effect of social interaction and children's relative inputs on older women's psychological well-Being. *Journal of Personality and Social Psychology* 41, pp. 1075–1086

Bian, F., J. Logan, and Y. Bian (1998), Intergenerational relations in urban China: proximity, contact, and help to parents. *Demography* 35 (1), pp. 115–129

Chen, J. (1998), Investment and support: A analysis of causality of intergenerational exchanges of urban residents. *Social Science in China* 6, pp. 131–145 (in Chinese)

Chen, C. (2000), Dependence is a primary issue in blest aging of elders in rural China. *Population Research* 24 (2), pp. 53–58 (in Chinese)

Deeg, D.J.H. and P.A. Bath (2003), Self-rated health, gender, and mortality in older persons: introduction to a special section. *The Gerontologist* 43 (3), pp. 369–371

Du, P. and C. Wu (1998), The primary financial resource of the Chinese elderly. *Population Research* 22 (4), pp. 51–57 (in Chinese)

Frankenberg, E., L. Lillard, and R.J. Willis (2002), Patterns of intergenerational transfers in southeast Asia. *Journal of Marriage and Family* 64, pp. 627–641

Guo, Z. and G. Chen (1998), The financial flows between the elders and their children. *Population Research* 22 (1), pp. 35–39 (in Chinese)

Hayward, M.D., E. Crimmins, T. Miles, and Y. Yang (2000), The significance of socioeconomic status in explaining the racial gap in chronic health conditions. *American Sociological Review* 65 (December), pp. 910–930

Helweg-Larsen, M., M. Kjoller, and H. Thoning (2003), Do age and social relations moderate the relationship between self-rated health and mortality among adult danes? *Social Science and Medicine* 57, pp. 1237–1247

Hermalin, A.I. (1995), Aging in Asia: Setting the research foundation. *Asia-Pacific Research Report* 4, p. 20s

Hermalin, A.I., M.B. Ofstedal, and M. Chang (1996), Types of support for the aged and their providers in Taiwan. In: T.K. Hareven (ed.): *Aging and generational relations over the life course: A historical and cross-cultural perspective.* New York: Walter de Gruyter, pp. 400–437

House, J.S. and R.L. Kahn (1985), Measures and concepts of social support. In: S. Cohen and L. Syme (eds.): *Social support and health.* Orlando, FL: Academic Press, pp. 83–108

Idler, E.L. (1993), Age differences in self-assessments of health: Age changes, cohort differences, or survivorship? *Journal of Gerontology: Social Sciences* 48, pp. S289–S300

Idler, E.L. (2003), Discussion: Gender differences in self-rated health, in mortality, and in the relationship between the two. *The Gerontologist* 43, pp. 372–375

Idler, E.L. and R.J. Angle (1990), Self-rated health and mortality in the NHAES-1 epidemiologic follow-up study. *American Journal of Public Health* 80, pp. 446–452

Idler, E.L. and Y. Benyamini (1997), Self-rated health and mortality: A review of twenty-seven community studies. *Journal of Health and Social Behavior* 38, pp. 21–37

Idler, E.L. and S. Kasl (1991), Health perceptions and survival: do global evaluations of health status really predict mortality? *Journal of Gerontology: Social Sciences* 46, pp. S55–S65

Kaplan, G.A. and T. Camacho (1983), Perceived health and mortality: A nine-year follow-up of the human population laboratory cohort. *American Journal of Epidemiology* 117, pp. 292–304

Katz, S., A.B. Ford, R.W. Moskowitz, B.A. Jackson, and M.W. Jaffe (1963), Studies of illness in the aged, the index of ADL: A standardized measure of biological and psychological function. *Journal of American Medical Association* 185, pp. 914–919

Kim, I.K. and C. Kim (2003), Patterns of family support and the quality of life of the elderly. *Social Indicators Research* 62,63, pp. 437–454

Krause, N. (1987), Satisfaction with social support and self-rated health in older adults. *Gerontologist* 27, pp. 301–308

Krause, N., J. Liang, and S. Gu (1998), Financial strain, received support, anticipated support and depressive symptoms in the people's republic of China. *Psychological Aging* 13, pp. 58–68

Lee, Y., W.L. Parish, and R.J. Willis (1994), Sons, daughters and intergenerational support in Taiwan. *American Journal of Sociology* 99(4), pp. 1010–1041

Liu, X., J. Liang, and S. Gu (1995), Flows of social support and health status among older persons in China. *Social Science and Medicine* 41, pp. 1175–1184

Litwin, H. (1998), Social network type and health status in a national sample of elderly Israelis. *Social Science and Medicine* 46, pp. 599–609

Logan, J.R., F. Bian, and Y. Bian (1998), Tradition and change in the urban Chinese family: The case of living arrangements. *Social Forces* 76 (3), pp. 851–882

Mangen, D.J., V.L. Bengtson, and P.H. Landry (1988), *The measurement of intergenerational relations*. Beverly Hills, CA: Sage Publications.

Maslow, A.H. (1970), *Motivation and personality*. Harper, New York.

McGee, D.L., Y. Liao, G. Cao, and R.S. Cooper (1999), Self-report health status and mortality in a multiethnic US cohorts. *American Journal of Epidemiology* 149, pp. 41–46

Morgan, L. and S. Kunkel (1998), *Aging: The social context*. Thousand oaks, California: Pine Forge Press.

Ofstedal, M.B., J. Knodel, and N. Chayovan (1999), Intergenerational support and gender: A comparison of four Asian countries. *Southeast Asian Journal of Social Sciences* 27 (2), pp. 21–42

Penning, M.J. and L. Strain (1994), Gender differences in disability, assistance, and subjective well-being in later life. *The Journal of Gerontology: Social Sciences* 49 (4), pp. S202–S208

Pijls, L.T., E.J.M. Feskens, and D. Kromhout (1993), Self-rated health, mortality and chronic diseases in elderly men: The Zutphen Study, 1985–1990. *American Journal of Epidemiology* 138, pp. 840–848

Qi, Y. (1998), *The life status and life quality of the elderly in China backland and in Hong Kong*. Beijing: Peking University Publishing House (in Chinese).

Radloff, L. (1977), The CES-D scale: A self-report depression scale for research in the general population. *Applied Psychological Measurement* 1, pp. 385–401.

Rahman, M.O. and J. Liu (2000), Gender differences in functioning for older adults in rural Bangladesh: the impact of differential reporting? *Journal of Gerontology: Medical Sciences* 55A (1), pp. M28–M33

Seeman, T.E., M.L. Bruce, and G.J. McAvay (1996), Baseline social network characteristics and onset of ADL disability: MacArthur studies of successful aging. *Journals of Gerontology: Social Sciences* 51b, pp. S191–S200

Shi, L. (1993), Family financial and household support exchange between generations: A survey of Chinese rural elderly. *The Gerontologist* 33 (4), pp. 468–480

Silverstein, M. and V.L. Bengston (1994), Does intergenerational social support influence the psychological well-Being of older parents? The contingencies of declining health and widowhood. *Social Science and Medicine* 38, pp. 943–957

Silverstein, M., T. Seeman, S. Kasl, and L. Berkman (1997), Gender differences in the comparison of self-reported disability and performance measures. *Journal of Gerontology* 52A (1), pp. M19–M26.

Smith, J.P. (1999), Healthy bodies and thick wallets: The dual relation between health and economic status. *Journal of Economic Perspectives* 13 (1), pp. 145–166

Sugisawa, H., J. Liang, and X. Liu (1994), Social networks, social support, and mortality among older people in Japan. *Journal of Gerontology: Social Sciences* 49, pp. S3–S13

Wang, J. (1990), New features of family conflicts in urban China. *Zhejiang Xuekan* 6, pp. 146 (in Chinese)

Wang, S. and J. Ma (2002), The new trend of intergenerational relationships in the process of population aging. *Population and Economics* 133, pp. 15–21 (in Chinese)

World Bank (1997), *Sharing rising incomes: Disparities in China*. Washington, DC: World Bank.

Xiong, Y. (1998), Intergenerational relation and old-age care in urban China, *Chinese Journal of Population Science* 69, pp. 15–21 (in Chinese)

Xu, Q. and Y. Yuan (1997), The role of family support in the old-age security in China. In: China Population Association (eds.): *23rd IUSSP general population conference: Symposium on demography of China*. Beijing: Xin Hua Press, pp. 265–273

Yao, Y. (1996), Old support: Special component of the traditional culture. *Population Research* 20, pp. 30–35 (in Chinese)

Yao, Y. (2001), A review of the researches on providing for the aged in household in China. *Population and Economics* 124, pp. 33–43 (in Chinese)

Zeng, Y., Y. Liu, Z. Xiao, and C. Zhang (2004), The socio-demographic and health status of the oldest-old in China. *Chinese Journal of Population Science* Supplement, pp. 4–13 (in Chinese)

Chapter 15
The Effects of Adult Children's Caregiving on the Health Status of Their Elderly Parents: Protection or Selection?

Zhen Zhang

Abstract The objective of this chapter is to re-examine the protective and selective effects of children's caregiving to their elderly parents. Although the protective effect is widely recognized, the selection of care provision necessitates a comprehensive investigation into elderly caregiving. We design a joint model to specify the two effects. The model is applied to dyadic data from two large-scale surveys conducted in China in 2002. We found that protection is seriously misestimated when selection is ignored. After having successfully identified the selective effect, care provision to the elderly is found to be far more beneficial than claimed in previous studies.

Keywords Adult children, Caregiving, Care provision, China, Dyadic data, Elder parents, Family support, Intraclass correlation, Joint model, Need-based support, Protection, Selection, Self-rated-health, Social support, Survey of Family Dynamics of the Elderly's Children, Unobserved heterogeneity, Unobserved propensity

15.1 Introduction

Many studies of children's caregiving to their elderly parents have found that there are reciprocal effects of caregiving and elderly health, namely, the protective effect and the selective effect. Caregiving protection refers to the beneficial effect of care to the elderly, that is, the more caregiving the elderly receive, the better their health (Cornell 1992; Rogers 1996; Silverstein and Bengtson 1991; Stoller, 1985; Zunzunegui et al. 2001). The selective effect implies that care by children is more likely to be provided to elderly parents who are in relatively poor health (Bian et al. 1998; Hermalin et al. 1996; Lee et al. 1994; Lee and Xiao 1998; Lin et al. 2003; Logan et al. 1998).

Z. Zhang
Max Planck Institute for Demographic Research Konrad-Zuse-Str. 1, 18057 Rostock, Germany
e-mail: zhang@demogr.mpg.de

An example will illustrate the two concepts and their effects. Consider an older person who is able to go shopping without assistance from anyone. Even though the older person is not in need of help, he/she would be more likely to forego unpredictable risks, such as falling down, if accompanied by one of his/her children. Consider another older person who is unable to go shopping by himself/herself due to some functional impairment. The older person needs to be accompanied by someone, e.g., one of his/her children. In neither case is there an essential difference in the activities of the accompanying child; the child just accompanies the parent while the parent is shopping. However, the motivation behind the care and the effect of the care provision are different. In the former case, the care given by the child is much more protective, but in the latter one, it is selective.

The protective and selective effects of caregiving on elderly health are not mutually exclusive, but are two sides of the same coin. However, the sides we observe in a study will largely depend on the context in which the study is carried out, and the methods of analysis used to investigate such effects. Different societies have different motives behind and patterns of caregiving, resulting in different effects of care on elderly health. Traditionally, adult children dutifully provide care to their elderly parents in need of help. In western countries, today's elderly may be more financially independent than their predecessors, and social support may play a more important role than family support, such as care by their children, in their elderly lives. In this case, care provided by children to their elderly parents who are not in need of care may yield additional benefits in terms of the health of the elder. Hence, we may find protective care by adult children in these developed societies.

In contrast, in many developing countries, such as China, the elderly depend heavily on their family, mainly their adult children, to provide a variety of help, including financial and emotional support and caregiving. Hence, the support the children provide to their elders is called "need-based support" (Lee and Xiao 1998). Moreover, the duty the offspring have towards their elders in China is further reinforced by laws and regulations, including the Chinese Constitution (Bartlett and Phillips 1997; Lee and Xiao 1998; Leung 1997; Wu 1994). If care of elderly parents is needed, their adult children are supposed to provide such support; this is the minimum requirement imposed on the family in terms of old-age support. Moreover, due to lack of social support for the elderly, adult children have to be the primary caregivers, and the selection of caregiving is relatively strong. As a result, care provision due to the needs of elderly parents is more likely to be observed in China.

Ignoring selection will result in bias to some extent or even in an estimation that is wrong (Vaupel et al. 1979). Conventionally, when estimating the impact of care on elderly health, caregiving is usually thought of as one factor among others that plays a major part in the health of an elder, and care provision is often set as an independent variable in statistical modeling. It is difficult to specify the selection of caregiving in this way, however. An analysis on the impact of intergenerational support on the survival of the oldest-old (i.e. elderly aged 80 and over) in China found that elderly who received care from children have a greater likelihood of dying than those who do not have this care (Zhang 2002). Clearly, this result cannot be interpreted that the care provided by children can be harmful to their elderly parents. Such a result comes from the fact that the elderly who received care by their children had a higher mortality themselves because children selectively provided care to their frail elderly parents. Thus, a method that is able to model this selection effect is required.

Using dyadic data, the research reported in this chapter employs a joint model so that the two effects of care can be specified comprehensively. The dyadic approach has been used in studying relationships between family members (Lye 1996; Miller et al. 1982; Thompson and Walker 1982). The dyad is considered as the unit of analysis throughout all phases of the research process (Wampler and Halverson 1993). Theoretically, the dyadic approach is a good choice for studying care provided by adult children and the elderly in a family context (Allen et al. 2000). Employing the joint model developed by Lillard and Panis (2003), we estimate the protective and selective effects of caregiving simultaneously. The parent–child dyad survey conducted in China in 2002 provides the dyadic data. Considering that caregiving by children to their elderly parents is prevalent in China, the dyadic analysis should easily capture the selective effect of caregiving.

15.2 Data and Methods

15.2.1 Sample

The sample in this study is taken from the parent-child dyadic survey data, i.e. the 2002 wave of the Chinese Longitudinal Healthy Longevity Survey (CLHLS) and the 2002 Survey of Family Dynamics of the Elderly's Children (SFDC). Out of 22 provinces covered by the CLHLS, eight were set as the sample pool for the SFDC. In this pool, an adult child of each older person interviewed (aged 65+) in the CLHLS was randomly chosen according to the sex, age and rural/urban residence of the children. Some of the parent-child dyads (435 out of a total of 4,364) were removed for the following reasons. First, the age of some respondents was not within the age range used in our analysis (ages 35–65 for adult children and ages 65–105 for elderly parents). Second, a small proportion of cases had missing values in some important variables (e.g., Self-Rated Health of elderly parent) and therefore were deleted. The sample was thus reduced to 3,929 parent–child dyads. Note that in this study the data were not weighted.

15.2.2 Models

Much debate surrounds the question of how to conduct dyadic analyses (Maguire 1999 and references therein). Previous dyadic studies have mainly focused on a comparison between the similarities and differences of dyadic members (e.g., husband vs. wife, parent vs. child), utilizing methods such as intraclass correlation, ANOVA, and the hierarchical linear model. To our knowledge, no research uses dyadic data to specify the protective and selective effects of care by children on the health status of elderly parents.

Here, we design a joint model that simultaneously evaluates the protective and selective effects of caregiving. The model contains two sets of sub-models for the Self-Rated Health (SRH) of elderly parents and the care provided by their adult children. The SRH model on elderly parents is given by an ordinal probit model.

$$y = \begin{cases} 0 \text{ if } y^* < \tau_1 \\ 1 \text{ if } \tau_1 \leq y^* < \tau_2 \\ 2 \text{ if } \tau_2 \leq y^* < \tau_3 \\ \vdots \\ n \text{ if } \tau_n \leq y^* \end{cases}, \text{ where } y^* = \sum \alpha_1' X_p + \sum \alpha_2' H_p + \sum \alpha_3' C + \varepsilon. \quad (15.1)$$

In Eq. 15.1, y^* stands for a latent propensity dependent on independent variables, y for the SRH reported by elderly parents, X_p for the sociodemographic characteristics of the elderly parents, H_p for the health condition indicators of the elderly parents, C for caregiving to the elderly parents reported by the adult children and ε for the residual term for the SRH of the parents.

Another model is the caregiving model, characterized by a probit model that measures the occurrence of caregiving,

$$p = \begin{cases} 0 \text{ if } p^* < 0 \\ 1 \text{ if } p^* \geq 0 \end{cases}, \text{ where } p* = \sum \beta_1' X_c + \sum \beta_2' H_p + \delta. \quad (15.2)$$

In Eq. 15.2, p^* denotes a latent propensity dependent on independent variables, p an observed indicator variable, X_c the sociodemographic characteristics of the adult children, H_p the health condition indicators of the elderly parents and δ the residual term for the occurrence of caregiving by the adult children.

The residual terms ε and δ, are assumed to be jointly normally distributed, represented by

$$\begin{pmatrix} \varepsilon \\ \delta \end{pmatrix} \sim N\left(\begin{pmatrix} 0 \\ 0 \end{pmatrix}, \begin{pmatrix} \sigma_\varepsilon^2 & \sigma_{\varepsilon\delta} \\ \sigma_{\varepsilon\delta} & \sigma_\delta^2 \end{pmatrix}\right), \text{ where } \sigma_{\delta\varepsilon} = \rho_{\varepsilon\delta}\sigma_\delta\sigma_\varepsilon,$$

and $\rho_{\varepsilon\delta}$ is the correlation coefficient of residual terms ε and δ. The two models are estimated jointly at the same time. The potential correlation between the two dependent variables can be obtained through estimating the correlation coefficient of residual terms in the two models. The critical test of the simultaneous relationship between SRH and caregiving is whether $\rho_{\varepsilon\delta}$ is equal to zero. We adopted the following decision strategy: First, if $\rho_{\varepsilon\delta} = 0$, this means there is no relationship between the propensity of the elderly's SRH and that of care provided by their children. And if the caregiving protection in the SRH model significantly exists (i.e. the coefficient of caregiving in the SRH model is positive), then we accept this protection as the effect of the care provision for the elderly on their health status.

Second, a positive $\rho_{\varepsilon\delta}$ implies that the propensity that the elderly parents self-assess good health is positively related with the propensity that their children provide care, thus indicating the caregiving protection. And if the coefficient of care provision by children is significantly positive in the SRH model, then the caregiving protection will be specified and there is no caregiving selection. Third, if $\rho_{\varepsilon\delta} < 0$ and if the positive effect of caregiving on the SRH significantly exists, then both caregiving protection and caregiving selection are identified. On the one hand, the negative $\rho_{\varepsilon\delta}$ indicates the caregiving selection; that is, children are more likely to provide care to their elderly parents in poor health. On the other hand, the positive coefficient of caregiving in the SRH sub-model suggests the protective effect of such care.

The joint model enables the researcher to capture the protective and selective effects of care given to the elderly, owing to its following features. First, a latent propensity corresponding to SRH is set as a function of children's care to elderly parents and sociodemographic features, physical ability and morbidity of the elderly parents. Second, the likelihood of providing care to elderly parents is set as a function of the physical ability and morbidity of the elderly and demographic and socio-economic features of adult children. Hence, the explanatory model of caregiving consists of two parts, the parents' need of care, and the children's capability of providing this care. Considering that physical ability and morbidity are directly related to the need of care, selective care provision to the elderly can be reflected in the caregiving model (e.g., Lee and Xiao 1998). Besides, children's decision to give care to their elderly parents is influenced by their sociodemographic conditions, such as sex, age, socio-economic status (SES) and so on. More importantly, the SRH of elderly parents is also included in the caregiving model as an independent variable, while it is a dependent variable in the SRH model (see below).

Finally, the joint model takes into account the unobserved heterogeneity underlying the elderly's SRH and care provided by their children. Each of the two heterogeneity components, measured by the residual terms in Eqs. 15.1 and 15.2, represents the effects of unmeasured latent factors that are the source of correlation across the two equations. Table 15.1 summarizes the specification of both equations in this model.

15.2.2.1 The SRH Model for Elderly Parents

SRH is a broader measure of health than the presence or absence of disease and functional limitation (Idler and Benyamini 1997; Strawbridge and Wallhagen 1999). SRH has been shown to be a significant and independent predictor of mortality and of the declining physical function of the elderly (Benyamini et al. 1999; Idler and Benyamini 1997; Idler et al. 1990; Wolinsky and Johnson 1992). The SRH of the elderly was measured using a single question in the CLHLS survey, "How do you rate your health at present?" Six levels of coding were provided in terms of the respondent's health self-evaluation: "very poor," "poor," "fair," "good," "very good," and "cannot answer." The number of elderly who evaluated their health as "very poor" was small (less than 1 percent of the older subjects) so it was assigned

Table 15.1 Model specifications

	SRH model	Caregiving model
Parent's Information		
Sex	X	
Age	X	
Race	X	
Place of residence	X	
Living arrangement	X	
Marital status	X	
IADL	X	X
ADL	X	X
Serious disease	X	X
Child's Information		
Sex		X
Age		X
Urban/rural residence		X
Co-resident with parents		X
Number of sibling		X
Caregiving	X	

to the level "poor." There were 193 older respondents who "cannot answer"; they were disregarded because they did not provide any information about SRH.

In earlier studies, SRH has been shown to be related to physical ability, morbidity, social support and life satisfaction (Goldstein et al. 1984; Idler 1993; Markides and Lee 1990; Minkler and Langhauser 1988; Zhang 2004). Physical ability of the elderly is assessed by Katz's Index of Activities of Daily Living (ADL) and Instrumental Activities of Daily Living (IADL). Because of its easy administration, ADL is usually applied to the oldest old, those aged 80 and over in the CLHLS. In 2002, the sample was extended to the younger old aged, i.e., those aged 65–79. Performing ADL tasks does not show enough variation in the younger old so that the IADL was included to measure the physical ability of the younger old. The serious diseases from which the elderly suffered were associated with the elderly's SRH (Ilder and Benyamini 1997). Note that, ADL/IADL and serious diseases are different in terms of their effects on the elder's health, as well as care provision by children. Although ADL/IADL and serious diseases can affect each other (e.g. some serious diseases result in an impairment in ADL or IADL), they are relatively independent. In the survey, the elderly were asked how many times he or she suffered from serious diseases. Serious diseases may not be a longstanding illness, whereas functional limitations in terms of ADL/IADL often exist for a long time. Clearly, this difference between ADL/IADL and serious diseases is important to caregiving selection.

Relevant sociodemographic characteristics of elderly parents should be controlled; these include age, place of residence, marital status, and living arrangements. The elderly's SES, measured by education, occupation and income, is not included in our analysis because previous studies have shown that it does not have a significant impact on the SRH of the elderly (Zhang 2002, 2004). Instead, the large gap between rural and urban regions in China implies that place of residence is a good proxy for SES (Liang et al. 2000). Differences between marital status groups have been found in relation to the elderly's SRH (Joung et al. 1995; Ren 1997). Living arrangements have also been shown to be important factors that determine the elders' SRH (Zunzunegui et al. 2001). Because of the very small proportion of those living in nursing homes (less than 1 percent), we group them into "lives alone."

Besides the observed SRH, the unobserved propensity of SRH can be estimated. The residual term ε represents the parent-specific unobserved heterogeneity in the propensity of SRH.

15.2.2.2 The Caregiving Model for Adult Children

The outcome variable in our caregiving model is whether or not the children interviewed provided care to their elderly parents in the last year. In the survey, the exact amount of time spent in caregiving is ascertained through the question, "How many days did you spend on caring for your parents in the last year?" If children did not give any care, the answer is zero. The time spent on care is very informative in terms of the amount of care provided. In this research, however, we have to reduce this continuous variable to a binomial one for two reasons. First, we find that the self-reported time spent on care is heavily heaped. It is difficult to remember the exact days of caregiving except in the case of no care; instead it is much easier to have an approximate estimate by some time unit, such as a week, month, half a year or 1 year. Second, the distribution of time has a positive skew. More than half of the children surveyed did not provide care to their elderly parent, so the time is zero. If the time expenditure on caregiving is set as a multi-categorical variable, another half of the cases will be regrouped into several levels. As an example, we categorize the time expenditure of caregiving into five levels: zero (no caregiving), not-zero but less than 1 week, more than 1 week but less than 1 month, more than 1 month but less than half a year, and more than half a year but less than 1 year. In the SRH model, the number of combinations between the time expenditure and SRH will be $20 (= 4 \times 5)$.

Regarding the remaining covariates in the model, the number of cases in some cells of the contingency table will be too small to obtain a reliable estimation. Therefore, caregiving is defined as a binomial variable, giving care or not. Although the dichotomous variable implies the loss of information to some extent compared with time expenditure as a continuous or multi-categorical variable, we argue that its employment is preferred in this analysis.

The sociodemographic features of adult children, such as gender, age (Coward and Dwyer 1990; Lin et al. 2003) and race (Dilworth-Anderson et al. 2002; Dwyer and Coward 1991)[1] have been shown to contribute to their decision to provide

---

[1] Some studies claim that marital status and SES (Lin et al. 2003) of adult children is significantly related to the caregiving decision. However, a study based on the same dataset as this paper shows that the two factors do not significantly impact the decision of whether or not to give care to elderly parents (Zhang 2004).

care or not provide care. The children's place of residence is not necessarily the same as that of their parents because some children might have migrated from a rural to an urban area. The patterns of caregiving in rural areas may also change to some degree as some adult children have gradually adapted to an urban lifestyle.

With respect to the importance of elderly living arrangements in terms of care provision, we also take into account in the caregiving model whether or not their children are living with them. Clearly, children who are living near their parents are more likely to give care than children who are living far away from their parents (Dwyer and Coward 1991; Ingersoll-Dayton et al. 1996). Co-residence with an adult child is considered to be a central feature of the familial support system in much of the developing world (United Nations 2000). Although in general, children living with their elderly parents are more likely to be the primary caregiver, in China the children who do not live with their parents but live nearby tend not only to maintain frequent face-to-face contact with their parents but also to provide help to them (Bian et al. 1998). Hence, co-residence with parents is not a necessary condition for caregiving as far as China is concerned. In addition, the number of siblings of an adult child may influence the decision to provide care to their elderly parents. If an adult child has no siblings, he/she will have to bear the entire responsibility for taking care of the older parents. In contrast, if an adult has siblings, he/she may share the responsibility of caregiving with the siblings. In other words, for an adult, the decision to provide care to elder parents will be impacted by whether or not he or she has siblings.

IADL, ADL and serious diseases are also directly related to the care needs of the elderly. The care provision resulting from impairment in IADL/ADL of elderly parents or the elder suffering from serious diseases is mainly selective. That is, the health status of elderly parents will predict whether or not their adult children give them care. Thus, the caregiving model can specify the selection of caregiving.

Analogously, the residual term, δ, in Eq. 15.2 denotes the child-specific unobserved heterogeneity in the propensity to give care to elderly parents. As shown in the example mentioned at the beginning of this chapter, children accompanying the elderly parent in good health when going shopping might eliminate the potential risk for that elder. This implies the unobserved propensity of children's care provision to an elderly parent who is not in need of care. In the caregiving study, the elderly respondent is often asked whether he/she is able to go shopping alone, and if not, who will accompany him/her. The adult children are asked a similar question. In fact, these questions collect information about selective caregiving. However, if an older person is able to go shopping alone, he/she will have little chance to provide information about the provision of care by his/her children (e.g. accompanying him/her going shopping). As a consequence, such care provision may be ignored. In other words, the benefits of this kind of caregiving to the elder's health are hard to be observed in surveys because the observed selective caregiving tends to overshadow the protective caregiving. The unobserved propensity of caregiving can be estimated in the model, that is, the residual term δ.

15.3 Results

15.3.1 Preliminary

Table 15.2 describes the sample characteristics. The information about elderly parents is presented in the left panel in Table 15.2. Because of the sampling design of the CLHLS (Zeng et al. 2002 and Chap. 2 of this volume), the size of the female elder sample is roughly the same as that for males. The oldest old comprise more than half of the elderly (59 percent). Han Chinese, the largest nationality population in China, comprise the majority of the parent sample. Most of the elderly are not married (i.e. they are widowed, divorced or never married). In China, most elderly live with their family (not necessarily with adult children), while only a few live alone or in a nursing home. For each IADL and ADL, the elderly are thought of as independent if they have no functional limitation and as dependent otherwise. The distributions of dependency and non-dependency in IADL and ADL are adverse: 61 percent of the elderly are independent in terms of IADL, whereas 79 percent are dependent in terms of ADL. Moreover, only 18 percent of the elderly suffered from a serious disease at least once in the last 2 years before the conduct of the 2002 CLHLS.

The child variables are listed in the right panel in Table 15.2. Fifty-nine percent of children do not provide care to their parents. But this does not mean that an equal proportion of parents does not receive care. Rather, they may receive care from

Table 15.2 Sample characteristics

Older parent ($n = 3{,}929$)		Percent	Adult child ($n = 3{,}929$)		Percent
SRH	Poor	13.4	Caregiving	No	58.5
	Fair	38.5		Yes	41.5
	Good	35.5	Sex	Female	30.5
	Very good	12.6		Male	69.5
Sex	Female	53.0	Age	35–44	29.1
	Male	47.0		45–54	37.7
Age	65–79	40.5		55–65	33.2
	80–89	29.3	Place of residence	Rural	59.2
	90–105	30.2		Urban	40.8
Race	Minorities	9.1	Co-resident with parents	Yes	43.6
	Han	90.9		No	56.4
Place of residence	Rural	54.7	Number of siblings	Zero	21.1
	Urban	46.3		1+	78.9
Marital status	Unmarried	63.5			
	Married	36.5			
Living arrangement	Alone	14.2			
	With families	85.8			
ADL	Independent	79.0			
	Dependent	21.0			
IADL	Independent	38.8			
	Dependent	61.2			
Serious diseases	No	82.0			
	At lease once	18.0			

children who are not included in the SFDC. We also find that co-residence with a parent is not always related to caregiving. Forty-four percent of children live with their parents, but only 41 percent give care to their elderly parents. Among adult children, sons comprise the main part. There are more adult children who live in rural regions compared to those in urban regions. The proportion of elderly living in urban areas is slightly higher than that of adult children residing in urban areas, possibly because adult children from rural areas who have settled in cities have better opportunities to invite elderly parents to live with them. Nearly 80 percent of adult children have at least one sibling.

15.3.2 The SRH of the Elderly Parent

The bivariate and multivariate regressions are placed together under the SRH model. The bivariate analysis is carried out without controlling for any other variables, and each variable in the regression is the only predictor of SRH. Note that the scores of SRH rank from "poor" to "very good," and that the positive coefficient of a variable simply means that the elder-feature specified by the variable has a larger likelihood to report a better SRH than the reference level. The bivariate analysis shows that older persons or those with a functional impairment are more likely to self-rate themselves as being in relatively poor health. The married elderly are in better health in terms of SRH than the non-married. Caregiving is negatively associated with the elder's SRH; that is, those who receive care have a greater likelihood of reporting poor health.

In China, place of residence (rural or urban) is expected to be a strong predictor of the elder's SRH because urban residence is a prerequisite for an individual to access various public resources and medical services. However, in neither the bivariate nor the multivariate regression does the elder's place of residence have a significant effect on the elder's SRH. One possibility for this is that because each elder in this study has at least one child, they may be selected by marriage protection (Lillard and Panis 1996), and for the elderly with late childbearing, fertility protection (Müller et al. 2002) in the earlier life of the elderly. Note that a similar study based on data drawn from the CLHLS found in a multivariate regression that place of residence of the elderly is not significantly associated with the elderly's SRH (Liu and Zhang 2004). The effect of place of residence on the elder's SRH needs to be examined further in future analyses.

The differences between the bivariate and multivariate regressions are found in the effects of the elder's age and marital status. In general, dysfunction tends to accelerate with the age of the oldest old (Fries et al. 2000). Thus, a decline in SRH with age is expected. In a multivariate regression that considers the elder's physiological conditions, however, nonagenarians are more likely to report better health than octogenarians. Moreover, because in China very old people generally receive higher respect from society, they will tend to have a more optimistic assessment of their health[2] (Liu and Zhang 2004).

---

[2] An alternative explanation is also possible; it focuses on mortality selection (Vaupel et al. 1979).

We found it interesting that "caregiving receipt" has a negative effect on the elder's SRH, even after controlling for other covariates. Elders who receive care are more likely to report poorer SRH than those who do not receive care. What we observe here seems to be a selective effect of care provision, although the regression model is conventionally used for estimating the expected caregiving protection effects. As mentioned above, the superficial relationship between the receipt of care and the relatively poor health of the elder may be misestimated. We will discuss this further below, where we look at the caregiving model.

In addition, we account for parent-specific propensities in the multivariate regression to self-rate better or worse health through a variance of parent-specific unobserved attributes. The estimation of the variance shows that the unobserved attributes of elderly parents contribute to a better SRH ($\sigma_\varepsilon > 0$).

15.3.3 Caregiving by Children

With respect to the caregiving model, we do not find differences in patterns of impact for the other covariates, except for age, between the bivariate and multivariate regressions. In both regressions, adult children who are male, who are living in a rural region, who do not reside with their parents, and who have siblings are less likely to provide care. In rural China, the coverage of social support and health services is so minimal that the majority of rural elderly rely on their families (World Bank 1997). However, the bivariate regression shows that the children in rural areas are less likely to provide care to their elderly parents. A possible explanation may be that rural elderly prefer to be autonomous for as long as possible, and only when they face difficulties in performing activities of daily life, are they willing to accept care from their children (Liang 1999). The place of residence of children and the number of siblings they have lose significance in the multivariate regression, when other variables are controlled. It is consistent in both regressions that the health conditions of the elderly parents predict whether their children provide care or not.

Regarding the effect of the children's age on their decision to provide care, the bivariate analysis provides estimates that are different from the multivariate analysis. In the bivariate regression, children aged 45–54 are less likely to give care than the younger age group. Older adult children (aged 55–65) inversely intend to provide much more care. However, in the multivariate analysis, when other variables are introduced, a clear age pattern of adult children is shown with respect to providing care to their elderly parents; that is, the older the children, the more likely they are to give care to their parents. This is consistent with findings from previous studies (e.g., Marks and Lambert 1997) that show that caregiving reaches a peak in midlife (i.e. age 45–65).

That is, the frail elderly die out first while the robust remain. Hence, nonagenarians are on average more robust than octogenarians and thus report relatively better health. However, the cross-sectional data used in this study cannot examine such selection processes for mortality.

Our emphasis is on the set of variables for the elderly parents and that are shared by the SRH model and the caregiving model. The caregiving model shows caregiving selection, i.e., children are more likely to give care to their elderly parents who are IADL and ADL dependent and who have suffered from a serious disease at least once. In the SRH model, however, the negative coefficient for caregiving implies caregiving selection, too. But what about the protective effect of caregiving?

15.3.4 Joint Model

The protective effect of care provision to the elderly fails to be specified by the SRH model and the caregiving model. Hence, a joint model is used to identify the two effects of caregiving. The results are shown in the joint model column in Table 15.3.

The coefficient of care provision by children changes from a negative value in the SRH model to a positive one in the joint model. The positive coefficient of care provision to the elderly in the joint model suggests that there is caregiving protection, i.e., elderly who receive care are in good health. How can we confirm that the protective effect of caregiving estimated in the joint model is truly protective?

As stated earlier, in each model (one for the elder's SRH and one for the children's care provision), there is a residual term that represents the unobserved propensity of the corresponding dependent variable. For the dependent variable in the SRH model, this residual term (ε) represents the propensity of self-rating higher health. For the caregiving model, the residual term (δ) means the propensity of providing care to elderly parents. The correlation coefficient of residual terms ε and δ, $\rho_{\varepsilon\delta}$, indicates the potential correlation between the propensities of self-rating higher health and providing care. In the joint model, we find that the correlation coefficient between variances specific to SRH and caregiving is negative; $\rho_{\varepsilon\delta} = -0.43$. This suggests that the elderly with an above-average propensity to report better health do not tend to have an above-average propensity to receive care by their adult children; or that the elderly whose children have an above-average propensity to provide care to them are less likely to report better health. This is the selective effect of caregiving. Because the caregiving selection is successfully identified, we believe that the positive coefficient of the care provision in the joint model is estimated.

When the caregiving selection is identified, the effects of some variables on their corresponding dependent variables change. The variables shared by the two separate models, ADL–IADL and serious diseases are underestimated in the two separate models compared with the joint model. However, despite the underestimated coefficients, ADL–IADL and serious diseases are stable in predicting elderly SRH and their children's caregiving.

In addition, in the caregiving model, daughters are more likely than sons to provide care to their elder parents. However, in the joint model, it is the sons rather than the daughters who have a greater likelihood of care provision for their elderly parents. In the context of China, most daughters will leave their parents after marriage to live with the husband's family. Many elderly parents live with their married sons,

15 The Effects of Adult Children's Caregiving on the Health Status

Table 15.3 Joint model versus separate models for SRH and caregiving

	SRH model		Caregiving model		Joint model
	Bivariate	Multivariate	Bivariate	Multivariate	
τ_1	–	−1.905***			−2.196***
τ_2	–	−0.248*			−0.220
τ_3	–	1.404***			1.748***
Sex (female=0)					
Male	0.165***	0.103*			0.135**
Age (65–79=0)					
80–89	−0.166***	0.086			0.074
90–105	−0.133***	0.222***			0.198**
Race (minorities=0)					
Han	0.017	0.184*			0.188*
Residence (urban=0)					
Rural	−0.034	−0.080			−0.075
Marital status (non-married=0)					
Married	0.115***	−0.064			0.019
Living arrangement (alone=0)					
Live with families	0.048	0.085			−0.066
ADL (independent=0)					
Dependent	−0.562***	−0.646***			−0.869***
IADL (independent=0)					
Dependent	−0.513***	−0.683***			−0.975***
Serious diseases (No=0)					
At least once	−0.583***	−0.630***			−0.853***
Caregiving (No=0)					
Yes	−0.212***	−0.121**			0.722***
Intercept			–	−0.546***	−0.235**
Sex (Female=0)					
Male			−0.182***	−0.202**	0.539***
Age (35–44=0)					
45–54			−0.251***	0.383***	1.070***
55–65			0.149***	0.800***	−0.178*
Residence (urban=0)					
Rural			−0.235***	−0.125	−1.600***
Co-resident with parents (yes=0)					
No			−0.571***	−1.310***	0.044
Number of sibling (zero = 0)					
1+ siblings			−0.261***	−0.013	−0.787***
The health status of the elderly					
ADL (independent=0)					
Dependent			0.293***	0.471***	0.600***
IADL (Independent=0)					
Dependent			0.058***	0.755***	0.956***
Serious diseases (No=0)					
At least once			0.158***	0.674***	0.843***
σ_ε		1.051***		1.462***	
σ_δ				1.430***	1.968***
$\rho_{\varepsilon\delta}$					−0.423***
ln-L		−5,034.2		−2399.3	−7426.9

* $p<0.05$; ** $p<0.01$; *** $p<0.001$

especially in rural China. Hence, the likelihood of a daughter's providing care to her parents is much lower than that of a son. Despite this, daughters may still come back to take care of their parents when they are in poor health. In other words, care provision by daughters probably is corresponding to the poor health status of their elderly parents, and thus is selective. Considering the marriage patterns and living arrangements in China, selective caregiving is much stronger. Therefore, it is the sons who usually live with their parents, and who are the true primary caregivers rather than the daughters, who usually live far away from their elderly parents.

Similar changes are found in the age patterns of the adult children in providing care to their elders. The caregiving model shows that the older the children, the more likely they are to give care to their parents. This age pattern, however, disappears when we look at the joint model. This is because older adult children (aged 55–65) are less likely than younger adults to provide caregiving. A possible explanation is that adult children will experience dysfunction with age to different degrees. They are sometimes in need of care themselves. But as children, their responsibilities of filial piety cannot be substituted by someone else. As for the older adult children, if their elderly parents are in good health, their younger siblings or their children can assume their responsibility of taking care of their parents. Hence, the provision of care by older adult children probably implies that the elderly parents have relatively serious health problems. With the successful identification of selective caregiving, the true role older adult children play in the family's support emerges: younger adult children, rather than older adult children, are the primary care providers.

15.4 Discussion

Previous studies of the effect of children's caregiving on their elderly parents' health are limited because they fail to correctly identify caregiving selection. Most previous studies note that the care received by the elderly is to some extent conditional on the health of the elderly. The analyses have endeavored to control for this effect by taking into account the health status of the elderly. This method, however, proves difficult to identify caregiving. In the research reported in this chapter, the emphasis has been on distinguishing the protective effect of caregiving from the selective one, both theoretically and methodologically.

We have reached several conclusions. First, there are protective and selective effects of care provided by children to their elderly parents. Many previous studies have observed caregiving selection and caregiving protection. However, they have failed to distinguish the two effects of caregiving because they employed inappropriate methods. In this research, we have succeeded in identifying the two effects of caregiving through a joint model.

Second, the protective effects of caregiving can correctly be estimated only if the selection of caregiving is successfully identified. It is easy to understand the difference between the protective and selective effects of care provision for the elderly. But the conventional approaches to analysis, especially those based on selectively collected information about care to the elderly, make it difficult to specify the two effects. We addressed this issue with a joint model on the relationship between the health of the elderly and care provision to the elderly. In this model, several important variables, such as ADL-IADL and suffering from serious diseases, are shared by the SRH model and the caregiving model. Moreover, a correlation coefficient between the residual terms of the two models is used to represent the unobserved relationship between the propensities of self-rating better health and of giving care. We can confirm that the correct estimate of the protective effect of caregiving has been obtained since the negative correlation coefficient suggests the existence of caregiving selection.

Third, the joint model reveals the gender and sex patterns of caregiving. When the sex-specific selection of caregiving is removed, it turns out that sons who usually live with their elderly parents are the primary caregivers. This is inconsistent with the general expectation regarding the role gender plays in providing elder care. Note that unlike other types of old-age support (e.g. financial support), caregiving usually requires face-to-face contact. Clearly, adult children who live with elderly parents are at an advantage compared to those who do not with respect to providing care to their elderly parents. Daughters who usually do not reside with their elderly parents are in a disadvantageous position regarding their ability to provide care to their parents. Hence, care provision by daughters probably means the relatively poor health status of their elderly parents (at least for the short term). Similarly, identification of caregiving selection reveals the age pattern of the adult children who provide care. The primary care givers are the relatively younger adult children.

Finally, this research makes sense in terms of methodology. First, in both separate and joint models, the variables of one member of the parent-child dyad predict the outcome variable for another member. In the joint model, especially, the residual components assumed to be jointly normally distributed represent a good identification of this modeling. This means that the joint model succeeds in the conceptual specification of the complex interrelationship between the health status of the elderly and the care provided by their adult children. With appropriate modeling, the dyadic approach is able to do much more than comparing similarities or differences between two dyadic members.

However, we must note that this research is limited in several aspects. First, our analysis does not cover everything regarding the health status of elderly parents and their adult children. For instance, following the exchange motive of transfers (Lillard and Willis 1997), care provision for elderly parents may partly result from the financial support of parents to their children. But in our model, the analysis is restricted to considering support coming from children and provided to parents based on the concept of filial piety in the Chinese context. Moreover, we do not consider the interrelations among caregiving and other types of support provided by children, such as financial and emotional support. This oversimplifies the complicated structure of intergenerational support. For instance, it is found that children who do not live with their parents prefer to provide financial support, whereas children who live with their parents or live nearby are more likely to provide face-to-face care (Chi 2002). Although such simplification suffices for the goal of this research, i.e., identifying the protective and selective effects of caregiving, there is considerable room left for further elaboration. Second, due to the known limitations of cross-sectional data, our model is unable to examine the dynamics of changes in health status and in caregiving. Thus, we cannot answer or explore questions such as how children's caregiving changes with elderly parent's health, and what impact children's caregiving will have on family relationships. In an important sense, the research reported here is only the beginning.

Acknowledgments I am grateful for the comments and suggestions from two anonymous reviewers. Also, I am in debt to Susann Backer who provided much help with the English editing of this paper. This paper is based on part of my doctorial thesis research under the supervision of Professor Zeng Yi.

References

Allen, K.R., R. Blieszner, and K.A. Roberto (2000), Families in the middle and later years: A review and critique of research in the 1990s. *Journal of Marriage and the Family* 62 (4), pp. 911–926

Bartlett, H. and D.R. Phillips (1997), Ageing and aged care in the people's republic of China: National and local issues and perspectives. *Health and Place* 3 (3), pp. 149–159

Benyamini, Y., E.A. Leventhal, and H. Leventhal (1999), Self-assessments of health: What do people know that predicts their mortality? *Research on Aging* 21 (3), pp. 477–500

Bian, F., J.R. Logan, and Y. Bian (1998), Intergenerational relations in urban China: Proximity, contact and help to parents. *Demography* 35 (1), pp. 115–124

Chi, I. (2002), Old age support in contemporary urban China from both parents' and children's perspectives sun. *Research on Aging* 24, pp. 337–359

Cornell, L.L. (1992), Intergenerational relationships, social support and mortality. *Social Forces* 71 (1), pp. 53–62

Coward, R.T. and J.W. Dwyer (1990), The association of gender, sibling network composition and patterns of parent care by adult children. *Research on Aging* 12 (2), pp. 158–181

Dilworth-Anderson, P., I.C. Williams, and B.E. Gibson (2002), Issues of race, ethnicity and culture in caregiving research: A 20-year review (1980–2000). *The Gerontologist* 42 (2), pp. 237–272

Dwyer, J.W. and R.T. Coward (1991), A multivariate comparison of the involvement of adult sons versus daughters in the care of impaired parents. *Journal of Gerontology: Social Sciences* 46 (5), pp. S259–S269

Fries, B.E. J.N. Morris, K.A. Skarupski, C.S. Blaum, A. Galecki, F. Bookstein, and M. Ribbe (2000), Accelerated dysfunction among the very oldest-old in nursing homes. *The Journal of Gerontology Series A: Biological Sciences and Medical Sciences* 55A (6), pp. M336–M341

Goldstein, M.S., J.M. Siegel, and R. Boyer (1984), Predicting changes in perceived health status. *American Journal of Public Health* 74 (6), pp. 611–615

Hermalin, A.I., M.B. Ofstedal, and M. Chang (1996), Types of support for the aged and their providers in Taiwan. In: T.K.Hareven (eds.). *Aging and generational relations over the life course*. New York: Walter de Gruyter, pp. 179–215

Idler, E.L. (1993), Age difference in self-assessments of health: Age changes, cohort differences, or survivorship? *Journal of Gerontology: Social Sciences* 48 (6), pp. S289–S300

Ilder, E.L. and Y. Benyamini (1997), Self-rated health and mortality: A review of twenty-seven community studies. *Journal of Health and Social Behavior* 38 (1), pp. 21–37

Idler, E.L., S.V. Kasl, and J.H. Lemke (1990), Self-evaluated health and mortality among the elderly in New Heaven, Connecticut and Iowa and Washington counties, Iowa, 1982–1986. *American Journal of Epidemiology* 131 (1), pp. 91–103

Ingersoll-Dayton, B., M.E. Starrels, and D. Dowler (1996), Caregiving for parents and parents-in-law: Is gender important? *The Gerontologist* 36 (4), pp. 483–491

Joung, I.M., K. Stronks, H. van de Mheen, and J.P. Mackenbach (1995), Health behaviors explain part of the differences in self reported health associated with partner/marital status in the Netherlands. *Journal of Epidemiology and Comunity Health* 49 (5), pp. 482–488

Lee, Y. and Z. Xiao (1998), Children's support for elderly parents in urban and rural China: Results from a national survey. *Journal of Cross-Cultural Gerontology* 13 (1), pp. 39–62

Lee, Y., W.L. Parish, and R.J. Willis (1994), Son, daughters and intergenerational support in Taiwan. *American Journal of Sociology* 99 (4), pp. 1010–1041

Leung, J.C.B. (1997), Family support for the elderly in China: Issues and challenges. *Journal of Aging and Social Policy* 9 (3), pp. 87–101

Liang, H. (1999), The study on capability of self-dependence and self-autonomous of the elderly people in rural China. *Population and Economy* 20 (4), pp. 21–25 (in Chinese)

Liang, J., J.F. McCarthy, A. Jain, N. Krause, J.M. Bennett, and S. Gu (2000), Socioeconomic gradient in old age mortality in Wuhan, China. *The Journals of Gerontology: Social Sciences* 55 (B), pp. S222–S233

Lillard, L.A. and C.W.A. Panis (1996), Marital status and mortality: The role of health. *Demography* 33 (3), pp. 313–327

Lillard, L.A. and C.W.A. Panis (2003), aML multilevel multiprocess statistical software, Version 2. Los Angeles: EconWare

Lillard, L.A. and R.J. Willis (1997), Motives for intergenerational transfers: Evidence from Malaysia. *Demography* 34 (1), pp. 115–134

Lin, I.-F., N. Goldman, M. Weistein, Y.-H. Lin, T. Gorrindo, and T. Seeman (2003), Gender differences in adult children's support of their parents in Taiwan. *Journal of Marriage and the Family* 65 (1), pp. 184–200

Liu, G. and Z. Zhang (2004), Sociodemographic differentials of the self-rated health of the Chinese oldest-old. *Population Research and Policy Review* 23 (2), pp. 117–133

Logan, J.R., F. Bian, and Y. Bian (1998), Tradition and change in the urban Chinese family: The case of living arrangements. *Social Forces* 76 (3), pp. 851–882

Lye, D.N. (1996), Adult child–parent relationships. *Annual Review of Sociology* 22, pp. 79–102

Maddox, G.L. and E.B. Douglass (1973), Self-assessment health: A longitudinal study of elderly subjects. *Journal of Health and Social Behavior* 14 (1), pp. 87–93

Maguire, M.C. (1999), Treating the dyad as the unit of analysis: A primer on three analytic approaches. *Journal of Marriage and the Family* 61 (1), pp. 213–223

Markides, K.S. and D.J. Lee (1990), Predictors of well-being and functioning in older Mexican Americans and Anglos: An eight-year follow-up. *Journals of Gerontology: Psychological Sciences and Social Sciences* 45 (2), pp. S69–S73

Marks, N.F. and J.D. Lambert (1997), Family caregiving: Contemporary trends and issues. NSFH Working Paper, No. 78, Child and Family Studies, University of Wisconsin-Madison

Miller, B.C., B.C. Rollins, and D.L. Thomas (1982), On methods of studying marriages and families. *Journal of Marriage and the Family* 44 (4), pp. 983–998

Minkler, M. and C. Langhauser (1988), Assessing health differences in an elderly population: A five-year follow-up. *Journal of the American Geriatrics Society* 36 (2), pp. 113–118

Müller, H.-G., J.-M. Chiou, J.R. Carey, and J.-L. Wang (2002), Fertility and life span: Late children enhance female longeivty. *The Journals of Gerontology: Biological Sciences and Medical Sciences* 57A (5), pp. B202–B206

Ren, X.S. (1997), Marital status and quality of relationships: The impact on health perception. *Social Science and Medicine* 44 (2), pp. 241–249

Rogers, R.G. (1996), The effects of family composition, health and social support linkages on mortality. *Journal of Health and Social Behavior* 37 (4), pp. 326–338

Silverstein, M. and V.L. Bengtson (1991), Do close parent–child relations reduce the mortality risk of elderly parents? *Journal of Health and Social Behavior* 32 (4), pp. 382–395

Stoller, E.P. (1985), Self-assessments of health by the elderly: The impact of informal assistance. *Journal of Health and Social Behavior* 25 (3), pp. 260–270

Strawbridge, W.J. and M.I. Wallhagen (1999), Self-rated health and mortality over three decades. *Research on Aging* 21 (3), pp. 402–416

Thompson, L. and A.G. Walker (1982), The dyad as the unit of analysis: Conceptual and methodological issues. *Journal of Marriage and the Family* 44 (4), pp. 889–900

United Nations (2000), *United Nations technical meeting on population ageing and living arrangements of older persons: Critical issues and policy responses.* New York: Department of Economics and Social Affairs, Population Division, Author

Vaupel, J.W., K.G. Manton, and E. Stallard (1979), The impact of heterogeneity in individual frailty on the dynamics of mortality. *Demography* 16 (3), pp. 439–454

Wampler, K.S. and C.F. Halverson (1993), Quantitative measurement in family research. In: P.G. Boss, W.J. Doherty, R. LaRossa, W.R. Schumm, and S.K. Steinmetz (eds.): *Sourcebook of family theories and methods: A contextual approach.* New York: Plenum Pub Corp, pp. 181–194

Wolinsky, F.D. and R.J. Johnson (1992), Perceived health status and mortality among older men and women. *Journal of Gerontology: Social Sciences* 47B (6), pp. S304–S312

World Bank (1997), *Sharing rising income.* Washington, DC: Author

Wu, C. (1994), The aging process and incomes security of the elderly under reform in China. In: United Nations (eds.): *The aging of Asian populations: Proceedings of the United Nations round table on ageing of Asian populations*, Bangkok, May, 1992. New York: United Nations, pp. 58–65

Zeng, Y., J.W. Vaupel, Z. Xiao, C. Zhan, and Y. Liu (2002), Sociodemographic and health profiles of oldest old in China. *Population and Development Review* 28 (2), pp. 251–273

Zhang, Z. (2002), The impact of intergenerational support on mortality of the oldest old in China. *Population Research* 26 (5), pp. 55–62 (in Chinese)

Zhang, Z. (2004), The impact of intergenerational support on healthy longevity of the elderly in China. Unpublished doctoral thesis (in Chinese). Institute of Population Research, Peking University.

Zunzunegui, M.V., F. Béland, and A. Otero (2001), Support from children, living arrangements, self-rated health and depressive symptoms of older people in Spain. *International Journal of Epidemiology* 30, pp. 1090–1099

Chapter 16
The Challenge to Healthy Longevity: Inequality in Health Care and Mortality in China

Zhongwei Zhao

Abstract China made great progress in lowering mortality in the second half of the twentieth century. Its life expectancy increased from approximately 35 years to more than 65 years between 1950 and 1980. China's recent socio-economic reforms started in the late 1970s and has brought about both positive and negative changes in social security and health care systems. Some of the changes have created difficulties for the further improvement of public health. While China's mortality has continued to fall in the last 25 years and life expectancy has now reached 72 years, it also faces serious challenges in further improving healthy longevity. This chapter first reviews the epidemiological transition and increasing longevity in the world. It then discusses China's mortality decline and some major changes in health care taking place in recent decades. Following that, it examines the increasing inequality in health care and mortality across different areas and population groups. Finally, it comments on major challenges in raising healthy longevity and some related issues.

Keywords Accessibility of health services, Advanced areas, Age reporting, Ageing, China, Cause of death, Cooperative Medical System, Disparity, Epidemiological Transition, Gini index, Government Insurance Scheme, Health care, Health care coverage, Health care system, Health expenditure, Health services, Healthy longevity, Health workers, Income distribution, Inequality, Labour Insurance Scheme, Large cities, Less developed areas, Life expectancy, Mortality decline, Mortality differential, Mortality transition, Regional variation, Rural areas, Sanitary conditions, Standards of living

China has made significant progress in reducing mortality in the second half of the twentieth century. Between 1950 and 1980, its life expectancy at birth has increased from approximately 35 to more than 65 years. This achievement and the successful experiences of lowering mortality in Sri Lanka, Costa Rica and some

---

Z. Zhao
The Australian Demographic and Social Research Institute,
The Australian National University, Canberra, Australia
e-mail: zhongwei.zhao@anu.edu.au

other populations have been widely regarded as "routes to low mortality in poor countries" (Caldwell 1986: 171). Radical socio-economic reform and rapid economic growth have occurred in China since the late 1970s. This profound socio-economic transformation has brought about both positive and negative changes in social security and health care systems. Some of them, for example, the collapse of the Cooperative Medical System in many rural areas, have created difficulties for the further improvement of public health. Growing inequality in income distribution has also made the reduction of mortality difficult to achieve in poor areas and among disadvantaged population groups. Although China's mortality has continued to fall in the last 25 years, and life expectancy has now reached 72 years, the country also faces great challenges in the further development of healthy longevity.

This chapter begins with a brief review of the epidemiological transition and increasing longevity in the world. It then discusses the mortality decline that has been observed in China in the last 50 years and some of the major changes in health care brought about by recent socio-economic reforms. Then the chapter examines the increasing inequality in health care and mortality across different areas and population groups. It next comments on major challenges in raising healthy longevity and some of the related issues. The chapter concludes with a brief summary.

16.1 Epidemiological Transition, Increasing Longevity and the Emergence of the Ageing Society

Although this chapter is about inequality in health care and healthy longevity in China, we begin with a brief review of the epidemiological transition, increasing longevity, and the emergence of the ageing society in the world. This discussion will help us place China's mortality decline in the context of world demographic changes and highlight China's major challenges in further improving public health and longevity.

Mortality was high throughout most of human history. Although considerable fluctuations in mortality were recorded in the past, its continuous and long-term decline did not begin until the second half of the eighteenth century (Laslett 1995). The mortality decline has been closely related to changes in major causes of death. They together are frequently referred to as the epidemiological transition.

According to Omran, before the start of the epidemiological transition, people lived in the "age of pestilence and famine" when mortality fluctuated around a high level. Average life expectancy at birth was low, ranging between 20 and 40 years. During this time, the major cause of death was infectious diseases occurring in pandemics. "The age of receding pandemics" started in some countries from the second half of the eighteenth century. During this transitional stage, the threat of infectious diseases decreased gradually, and mortality fell to a lower level. Life expectancy at birth rose from about 30 years to about 50 years. This led to "the age of degenerative and man made diseases," when deaths were largely caused by chronic degenerative diseases such as cardiovascular diseases and cancers. Life expectancy at birth further increased to 50 years and above (Omran 1971, 1983).

When Omran first proposed this theory, death rates due to some of the degenerative diseases had already fallen in some developed countries, and they have continued to decline further. This has resulted in a continuous increase in life expectancy. Olshansky and Ault accordingly revised the epidemiological transition theory by adding a fourth stage, namely, "the age of delayed degenerative diseases." During this stage, the major degenerative causes of death remain as the major killers, "but the risk of dying from these diseases is redistributed to older ages" (Olshansky and Ault 1986: 361). According to this theory, most of the developed countries in the world have already entered the fourth stage of the epidemiological transition, where the major causes of deaths are degenerative diseases, and mortality has fallen to a very low level.[1]

One of the direct outcomes of the epidemiological transition is the persistent rise in life expectancy and the notable increase in the recorded maximum length of the human life span. Oeppen and Vaupel (2002) studied mortality changes in the world and found that life expectancy in record-holding countries (i.e. countries with the lowest mortality in a given time) has risen steadily since 1840. In the female population, for example, life expectancy has increased in an extraordinary linear fashion at a rate of almost 3 months per year. There is no sign that this trend will stop soon (Oeppen and Vaupel 2002). Because of this change, most of the developed countries now have a life expectancy at birth of around 80 years.

In the age of pestilence and famine, it was very difficult for an individual to survive to advanced ages. When the average life span was only some 30 years or less people rarely lived to 70 years, as suggested by a Chinese proverb, let alone a hundred years. A statistical analysis conducted by Wilmoth also shows that it was very difficult for people to live to 100 years, and it was almost certain that no people survived to 110 before the year of 1800 (Wilmoth 1995). The rapid increase in the number of very old people is largely a phenomenon of the twentieth century. As indicated by the data assembled by Kannisto and other researchers, in twelve selected developed countries (Austria, Belgium, Denmark, England and Wales, Finland, France, West Germany, Italy, Japan, Norway, Sweden and Switzerland) with the most accurate population data, the number of people aged 80–89 increased by four times during the period from 1950 to 2000. But those aged between 90 and 99 increased 15 times, and those aged 100 and over increased almost 50 times from just over 800 to about 40,000 people (Kannisto 1994; University of California at Berkeley and Max Planck Institute for Demographic Research 2006).

These changes have led to the emergence of an ageing society, or what Laslett called "the emergence of the third age" (Laslett 1989). The most important demographic feature of such a society is the increase in the proportion of old people and the gradual increase in the age of the population. This chapter is not about the ageing society itself, and a detailed discussion of this topic is beyond its scope. However, it is useful to briefly mention some characteristics which have been widely observed

---

[1] For recent discussion on the epidemiological transition, see Olshansky et al. (1998) and Vallin (2005).

in developed countries and are most relevant to the issues that will be examined in the remaining part of this chapter.

In most developed countries such as the twelve mentioned above, demographic changes and the epidemiological transition started earlier than in other parts of the world. By the year 1900, mortality in many of these countries had already declined to a relatively low level, and life expectancy at birth had reached around 50 years. People in most of these countries have enjoyed a high standard of living for quite a long time. Per capita Gross Domestic Product at purchasing power parity is high in all these countries. Their human development indices are at the top of the list in the world. The level of socio-economic equality is also relatively high in most of these countries, where fairly good support is available for poor and disadvantaged peoples. Most of these countries have well-established health care systems and social security systems. According to an evaluation made by the World Health Organization (WHO) in a report published in 2000, the level of fairness in financial contribution to health care is high in all these countries, and regional variations in public health and mortality are relatively small. In most of these countries, government spending on health care is relatively large and has been increasing in recent years. All these are obviously some of the major contributing factors that have led to mortality decline, and they are likely to be important conditions for further increases in life expectancy.

16.2 China's Mortality Decline and Recent Changes in Health Care

Levels of mortality level were high in historical China (Barclay et al. 1976; Zhao 1997; Lee and Wang 1999). While survivorship improved in some areas in the first half of the twentieth Century, the crude death rate for the national population "was close to 40 per thousand population in 1930 and stayed that high or higher during most of the 1930s and 1940s" (Banister 1987: 80; Campbell 2001). Life expectancy at birth was likely less than 35 years. China made very impressive progress in reducing mortality between the early 1950s and the late 1970s, although it was interrupted severely first by the Great Famine and then by the social upheavals in the early years of the Cultural Revolution. According to low but perhaps more reliable estimates, China achieved great success in lowering mortality. Life expectancy at birth increased to 50 years in 1957, to 61 in 1970 and to 65 in 1981, a gain of about 10 years per decade for some 30 years (Banister 1987: 116).[2] This is not only faster

---

[2] The most rapid improvement was recorded between 1950 and 1957. During this period, the fall in mortality was dramatic and the life expectancy at birth rose between 15 and 20 years. Data provided by China's Population Information and Research Center indicate that life expectancy at birth for the national population rose to 56 in 1957, to 64 in the early 1970s and to 68 in 1981 (Huang and Liu 1995). These figures likely under-represent actual mortality levels because of the under-registration of deaths.

than that recorded in Europe, but also surpasses what was observed in Japan and Korea in recent decades (Zhao and Kinfu 2005).

Rapid socio-economic changes have occurred in China over the last 25 years. According to unadjusted (for inflation) figures, China's GDP rose about 32 fold from 362.4 billion *yuan* in 1978 to 11725.1 billion *yuan* in 2003. Results based on comparable/standard prices show a slower growth, but the resulting indices still increased from 100 to 940 during this time (National Bureau of Statistics (NBS) 2004). Thanks to this achievement, standards of living have greatly improved despite the fact that China's total population increased by some 300 million during this period, which is about the size of the population of the United States in 2007. According to recent estimates made by the World Bank, in China the number of people living under the poverty line of one dollar per day decreased by over 420 million, from 633.7 million in 1981 to 211.6 million in 2001. This played an important part in poverty reduction in the world where the number of people living under the poverty line fell by 390 million during the same period (Chen and Ravallion 2004). Because of these changes, the health status of the Chinese population as a whole has continued to improve. Unlike the republics of the former USSR and some of the Eastern European countries where mortality stopped declining or even increased during their recent reconstruction (Mesle 2004), China's life expectancy at birth has reached 72 years—a level well above the world average, but ten years lower than the world record (Population Reference Bureau 2004; Ren et al. 2004).

Because of these changes, the number of very old people increased rapidly in China. According to published census results, from 1953 to 2000 the number of people aged 80–89 increased by six times from less than two million to eleven million. The number of those aged 90–99 increased by 14 times from 68,000 to 953,000. (See the first chapter in this volume by Poston and Zeng for more discussion.) In these age groups, recorded mortality rates have also shown some decrease in recent years. According to the 2000 census, however, mortality rates for China's elderly population (up to age 94) are still notably higher than those recorded in many developed countries.

It is difficult to estimate the change in the number of centenarians, because this figure has been over-reported in China's early censuses. For example, the 1953 census reported 3,384 centenarians, which is four times the 809 centenarians recorded in the twelve developed countries mentioned earlier, although the number of Chinese who were aged 90–99 is less than half that of their counterparts. While considerable improvement has been made, age misreporting at the very old ages is still noticeable in recent censuses (Fan 1995; Zhuang and Zhang 2003).[3] This issue along with registration problems of other kinds may contribute to the relatively low mortality recorded among those aged 95 and above.

---

[3] For example, in the year 2000, China's mortality was notably higher and life expectancy was some eight years lower than the corresponding figures in the twelve developed countries mentioned earlier. However, the ratio of centenarians to people aged 90–99 was 0.019 and slightly higher than that in the 12 countries. This could be a result of age misreporting at the very old ages.

China's recent socio-economic reforms have significantly transformed Chinese society and made China a new powerhouse for economic development in the world. However, this transformation has also brought about some negative changes. They are particularly noticeable in the area of health care and are summarized in the next few paragraphs.

Firstly, China's health services have been commercialized considerably during its recent market oriented economic reforms and have increasingly become profit driven. As a result, the cost of health care has risen sharply. An estimate produced by the Ministry of Health (MOH) shows that in comparison with 1990, per capita income in 1999 increased by 288 percent and 222 percent in China's urban and rural areas, respectively. However, the cost of medical care grew at an even faster pace; the cost of visiting doctors (*menzhen*) rose by 625 percent and that of in-hospital treatments (*zhuyuan*) by 511 percent during the same period (Rao and Liu, 2004: 51).[4] The high costs have already become a major obstacle preventing many people from receiving needed medical treatment.

Secondly, many workers and peasants have lost their health care coverage because of recent changes in the health care system. In the late 1970s, more than 75 percent of urban employees and retirees were covered by the Government Insurance Scheme or Labour Insurance Scheme, and benefits were often extended to their dependents. Co-operative Medical Systems were set up in 90 percent of China's rural areas (Wu 2003: 234 and 244). According to the National Health Service Survey (NHSS) conducted by the MOH in 2003, the proportion of people covered by various health insurance schemes was moderate, even though it had increased from a lower level in the early 1990s. Even in large and medium cities, some 40 percent of people did not have any health care coverage. In rural areas the figure was almost 80 percent (Center for Health Statistics and Information (CHSI) of MOH 2004).

Thirdly, due partly to the commercialization of health care services, China's total health expenditures rose by 40-fold from 14.3 billion *yuan* in 1980 to 568.5 billion *yuan* in 2002, or from 3.2 to 5.4 percent of GDP. These increasing costs have fallen largely on individuals. Government health spending as a percentage of GDP actually decreased from 1.1 to 0.8 percent, but the total amount spent by individuals as a percentage of GDP increased from less than 0.7 to 3.2 percent. Measured in absolute terms, the health expenses paid by individuals increased from 3.0 billion *yuan* to 331.4 billion *yuan*, or about 110 times (Rao and Liu 2004: 37).[5] This pattern is radically different from that in most developed and many less developed countries where government expenditures on health are far greater than the contributions made by individuals (WHO 2004).

Fourthly, medical services in China's large cities and less developed rural areas have become more polarized than before. Medical facilities in large cities and developed areas have improved in recent years. The gap between these places and

---

[4] The rate of income increase would be somewhat different were it computed with data published by NBS, but would still be markedly smaller than the increase in the cost of visiting doctors and in-hospital treatments (NBS 2004).

[5] These figures are estimated from the data presented by Rao and Liu and not adjusted for inflation (Rao and Liu 2004: 37).

developed countries in the number of medical professionals and in types of health services they offer has become small in recent years. In contrast, government investment in health has been inadequate in less developed rural areas where the number of medical professionals actually decreased at the grass-root level. According to the figures reported by MOH, while the number of medical professionals working in county-level hospitals or health facilities of other kinds increased slightly from 1.3 million in 1977 to 1.5 million in 2003, the number of village doctors (formerly known as "barefoot doctors") decreased from 1.8 to 0.8 million and the number of health workers (*wei sheng yuan*) decreased from 3.4 to 0.8 million during the same period (Rao and Liu 2004: 35).

Fifthly, the inequality in income distribution has grown at a startling speed, and wealth has increasingly become controlled by a small number of people. The Gini index increased from 0.31 in 1985 to 0.42 in 2001, according to Wu and Perloff (2004). While others have made different estimates, they all generally agree that disparities in income and wealth have reached an alarming level (Meng 2004; China Daily 2004). This stark inequality has also been noted by the United Nations (2005). China's recent reforms have broken the "iron rice bowl."[6] Many poorly managed companies and factories have undergone bankruptcy during the last 20 years, and their former employees have lost their jobs.[7] Changes of a similar nature have also been observed in the rural areas. Unemployment, along with other factors, often leads to poverty. Accordingly, the rapid socio-economic changes have produced not only the new rich, but also the new poor.[8]

16.3 Inequality in Health Care and Variation in Morbidity and Mortality[9]

The changes discussed in the preceding sections have resulted in increases in the inequality in health care in China and have had a marked impact on public health and mortality improvement. Although impacts of this kind are often difficult to detect from health or mortality statistics aggregated at the national or provincial levels, they have already reached a critical level and threaten the potential of future economic growth, and to some extent the stability of the society. This section discusses some of these problems via an examination of data collected by the NHSS in 2003, the population census undertaken in 2000, and data gathered from National Diseases Surveillance Points in the late 1990s.

---

[6] This is a Chinese phrase often used to refer to the guaranteed job provided by the government and state.

[7] According to the NBS, registered urban unemployment rates have varied between 3 and 5 percent in most provinces in recent years (NBS 2004). The actual figures are likely higher.

[8] While the number of people living under the poverty line has decreased greatly in China since the start of the reforms, a recent health survey revealed that there were still 7.2 percent of surveyed households receiving financial support from governments (CHSI of MOH 2004).

[9] This section summarizes some of the research findings reported in Zhao (2006), where a more detailed discussion may be found.

16.3.1 Considerable Variation in the Availability of Health Services, Living Standards and Sanitary Conditions

According to the NHSS, there is still a huge gap in the availability of healthcare services, standards of living and sanitary conditions between China's advanced and less developed areas, as shown in Table 16.1.

The NHSS collected detailed data on the economic status, health conditions, participation in health care systems, use of health facilities, and other relevant information from 194,000 respondents and 57,000 households selected from 28 districts and 67 counties. These districts and counties, which had a total population of approximately 45 million at the time of the survey, were sampled from, and also grouped into, three types of cities and four types of rural areas (CHSI of MOH, 2004). The major criterion for grouping the cities is their population size. As in many other countries, the population size of Chinese cities is related to the level of socio-economic development.[10] Large cities generally are regional or national economic centers, while the medium-sized and especially the small cities tend to be in remote and less developed areas. Rural areas are classified into four categories according to their levels of socio-economic development. Type I includes China's advanced and rich rural areas; Type II areas are not as developed as the Type I areas, but peasants in these places can have a reasonably comfortable life; Type III areas consist of those where food, clothing and other living necessity have reached just adequate levels; and Type IV includes China's least developed and poor rural areas.

Table 16.1 Disparities in the availability of health care, living standards and sanitary conditions 2003

	Cities			Rural areas			
	Large	Medium	Small	Type I	Type II	Type III	Type IV
Number of doctors per 1000 population	5.8	4.4	1.7	1.3	1.0	0.8	0.6
Number of nurses per 1000 population	5.8	4.8	1.4	1.1	0.7	0.6	0.4
Proportion having no medical care coverage	38.5	41.2	55.0	67.8	80.7	88.6	70.8
Per capita income (*yuan*)	8292	6607	4589	3163	2187	1938	1187
Per capita expenditure (*yuan*)	6297	4791	3524	2466	1763	1666	1039
Proportion of households using tap-water	99.5	99.8	87.6	49.3	31.1	27.4	30.1
Proportion of households using flush toilets	86.1	93.5	57.6	13.5	4.1	2.1	1.2

Sources: CHSI of MOH (2004).

---

[10] Chinese cities are often divided into three groups according to their population size: cities with a population of less than 200,000 are defined as small cities; those with a population between 200,000 and 500,000 are defined as medium cities; and those with a population of over 500,000 are defined as large cities (The Standing Committee of the National People's Congress 1989).

In general, more people engage in agricultural production in the least developed rural areas as compared to the advanced rural areas.

Table 16.1 reveals great disparities between the advanced and less developed areas in the availability of health services, standards of living and sanitary conditions. In large cities, there are 5.8 doctors and 5.8 nurses for every 1,000 residents. In contrast, there is only 1 such medical professional (0.6 doctors and 0.4 nurses) per 1,000 population in Type IV rural areas. In large cities, a hospital serves an area of 1.4 square kilometres and 14,800 people on average, but in the least developed (or Type IV) rural areas, a hospital serves an area of 1,047 square kilometres and 103,000 people on average. This makes it very difficult for those with acute health conditions to get high quality and timely treatment.[11] In large and medium-sized cities more than 60 percent of the people are covered by some kind of health care scheme, but in most of the rural areas, people without any health care coverage account for about 80 percent of the population. Because of the shortage of health facilities and health professionals, people in the less developed rural areas are disadvantaged in their ability to access knowledge about health care and health services. For example, while in large and medium cities, 94 percent of respondents had some knowledge or had heard about HIV/AIDS, only 40 percent of respondents in Type IV rural areas had such knowledge.

There is also a large gap across these areas in people's standard of living and living environment. As shown in Table 16.1, in large cities average annual income and expenditures are seven times and six times, respectively, those for Type IV rural areas. In many rural areas, especially in Type IV rural areas, the income level is well below the internationally recognized poverty line of one dollar per day. Such a low living standard directly affects people's nutritional intake which in turn makes them vulnerable to many diseases. In China's poor rural areas, more than 20 percent of children below age 5 are under weight for their age, and infant mortality is high; this is a clear indication of such an impact (Rao and Liu 2004). In addition, people in less developed rural areas usually live in a hygienically inadequate environment. For example, compared with large and medium-sized cities where almost 100 percent of people use tap-water and around 90 percent of households have flush toilets, in Type III and IV rural areas, less than one-third of the people use tap-water and about 90 percent of toilets are regarded as unhygienic (CHSI of MOH 2004: 70). It is important to note that all of these factors are closely related to morbidity patterns and the level of mortality, which will be discussed later.

16.3.2 Increasing Inequality in the Accessibility of Health Services

Because of the increasing inequality in healthcare coverage and in the distribution of income and wealth, great intra-regional inequality in healthcare has become a major social problem. This has also had in recent years a considerable effect on the

---

[11] There are clinics and health stations in some of the poor areas, but they are usually not equipped to treat complex health problems.

improvement of public health. This is particularly the case with regard to patients' access to health services, as shown in Table 16.2.

According to the NHSS, about 28,000 out of nearly 200,000 respondents, reported having suffered from various types of illness during the two weeks before the time of the interview. Slightly more than half of them visited doctors for treatments, and a total of 26,000 visits were recorded. As indicated by Table 16.2, nearly half of those suffering from various kinds of health problems did not see doctors for treatment. Some people treated the illness themselves, and 13 percent did not have any treatment at all. In large and medium-sized cities and in Type I and Type II rural areas, about one-third of those who did not receive any treatment did not do so due to economic hardships. In small cities and in Type IV rural areas, the proportions were markedly higher and close to 50 percent. Poverty and inequality have noticeably influenced people's access to medical services. Even in large cities, 3 percent of all patients could not get the needed treatment because of economic difficulties. The impact was greater in the least developed rural areas where nearly 10 percent of patients were not treated because they did not have the financial resources.[12]

Table 16.2 Proportions of patients visiting doctors or getting in-hospital treatments

	Cities			Rural areas			
	Large	Medium	Small	Type I	Type II	Type III	Type IV
Number of person–time visits to doctors in the two weeks before the survey (per 1,000 people)	119.7	92.6	138.2	115.7	145.7	158.0	118.3
Proportion of patients who did not visit doctors for treatment	57.7	63.8	48.9	49.8	43.0	46.7	43.0
Proportion of patients who did not have any treatment	10.7	7.6	10.6	11.9	12.8	15.5	18.9
Proportion having not had any treatment because of financial difficulties	30.8	32.7	47.0	29.2	33.9	41.2	49.1
Number of person–time hospitalizations in the previous year (per 1,000 people)	40.3	46.0	41.6	34.2	30.1	35.7	36.5
Proportion of in-patients asking for early discharge from the hospital	32.6	31.8	38.8	39.0	45.2	53.5	47.5
Proportion asking for early discharge because of financial difficulty	40.8	50.8	65.1	66.8	62.9	68.2	73.2
Proportion of patients who ought to but did not have in-hospital treatment	23.3	31.3	28.7	23.2	27.1	35.8	31.2
Proportion not receiving in-hospital treatment because of financial difficulty	64.4	35.6	74.8	77.6	74.9	75.5	73.6

Sources: CHSI of MOH (2004).

---

[12] These figures are derived by multiplying the proportion of patients who did not have any treatment (figures in row three of Table 16.2) by the proportion not having had any treatment because of financial difficulty (the figures in row four of Table 16.2) and then dividing by 10,000.

Influence of this kind is more observable in people's accessibility to in-hospital treatment. The NHSS also recorded that nearly 6,000 patients were hospitalized for a total of about 7,000 times for treatment during the year prior to the survey. Because of the high cost of the in-hospital treatment, which almost equals a person's average annual income, 43 percent of in-patients asked to be released from hospitals before the date recommended by their doctors. This proportion is highly related to the level of socio-economic development, people's income and the percentage of respondents covered by various medical care schemes in each type of area.[13] As expected, the majority of those asking for an early discharge did so because of economic hardship. In large cities, 41 percent of patients who wanted to be released earlier did so due to financial difficulty, and in Type IV rural areas the figure was as high as 73 percent. In addition, there was also a large number of patients who should, but did not, have any in-hospital treatment at all, varying from 23 to 36 percent across the seven types of areas. Again, economic difficulty was the major determinant for such decisions. In small cities and in rural areas, three quarters of those who should, but did not, have the in-hospital treatment did not do so because of this reason. In medium-sized cities, the proportion of patients who did not receive in-hospital treatment because of financial difficulty was noticeably lower than in all other areas. This was likely caused by reporting or coding problems in one district (the Lubei District, Hebei Province) where 85 per cent of those who should, but did not, have in-hospital treatment reported that they felt the treatment was unnecessary and only 9 percent reported that they did not seek treatment due to financial difficulties. If this district is excluded, then in the medium-sized cities the proportion of patients not getting in-hospital treatment because of financial difficulties increases from 36 to about 63 percent, which is very close to the proportion in large cities, which is 64 percent.

16.3.3 *Great Disparities in Causes of Death and Mortality*

The striking gap in living conditions and in the availability and accessibility of health services directly results in great disparities in public health and in mortality patterns between the advanced and the less developed areas. As indicated by Table 16.3, there is great variation in morbidity patterns and major causes of death across regions with different levels of socio-economic development. The NHSS reveals that a larger proportion of urban than rural respondents reported that they suffered from cardiovascular diseases during the two weeks before the survey. Cardiovascular diseases comprise more than one third of all reported diseases in large cities, but account for less than one tenth of the diseases in Type IV rural areas.[14] In

---

[13] These conclusions are supported by the analysis of data recorded at the individual level, which was conducted by CHSI of MOH (2004).

[14] This may be partly related to the fact that in less developed rural areas there is a lack of accurate diagnosis, and many people are not aware of these and some other types of non-communicable diseases.

Table 16.3 Variations in mortality and causes of death

	Cities			Rural areas			
	Large	Medium	Small	Type I	Type II	Type III	Type IV
Proportion of deaths by							
Infectious and maternal diseases[a]	3.8	3.9	6.2	4.4	6.1	11.4	23.1
Non-communicable chronic diseases	84.2	80.8	74.7	80.9	78.6	70.3	60.6
Injury and poisoning	6.0	7.4	4.8	10.3	11.2	13.1	10.2
Unknown reasons	6.0	7.9	14.3	4.4	4.0	5.1	6.1
TB prevalence rate (per 100,000 population)	37.3	69.9	150.1	81.1	96.3	140.8	223.2
Average life expectancy at birth	77.7	77.7	75.7	73.8	73.0	71.3	65.2
Average infant mortality rate in 2000	6.0	8.6	14.5	14.1	24.2	30.6	54.0

Sources: Department of Control Disease of MOH and Chinese Academy of Preventive Medicine 1997 and 1998. The life tables for these districts and counties are provided by Yong Cai. CHSI of MOH, 2004.
[a] See the text for the classification of these categories.

contrast, the proportion of people who reported to have suffered infectious diseases, digestive diseases, and respiratory diseases was greater in the less developed rural areas than in the large and medium-sized cities. A notable example is the difference between China's advanced and less developed areas in their TB prevalence rates. According to the data collected by the NHSS, the TB prevalence rate is 37 per 100,000 in large cities, but is significantly higher at 223 per 100,000 in Type IV rural areas. These results are computed based on patients who coughed continuously for at least three weeks and were then diagnosed with TB in hospitals. The adjusted rates, which were estimated by the MOH and included patients who had the disease but had not been diagnosed in this way, were noticeably higher.[15]

The results are supported by the causal structure of deaths observed from China's disease surveillance points. As shown in Table 16.3, in China's large and medium-sized cities and in advanced rural areas, the major causes of death are very similar to those in developed countries. The overwhelming majority of deaths were caused by cancers, heart diseases, cerebrovascular diseases and other non-communicable diseases. The proportion of those dying from infectious diseases is low. In contrast, in Type III and IV rural areas, the proportion of deaths due to non-communicable chronic diseases is much lower; and deaths caused by infectious diseases, diseases of upper respiratory tract, pneumonia, influenza, maternal diseases, and diseases

---

[15] The MOH adjusted rate for the surveyed population is 153.4 per 100,000, which is notably higher than the 115.6 per 100,000 computed according to the above method (WHO 2004).

originating in the perinatal period are considerably higher.[16] This pattern is very similar to that found in many less developed countries.

The marked gap in public health and morbidity patterns also affects and is reflected in the level of mortality. There is a strong positive relationship between life expectancies at birth and levels of socio-economic development.[17] Life expectancy at birth in large cities is more than 12 years greater than in Type IV rural areas. When examined at the district and the county levels, the gap between the highest and lowest life expectancies is even larger and is close to 20 years. Great disparities of this kind also exist in infant mortality. According to data in the 2000 census, the recorded infant mortality rate in large cities was 6 per 1,000 and very close to the lowest in the world, but in Type IV rural areas it was 54 per 1,000, or nearly 10 times higher. High infant mortality in less developed rural areas is directly related to poor access to effective medical services. Data collected from China's disease surveillance points show that in rural areas in 1998 the infant mortality rate due to pneumonia was about 15 times that in urban areas (Department of Disease Control of MOH and the Chinese Academy of Preventive Medicine 1998). Had timely treatment been available or easier to access in less developed rural areas, many of these lives could have been saved.

16.4 Challenges to Healthy Longevity

The problems examined in the last two sections have already become major obstacles in improving public health. They are also major challenges for further increasing life expectancy and for understanding and promoting longevity, which are discussed below under two sub-headings.

16.4.1 Improving Healthy Longevity

A number of observations can be made from the evidence presented up to this point. Because of the deterioration of China's health care system in the 1980s and the recent spiralling increase in the cost of medical services, it has become difficult for poor or disadvantaged people to obtain adequate medical care. It is alarming that financial difficulties were the reason that in less developed rural areas more than half (52 percent) of the patients who needed to have in-hospital treatments either

---

[16] This is rather similar to the causal structure of death found in many less developed countries. These patterns are generally consistent with those described by epidemiological transition theory and observed in other populations (Omran 1971, 1983).

[17] These life expectancies are obtained from the life tables constructed by Yong Cai on the basis of unadjusted mortality data gathered in the 2000 census. Studies have shown that the 2000 census under-recorded deaths (Li and Sun 2002). However, because detailed information is not available, we are not able to adjust recorded mortality rates. While the recorded life expectancies might be slightly higher than the actual ones, the marked difference in mortality among the different areas as indicated by these data is reliable.

could not have them or had not completed the in-hospital treatment recommended by doctors.[18] The problem has become so severe that some Chinese officials have recently suggested that China's healthcare reform is largely a failure (Bai 2005). The impact of increasing inequality in income distribution and healthcare on population health is also indicated by the fact that between 1981 and 2000, infant mortality fell by 59 percent in large and in medium-sized cities, but the reduction was only about 30 percent in small cities and in poor rural areas. The disparity in mortality between advanced and less developed areas has actually increased (Zhao 2006).

Evidence also shows that because of the changes and problems examined above, there have been some downward trends in improving public health. In addition to the great reduction in the number of village doctors, the proportion of children receiving immunizations or vaccinations has decreased in recent years (CHSI of MOH 2004; Riley 2004). The prevalence rate of TB has remained high and has increased in some areas. This is particularly the case among the poor and among temporary migrants who often live under crowded and poor living conditions. TB kills many more people each year than most other infectious diseases.[19] The increase in HIV/AIDS, in poor and in less developed areas in particular, is another example. According to data published by official reports, more than 800,000 people have been infected by the virus since it was first recorded in China, and AIDS may become a major killer in China if it can not be controlled effectively (State Council AIDS Working Committee Office and UN Theme Group on HIV/AIDS in China 2004). Partly due to these changes, China's recent mortality decline has become slower in comparison with that recorded in earlier periods, and its mortality advantage over countries with similar levels of development, which greatly impressed the world in the 1970s and 1980s, has largely disappeared.

Poverty is an important reason for poor health, and this has been illustrated by the research findings reported in the last section. Poor health, especially in a society where an effective healthcare system does not exist, could also lead to poverty. This is readily observable in China's less developed areas and disadvantaged population groups. According to a survey conducted in China's poor rural areas and reported by Rao and Liu (2004), high medical costs have already become a heavy financial burden and a major cause of poverty in many families. The medical expenditures in about 18 percent of the surveyed families were actually greater than their total incomes. Nearly a quarter of the 11,353 surveyed households borrowed money, and 5.5 percent of them sold their properties or belongings to pay their medical bills. According to the 1998 NHSS, after paying their medical expenses, the proportion of poor people in rural areas increased from 7.2 to 10.5 percent (Rao and Liu 2004:

---

[18] In another study, Rao and Liu reported that in China's rural areas, 37 per cent of the patients ought to, but did not, see doctors for treatment, and 65 per cent of the patients should, but did not, have in-hospital treatments (Rao and Liu 2004: 47).

[19] According to the NHSS, the MOH adjusted rate is 153.4 per 100,000 in the surveyed population. According to a recent WHO health report, China's TB prevalence rate was 250 per 100,000 in the year 2000, which is noticeably higher than the rate published by the Chinese government (WHO 2004).

51–52). Poverty thus exerts a further impact on the health of these impoverished people, and a new cycle begins again.

Old people, especially those without a pension and healthcare coverage, are particularly vulnerable to the influence of both poor health and poverty. For example, while it is less observable in cities, healthcare provided to old people is often less adequate than that provided to the general population in rural areas. This is indicated by the fact that in rural China, 45.8 percent of people of all ages who suffered from various diseases during the two weeks before the 2003 NHSS did not visit doctors for treatment, compared to 51.6 percent who did not do so among those aged 65 and above. Similarly, among those who should have in-hospital treatment, 30.3 percent did not have such treatment in the population of all ages, but 44.4 percent did not have the required treatment among people aged 65 and over. Restricted medical care of this kind greatly prevents older people from maintaining good health and could reduce their life span considerably.

To stop the negative trends discussed above and to meet the challenges in improving healthy longevity, there is a need to control the rapid increase in the price of medical services. This is a relatively easy step in improving accessibility to medical care. The government needs to invest more in public health, especially in providing help to less developed areas and disadvantaged people. After a quarter of a century of successful reform and economic growth, the government is now in a much better position to provide such support than was the situation during the pre-reform period. Moreover, further actions need to be taken to re-establish and consolidate a nationwide health insurance system. The current health care coverage simply cannot meet people's immediate and, especially, long-term health needs. China has made some progress in recent years in developing a medical care system, but further effort is crucial.

16.4.2 Understanding the Process of Increasing Longevity

The discussions presented in this chapter and elsewhere in this book also raise another question which is of considerable importance for the study and our understanding of longevity. In comparison with countries such as the 12 developed countries mentioned earlier, China's censuses and some surveys often recorded a relatively large number of very old people, especially centenarians. These centenarians also tended to concentrate in some of the less developed areas. For example, a large number of centenarians were recorded in Xinjiang and Guangxi. In these two autonomous regions, many centenarians were reportedly living in the counties of Hotan and Bama, and both places have thus been regarded as the "home of longevity" (Xing and Zhang 2005; People's Daily 2003). However, both the Hotan and Bama Counties were also defined by the Chinese government as poor counties where per capita annual income was below 700 yuan in 1994. The two counties are included in China's least developed areas. According to unadjusted census data, life expectancy at birth in 2000 was only about 65 for Hotan County and 70 for Bama County, lower than the national average.

The following example further illustrates this rather unusual situation. Sweden and Japan are two very wealthy countries in the world where life expectancies at birth have reached more than 80 years. In 1998 there were 15 and 17 centenarians per 1,000 people aged 90 and above in the two countries respectively. In contrast, the standards of living and healthcare availability levels were much lower in China's Guangxi Autonomous Region where life expectancy at birth was about 10 years lower. Despite such huge gaps, however, there were 40 reported centenarians for every 1,000 people aged 90 plus in Guangxi in 1990—a ratio that is far greater than those recorded in Sweden and Japan.

Here we have a paradoxical situation. The brief review presented in the first section of this chapter suggests that while it was possible in the past for a small number of people to survive to an advanced age or to even become centenarians when mortality was relatively high, a large increase in the proportion of very old people, centenarians in particular, is a relatively new phenomenon taking place largely in the second half of the twentieth century. This was first observed in developed countries where standards of living and levels of healthcare are both high, which are important contributing factors for lowering mortality. The reduction of mortality across all age groups is an important condition for the increase in the number of the oldest old. If these suggestions are true, why have so many centenarians been reported in China where levels of socio-economic development and health care are much lower and where overall mortality is considerably higher than in most developed countries? Why have many of China's centenarians been reported in its less developed areas where standards of living are still very low?

An important step in solving this puzzle is to find out whether there were really as many centenarians as reported in places like Xinjiang and Guangxi. If they are indeed "homes of longevity" and have considerably higher proportions of centenarians, we not only need to find out the major factors or conditions that promote healthy longevity, but also need to reassess existing theories of epidemiological transition and increasing longevity such as those summarized earlier. It becomes important, for example, to address these questions: (1) to what extent do lowering mortality in the whole population and improving public health contribute to increasing healthy longevity; and, (2) what are the reasons that make China's old people (compared with old people in other national populations) vulnerable to the risk of death when they are relatively young but more invincible after age 95 or 100.

If however the relatively large number of centenarians is not genuine but a result of misreporting, it is also important to assess the impact of such problems on our understanding of the process of increasing longevity and on the conclusions drawn from these data. Studies have already shown that a large number of centenarians recorded in Xinjiang in 1982 was largely a result of age mis-reporting (Coale and Li 1991). Whether problems of the same nature existed in Guangxi (which has also claimed to be the "home of longevity") or other places needs further investigation. The outcome of such investigations will have important policy implications. If the number of centenarians in places like Xinjiang and Guangxi is not as large as is being reported, and if lowering the mortality of the entire population and improving public health are still vital conditions for increasing levels of longevity of the oldest old, then the effort, especially that made by the government, perhaps should be directed to the promotion of public health and equality in healthcare rather than to the promotion of various "homes of longevity." The objective of such efforts is not finding a large number of centenarians, but rather finding ways to help the majority of the population gain a longer and healthier life. This would be very difficult to achieve if the problems addressed in this chapter could not be solved effectively.

16.5 Conclusion

China made very impressive progress in reducing mortality in the last half century. However, further improving healthy longevity in the population is a serious challenge. In addition to raising the standard of living, which China has done remarkably well during its recent economic reforms, it is also essential that China improves its healthcare system and reduces the inequality in the availability of, and accessibility to, healthcare services. It is important to recognize the fact that compared with many other countries in the world, China's mortality level is still relatively high; there are still many people who die of infectious diseases; regional variations in causes of death and mortality patterns are still very pronounced; and the level of healthcare and services is still low. Further efforts need to be made in these areas if China wants to maintain its success in lowering mortality and to continue to improve public health. In a population where a considerable number of people live under the poverty line, the majority do not have any healthcare coverage and many people can not obtain the requested medical treatments because of financial difficulties, increasing healthy longevity will be difficult to achieve.

Acknowledgments The author would like to thank Cai Yong and Li Li for their help. The author would also like to thank University of Cambridge and Pfizer for their support of this research.

References

Bai, J. (2005), Ge Yanfeng: Rethinking China's health reform. *China Economic Times* (in Chinese), http://business.sohu.com/20050615/n225953302.shtml, accessed on 11 August, 2005
Banister, J. (1987), *China's changing population*. California: Stanford University Press
Barclay, G.W., A.J. Coale, M.A. Stoto, and T.J. Trussell (1976), A reassessment of the demography of traditional rural China. *Population Index* 42, pp. 606–635
Caldwell, J.C. (1986), Routes to low mortality in poor countries. *Population and Development Review* 12, pp. 171–219
Campbell, C. (2001), Mortality change and the epidemiological transition in Beijing, 1644–1990. In: T. Liu, J. Lee, D.S. Reher, O. Saito, and W. Feng (eds.): *Asian population history*. Oxford: Oxford University Press, pp. 221–247
Coale, A. and S. Li (1991), The effect of age misreporting in China on the calculation of mortality at very high ages. *Demography* 28, pp. 293–301
Center for Health Statistics and Information of MOH (2004), *An analysis report of national health services survey in 2003*. Beijing: Zhongguo Xiehe Yike Daxue Chuban She (in Chinese)

Chen, S. and M. Ravallion (2004), How have the world's poorest fared since the early 1980s? *The World Bank Research Observer* 19, pp. 141–169

China Daily (2004), China's income distribution policy urged to be adjusted. (In Chinese) At http://www.chinadaily.com.cn/english/doc/2004-03/18/content_315965.htm, accessed on 11 August 2005

Department of Control Disease of MOH and Chinese Academy of Preventive Medicine (1997), *1997 annual report on Chinese diseases surveillance*. Beijing: Beijing Yike Daxue Chuban She

Department of Control Disease of MOH and Chinese Academy of Preventive Medicine (1998), *1998 annual report on Chinese diseases surveillance*. Beijing: Department of Control Disease of MOH and Chinese Academy of Preventive Medicine (in Chinese)

Fan, J. (1995), *China population structure by age and sex*. Beijing: Data User Service, CPIRC (in Chinese)

Huang, R. and Y. Liu (1995), *Mortality data of China population*. Beijing: Data User Service, CPIRC

Kannisto, V. (1994), *Development of oldest-old mortality, 1950–1990: Evidence from 28 developed countries*. Odense: Odense University Press

Laslett, P. (1989), *A fresh map of life: The emergence of the third age*. London: Weidenfeld and Nicolson

Laslett, P. (1995), Necessary knowledge: age and aging in the societies of the past. In: D. Kertzer and P. Laslett (eds.): *Aging in the past: Demography, society, and old age*. Berkeley: University of California, pp. 3–79

Lee, J.Z. and F. Wang (1999), *One quarter of humanity: Malthusian mythology and Chinese realities, 1700–2000*. Cambridge: Harvard University Press

Li, S. and F. Sun (2002), A preliminary analysis of mortality level in China in 2000. Unpublished manuscript

Meng, X. (2004), Economic restructuring and income inequality in urban China. *Review of Income and Wealth* 50, pp. 357–379

Mesle, F. (2004), Mortality in Central and Eastern Europe: Long-term trends and recent upturns. *Demography Research* (electronic journal), Special collection 2, pp. 45–70

National Bureau of Statistics (2004), *China statistics yearbook*. Beijing: China Statistics Press

Oeppen, J. and J. Vaupel (2002), Broken limits to life expectancy. *Science* 296, pp. 1029–1031

Omran, A.R. (1971), The epidemiological transition: A theory of the epidemiology of population change. *Milbank Memorial Fund Quarterly* 49, pp. 509–538

Omran, A.R. (1983), The epidemiological transition theory: A preliminary update. *Journal of Tropical Pediatrics* 29, pp. 305–316

Olshansky, S.J. and B. Ault (1986), The fourth stage of the epidemiologic transition: the age of delayed degenerative diseases. *The Milbank Quarterly* 64, pp. 355–391

Olshansky, S.J., B.A. Carnes, R.G. Rogers, and L. Smith (1998), Emerging infectious diseases: The fifth stage of the epidemiological transition? *World Health Statistics Quarterly* 51, pp. 207–217

People's Daily (2003), Facing the challenge of population ageing: Discovering the secret of Bama—the home of longevity. (In Chinese). At http://www.chinapop.gov.cn/rkxx/ztbd/t20040326_10985.htm, accessed on 15 August 2006

Population Reference Bureau (2004), *World population data sheet*. Washington: Population Reference Bureau

Rao, K. and Y. Liu (2004), Study of health care systems and related policies in rural China. In: Center for Health Statistics and Information of MOH (eds.): *Research on health reform issues in China, 2003*. Beijing: Zhongguo Xiehe Yike Daxue Chuban She, pp. 34–81 (in Chinese)

Ren, Q., Y. You, X. Zheng, X. Song, and G. Chen (2004), The levels and patterns of mortality and their variations in China since the 1980s. *Chinese Journal of Population Science* 3, pp. 19–29 (in Chinese)

Riley, N. (2004), China's population: new trends and challenges. *Population Bulletin* 59 (2), pp. 3–36

State Council AIDS Working Committee Office and UN Theme Group on HIV/AIDS in China (2004), *A joint assessment of HIV/AIDS prevention, treatment and care in China.* At http://www.casy.org/engdocs/JAREng04.pdf, accessed on 11 August 2005

The Standing Committee of the National People's Congress (1989), *The urban planning law of the People's Republic of China* (in Chinese). At http://www.cin.gov.cn/law/main/law023.htm, accessed on 20 July 2006

United Nations (2005), *The inequality predicament.* New York: United Nations

University of California at Berkeley and Max Planck Institute for Demographic Research. The Human Mortality Database. At http://www.mortality.org, accessed on 10 January 2006

Vallin, J. (2005), Diseases, deaths, and life expectancy. *Genus* LXI, pp. 279–296

WHO (2000), *The world health report 2000.* Geneva: World Health Organization

WHO (2004), *The world health report 2004.* Geneva: World Health Organization

Wilmoth, J. (1995), The earliest centenarians: A statistical analysis. In: J. Bernard and J.W. Vaupel (eds.): *Exceptional longevity: From prehistory to the present.* Odense: Odense University Press, pp. 125–169

Wu, X. and J.M. Perloff (2004), *China's income distribution over time: Reasons for rising inequality.* Department of Agriculture and Resource Economics, University of California, Berkeley

Wu, R. (2003), *Medical insurance systems: An international comparison.* Beijing: Huaxue Gongye Chuban She (in Chinese)

Xing, J. and L. Zhang (2005), From average life span of less than 30 years to the home of longevity. (In Chinese). At http://www.chinaxinjing.cn/news/xjxw/shjj/t20050929_59840.htm, accessed on 20 August 2006

Zhao, Z. (1997), Demographic systems in historic China: Some new findings from recent research. *Journal of the Australian Population Association* 14, pp. 201–232

Zhao, Z. (2006), Income inequality, unequal health care access, and mortality in China. *Population and Development Review* 32, pp. 461–483

Zhao, Z. and Y. Kinfu (2005), Mortality transition in East Asia. *Asian Population Studies* 1, pp. 3–30

Zhuang, Y. and L. Zhang (2003), *Basic data of China population since 1990.* Beijing: China Population Publishing house (in Chinese)

Part IV
Subjective Wellbeing and Disability

Introduction *by Denese Ashbaugh Vlosky*

The CLHLS is an important dataset for examining the impact of various dimensions of subjective well-being on a wide variety of outcomes. These impacts have not been adequately investigated in previous studies because of data limitations. For example, studies focused mainly on the young-old, were of limited sample size, and were conducted primarily in Western countries with advanced medical treatments and interventions, and where the population and therefore subjects had higher levels of SES. The CLHLS rectifies this imbalance in data by offering us a dataset of the oldest-old from a developing country in the East. The CLHLS also allows us the opportunity to utilize data collected in a country that has very different cultural norms that influence both the perceptions of quality of life, and the support and care provided to their elderly population, also known to impact quality of life. In addition, the very size of the CLHLS provides great opportunities to examine gender and cohort specific differences in the meaning of the constructs as well as in their impacts. In all, the CLHLS allows previous studies to be extended and comparisons to be made that can highlight potentially new and useful ways to enable the elderly to age more successfully. These findings will surely have important policy implications for planning purposes for countries worldwide. The chapters in Part IV are all excellent examples of how the CLHLS has been utilized to extend and conduct research and discover new knowledge in this important area of well-being.

Various dimensions of "subjective well-being" or quality of life were addressed in the chapters contained in Part IV. Moran, Sihan and Chen in Chap. 18 and Yun Zhou and Zhenzhen Zheng in Chap. 19 focus on the physical limitation (disability) aspect of subjective well-being. Moran and colleagues compared ADL impairments between the Chinese and American oldest-old, and how the impairments were moderated through demographic characteristics, attitudes, behaviors and coping ability, as well as through medical care and environmental factors. They found that Chinese

---

D. A. Vlosky
Office of Social Service Research & Development, School of Social Work, The Louisiana State University
denese@lsu.edu

elders showed significant, and in the case of transferring, dramatically lower odds of functional impairment after adjusting for known confounders. Their important findings pave the way for further studies into behavioral, environmental or lifestyle factors that may moderate and reduce disability levels, and thereby improve quality of life.

Yun Zhou and Zhenzhen Zheng, on the other hand, use the CLHLS to investigate a relatively new research area in China, tooth loss and the denture wearing status of the oldest-old and how these impact quality of life. The objective of their research was to describe the dentate, edentulous and denture wearing status of the older Chinese, to highlight the discrepancy between reality and the goal of the "WHO 80/20 plan," (i.e., having 20 teeth at 80 years of age), and to add more knowledge and information about oral hygiene among Chinese elderly. They found that older elders, rural residents, those of lower SES, and females, all had fewer teeth, although females lived longer. Unfortunately, tooth loss did not equate with denture wear, an important finding with regard to the policy addressing the dentate health and quality of life of the Chinese oldest-old.

Jiajian Chen and Zheng Wu in Chap. 24, and Quiang Li and Yuzhi Liu in Chap. 23, look at gender differences in the relationship between Self Reported Health (SRH) and mortality. Jiajian Chen and Zheng Wu investigate gender differences in the SRH–mortality relationship and the factors that tend to mediate the relationship among the oldest old in China. The authors ask if SRH is predictive of oldest-old mortality in cohort groups with low SES, whether risk factors moderate the association between SRH and mortality, and whether SRH is a valid evaluation of health among the Chinese oldest old. They find that self-rated good health is consistently related to longer life after controlling for the effects of socio-demographic and physical health conditions among men and women. Education (as a measure of SES) and psychosocial factors are found to be independently predictive of mortality and appear to modify longevity. These findings suggest that self-rated health status is an important health indicator and mortality predictor for the oldest-old men and women in China, an important finding in a part of the world where low-cost diagnostic tools are much valued.

Quiang Li and Yuzhi Liu also looked at gender differences in the SRH–mortality relationship with the intent of teasing out the mechanisms that underlie it. They utilized the 1998–2002 waves of the CLHLS, focusing on the 1998 wave, which was comprised of a cohort with much lower levels of education and a much higher likelihood of widowhood because these factors may have influenced self assessments of health which in turn may have influenced mortality. The CLHLS' large sample size also allowed the authors to investigate changes in SRH overtime and the impact of SRH missing values on mortality. The results of their Cox Hazard Model show that SRH was significantly associated with mortality in the oldest-old even when confounding variables such as sociodemographic, health and behavioral factors were controlled. The effects were stronger in men than in women, and the modifying effects of some factors varied by gender. Missing values predicted higher mortality but were explained by physical and cognitive functioning.

Part IV Subjective Wellbeing and Disability 291

In a related piece, Jacqui Smith, Quiang Li and Denis Gerstorf, in Chap. 20, utilize the large CLHLS database to investigate psychological resources for well-being among three groups, octogenarians, nonagenarians and centenarians. The authors compare those who survived an additional 2 years and those who did not. Although previous studies had been done on contemporary generations of the young-old in many countries, little was known about the constellations of various characteristics promoting well-being in the oldest old (aged 80+), particularly those living in a developing country. The authors found that despite constraints in objective life conditions, long-lived individuals tend to show reasonably high levels of psychological resources for well-being. Most of the variance was explained by selective mortality and individual differences in life history and context. In fact age-cohort differences were small. Surprisingly, the authors note that their findings were very similar to studies in other countries. They state that "the CLHLS provides a wealth of opportunities for future analyses to test hypotheses about determinants of longevity and life quality in the oldest old and also to distinguish sources of heterogeneity in this period of life (e.g., biological, social, age-related, and death-related). Such future work will further our understanding of how individuals age differently at the end of life."

Min Zhou and Zenchao Qian, in Chap. 22, use the first wave of the CLHLS to break new ground examining social support and quality of life in an oldest old population in a developing country where declining fertility will mean fewer children to care for and live with their parents. They utilize objective measures of social support by examining living arrangements, children's visits and perceived social support; and they evaluate subjective measures through self reports of quality of life. They found that all sources of social support were important to overall quality of life. Elders living alone reported the lowest quality of life, and surprisingly those living in nursing homes reported the highest qualify of life. The authors offer several explanations for their unexpected findings, including the importance of possible peer relationships. Probably most importantly, this study provides fodder for the further examination of how living arrangements in conjunction with other factors impact quality of life for the oldest-old in China.

Du Peng, in Chap. 17, uses the large CLHLS database to further the study of "successful aging" in China as defined by predetermined criteria in biomedical, physiological, psychological and social functioning. Former studies originated in the developed world, beginning with the MacArthur Studies in the United States and extending to the Australian Longitudinal Study of Ageing in Australia. Subsequent work included the Beijing Multidimensional Longitudinal Study of Ageing (BMLSA) in Beijing, although the sample was geographically limited and homogeneous. With the CLHLS, the author was able to extend study results to a representative sample of the oldest-old in China. He found that successful aging in Australia and in the United States, for the entire age range of the elderly population was associated with more years of education, and with more household assets or monthly income (higher SES). In China successful aging was associated with more years of education and with marital status. Those with positive attitudes on life, who exercised more often and who had better physical performance were more successful in their old age and had a higher survival rate.

Finally, Deming Li, Tianyong Chen, and Zhenyun Wu in Chap. 21 utilize the extensive data in the CLHLS to examine the subjective well-being of the Chinese oldest-old. Prior studies were mainly conducted with the young-old, and they failed to fully explore whether subjective well-being remains stable overtime at the oldest ages. A few studies were conducted in western nations but were never extended to the oldest old in developing countries. Li and colleagues found that life satisfaction and affective experience were well above the neutral level among the oldest-old in China, despite declines in physical and cognitive areas, and social losses. Self-reported life satisfaction remained constant or slightly increased with age, while scores of affective experience decreased with age. Subjective well being was influenced by the expected demographic variables (sex, education and SES, for example) and by social supports, including family, friends and surprisingly social workers. These results may well become the basis of important policies in China that reinforce the role and importance of social workers in the lives of those elderly who do not have other sources of social support.

Chapter 17
Successful Ageing of the Oldest-Old in China

Peng Du

Abstract Earlier case studies in Australia, the United States and China show that successful ageing is associated with more years of education, and more household assets or monthly income. The influencing factors are not limited specifically to a certain age group; they tend to be significant for the entire age range of the elderly population. This chapter examines these findings in broader geographic settings and advanced age groups. Using data from the Chinese Longitudinal Healthy Longevity Study, we extend the analysis of successful ageing to the oldest-old in China. The results confirm that successful ageing of the oldest-old in China is associated with more years of education and marital status. The oldest-old with positive attitudes on life, and who both exercise more often and have better physical performance are more successful in their old age and have a higher survival rate. Analysis of the substantial heterogeneity of the elderly may lead to both meaningful policies on ageing, and the promotion of a better quality of life.

Keywords Activities of daily living, Chinese oldest-old, Cognitive function, Functional classification, Longitudinal study, Medical conditions, Minimal functional limitations, Mortality, Normal ageing, Successful ageing

17.1 Introduction

Research on ageing has long emphasized average-related losses and has neglected the substantial heterogeneity of older persons (Rowe and Kahn 1987). It has thus become common in ageing research to group the elderly into the dichotomous categories of impaired or normal; but this classification conceals the vast heterogeneity of the elderly who are not impaired. Theoretically this dichotomous classification is not very helpful for discussions of policies on ageing, because it limits our research about the most psychologically and physically healthy group and their characteristics.

P. Du
Professor, Director, Gerontology Institute, Renmin University of China, Beijing 100872, China
e-mail: dupeng415@yahoo.com.cn

In 1987, Rowe and Kahn suggested that within the category of normal ageing, a distinction can be made between usual ageing and successful ageing (Rowe and Kahn 1987). Since then many researchers have been focusing on what successful and usual aging means. Several studies have sought to identify subgroups of old people in the population who exhibit minimal functional limitations, using a variety of approaches (Berkman et al. 1993; Garfein and Herzog 1995; Guralnik and Kaplan 1989; Harris et al. 1989; Jorm et al. 1998; Strawbridge et al. 1996; Suzman et al. 1992).

In these studies, successful ageing has often been defined as living in the community, without restriction on activities of daily living, no serious difficulties on gross mobility and physical performance, a high score on a cognitive screening test, and excellent or good self-rated health. Usually, the older men and women were classified into three functioning groups: high, medium and impaired (or low) (Andrews et al. 2002). In some studies they were directly labelled as successful, usual and diseased. The criteria by which *successful or usual* ageing was distinguished were often acknowledged as arbitrary.

One reason for differentiating among these three groups is to determine the range of complex physical and cognitive abilities of older men and women functioning at high, medium and impaired ranges, and to ascertain the psychosocial and physiological conditions that discriminate among those in the high functioning group from those functioning at middle or impaired ranges (Berkman et al. 1993).

There is some research on successful ageing based on longitudinal data in developed countries. Examples include the MacArthur study of ageing in the United States, first reported by Berkman et al. (1993), and two Australian studies (Jorm et al. 1998; Andrews et al. 2002). The MacArthur studies examined data from participants 70–79 years old drawn from three community-based populations (East Boston MA, New Haven CT, and Durham County NC) within the Established Populations for the Epidemiologic Studies of Elderly cohorts (EPESE; Cornoni-Huntley et al. 1993). High, medium, and low functioning subgroups were defined on the basis of predetermined criteria of physical and cognitive functioning; significant differences were identified among these three subgroups in biomedical, physiological, psychological and social functioning (Berkman et al. 1993). These findings have been replicated and extended in a series of subsequent MacArthur studies (Cook et al. 1995; Glass et al. 1995; Guralnik et al. 1994; Inouye et al. 1993; Seeman et al. 1994, 1995, 1996, 1997, 1999; Wallsten et al. 1995). Major objectives of the study by Andrews and his colleagues using data from the Australian Longitudinal Study of Aging (ALSA) data were to examine the applicability of the MacArthur model for an analysis of successful ageing in a setting removed both geographically and culturally from the USA, to determine whether subgroups differed from each other across a range of domains, and to signal key markers of successful ageing. Their findings suggested that indeed people age with differing degrees of success and those ageing most successfully are likely not only to live longer, but also to experience a better quality of life (Andrews et al. 2002).

In 2003, Du and Andrews extended the application of the MacArthur Model to the Beijing Multidimensional Longitudinal Study of Ageing (BMLSA). The BMLSA used the same indicators as in ALSA; thus it was possible to examine cultural differences and socio-economic differences between China and Australia. The comparison among United States, Australia and China proved useful in the analysis of successful ageing (Du and Andrews 2003). The similarities and differences of the findings between the developed and developing countries pointed to the need to undertake further work along these lines, both to identify some of the more universal patterns of ageing, and to assess variations that may arise from the complex complement of culturally distinct circumstances and experiences (Andrews et al. 2002).

As the capital of China, Beijing is a well developed area. Thus, research results from the BMLSA, although based on 3,000 Beijing elderly, may not have been representative of the Chinese elderly as a whole. However, data from the Chinese Longitudinal Healthy Longevity Study (CLHLS) conducted in 1998 and 2000 covered 22 provinces and 10,000 elderly aged 80 and over. These data enable us to examine previous findings from the Beijing study, and to extend an analysis of successful ageing to a nationally representative sample of oldest-old in China. This is important because according to the latest 1 percent National Survey data of 2005, China now has more than 13 million oldest old aged 80 and over, and projections indicate that there will be more than 100 million oldest old by 2050 (for more discussion, see also the first chapter in this book by Poston and Zeng). The baby-boomers of the 1950s and the 1960s will reach their advanced ages in the first half of the new century. Research findings about the elderly population may well help the development of policies to improve their active quality of life.

17.2 Design and Methods

17.2.1 Participant and Procedure

The data of the Chinese oldest old aged 80 and over to be used here are based on the Chinese Longitudinal Healthy Longevity survey, which was implemented by the Center for Healthy Ageing and Family Studies at Peking University (CHAFS) and the China Research Center on Ageing (CRCA) (Zeng et al. 2004). We used the data from the baseline survey in 1998 and the first follow-up in 2000 which covered 22 provinces. Data were collected through home interviews as well as physical examinations. In 1998, 8,805 persons aged 80–105 were interviewed. For the purposes of this study, we only choose respondents for whom necessary information was available for an analysis of successful ageing. This means that the oldest old had to live in the community, answer the questions on the cognitive and self-rated health instruments, and participate in a physical test. This reduced the available cohort size for the study to 7,737.

17.2.2 Measurements and Functional Classification

Self-reports of health were rated from excellent (1) to poor (5). Medical conditions were obtained by asking participants to indicate from a comprehensive list of conditions those they had ever experienced. Cognitive functioning was assessed using the Mini Mental State Examination-Modified Chinese version (MMSE-MC) (Meng and Meng 1991) with scores ranging from 0 to 23. The activities of daily living (ADL) include six categories, namely, bathing, dressing, toileting, transferring, feeding and continence. Questions on physical exercise dealt with whether the respondent had no exercise or often exercised.

Participants were classified as higher, intermediate or lower functioning. Cognitive functioning was divided into three groups according to the cognitive score of the respondent. If the elderly answered all questions correctly or gave a wrong answer to only one question, they were classified into the higher functioning group. If there was at least one question that the elderly could not answer or they gave incorrect answers to five or more questions, they were classified into the lower functioning group. The rest were placed in the intermediate group.

Individuals were classified as higher functioning ($N = 1,161$ or 15.0 percent of the total) if they fulfilled all of the following five criteria: 1. grouped as successful according to their Mini-Mental State Examination (MMSE-MC) score; 2. had good or very good self-rated health; 3. had no disability in the six activities of daily living; 4. could stand up from a chair; and 5. were able to pick up a book from the floor.

Individuals were classified as lower functioning ($N = 3,222$, or 41.6 percent) if they fulfilled any of the following four criteria: 1. there was at least one question that they could not answer, or they gave wrong answers to five or more questions; 2. they had poor or very poor self-rated health; 3. they had one or more disabilities among the activities of daily living; and 4. they had 1one or more disabilities in physical performance. The remaining individuals were classified as intermediate in functioning ($N = 3,354$ or 43.4 percent).

17.2.3 Analytic Approach

We evaluated the relative importance of a large number of potential risk or protective factors for successful ageing, using a two-stage process. First, descriptive analyses were undertaken regarding the relationship between each predictor and the functional classification of successful ageing. Second, a logistic regression model was used that controlled for the effects of age, sex, education, urban/rural residence and marital status (see Table 17.1 for categories, reference category, and descriptive statistics). Groups of conceptually related predictors were entered as a block to correct for their mutual associations. In all analyses assessing factors associated with successful ageing, the higher functioning group was designated as the reference group.

17.3 Results

17.3.1 Control Variables

As seen in Table 17.1, the results of the descriptive analyses show that each of the control variables distinguished among members of the three functional categories. Successful ageing was associated with lower age, male, more years of education,

Table 17.1 Control variables by level of function

Variable	High		Intermediate		Low		x^2	df
	No.	Percent	No.	Percent	No.	Percent		
Age group								
80–84	504	43.4	768	22.9	486	15.1	711.747***	8
85–89	291	25.1	672	20.0	488	15.1		
90–94	208	17.9	688	20.5	576	17.9		
95–99	99	8.5	513	15.3	613	19.0		
100+	59	5.1	713	21.3	1059	32.9		
Gender								
Male	725	62.4	1465	43.7	1023	31.8	342.458***	2
Female	436	37.6	1889	56.3	2199	68.2		
Education								
0 year	513	44.3	2152	64.4	2428	75.7	384.223***	4
1–6 years	460	39.8	871	26.1	587	18.3		
7+ years	184	15.9	320	9.6	192	6.0		
Residence								
Urban	508	43.8	1239	36.9	1034	32.1	52.981***	2
Rural	653	56.2	2115	63.1	2188	67.9		
Marriage								
Married	365	31.4	613	18.3	398	12.4	218.578***	4
Widowed	759	65.4	2657	79.2	2739	85.0		
Other	37	3.2	84	2.5	85	2.6		

*** $p < 0.001$

living in an urban area and being married. These findings are consistent with previous findings in the Beijing case study regarding the elderly aged 60 and over.

17.3.2 Health and Medical Conditions

Thirteen medical conditions were of sufficient prevalence to permit meaningful statistical analyses. The results of the descriptive analysis shown in Table 17.2 indicate that all of these conditions discriminated among the three functional categories. In nine of the thirteen conditions, the presence of a medical condition was associated with a poorer functional classification.

These associations were explored in more detail in the logistic regression (see Table 17.3) in which the morbid conditions were entered as a block and adjusted for with the control variables. Only heart disease, bronchitis, cataract and glaucoma emerged as significant risk factors for being in the intermediate functioning group relative to the higher functioning group. That finding is slightly different from the significant risk factors in the same comparison in the Beijing study of the elderly aged 60 and over, in which only hypertension and coronary heart disease were significant factors. For the Chinese oldest-old, heart disease, stroke, bronchitis, cataract and glaucoma were all associated with an increased risk of being in the lower functioning group relative to the higher functioning group; the risk was

Table 17.2 Medical conditions by level of function

Variable	High		Intermediate		Low		χ2
	No.	Percent	No.	Percent	No.	Percent	
Hypertension	169	14.6	485	14.5	383	11.9	29.65***
Diabetes	8	0.7	32	1.0	22	0.7	30.75***
Heart disease	60	5.2	246	7.3	279	8.7	36.33***
Stroke or CVD	19	1.6	75	2.2	129	4.0	48.10***
Bronchitis	108	9.3	427	12.7	432	13.4	33.23***
Tuberculosis	6	0.5	31	0.9	30	0.9	25.81***
Cataract	133	11.5	595	17.7	697	21.6	81.92***
Glaucoma	9	0.8	72	2.1	98	3.0	48.87***
Cancer	5	0.4	10	0.3	22	0.7	27.77***
Prostate tumour	52	4.5	131	3.9	113	3.5	36.20***
Gastric or duodenal ulcer	30	2.6	118	3.5	107	3.3	27.88***
Parkinson's disease	6	0.5	26	0.8	44	1.4	35.08***
Bedsore	2	0.2	20	0.6	35	1.1	38.88***

*** $p < 0.001$

particularly high for stroke; and heart disease, cataract and glaucoma at least doubled the risk.

17.3.3 Activity, Physical Performance and Health Indicators

Our analyses also showed that a range of activities, physical performance and health indicators discriminated among the three functional categories (see Table 17.4).

Table 17.3 Logistic regression summary: level of function by medical conditions entered as a block

	High vs intermediate		High vs low	
	AOR[a]	95%CI[b]	AOR	95%CI
Hypertension	1.08	0.88–1.32	0.75*	0.60–0.94
Diabetes	1.61	0.71–3.65	1.31	0.53–3.25
Heart disease	1.56**	1.14–2.14	2.47***	1.76–3.47
Stroke or CVD	1.67	0.98–2.85	4.05***	2.37–6.92
Bronchitis	1.54***	1.22–1.95	1.82***	1.41–2.34
Tuberculosis	1.66	0.64–4.30	1.64	0.59–4.53
Cataract	1.47***	1.18–1.83	1.57***	1.25–1.98
Glaucoma	2.54*	1.23–5.28	3.67***	1.72–7.82
Cancer	0.65	0.20–2.14	1.31	0.40–4.25
Prostate tumour	1.08	0.75–1.56	1.18	0.78–1.77
Gastric or duodenal ulcer	1.44	0.93–2.21	1.23	0.77–1.97
Parkinson's disease	1.28	0.50–3.33	2.36	0.90–6.20
Bedsore	2.61	0.58–11.74	3.74	0.83–16.74

[a]Adjusted odds ratio, where odds ratio are adjusted for age, gender, education, place of residence and marital status
[b]95% confidence interval
*$p < 0.05$ **$p < 0.01$ ***$p < 0.001$

Table 17.4 Activity, physical performance, health and psychological indicators by level of function

Variable	High		Intermediate		Low	
	Percent	n	Percent	n	Percent	n
Intensity of exercise						
Often	48.1	559	31.3	1,051	17.6	567
No	51.9	602	68.7	2,303	82.4	2,655
Self-rated health						
Excellent/good	100.0	1,161	54.7	1,833	44.3	1,428
Fair/poor	0.0	0	45.3	1,521	55.7	1,794
Able to pick up a book from the floor?						
Yes	100.0	1,161	71.0	2,383	47.0	1,514
No	0.0	0	29.0	971	53.0	1,708
Able to stand up from a chair?						
Yes	100.0	1,161	74.4	2,494	51.6	1,661
No	0.0	0	25.6	860	48.4	1,561
Look on the bright side of things						
Always/often	100.0	1,161	100.0	3,354	83.4	2,687
Sometimes/never	0.0	0	0.0		16.6	535
Keep my belongings neat and clean						
Always/often	92.0	1,068	83.2	2,792	74.9	2,413
Sometimes/never	8.0	93	16.8	562	25.1	809
Feel fearful or anxious						
Always/often	9.5	110	11.2	376	13.6	437
Sometimes/never	90.5	1,051	88.8	2,978	86.4	2,785
Feel lonely and isolated						
Always/often	9.3	108	13.2	444	18.5	595
Sometimes/never	90.7	1,053	86.8	2,910	81.5	2,627
Make own decision						
Always/often	71.5	830	59.4	1,993	49.3	1,590
Sometimes/never	28.5	331	40.6	1,361	50.7	1,632
Feel useless with age						
Always/often	24.1	280	35.0	1,173	44.2	1,424
Sometimes/never	75.9	881	65.0	2,181	55.8	1,798
Be happy as younger						
Always/often	65.3	758	48.3	1,620	39.4	1,270
Sometimes/never	34.7	403	51.7	1,734	60.6	1,952

*** $p < 0.001$

Lower levels of activity, physical performance and health were consistently associated with a poorer functional classification. The oldest-old who fell into the lower functioning group were usually those who reported their self-rated health as poor, did not exercise often, whose physical performance proved to be more difficult, and whose attitude about life tended to be negative. The higher functioning group reported higher morale and took a more active attitude toward life.

Table 17.5 Survival rate of the oldest-old by level of function

	High		Intermediate		Low	
	Percent	n	Percent	n	Percent	n
Lost to follow-up	11.4	133	9.1	306	9.4	303
Still alive at 2000 survey	72.4	840	57.4	1,925	46.5	1,498
Died before 2000 survey	16.2	188	33.5	1,123	44.1	1,421
Total	100	1,161	100	3,354	100	3,222

17.3.4 Mortality

Out of 7,737 oldest-old persons in the 2000 follow-up survey, 6,995 cases were located; another 742 cases were missing because they had moved away from the area, were away from home during the time of the field work, or refused to participate. There were a total of 2,732 deaths (39.1 percent) among the traced 6,995 cases. There were 188 deaths (16.2 percent) in the higher functioning group, 1,123 (33.5 percent) in the intermediate functioning group, and 1,421 (44.1 percent) in the lower functioning group. Death was strongly related to functional classification ($\chi^2 = 308.21$, $p < 0.001$) (Table 17.5).

17.4 Discussion

Research findings on successful ageing are usually based on the data from developed countries. Even the previous study in China was limited to the city of Beijing. The research reported in this chapter used data from the CLHLS and is thus based on a more representative sample of the Chinese oldest-old.

Our findings confirm that place of residence, education and marital status are the factors associated with the successful ageing of the oldest-old. Successful ageing is associated with more years of education and being married.

In the MacArthur studies (with subjects aged 70–79), the ALSA study (which further extended the age range to 70 and over), and in the present China study (with the age range further extended to 80 and over), the results indicate that general influencing factors on successful ageing are not limited to a certain age group, but are significant for the whole age range of the elderly population.

The interconnection among physical, psychological and social functioning with ageing is evident. As with the American and Australian research findings, the Chinese elderly studied here could be distinguished from each other on additional measures of physical functioning, health and psychological status; and key indicators remained after controlling for the effects of age, gender, and education.

In the ALSA study, the mortality data showed that death in the intervening years was more likely in those originally classified as ageing less successfully. The authors suggested that indeed people age with differing degrees of success, and those ageing most successfully are likely not only to live longer, but to experience a better quality of life (Andrews et al. 2002). The Chinese study confirms their findings.

On the other hand, many differences remain, and these may be due to economic and cultural differences. These remind us to pay attention to possibly revising questions when they are used in different cultural backgrounds.

First, due to differences in epidemiology which are influenced by socioeconomic development and medical conditions, risk factors for successful ageing are different in China compared to the United States and Australia. In the ALSA study, lower functioning is especially found following the major disabling effects of stroke and a fractured hip, while arthritis, a slipped disc and asthma at least doubled the risk. In the present China study the highest risk was for heart disease, stroke, bronchitis, cataract and glaucoma, with heart disease, cataract and glaucoma at least doubling the risk. Some of these conditions have a very low prevalence in developed countries.

Secondly, notice should be taken regarding the differences in the classification of some indicators since these may well impact our projection of future trends. For example, in the MacArthur and ALSA studies, the educational level of the elderly is classified into two groups: left school younger than age 15, and left school at the age of 15 years or more. However, the Chinese oldest-old have very high illiteracy. Amongst those aged 80 and over in the Chinese Healthy Longevity Study, more than 66 percent of them never attended school; the same situation is found in most developing countries. Therefore, we used years attending school as the measure. The study by Jorm and colleagues found that better educated cohorts perform better on cognitive tests (Jorm et al. 1998). With the education systems in developing countries such as China experiencing significant improvement since the 1950s, positive effects on the prevalence of successful ageing may be expected when these better educated cohorts become old.

In conclusion, our research results on successful ageing using data from the China Longitudinal Healthy Longevity Study may be compared with results from studies conducted in the United States, Australia and Beijing. Analysis of the substantial heterogeneity of the elderly can lead to very meaningful policies on ageing, which will hopefully promote a better quality of life. The similarities and differences of findings between studies in developed and developing countries point to the need to undertake further work along these lines to identify not only some of the more universal patterns of ageing well, but also variations that may arise from the complex mix of culturally distinct circumstances and experiences. It is hoped that policy suggestions could then be formed to promote world-wide successful ageing.

References

Andrews, G., M. Clark, and M. Luszcz (2002), Successful aging in the Australian longitudinal study of aging: Applying the MacArthur model cross-nationally. *Journal of Social Issues* 58 (4), pp. 749–765

Berkman, L.F., T.E. Seeman, M. Albert, D. Blazer, R. Kahn, R. Mohs, C. Finch, E. Schneider, C. Cotman, G. McClear, J. Nesselroade, D. Featherman, N. Ganmezy, G. McKhann, G. Brim,

D. Prager, and J. Rowe (1993), High, usual and impaired functioning in community-dwelling older men and women: Findings from the MacArthur Foundation Research Network on Successful Ageing. *Journal of Clinical Epidemiology* 46, pp. 1129–1140

Cook, N.R., M.S. Albert, L.F. Berkman, D. Blazer, J.O. Taylor, C.H. Hennekens (1995), Interrelationships of peak expiratory flow rate with physical and cognitive function in the elderly: MacArthur Foundation studies of ageing. *Journal Gerontology: Medical Sciences* 50A, pp. M317–M323

Cornoni-Huntley, J., A.M. Ostfeld, J.O. Taylor, R.B. Wallace, D. Blazer, L.F. Berkman, D.A. Evans, F.J. Kohout, J.H. Lemke, P.A. Scherr, and S.P. Korper (1993), Established populations for the epidemiological studies of the elderly: Study design and methodology. *Ageing: Clinical and Experimental Research* 5, pp. 27–37

Du, P. and G.R. Andrews (2003), Successful aging: A case study on Beijing elders. *Population Research* 3, pp. 4–11 (in Chinese)

Garfein, A.J. and A.R. Herzog (1995), Robust ageing among the youngold, oldold, and oldestold. *Journal of Gerontology: Social Sciences* 50B, pp. S77–S87

Glass, T.A., T.E. Seeman, A.R. Herzog, R. Kahn, and L.F. Berkman (1995), Change in productive activity in late adulthood: MacArthur studies of successful ageing. *Journal of Gerontology: Social Sciences* 50B, pp. S65–S76

Guralnik, J.M. and G.A. Kaplan (1989), Predictors of healthy ageing: Prospective evidence from the Alameda County Study. *American Journal of Public Health* 79, pp. 703–708

Guralnik, J.M., T.E. Seeman, M.E. Tinetti, M.C. Nevitt, and L.F. Berkman (1994), Validation and use of performance measures of functioning in a non-disabled older population: MacArthur studies of successful ageing. *Ageing: Clinical and Experimental Research* 6, pp. 410–419

Harris, T., M.G. Kovar, R. Suzman, J.C. Kleinman, and J.J. Feldman (1989), Longitudinal study of physical ability in the oldestold. *American Journal of Public Health* 79, pp. 698–702

Inouye, S.K., M.S. Albert, R. Mohs, K. Sun, and L.F. Berkman (1993), Cognitive performance in a high-functioning community-dwelling elderly population. *Journal of Gerontology: Medical Sciences* 48A, pp. M146–M151

Jorm, A.F., H. Christensen, S. Henderson, P.A. Jacomb, A.E. Korten, and A. Mackinnon (1998), Factors associated with successful aging. *Australian Journal on Aging* 17, pp. 33–37

Meng, C. and J. Meng (1991), The application of MMSE in ordinary population group. *Journal of Gerontology* 11, pp. 203–208 (in Chinese)

Rowe, J.W. and R.L. Kahn (1987), Human ageing: Usual and successful. *Science* 23, pp. 143–149

Seeman, T.E., P.A. Charpentier, L.F. Berkman, M.E. Tinetti, J.M. Guralnik, M. Albert, D. Blazer, and J.W. Rowe (1994), Predicting changes in physical performance in a high-functioning elderly cohort: MacArthur studies of successful ageing. *Journal of Gerontology: Medical Sciences* 49A, pp. M97–M108

Seeman, T.E., L.F. Berkman, P.A. Charpentier, D.G. Blazer, M.S. Albert, and M.E. Tinetti (1995), Behavioral and psychosocial predictors of physical performance: MacArthur studies of successful ageing. *Journal of Gerontology: Medical Sciences* 50A, pp. M177–M183

Seeman, T.E., M.L. Bruce, and G.J. McAvay (1996), Social network characteristics and onset of ADL disability: MacArthur studies of successful ageing. *Journal of Gerontology: Social Sciences* 51B, pp. S191–S200

Seeman, T.E., B.H. Singer, J.W. Rowe, R.I. Horwitz, and B.S. McEwen (1997), Price of adaptation—allostatic load and its health consequences: MacArthur studies of successful ageing. *Archives of Internal Medicine* 157, pp. 2259–2268

Seeman, T.E., J.B. Unger, G. McAvay, and C.F. Mendes de Leon (1999), Self-efficacy beliefs and perceived declines in functional ability: MacArthur studies of successful ageing. *Journal of Gerontology: Psychological Sciences* 54B, pp. P223–P230

Strawbridge, W.J, R.D. Cohen, S.J. Shema, and G.A. Kaplan (1996), Successful ageing: Predictors and associated activities. *American Journal of Epidemiology* 144, pp. 135–141

Suzman, R.M., T. Harris, E.C. Hadley, M.G. Kovar, and R. Weindruch (1992), The robust oldest old: Optimistic perspectives for increasing healthy life expectancy. In: Suzman, R.M.,

D.P. Willis, and K.G. Manton (eds.): *The oldest old.* New York: Oxford University Press, pp. 341–358

Wallsten, S.M., R.J. Sullivan, J.T. Hanlon, D.G. Blazer, M.J. Tyrey, and R. Westlund (1995), Medication taking behaviors in the high- and low-functioning elderly: MacArthur field studies of successful ageing. *Annals of Pharmacotherapy* 29, pp. 359–364

Zeng Y., Y. Liu, C. Zhang, and Z. Xiao (2004), *Analyses of the determinants of healthy longevity.* Beijing: Peking University Press (in Chinese)

Chapter 18
Impairments and Disability in the Chinese and American Oldest-Old Population

William P. Moran, Sihan Lv and G. John Chen

Abstract This study used data from the 1998 Chinese Longitudinal Healthy Longevity Survey (CLHLS) and the 1998 Medicare Current Beneficiary Survey (MCBS) to compare impairment and disability in the Chinese and American Oldest-Old Populations. The standard measures of activities of daily living (ADL) were utilized for assessing functional impairment in the study subjects. Descriptive and multivariate regression analyses were undertaken to examine differences in functional impairments between Chinese and US oldest-old populations. It was found that the oldest old Chinese were less likely to have difficulties in bathing (OR = 0.748), dressing (OR = 0.555), toileting (OR = 0.838), and transferring (OR = 0.319) compared to the oldest old Americans ($p < 0.01$), after adjusting for age, gender and self-perceived health status. The oldest-old Chinese were more likely to have difficulty in feeding (OR = 1.12) than their counterparts but this finding was not statistically significant ($p > 0.05$).

Keywords Activities of daily living, Age difference, American oldest-old, Center for Medicare and Medicaid Services, Chinese oldest-old, Comparison, Disability, Gender difference, Impairment, Multivariate analysis, Self-perceived health status, Univariate analysis

18.1 Introduction: Disability and Ageing

As the population ages, many elders develop a functional impairment or disability, defined as an inability to perform activities of daily living without assistance. With an ever growing elderly population, functional impairment is becoming a significant public health concern. The rate of impairment and disability rises steeply with age and is especially high for persons aged 80 and over, one of the most rapidly growing

---

W.P. Moran
Professor of Medicine, Director, Division of General Internal Medicine and Geriatrics, Medical University of South Carolina, Rutledge Tower, 12 Floor, 135 Rutledge Avenue, PO Box 250591, Charleston, SC 29425-0591, USA
e-mail:moranw@musc.edu

age groups in developed countries. Disability at older ages is frequently the result of chronic health conditions such as diabetes, cardiovascular disease, stroke, or movement disorders (Fried and Guralnik 1997).

A conceptual model for the relationship among disease, functional impairment and disability and the Disablement Process has been proposed by Verbrugge and Jette (1994), and is illustrated in Fig. 18.1. The model posits that disease or pathology causes impairment, leading to functional impairment. Disability results when the impairment exceeds the ability of the elder to function independently in that task. The model illustrates important mediators and moderators of the disablement process; these include individual factors such as demographics and genetics, attitude, behavior and coping ability, as well as medical care and environmental factors. The model implies that reducing the population prevalence or impact of chronic illness by behavioral or environmental interventions may result in reductions in disability. In fact data from the National Longitudinal Survey suggest that age-adjusted disability rates for elders in the United States are falling, suggesting that disability rates in the U.S. have been influenced by various factors (Manton et al. 1997; Singer and Manton 1998), and there is evidence that chronic disability is overestimated (Gill and Gahbauer 2005).

One standard measure for functional impairment in the elderly population is activities of daily living (ADL) and instrumental activities of daily living (IADL) (Katz et al. 1970). IADLs assess performance of instrumental activities such as the ability of the elder to use the telephone, go shopping, prepare meals, walk long distances, and manage finances. ADLs assess more basic tasks of everyday life such as the ability to independently eat, bathe, dress, toilet, and transfer from bed to chair. ADL impairments reflect more severe physical or cognitive impairments and are a threat to the elders' ability to live independently. ADL impairments require the elder to employ a compensatory strategy such as assistive devices (e.g., a walker), environmental modifications (e.g., adjustments made in the bathroom) or personal assistance (e.g., family member or paid worker) to maintain maximal functioning. Importantly, ADL impairment is also a marker for increased risk of utilization of acute services such as emergency department services and hospitalization. Thus the rate and degree of functional impairment in aging populations carry significant implications for policy planning and cost estimation for services to support elders (Fried and Guralnik 1997).

The purpose of the research we report in this chapter was to compare ADL impairments between the Chinese and American oldest-old population. To our knowledge, data directly comparing these populations are not available. We chose two data sets for which sufficient sample size and individual variables were available to make meaningful comparisons.

18.2 Data and Methods

This study used data from the 1998 Chinese Longitudinal Healthy Longevity Survey (CLHLS) and the US 1998 Medicare Current Beneficiary Survey (MCBS). The data provide point-in-time estimates of the outcomes of interest.

EXTRA-INDIVIDUAL FACTORS

MEDICAL CARE & REHABILITATION

MEDICATIONS & OTHER THERAPEUTIC REGIMENS

EXTERNAL SUPPORTS

BUILT, PHYSICAL, & SOCIAL ENVIRONMENT

↓

THE MAIN PATHWAY			
PATHOLOGY →	IMPAIRMENTS →	FUNCTIONAL LIMITATIONS →	DISABILITY
(diagnoses of disease, injury, congenital/ developmental condition)	(dysfunctions and structural in specific body systems: musculo- skeletal, cardio- vascular, neuro- logical, etc.)	(restrictions in basic physical and mental actions: ambulate, reach, stoop, climb stairs, produce intelligible speech, see standard print, etc.)	(difficulty doing activities of daily life: job, house- hold management, personal care, hobbies, active recreation, clubs, socializing with friends and kin, childcare, errands, sleep, trips, etc.)

RISK
FACTORS
Age, Gender

INTRA-INDIVIDUAL FACTORS

LIFESTYLE & BEHAVIOR CHANGES

PSYCHOSOCIAL ATTRIBUTES & COPING

ACTIVITY ACCOMMODATIONS

Fig. 18.1 The disablement process. Verbrugge and Jette (1994)

The CLHLS was designed to examine the determinants of healthy longevity in the Chinese elderly population; it is an observation cohort study started in 1998. The survey was conducted in randomly selected counties and cities of 22 provinces in China. An interview and a basic health examination were conducted at the inter- viewee's home. Data were collected on family structure, living arrangements and proximity to children, self-rated health, self-evaluation on life satisfaction, chronic diseases, use of medical care, social activities, diet, smoking and alcohol drinking, psychological characteristics, economic resources, caregivers and family support, nutrition and other health-related conditions in early life (childhood, adulthood, and around age 60), activities of daily living (ADL),[1] physical performance capacity, and cognitive functioning. The questionnaire design was based on international standards and was adapted to the Chinese cultural/social context and carefully tested by pilot studies/interviews (CLHLS 2006).

MCBS is a continuous, multi-purpose survey of a representative sample of the US Medicare population, conducted by the Center for Medicare and Medicaid Services (CMS). The data collected include the study subjects' demographics, self-reported health, medical conditions, ADL and IADL, use and cost of medical care; these individual survey data were cross-linked with their medical claims (Adler 1994).

Study subjects aged 80 or over in the CLHLS and the MCBS were included in our study. Measures of ADL impairments were elders' self-reports of any individual difficulty in eating, bathing, dressing, toileting, and transferring (yes/no), as well as the level of ADL impairments measured by the number of ADL difficulties (none, 1–2, or ≥ 3). Descriptive and multivariate regression analyses were undertaken to examine differences in ADL impairments between Chinese and the US oldest-old populations.

18.3 Results

18.3.1 Self-perceived Health Status and ADL by Age Group

Across all age groups, the oldest-old Americans consistently reported having "Good" self-perceived health status relative to their Chinese counterparts, as seen in Table 18.1. However, we found that relatively higher proportions of the oldest-old Chinese in all age groups were reported as having no difficulties in ADL impairments, as measured by the level of difficulty. For example, 86.17 percent of the Chinese had "none" or no ADL difficulties compared to 74.28 percent of the Americans. The oldest-old Americans showed higher proportions of difficulties in bathing, dressing, and transferring than the Chinese in all the age groups. There were higher proportions of Chinese in the 90+ age group who had difficulties in toileting and feeding than Americans in the same age group.

18.3.2 Self-perceived Health Status and ADL by Age Group and Gender

Across all the age groups, oldest-old American men consistently reported having "Good" self-perceived health status relative to their Chinese counterparts, as seen

---

[1] Instrumental Activities of Daily Living (IADL) questions were not included in the 1998 baseline and 2000 follow-up CLHLS surveys because the 1998 and 2000 waves interviewed the oldest-old only and the Chinese oldest-old are generally limited in IADL. The IADL questions were added in the 2002 and 2005 CLHLS surveys when the CLHLS study was expanded to cover both the oldest-old and the younger elderly aged 65-79 (for more discussion, see Chapter 2 in this Volume).

Table 18.1 Self-reported health status and ADL between Chinese and American oldest old

	China			USA		
	80–85 n = 2271 (%)	86–90 n = 1629 (%)	90+ n = 5059 (%)	80–85 n = 2139 (%)	86–90 n = 867 (%)	90+ n = 386 (%)
Health Status						
Good	60.48	57.62	55.12	70.83	72.22	72.21
Fair	31.42	33.06	35.50	20.65	18.52	20.26
Poor	8.10	9.32	9.39	8.52	9.26	7.53
Number of difficulties in ADL						
None	86.17	77.14	50.49	74.28	65.63	45.08
1–2	9.64	14.69	23.22	16.93	20.30	28.24
≥ 3	4.18	8.17	26.29	8.79	14.07	26.68
Any difficulty with (yes)						
Bath	12.17	20.48	44.39	18.38	25.37	42.75
Dress	4.89	8.73	26.63	9.73	17.07	29.79
Toilet	5.37	9.96	28.88	8.47	12.57	24.09
Transfer	4.33	7.89	25.03	15.76	22.26	38.86
Feed	2.64	5.53	18.35	3.60	5.42	11.92

in Table 18.2. A similar pattern may be observed for women (Table 18.3). In terms of having difficulties in ADLs, however, higher proportions of Chinese men and women relative to American men and women had no difficulties (Tables 18.2 and 18.3). Lower proportions of Chinese men were consistently seen as having difficulties in bathing, dressing, toileting, and transferring than American men in all the age groups. The Chinese men in the 90+ age group showed a higher proportion having difficulty in feeding compared to American men (12.34 vs. 10.58 percent). In the 80–85 and 86–90 age groups, it seems that Chinese women were doing much better than American women in bathing, dressing, toileting, and transferring (Table 18.3), except that in the 86–90 age group, there were more Chinese women having difficulty in feeding (6.94 vs. 4.46 percent) than American women. However, the Chinese women in the 90+ age group had higher proportions having difficulties in bathing (49.32 vs. 47.16 percent), toileting (33.30 vs. 28.01 percent), and feeding (21.14 vs. 12.41 percent) than their US counterparts.

18.4 Multivariate Regression Analysis

To control for the potential confounding effects of age, gender, and health status, multivariate logistic regression analysis was undertaken. As shown in Table 18.4, the oldest old Chinese were less likely to have difficulties in bathing (OR = 0.748), dressing (OR = 0.555), toileting (OR = 0.838), and transferring (OR = 0.319) compared to the oldest old Americans ($p < 0.01$), after adjusting for age, gender and self-perceived health status. The oldest-old Chinese were more likely to have

Table 18.2 Male: self-reported health status and ADL between Chinese and American oldest old

	China			USA		
	80–85 n = 1153 (%)	86–90 n = 808 (%)	90+ n = 1606 (%)	80–85 n = 809 (%)	(86–90) n = 307 (%)	90(+) n = 104 (%)
Health status						
Good	62.00	59.75	60.83	70.92	67.10	73.79
Fair	30.47	31.90	32.49	20.42	21.82	16.50
Poor	7.53	8.35	6.68	8.66	11.07	9.71
Number of difficulties in ADL						
None	87.68	82.01	61.39	78.12	68.08	55.77
1–2	8.15	11.79	20.73	13.84	17.59	26.92
≥3	4.16	6.20	17.87	8.03	14.33	17.31
Any difficulty with (yes)						
Bath	10.77	16.13	33.79	14.22	22.48	30.77
Dress	4.86	7.07	19.80	9.27	18.89	21.15
Toilet	5.12	7.44	19.38	6.92	12.05	13.46
Transfer	4.26	5.84	16.47	14.38	21.82	32.69
Feed	2.35	4.09	12.34	3.58	7.17	10.58

Table 18.3 Female: self-reported health status and ADL between Chinese and American oldest old

	China			USA		
	80–85 n = 1118 (%)	86–90 n=821 (%)	90+ n = 3453 (%)	80–85 n = 1330 (%)	86–90 n = 530 (%)	90+ n = 282 (%)
Health status						
Good	58.91	55.51	52.27	70.78	75.04	71.63
Fair	32.40	34.21	37.00	20.78	16.70	21.63
Poor	8.69	10.28	10.73	8.43	8.26	6.74
Number of difficulties in ADL						
None	84.62	72.35	45.42	71.93	64.29	41.13
1–2	11.18	17.54	24.38	18.81	21.79	28.72
≥3	4.20	10.11	30.20	9.26	13.93	30.14
Any difficulty with (yes)						
Bath	13.61	24.76	49.32	20.92	26.96	47.16
Dress	4.92	10.35	29.42	10.01	16.07	32.98
Toilet	5.64	12.42	33.30	9.41	12.86	28.01
Transfer	4.39	9.91	29.02	16.33	22.50	41.13
Feed	2.95	6.94	21.14	3.61	4.46	12.41

Table 18.4 Results from multivariate logistic regression models

	Bathing	Dressing	Toileting	Transferring	Feeding
					OR
					95% CI
					P-value
China	0.748	0.555	0.838	0.319	1.12
	(0.669–0.836)	(0.483–0.637)	(0.727–0.966)	(0.279–0.364)	(0.926–1.355)
	$P < 0.001$	$P < 0.001$	$P < 0.05$	$P < 0.001$	$P = 0.2441$
female	1.699	1.411	1.766	1.592	1.581
	(1.553–1.859)	(1.264–1.574)	(1.579–1.974)	(1.428–1.774)	(1.376–1.815)
	$P < 0.001$	$P < 0.001$	$P < 0.001$	$P < 0.001$	$P < 0.001$
Age 85	1.667	1.836	1.687	1.594	1.761
	(1.463–1.898)	(1.544–2.183)	(1.414–2.013)	(1.357–1.872)	(1.377–2.251)
	$P < 0.001$	$P < 0.001$	$P < 0.001$	$P < 0.001$	$P < 0.001$
Age 90	5.022	6.196	5.823	5.612	6.343
	(4.479–5.630)	(5.324–7.210)	(5.015–6.762)	(4.853–6.488)	(5.185–7.761)
	$P < 0.001$	$P < 0.001$	$P < 0.001$	$P < 0.001$	$P < 0.001$
Fair	1.656	1.646	1.697	1.712	1.443
	(1.509–1.817)	(1.471–1.842)	(1.517–1.894)	(1.533–1.913)	(1.252–1.657)
	$P < 0.001$	$P < 0.001$	$P < 0.001$	$P < 0.001$	$P < 0.001$
Poor	4.648	5.394	5.217	5.141	4.313
	(4.029–5.363)	(4.638–6.273)	(4.478–6.078)	(4.429–5.967)	(3.622–5.137)
	$P < 0.001$	$P < 0.001$	$P < 0.001$	$P < 0.001$	$P < 0.001$
Model fit: −2Log likelihood	13196.569	11122.747	11344.519	11471.851	8080.024

OR: Odds ratio; CI: Confidence interval

difficulty in feeding (OR = 1.12) than their counterparts, but this finding was not statistically significant ($p > 0.05$).

18.5 Discussion

This is the first comparative analysis of functional status of which we are aware among the oldest old individuals in China and the US Demographic data for the populations of both countries show rapidly expanding elder populations; unfortunately, the rates of functional impairment and disability parallel this increase. The sampling frame of the Chinese Longitudinal Healthy Longevity Survey reflects a focus on the oldest members of a family, compared to the random sampling methodology of the MCBS, which is reflected in the large differences in percentages of age categories. As a result, the Chinese sample is skewed toward older individuals; thus the univariate comparisons must be interpreted with caution. Nonetheless, both populations analyzed show the well-accepted increase in disability with advancing age. Furthermore, the higher rate of disability in females is consistent with prior studies in the US (Arbeev et al. 2004). Reflecting the age skew, the Chinese elders reported overall lower global health status and more disability reflected by relatively larger proportions of elders with three or more ADL impairments. Chinese elders also report higher rates of individual impairments with the exception of transferring, where US elders report more impairments.

The stratified analyses are very informative. Despite a higher proportion of Chinese elders reporting fair or poor health, Chinese elders report the same or fewer aggregate ADL impairments compared to US elders, with the exception of toileting and eating, where 90+ year old Chinese elders report higher rates of impairments. We can only speculate that impairments in these areas reported by Chinese elders reflect environmental differences in bathroom and toilet design (e.g. floor level toilets), or eating utensils (e.g. chopsticks). The higher level of transfer impairments in the US population is striking, perhaps reflecting characteristics prevalent in US populations associated with functional impairments such as higher weight, lower levels of precedent physical activity, or higher rates of chronic disease-associated impairments such as arthritis (Nagamastu et al. 2003; Houston et al. 2005; Dunlop et al. 2005). Although there are higher overall rates of impairments among the female elders in both populations, the age-gender stratified analysis demonstrates the persistence of differences in impairments between the Chinese and US populations.

In the multivariate analysis adjusting for age, self-reported health status and gender, Chinese elders consistently show lower rates of individual functional impairments. The exception is eating, where the difference noted in the stratified analysis is no longer significant. All covariates such as older age, female gender and fair or poor self reported health status show the expected statistically significant association with greater impairments. Thus, if we assume we have overcome the differences in sampling frames, Chinese elders show significant, and in the case of transferring, dramatically lower odds of functional impairments after adjusting for the known confounders.

The relationship between lifestyle, behaviors, sensory impairment, and acute and chronic diseases could not be explored in this study. Manton and his colleagues (1997) have suggested that US elders may experience lower than expected rates of disability due to changes in many of these factors. Detailed additional risk factor data, including IADL, falls, mobility measures and affective disorders, could help policy makers in understanding the prevalence of such factors for functional impairments and the potential impact of risk factor reduction interventions (Jette 1997; Chu et al. 2005) and implementing intervention programs. Given the magnitude of the aging population of China and the large and predictable direct and indirect costs of disability in the elderly, longitudinal data now being collected in the CLHLS could inform policy on prevention of disability, as has been the case recently in Europe (van Gool et al. 2005).

Clearly the results of this exploratory study must be interpreted with caution. Alternative US data sets could be used for future comparisons with Chinese elders. This analysis is based on cross-sectional data, which limits inference. For example, one could hypothesize that behaviors of Chinese elders such as exercise, diet, and weight control are associated with lower impairment rates which could lead to the observed differences, as has been seen in other populations (Ohmori et al. 2005). Alternatively, one could explain such differences by prolonged survival of chronically ill elders in the US population such that higher rates of functional impairments reflect the survival advantage conveyed by an aggressive US medical care system. Nonetheless, the differences bear closer study in the future, and the continued longitudinal study of Chinese elders in the CLHLS will be very informative.

References

Adler, G.S. (1994), A profile of the Medicare current beneficiary survey. *Health Care Financing Review* 15 (4), pp. 153–163

Aijanseppa, S., I.L. Notkola, M. Tijhuis, W. Van Staveren, D. Kromhout, and A. Nissinen (2005), Physical functioning in elderly Europeans: 10 year changes in the north and south: the HALE project. *Journal of Epidemiology and Community Health* 59, pp. 413–419

Arbeev, K.G., A.A. Butov, K.G. Manton, I.A. Sannikov, and A.I. Yashin (2004), Disability trends in gender and race groups of early retirement ages in the USA. *Sozial-Und Proventivmedizin* 49 (2), pp. 142–151

CLHLS: http://www.pubpol.duke.edu/centers/pparc/research/data/china/details.php. Accessed 12-10–2006

Chu, L.W., I. Chi, and A.Y.Y. Chiu (2005), Incidence and predictors of falls in the Chinese elderly. *Annals of the Academy of Medicine Singapore* 34, pp. 60–72

Dunlop, D.D., P. Semanik, J. Song, L.M. Manheim, V. Shih, and R.W. Chang (2005), Risk factors for functional decline in older adults with arthritis. *Arthritis & Rheumatism* 52 (4), 1274–1282

Fried, L.P. and J.M. Guralnik (1997), Disability in older adults: evidence regarding significance, etiology, and risk. *Journal of the American Geriatric Society* 45 (1), pp. 92–100

Gill, T.M. and E.A. Gahbauer (2005), Overestimation of chronic disability among elderly persons. *Archives of Internal Medicine* 165, pp. 12–26

Houston, D.K., J. Stevens, and J. Cai. (2005), Abdominal fat distribution and functional limitations and disability in a biracial cohort: the atherosclerosis risk in communities study. *International Journal of Obesity* 12, pp. 1457–1463

Jette, A.M. (1997), Disablement outcomes in geriatric rehabilitation. *Medical Care* 35 (6 suppl), pp. JS28–JS37

Katz, S., T.D. Downs, H.R. Cash, and R.D. Frotz (1970), Progress in development of the index of ADL. *The Gerontologist* 10, pp. 20–30

Manton, K.G., L. Corder, and E. Stallard (1997), Chronic disability trends in elderly United States populations: 1982–1994. *Proceedings of the National Academy of Sciences of the United States of America* 94, pp. 2593–2598

Nagamastu, T., Y. Oida, Y. Kitabatake, H. Kohno, K. Egawa, N. Nezu, and T. Arao. (2003), A 6-year cohort study on relationship between functional fitness and impairment of ADL in community-dwelling older persons. *Journal of Epidemiology* 12 (3), pp. 142–148

Ohmori, K., S. Kuriyama, A. Hozawa, T. Ohkubo, Y. Tsubono, and I. Tsuji (2005), Modifiable factors for the length of life with disability before death: Mortality retrospective study in Japan. *Gerontology* 51, pp. 186–191

Singer, B.H. and K.G. Manton (1998), The effects of health changes on projections of health services needs for the elderly population of the United States. *Proceedings of the National Academy of Sciences of the United States of America* 95, pp. 15618–15622

van Gool, C.H., G.I. Kempen, B.W. Penninx, D.J. Deeg, A.T. Beekman, and T.J. van Eijk (2005), Impact of depression on disablement in late middle aged and older persons: Results from the longitudinal aging study Amsterdam. *Social Science and Medicine* 60 (1), pp. 25–36

Verbrugge, L.M. and A.M. Jette (1994), The disablement process. *Social Science and Medicine* 38 (1), pp. 1–14

Chapter 19
Tooth Loss Among the Elderly in China

Yun Zhou and Zhenzhen Zheng

Abstract Tooth loss among the elderly is an important research and policy issue. The study we report in this chapter analyzed the edentulous and denture wearing status of 15,766 Chinese elderly (65–100+), using data from the 2002 CLHLS survey. Our results showed that age played an important role in tooth loss and denture wear. The older the elder, the fewer the teeth. There were gender and residence differences among those with tooth loss and those who were denture wearers. Males had more teeth than females, while life expectancy of females was longer than that of males. Rural elders in general lost more teeth than their counterparts in urban settings. Denture wear among the population was disproportionate to the level of edentulousness. Socio-economic factors also affected the patterns of dentate status of the elderly.

Keywords Age pattern, The Chinese elderly, Dental care, Dentate status, Denture, Edentulism, Gender difference, Natural teeth, Oral disease, Oral hygiene, Residence difference, Second National Epidemiological Survey on Oral Health, Survival curve, Tooth loss

19.1 Introduction

Only recently have dental care and oral health, especially among the Chinese elderly, attracted the attention of the public and researchers in China and throughout the world. For example, in the *Journal of Dental Research*, only 18 articles on oral health issues in China were published between 1919 and 2006. The earliest research on the Chinese population was published in 1932 by Anderson. But most of the articles were published in a special section of the journal in 2001, followed by a large-scale oral epidemiological study in Southern China, including elderly who were up to 74 years old. There are reasons for limited dental

Y. Zhou
Institute of Population Research, Peking University, Beijing, China
e-mail: zhouyun@pku.edu.cn

research on the elderly in China. First, rapid demographic changes in China in recent decades have resulted in a growing number of elderly; thus there have been increased concerns regarding their health and quality of life. Second, limited research resources have been committed to oral health studies in China; this has led to less information available about the oral health status of the elderly.[1] A review article by Lin and Schwarz (2001: 324), based on information obtained via searches of Medline and other Chinese medical and technical materials, concluded that "surveys focusing on tooth loss and prosthetic status were uncommon and mainly conducted in urban areas...." And third, only until a society reaches certain levels of social and economic development will it have the energy and resources to pay more attention to factors that less immediately affect an individual's life.

Dental problems do not usually threaten an individual's life directly. For many Chinese, a quick solution to a prolonged and uncontrollable toothache is the extraction of the troubled tooth; loss of one or even a few teeth is not considered a problem. However, tooth loss will not only affect the appearance of an individual, it will also affect the kinds and varieties of foods an individual may eat, thus affecting the balance of his/her in-take nutrition. For example, in 1991, Gershen reported that partial or complete loss of teeth can result in impairment of the masticatory function, and thus the ability to consume a well-balanced nutritional diet. Joshipura et al. (1996) considered that a reduction in number of teeth may cause deficiencies of various micronutrients which in turn will compromise the persons' immune status. Other scholars found that poorer dentition status, especially edentulousness without dentures, may be related to deterioration in the systemic health of the elderly (Shimazaki et al. 2001). Takata et al. (2004) concluded in their study that there was no significant relationship between the number of intact teeth and activities of daily living (ADL) status in a sample of 823 elderly 80+ years olds; it might be that chewing ability rather than number of intact teeth affected the ADL status of the elderly. In their study of Japanese elderly, Nasu and Saito (2006) concluded that maintenance or recovery of sufficient chewing ability was related to a longer total life expectancy and was even more related to a longer active life expectancy. Although research has been conducted on dental factors that may contribute to the specific types of health statuses among elderly, according to the World Health Organization, the interrelationship between oral health and general health is more pronounced among older people; in this population, poor oral health can increase the risks to general health (www.who.int/oral_health).

---

[1] Up to the present China has carried out three waves of oral epidemiological surveys in 1983, 1995, and 2005. Each survey has its own goal and coverage. The first covered only children; the second and third surveys included individuals between 5 and 74 years of age. The surveys provided very useful information about the oral health of the Chinese, but less information has been available on the elderly. Although the surveys followed international guidelines and standards of other oral epidemiological surveys, they ignored the oral health status of the oldest-old in the population, tending to assume that the oral health status of the oldest-old population was similar to that of young-old population.

Tooth loss among the elderly is an important research topic because it leads to many adverse consequences. Representative studies, however, are limited. Most researched different subgroups of the elderly population; the results tend to vary with age, residence, gender and other variables regarding tooth loss or denture wear. For example, a study in Kitakyushu Japan found that among the institutionalized elderly, males lost more teeth than females, and edentulism was also more likely in males than in females (Shimazaki et al. 2003). Among 338 elderly in Israel, 54 percent of them were edentulous; higher rates of edentulism were found among subjects living in urban areas compared to subjects living in rural areas (Adut et al. 2004). Among the 65–74 years old population, 4.4 percent of the urban and 3.4 percent of the rural elderly were edentulous in southern China, even though the number of missing teeth was not significantly different between urban and rural residents (Lin et al. 2001). Being older or being female in the older Mexican American population (65–99 years old) was significantly associated with tooth loss (Randolph et al. 2001). In a study of health in Pomerania, researchers found that age, low income, low educational level, smoking and alcohol abuse seemed to be risk markers for edentulism; whereas the number of diseases, diabetes, and gender were not (Mack et al. 2003). Shah's (2003) study revealed that elderly men had a higher percentage of filled teeth and denture wear compared to elderly women.

Due to a lack of awareness of tooth health and poor oral hygiene in China, many elderly suffer from notable tooth loss and severe periodontal conditions. The Chinese government has recognized this problem, as well as the relationship between oral diseases and quality of life among the population (including the elderly). In 1989 the government designated September 20th as "National Teeth-Loving-Day" (*Quan-guo Ai-ya-ri*) with each year having a special theme (Zhang 1999: 54).[2] In 2001, WHO proposed the "80/20 plan" which encouraged individuals to take care of their teeth for the sake of their own health and to achieve the goal of having 20 teeth at the age of 80. The objective of the research we report in this chapter is to describe dentate status among older Chinese (65+ years old), and the edentulous and denture wearing status among the population. We hope to identify the discrepancy between reality and the goal of the "80/20 plan," as well as to add more knowledge and information about oral hygiene among the Chinese elderly.

19.2 Materials and Methods

Data for this study were collected by the Chinese Longitudinal Healthy Longevity Survey "CLHLS." This is an ongoing project covering the elderly population in 22 provinces in China. Counties and cities in the 22 provinces were randomly selected for the survey. The project has conducted four waves of the survey (in 1998, 2000, 2002, and 2005). The initial survey included elderly 80+ years only; 65+ year olds

---

[2] For example, the theme in 1989 was "brush teeth and oral health," in 1999 it was "oral heath care for the elderly," and in 2005, "oral health care during pregnancy."

were added to the survey in 2002. We have used data from the 2002 survey wave because the population was expanded to 65+ years in that year and those data were the latest available at the time of our research. The survey covered various aspects regarding the lives of the elderly, including general information, physical ability, life style, personal background, family structure, and a basic physical examination. This was not a survey specifically designed to study dental health among the elderly, however. With regard to oral health status, the survey contained two questions directly related to the tooth status of the elderly: "How many teeth do you have (not including dentures)?" and "Do you wear dentures?" These questions, as well as socio-demographic information (i.e., age, gender, residence, schooling and pension status) were used to analyze the very basic condition of oral health among the elderly, including the oldest-old (80+), about whom we have never had any general dental information.

For our study we relied on the reports of oral health status among the 15,766 elderly in the survey; as already noted, this information included number of remaining teeth and denture wearing status. Table 19.1 shows the general characteristics of the population by age, sex, and residence; our analyses and conclusions were drawn from this sample population. The overall quality of the survey was considered generally good, but answers to some questions (especially personality-related items) should be used with caution, as discussed by Zeng et al. in 2001 and in 2002.

The results of this survey show that the number of teeth left (18) among the younger elderly is similar to that in the Second National Epidemiological Survey on Oral Health conducted in 1995 (National Supervising Committee on Oral Disease Prevention, 1999). This leads us to believe that the results of this survey represent the general patterns of tooth loss among the elderly in China.

The statistical methods we used were descriptive, bivariate, and multivariate analyses. The descriptive analysis provided information about the general status of oral health, the bivariate analysis checked the significance of the differences between

Table 19.1 Number of cases by age group, sex, and urban/rural residency

	Total	Urban		Rural	
		Male	Female	Male	Female
65–69	1,606	202	197	613	594
70–74	1,669	209	204	632	624
75–79	1,565	186	185	593	601
80–84	2,105	311	278	784	732
85–89	2,126	249	256	779	842
90–94	2,321	235	285	803	998
95–99	1,419	121	162	422	714
100+	2,955	163	535	471	1,786
Total	1,5766	1,676	2,102	5,097	6,891

The data is from the 2002 survey, with 15,798 unweighted cases, there were 32 cases with missing or invalid tooth information. The following analyses are based on cases described in this table

19.3 Results

19.3.1 General Trends

According to physiology, an adult should have 32 natural teeth, of which 28 have different functions for daily life. Among the elderly (age 65 and over), the average number of teeth remaining at the time of survey (2002) was 9.1; males had more teeth than females (11.0 vs. 7.6) (Fig. 19.1). Even though loss of teeth does not necessarily increase with age as does human mortality, the trend of the number of teeth remaining in this population indicates a decline in the number of remaining teeth as an individual ages. If we evaluate the speed of the decline by the steepness of the line of the mean number of teeth left in Fig. 19.1 among different age groups, the elderly in age group 85 and over were starting to experience a slower tooth loss than that experienced by the younger age groups. This trend persisted in the experience of male and female elderly and is shown in Fig. 19.1. Tooth loss among residents in urban and rural areas followed a similar pattern and only differed in degree of the number of teeth lost. Reasons behind this pattern are not clear, but the less teeth remaining, the less chance an individual will loose more teeth.

While the mean number of teeth remaining among older Chinese decreased with age, the percentage of edentulousness increased with age (Table 19.2). For both males and females, the percentage of edentulous elderly increased with age;

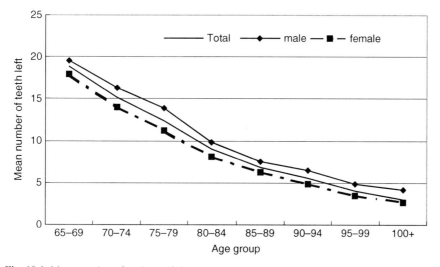

Fig. 19.1 Mean number of teeth remaining by age and sex, 2002

Table 19.2 Percentage of edentulousness among elderly in China, 2002

Gender	Age	Urban	Rural	Total
Male	65–69	9.9	8.0	8.4
	70–74	12.0	13.0	12.7
	75–79	18.3	15.2	15.9
	80–84	26.0	25.4	25.6
	85–89	37.8	29.2	31.2*
	90–94	41.1	31.3	33.6**
	95–99	47.9	41.1	42.8
	100+	46.6	44.1	44.7
Female	65–69	11.2	11.3	11.3
	70–74	23.0	16.5	18.1*
	75–79	33.0	21.1	23.9**
	80–84	33.5	30.7	31.4
	85–89	36.6	35.2	35.5
	90–94	46.9	40.7	42.1
	95–99	59.3	51.0	52.6
	100+	61.2	55.5	56.8*

Urban–rural difference was significant at *$p < 0.05$, **$p < 0.01$

however, the percentage of edentulous females was higher than that of males. At age 95 and over, more than half the females had lost all their teeth.

19.3.2 Gender and Residential Differences in Tooth Loss

Based on the information in Table 19.3, there were differences in the number of teeth remaining between male and female elderly in urban and rural China. In general, urban male elderly have more teeth remaining than do rural male elderly, although the pattern is reversed in the 95–99 and 100+ age groups. Urban female elderly have more teeth than rural elderly women in only three of the eight age groups. Gender differences in the number of teeth remaining among the urban elderly were greater than that of the rural elderly, while general tooth loss status among rural males and

Table 19.3 Mean number of teeth remaining and standard deviation (SD) by gender and residence (2002)

Age group	Urban total	Rural total	Urban male	(SD)	Urban female	(SD)	Rural male	(SD)	Rural female	(SD)
65–69	20.2	18.3	20.2	10.5	20.2	11.0	19.4	10.4	17.1	10.3
70–74	16.7	14.7	18.2	10.8	15.1	11.7	15.7	10.7	13.6	10.4
75–79	12.7	12.4	14.8	10.9	10.7	10.9	13.7	10.3	11.3	10.2
80–84	10.1	8.6	11.2	10.3	8.8	9.5	9.3	9.1	7.9	8.5
85–89	7.2	6.7	8.3	9.4	6.1	7.7	7.2	8.0	6.4	7.9
90–94	5.7	5.5	6.8	8.7	4.8	7.1	6.4	7.6	4.8	6.7
95–99	3.9	4.1	4.8	7.2	3.3	6.5	4.9	7.1	3.5	5.9
100+	2.6	3.1	3.6	5.2	2.3	4.9	4.3	6.9	2.7	5.3
Total	9.8	8.8	11.8		8.2		10.8		7.4	

females was worse than that of urban males and females. Note that the standard deviation is larger for the 85+ age groups, implying a more severe heterogeneity in each age group.

One's oral health may be intertwined with his/her health status. Bad health may result in more tooth loss, and a decreasing number of teeth may lead to a change in eating habits or cause problems in digestion and malnutrition. In multivariate analyses we focused mainly on socio-economic status as an independent variable because it is often used to examine causal relationships between oral health and sex, age, place of residence, educational attainment, and occupation. Our statistical analysis showed that age was the most influential factor regarding the number of teeth remaining in this population; the older the people, the less remaining teeth. Gender and place of residence were also significant determinants as shown in the first model in Table 19.4. For the same age group, males tended to have more teeth than females, and those who lived in urban areas had more teeth than those who lived in rural areas. Model 2 adds two more variables: if the individual had a pension, and years of schooling. The more wealthy the individual, the more he or she may spend on dental care. Educational attainment may improve knowledge of behavior to improve oral health status. Although age still played an important role in Model 2, place of residence became insignificant once pension and schooling variables were introduced. Both variables were positively correlated with number of teeth.

19.3.3 Denture Wear

Number of teeth (remaining) affects an individuals' daily life. We do not have information on the exact number and position of teeth remaining, or their relationship to the elder's functioning in daily life, or the impact of remaining teeth on health status. However, when the number of teeth was reduced to a point that hindered the elder's ability to eat and enjoy certain types of food, he or she may well wear dentures to minimize the problem. Among the elderly interviewed in the CLHLS, about 26 percent on average wore dentures (Table 19.5). Urban and rural difference in denture wear was obvious; more urban elderly wore dentures than rural elderly (38 vs. 22 percent). The younger the age group, the less difference there was in denture wear

Table 19.4 Parameter estimate by linear regression

Variables	Model 1		Model 2	
	Standardized B	p-value	Standardized B	p-value
Age	−0.492	< 0.001	−0.485	< 0.001
Sex: male	0.071	< 0.001	0.059	< 0.001
Residence: urban	0.024	< 0.001	0.005	0.517
Have pension: yes			0.039	< 0.001
Years of schooling			0.019	0.008
Adjusted R^2	0.261		0.262	

Dependent variable: Number of natural teeth left; the parameters were estimated by ordinary least squares method

Table 19.5 Percentages of elderly wearing dentures

Age group	Total	Male	Female	Urban total	Rural total
65–69	26.7	24.2	29.5	30.8	25.3
70–74	31.4	28.6	34.3	41.0	28.4
75–79	32.2	30.9	33.3	48.7	27.5
80–84	32.0	33.9	29.9	44.9	27.4
85–89	28.7	31.7	26.2	44.0	23.4
90–94	23.8	25.8	22.5	37.4	20.1
95–99	20.8	26.9	16.3	35.8	16.9
100+	15.7	22.4	13.9	29.3	11.3
Total	25.9	28.3	24.0	38.0	22.0

between urban and rural elderly. However, the difference in the percentage of those wearing dentures between urban and rural elderly increased after age 75.

Although women in general have fewer teeth left and a higher percentage of edentulousness than men, fewer wore dentures; elderly in rural areas had fewer teeth than urban elderly, but fewer were edentulous and less wore dentures (Tables 19.2, 19.3, and 19.5). Among the elderly younger than 80 years of age, more females wore dentures than males. Compared to the number of teeth remaining among urban and rural elderly in Table 19.3, those having more teeth in urban areas were more likely to wear dentures than those having less teeth but living in rural areas. This difference existed in almost all age groups and between males and females in urban and rural areas.

A further analysis indicated that among those elderly without a single tooth remaining, the proportion wearing dentures increased compared with those who still had a few remaining teeth (Table 19.6). For example, among urban residents, the percentage of denture wearers increased from 38 percent (Table 19.5) for the general elderly population to 61.2 percent among those who had lost all their teeth; in rural areas the increase was from 22 to 35 percent. The extent of the increase was larger in urban areas (23 percentage points) than in rural areas (13 percentage points), indicating both the need for dentures to assist in eating, and the active action taken to acquire dentures. The analysis also showed that more males than females wore dentures and that more urban than rural elderly wore dentures. Both comparisons were statistically significant.

Model 1 in Table 19.7 includes all the respondents and Model 2 in the same table only includes those who do not have any teeth. Results from the logistic regression on denture wear showed that gender was no longer significant when other variables were controlled. Urban residents were more likely to wear dentures than rural

Table 19.6 Proportion of denture use among those having no teeth (%)

Urban			Rural		
Male	Female	Urban total	Male	Female	Rural total
72.8	54.8	61.2	48.2	28.5	35.0

The difference between males and females as well as the difference between urban and rural are all statistically significant ($p < 0.001$)

Table 19.7 Parameter estimates of logistic regression on denture wear (odds ratio)

	Model 1	Model 2
	$n = 15,747$	$n = 5,226$
Sex: male	1.073	0.899
Residence: urban	1.831***	2.525***
Has pension: yes	2.175***	2.248***
Has schooling: yes	1.574***	1.800***
Age group		
65–69	7.060***	13.090***
70–74	6.457***	14.471***
75–79	5.166***	8.253***
80–84	3.584***	6.141***
85–89	2.634***	3.359***
90–94	1.786***	2.243***
95–99	1.437***	1.561***
100+	(reference)	
Number of teeth left	0.895***	–
Constant	0.154***	0.172***
Nagelkerke R^2	0.236	0.318
Correct classification	78.8%	73.2%

*$p < 0.05$, **$p < 0.01$, ***$p < 0.001$

residents, given the same age, education level, and number of teeth left. Younger elderly were more likely to wear dentures than older elders; there is a clear linear trend of a decrease in the odds ratio associated with the increase in age groups. Both pension status and place of residence were statistically significant in the models, implying that denture wear was not only related to the place of residence but also to the former occupation of the respondent (which was also a measure of financial dependency).

19.3.4 Survival of Teeth Among Elderly

Instead of cohort data, our research used cross-sectional data to build a hypothetical survival pattern of teeth. Survival of teeth among the elderly was calculated according to the status of having teeth (no matter how many), and the loss of all teeth (edentulousness). The survival curve of teeth in Fig. 19.2 explains the loss of all teeth among the elderly surveyed. First, total loss of teeth increased with age. In the beginning age group, only about 1 percent of the elderly lost all their teeth. Toward the higher age group, e.g. 100–104, about 81 percent of them were edentulous. Second, as indicated in the "general trend" section of dentate status among the elderly in this chapter (which was explained largely by the mean number of teeth remaining), there was a gender difference in the survival rate of teeth. While the mean number of teeth remaining among males was more than that of females (Fig. 19.1), more males lost all their teeth compared to females (Fig. 19.2). And third, unlike mortality where all human beings will die eventually, teeth often remain well after the death of an individual; in other words loss of life does not mean the loss of all or some teeth.

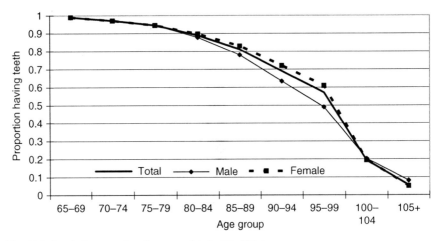

Fig. 19.2 Survival curve of teeth among elderly (65–100+)

Based on this premise, the 80/20 WHO goal is possibly attainable, and China should be able to achieve good dental care in both the young and older population in the future.

19.4 Discussion

Age is related to the loss of teeth in all populations. In this large sample of elderly in China, age is shown to play an important role in tooth loss. The older the age, the less the teeth the elderly have. On average, elderly between 65 and 100+ have 9.1 teeth. The number of teeth remaining between the ages of 65–74 (18) is the same as that from the Second National Epidemiological Survey on Oral Health (18 natural teeth; National Supervising Committee on Oral Disease Prevention 1999: 55); however centenarians had less teeth than in the Bama (Guanxi province) study, which showed that only 20 percent of the elderly lost all their natural teeth (Xiao et al. 1996). As to the population of those 60+ in Beijing, a survey found that the mean number of missing teeth was 11, excluding third molars (Cooperation Group for Beijing Elderly Oral Health Survey 1988); a similar finding of elderly in Chengdu showed that on the average, each person had lost 14.2 teeth, excluding third molars (Chen et al. 1985). Our research showed that the number of teeth remaining and the percentage of edentulous elders may differ from other populations. For example, a study in the 1990s of elders aged 70 and over in six New England States (NEEDS, the New England Elders Dental Study) showed that about 38 percent of the elderly surveyed lost all their natural teeth, and the percentages among male and female elders were similar (Douglass et al. 1993). In Japan, elderly over 80 years of age had 6.2 natural teeth on average, and less than 15 percent of them had 20 natural teeth (Ministry of Health and Welfare 2000). However, all this information among different groups in the Chinese population and the population in other countries indicates the deterioration of dentate status as people age.

Gender differences in tooth loss are also important, for example for preventive, treatment and socio-demographic research and intervention programs. From a medical point of view, it is necessary to explore the physiological reasons behind a possible gender difference in loss of teeth. The results of such research may well contribute to the prevention of tooth loss, especially among the older population, either by intervention in nutrition in-take, supplements of calcium and other necessary nutritional elements, or through improvement of oral hygiene. Socio-demographic research will add more information, such as the characteristics of individuals who have more (or less) teeth than others in a population of the same age. This information is useful for targeting prevention and treatment programs which aim at the improvement of oral health, retaining more teeth during aging, and ultimately improving the quality of life among the elderly.

In our study, there are gender and residence differences in tooth loss, even after controlling for age. Males have more teeth than females, but the life expectancy of females is longer than that of males; however, a gain in life expectancy among females may be compromised by their suffering from other health conditions; for example loss of teeth is more serious among females than males. Also rural elders in general were shown to have lost more teeth than their counterparts in urban settings. It is important to examine the reasons that lead to urban and rural differences in tooth loss. For example, different dietary styles, types of food, or the hardness of food consumed during life or later life may well be related to tooth loss, especially at the older ages. However, we suspect that oral hygiene may also be an important factor affecting tooth loss among the population we surveyed. Results from the National Oral Epidemiological Survey in China in 1995 showed that 72 percent of elderly between 65 and 74 years of age brushed their teeth 1–2 times a day; and the percentage doing so in urban areas was higher than in rural areas (90 vs. 53 percent; National Supervising Committee on Oral Disease Prevention 1999: 607).

An increase in the percentage of denture wearers may indicate that serious tooth loss has affected an individual's life to an unbearable point; or the increase may be due to an increased awareness of oral health, an increased ability to afford dentures, or an increase in dental care services in China. Considering the age of and the dental services available to this population in the past and the present, we believe that the last three aforementioned reasons have influenced this changing pattern of denture use. However, male and female differences in denture wear may be caused by the degree of edentulous status among males and females (though more females lost their teeth than did males), or by other socio-cultural factors (e.g., more males in that age group worked at a young age, thus had medical insurance to cover the cost of dentures; or there might be gender differences in attitudes toward dentures). It seems that denture wear is disproportionate to the level of edentulousness among the population.

There are limitations in this research. First, the CLHLS was not specifically designed for dentate studies, and information available on oral health is hence limited. A detailed analysis of tooth loss and the factors affecting it are not possible with these data. Second, information on teeth retained and denture wear is not detailed (such as position and quality of the teeth remaining and size and types of dentures) which also prevents further oral health analysis. And third, although the CLHLS provides more information about the socio-demographic and health status of the elderly, which was lacking in other surveys, our pilot analysis of data from the 2002 survey included only a few variables that we thought might affect the immediate dental status of the elderly. Even though there are limitations on data availability and the resulting analyses, our research does provide a basic albeit crude dental picture of the elderly in China which is useful for further research. For example, further research might explore reasons behind gender differences in tooth loss in reference to childbirth history, socio-cultural and economic differences, and urban and rural differences in diet in relation to tooth loss.

The policy implications of this study are significant. With the increasing size and percentage of the older population, especially the oldest-old (80+) in China—concurrent with economic growth and improvements in the standard of living—oral health will attract more attention from the government (to provide more services to the population needed) and from individuals (to change their oral hygiene and to keep more teeth as long as possible as they age). From the results of our study, we are able to demonstrate a general trend of the extent and the severity of tooth loss and edentulism among the elderly. While attention is increasing in more developed countries on oral health status and factors that affect these outcomes, we also need more research in developing countries to accumulate information on oral health status, services availability and services needed. Research presented in this chapter is one example of an attempt to better understand oral health in a developing country. For further more-detailed and in-depth research, there is a need to carefully construct the relationship between oral health and other health statuses, with a longer period of follow-up observations. More surveys and other information, specifically on oral health, need to be conducted for a better understanding of oral hygiene among the elderly. With more information, we will likely have more effective and specific interventions to improve the oral health in the Chinese population at large.

References

Adut, R., J. Mann, and H.D. Sgan-Cohen (2004), Past and present geographic location as oral health markers among older adults. *Journal of Public Health Dentistry* 64 (4), pp. 240–243

Anderson, B.G. (1932), An endemic center of mottled enamel in China. *Journal of Dental Research* 12, pp. 591–593

Chen, H.M., J.Z. Zhang, Y.X. Quan, and J. Pu (1985), The situation of senile anodontia of 926 old people in Chengdu. *West China Journal of Stomatology* 3, pp. 21–124 (in Chinese)

Cooperation Group for Beijing Elderly Oral Health Survey (1988), A survey of oral health in 2191 elderly inhabitants in Beijing. *China Journal of Stomatology* 23, pp. 29–32 (in Chinese)

Douglass, C.W., A.M. Jette, C.H. Fox, S.L. Tennstedt, A. Joshi, H.A. Feldman, S.M. McGuire, and J.B. McKinlay (1993), Oral health status of the elderly in New England. *Journal of Gerontology, Medical Science*, 48 (2), pp. M39–M46

Gershen, J.A. (1991), Geriatric dentistry and prevention: Research and public policy. *Advances in Dental Research*, 5, pp. 69–73

Joshipura, K.J., W.C. Willet, and C.W. Douglass (1996), The impact of edentulousness on food and nutrient intake. *Journal of American Dental Association* 127, pp. 459–467

Lin, H.C. and E. Schwarz (2001), Oral health and dental care in modern-day China. *Community Dentistry and Oral Epidemiology* 29 (5), pp. 319–328

Lin, H.C., E.F. Corbet, E.C.M. Lo, and H.G. Zhang (2001), Tooth loss, occluding pairs, and prosthetic status of Chinese adults. *Journal of Dental Research* 80 (5), pp. 1491–1495

Mack, F., T. Mundt, P. Mojon, E. Budtz-jorgensen, C. Schwahn, O. Bernhardt, D. Gesch, U. John, T. Kocher, and R. Biffar (2003), Study of Health in Pomerania (SHIP): Relationship among socioeconomic and general health factors and dental status among elderly adults in Pomerania. *Quintessence International* 34 (10), pp. 772–778

Ministry of Health and Welfare (2000), *Annual report of health and welfare*. Tokyo: Ministry of Health and Welfare (in Japanese)

Nasu, I and Y. Saito (2006), Active life expectancy for elderly Japanese by chewing ability. *Japanese Journal of Public Health* 53 (6), pp. 411–423 (in Japanese)

National Supervising Committee on Oral Disease Prevention (1999), *Second oral health epidemiological surveys in China*. Beijing: People's Health Press (in Chinese)

Randolph, W.M., G.V. Ostir, and K.S. Markides (2001), Prevalence of tooth loss and dental service use in older Mexican Americans. *Journal of American Geriatrics Society* 49 (5), pp. 585–589

Shah, N. (2003), Gender issues and oral health in elderly Indians. *Internal Dental Journal* 53 (6), pp. 475–484

Shimazaki, Y., I. Soh, T. Saito, Y. Yamashita, T. Koga, H. Miyazaki, and T. Takehara (2001), Influence of dentition status on physical disability, mental impairment, and mortality in institutionalized elderly people. *Journal Dental Research* 80 (1), pp. 340–345

Shimazaki, Y., I. Soh, T. Koga, H. Miyazaki, and T. Takehara (2003), Risk factors for tooth loss in the institutionalized elderly: A six year cohort study. *Community Dental Health* 20 (2), pp. 123–127

Takata, Y., T. Ansai, S. Awano, K. Sonoki, M. Fukuhara, M. Wakisaka, and T. Takehara (2004), Activities of daily living and chewing ability in an 80-year-old population. *Oral Diseases* 10 (6), pp. 365–368

Xiao, Z., Q. Xu, and Y. Yuan (1996), Bama centenarians and their secrets of longevity. *Chinese Journal of Population Science* 3, pp. 29–32

Zeng, Y., J.W. Vaupel, Z. Xiao, C. Zhang, and Y. Liu (2001), The healthy longevity survey and the active life expectancy of the oldest-old in China. *Population: An English Selection* 13 (1), pp. 95–116

Zeng, Y., J.W. Vaupel, Z. Xiao, C. Zhang, and Y. Liu (2002), Sociodemographic and health profiles of the oldest old in China. *Population and Development Review* 28 (2), pp. 251–273

Zhang, B.X. (eds.) (1999), *A decade prevention of oral disease in China*. Beijing: Beijing Medical University Publisher (in Chinese)

Chapter 20
Psychological Resources for Well-Being Among Octogenarians, Nonagenarians, and Centenarians: Differential Effects of Age and Selective Mortality

Jacqui Smith, Denis Gerstorf and Qiang Li

Abstract Research on the young old indicates that psychological processes associated with the maintenance of subjective well-being are effective despite declining health and age-related social losses. In this chapter, we examine the robustness of this system in the oldest old. We divided the first wave cross-sectional sample of the Chinese Longitudinal Healthy Longevity Study (CLHLS) into two subsamples: 2-year survivors ($N = 4,006$) and 2-year drop-outs ($N = 4,799$). Psychological resources for well-being were measured by seven items (5-point response scale). Selectivity analyses and multiple regression analyses were conducted. Despite constraints in objective life conditions, long-lived individuals showed reasonably high levels of psychological resources for well-being. Age-cohort differences were small. Selective mortality and individual differences in life-history and life-context factors accounted for substantial amounts of variance. Individual differences were primarily associated with engagement in life, cognitive functioning, and health. The efficacy of this psychological system is vulnerable to losses and is associated with survival in the oldest old.

Keywords Age difference, Attrition sample, Centenarian, Chinese oldest old, Cohort difference, Cross-sectional sample, Engagement in life, Fourth age, Mortality, Nonagenarian, Objective life conditions, Octogenarian, Psychological resources, Sample selectivity, Structural equation, Subjective well-being, Successful aging, Survivor sample

---

J. Smith
Department of Psychology and Institute for Social Research, University of Michigan, 426 Thompson Street, Ann Arbor MI 48106-1248, USA
e-mail: smitjacq@isr.umich.edu

20.1 Introduction

Gerontologists have long been interested in processes linked to the malleability of aging and the determinants of survival into very old age. Research has tended to focus primarily on biogenetic mechanisms and risk factors, but there is increasing consideration given to the interactive role of social, environmental, behavioral, and psychological factors (e.g., Seeman et al. 2004; Vaupel et al. 1998). Such a multilevel and multifaceted perspective on longevity coincides with the proposals of several models of successful aging (e.g., Baltes 1987; Baltes and Baltes 1990; Rowe and Kahn 1987, 1997; Strawbridge et al. 2002). Although these models suggest slightly different sets of components and processes, there is general agreement that two central outcomes of successful aging are a sense of personal well-being and a healthy long life.

Much is known about the characteristics and processes of aging successfully among individuals aged 60–80 years (e.g., Baltes and Smith 2003; Rowe and Kahn 1997). Among contemporary generations of the young old in many countries, there are high levels of physical and mental health, cognitive fitness, and engagement in productive, social, and solitary leisure activities, healthy lifestyles and subjective well-being (e.g., Antonucci et al. 2002). Little is known about the constellations of these characteristics in the oldest old (aged 80+). Advanced old age (the Fourth Age) has been described as a phase of life unlike earlier periods of the lifespan (e.g., Baltes and Smith 2003; Smith 2001; Suzman et al. 1992). It is associated with high levels of comorbidity as well as increased risks of dementia, need-for-care, and institutionalization. It is thus important to ask whether very long-lived persons have the psychological capacity to sustain a sense of personal well-being and, hence, to age successfully. Furthermore, are there differences in the profile of psychological resources for well-being observed in octogenarians, nonagenarians, and centenarians?

We examine these questions in the context of a cross-sectional sample from the first wave of the Chinese Longitudinal Healthy Longevity Study (CLHLS: Zeng et al. 2002). This study is unique in that it consists of a representative sample of the oldest old stratified by age and gender, meaning that it provides a context for age- and gender-comparative analyses within the oldest old. Furthermore, each participant was assessed individually with a standardized protocol in a face-to-face interview.

20.2 Psychological Resources for Well-Being

Depending on the researcher's method and theoretical stance, data about personal well-being can serve as an indicator of perceived current life status, an evaluation of life up-to-the-present, and/or as an estimate of the psychological resources that an individual could use to adapt to future challenges (e.g., Diener et al. 1999; Ryff 1995; Lawton 1991). In the research we report in this chapter, we adopt the latter approach.

We examine two sets of psychological resources, which together facilitate and constrain the maintenance of well-being and thriving in old age and so contribute to healthy longevity. One set involves resources that enhance positive well-being (e.g., optimism, a sense of personal control, conscientiousness, and positive feelings about aging). The other set is linked to negative aspects of well-being (e.g., loneliness, negative emotions such as anxiety, and associating aging with a loss of self worth or competence). Each of these psychological resources has been shown to be a predictor of mortality in old age (e.g., Berkman 1988; Friedman et al. 1995; Seeman et al., 1987; Swan and Carmelli 1996). Maier and Smith (1999) and Levy et al. (2002), for example, reported that older individuals who evaluated their own experiences of aging in a positive way and maintained a high sense of self worth also lived longer. These authors assessed positive feelings about aging using a subset of items from the PGCMS (Lawton 1975). In both studies, the predictive effects of these items remained after statistically controlling for other indicators associated with mortality in old age (e.g., age, gender, SES, and health).

A well-functioning psychological system that is capable of adapting to new challenges should be characterized by a pattern of higher levels (i.e., maximization) of resources associated with positive well-being, and relatively low levels (i.e., minimization) on indicators of negative well-being (e.g., Taylor 1991; Kahneman et al. 1999; Baltes and Baltes 1990). Other patterns across these resources are indicative of acute or chronic stress and of less effective functioning. An individual adjusts his or her level of aspirations to the reality of present life conditions in order to protect the self against a loss of well-being and to maintain a sense of purpose in life. Together, these psychological processes contribute to a positive aura of well-being and to seemingly paradoxical observations that some subgroups of individuals report high life satisfaction in contexts of relatively poor objective life circumstances (i.e., the well-being paradox). To the extent to which a sense of well-being contributes to a long life, these psychological resources and processes of evaluating life experiences play a critical role.

Theoretically, as a function of enduring personality dispositions and other psychological resources, an individual's level of well-being is expected to be generally stable across the lifespan, with short-term fluctuations contingent on acute negative events. However, research has revealed that some components of well-being show different age associations from age 20–75 years (e.g., Diener et al. 1999; Mroczek and Kolarz 1998). Feelings of happiness show negative age correlations, while reports of life satisfaction either reveal no age trends or a small increase with age. In very old age, it is suggested that the increased risk of frailty, the accumulation of debilitating health conditions, functional impairments, and personal losses may increasingly place constraints on life satisfaction (Isaacowitz and Smith 2003; Kunzmann et al. 2000; Smith et al. 2002). Beyond age and health, other factors such as life history and social embeddedness as well as activities and engagement in life are expected to contribute to differences between individuals (George 2000; Kahneman et al. 1999; Lennartsson and Silverstein 2001; Menec 2003).

Using cross-sectional data from the first wave of the CLHLS, we examine age cohort and selectivity effects in the levels of psychological resources associated with maximizing positive and minimizing negative well-being as well as predictors of individual differences in resource availability. Examination of age-related differences in the period of very old age is conceptually and methodologically complex because components of heterogeneity differ across age groups. Subgroups representing each decade (octogenarians, nonagenarians, and centenarians) reflect cohort differences in life history, differential amounts of cumulative age-related change, differential impact of selective mortality, and differential distance from death (e.g., Manton 1990). We adopt a strategy that provides partial insight into the possible effects of distance from death on functioning within and across age-cohort groups. To do this, we compare levels of functioning within age-cohort groups between those individuals who subsequently survived a further 2 years and continued to participate in the study with those who did not survive. Because the likelihood of 2-year survival is higher for octogenarians than for centenarians, we argue that selection effects found for centenarians provide strong evidence for the important role of psychological resources for aging successfully in the Fourth Age. Observed levels of psychological resources in centenarians reflect multiple selection effects. For some (unknown) reasons they are the positive outliers in terms of survival in their birth cohort. Observations among octogenarians are probably less select on these factors given that we do not know how many of them will live until age 100.

Based on the assumption that psychological resources contribute to healthy longevity, we hypothesize that we should observe selectivity differences in this sample such that individuals who subsequently survive for 2 years and continue in the second wave of data collection would show higher levels on psychological resources that maximize positive and minimize negative well-being. The effects are expected to be larger in centenarians than in octogenarians. Within the positively selected group of "2-year survivors," we expect to observe only minimal age cohort differences but substantial individual differences reflecting diversity in life history and present life conditions that are also linked to well-being.

20.3 Methods

20.3.1 Sample

We compare two nested subsamples from the first wave of the Chinese Longitudinal Healthy Longevity Study (CLHLS; $N = 8,805$): One subgroup survived for a further 2 years after baseline assessment and continued in the longitudinal study (1998–2000: $N = 4,006$), whereas the other subgroup dropped out after baseline ($N = 4,799$). The major reason underlying the definition of the two groups and then concentrating on the 2-year survivor subsample was to shed some light on the effects of sample selectivity on research findings in very old age.

Detailed information about the assessment battery of the total cross-sectional CLHLS sample is reported in Zeng and Vaupel (2002), and in Chap. 2 in this volume. The survey was conducted in 631 randomly selected counties and cities of 22 provinces in which Han Chinese predominate. These provinces covered 85 percent of the total population in China (985 million persons). All centenarians from the selected areas who agreed to participate were included in the study. Based on gender and place of residence (i.e., living in the same street, village, city, or county) for a given centenarian, randomly selected octogenarians and nonagenarians were also sampled. This matched-recruitment procedure resulted in an over-sampling of the oldest old and older men at baseline. Records of the age of Han Chinese have been verified as accurate for cohorts born after 1893 (see Coale and Li 1991; Zeng et al. 2002). Interviews and basic health examinations were carried out at the participant's place of residence (i.e., private household or institution) by a doctor, nurse, or medical student.

Descriptive information for the 2-year survivor and attrition subsamples analyzed in this chapter is provided in Table 20.1. To be included in the 2-year survivor subgroup, participants had to provide valid data on at least one indicator of a resource for positive and negative well-being at baseline and the subsequent follow-up in 2000. The attrition sample consisted of participants who were available for testing at baseline only (primarily due to mortality) as well as those who provided missing psychological data at either occasion (baseline: 3–7 percent; 2000: 9–15 percent). Missing data were primarily due to poor hearing and vision, and severe cognitive impairment.

Table 20.1 Differences on demographic, physical-functioning, and psychological characteristics for subsamples of the CLHLS participants who survived two years after baseline (S; $N = 4006$) or dropped out (A; $N = 4799$) across the three age cohorts

	80–89 years		90–99 years		100–105 years	
	S	A	S	A	S	A
N	2,239	1,289	1,216	1,797	551	1,713
% Women	51^a	47^a	55^a	58^a	81^a	79^a
% City	43^a	49^b	38^a	36^a	24^a	29^a
% No school education	56^a	53^a	66^a	70^a	82^a	85^a
% No spouse	69^a	73^b	87^a	90^b	98^a	97^a
Number of children alive	5.23^a	4.89^a	5.27^a	5.20^a	5.32^a	5.62^a
% Poor hearing	3^a	11^b	10^a	29^b	18^a	47^b
% Poor vision	7^a	12^b	12^a	26^b	23^a	42^b
ADL (max = 12)	11.68^a	10.99^b	11.18^a	9.93^b	10.62^a	8.53^b
Engagement in life (max = 16)	4.01^a	3.29^b	2.78^a	1.83^b	2.03^a	1.12^b
MMSE (max = 23)	20.17^a	18.85^b	18.27^a	15.28^b	15.99^a	12.15^b
Word fluency (foods)	11.29^a	10.52^b	9.39^a	7.26^b	7.32^a	5.47^b
Life satisfaction (max = 5)	3.90^a	3.88^a	3.92^a	3.83^b	3.93^a	3.87^a
Self-rated health (max = 5)	3.74^a	3.54^b	3.70^a	3.48^b	3.70^a	3.46^b

Total $N = 8805$. S = 2-year survivor subsample, A = attrition (drop-out sample). ADL, activities of daily living; MMSE, mini-mental state examination. Higher scores on ADL, engagement in life, MMSE, and word fluency indicate higher functioning. Within age cohorts, indices with different superscripts are significantly different between the samples at $p < .01$ or below. For statistically significant differences between the age cohorts, see text.

20.3.2 Measures

Psychological resources. Our research includes indicators of psychological resources associated with positive and negative aspects of well-being. Resources for positive well-being were measured using four items: *Optimism* ("I always look on the bright side of things"), *Conscientiousness* ("I like to keep my belongings neat and clean"), *Sense of personal control* ("I can make my own decisions concerning my personal affairs"), and *Positive feelings about aging* ("I am just as happy now as when I was younger"; item from the PGCMS; Lawton 1975). To assess negative aspects of well-being, the following three items were used: *Neuroticism* ("I often feel fearful or anxious"), *Loneliness* ("I often feel lonely or isolated"), and *Perceived loss of self-worth* ("The older I get, the more useless I feel"; item from the PGCMS; Lawton 1975). Responses were recorded on a 5-point scale (1—describes me very well; 5—does not describe me at all). To have both sets of resources scored in the same direction, all responses on items for positive well-being were reverse coded. As a consequence, high scores on resources against negative well-being reflect low neuroticism, low loneliness, and high self-worth.

Structural characteristics of the resource indicators of maximizing positive and minimizing negative well-being were evaluated using structural equation modeling techniques. Table 20.2 contains the standardized factor loadings, their levels of statistical significance and standard errors, as well as the communalities for the measurement model of well-being. This model produced bivariate correlations

Table 20.2 Standardized factor loadings and communalities for the measurement model of psychological resources for well-being

Indicator	Factor loading	T	SE	R^{2a}
Resources for positive well-being				
1. I always look on the bright side of things.	0.64 (0.65)	32.24 (48.74)	0.02 (0.01)	0.41 (0.42)
2. I like to keep my belongings neat and clean.	0.53 (0.53)	27.38 (40.85)	0.01 (0.01)	0.28 (0.28)
3. I can make my own decisions concerning my personal affairs.	0.45 (0.42)	23.58 (32.55)	0.02 (0.01)	0.21 (0.18)
4. I am just as happy now as when I was younger.	0.54 (0.54)	27.91 (41.63)	0.02 (0.01)	0.29 (0.29)
Resources against negative well-being				
1. I often feel fearful or anxious.[+]	0.62 (0.61)	29.93 (43.96)	0.02 (0.01)	0.38 (0.37)
2. I often feel lonely or isolated.[+]	0.75 (0.75)	33.40 (49.64)	0.02 (0.01)	0.56 (0.56)
3. The older I get, the more useless I feel.[+]	0.43 (0.43)	23.05 (33.67)	0.02 (0.01)	0.19 (0.18)

$N = 4006$. In parentheses, indices for the total sample ($N = 8805$).

[+] Scores were recoded so that high scores represent higher well-being-associated resources.

[a] Communality (R^2) = 1—standardized residual variance. Communality indicates the proportion of variance each single indicator explains of its associated latent factor: Squared multiple correlations.

between the positive and negative factors of well-being and found them to be of moderate size both in the total cross-sectional CLHLS sample ($N = 8,805$: $r = 0.34$) and in the restricted subsample of 2-year survivors ($N = 4,006$: $r = 0.33$). This specified model showed acceptable fit with the data in the total sample ($N = 8,805$: RMSEA = 0.061, NFI = 0.94, CFI = 0.94) as well as in the 2-year survivor subsample ($N = 4,006$: RMSEA = 0.054, NFI = 0.95, CFI = 0.95). As can be seen in Table 20.2, all factor loadings were reasonable, which also indicates that the fit between the model specified and the current data set is acceptable.

Individual difference correlates. To examine cross-disciplinary correlates of individual differences in the psychological resources for well-being, we entered six sets of measures into regression models. A first set contained sociodemographic characteristics including gender (1 = men, 2 = women), education (0 = no education, 1 = attended school), and place of residence (1 = urban, 2 = rural). A second set of correlates included in the analyses comprised measures of functional health such as Activities of Daily Living (ADL) and sensory functioning. ADL represents the number of basic activities (i.e., getting out of bed, dressing, toileting, bathing, and eating) in which participants' reported needing assistance (max 12 = no assistance needed on six activities; Katz et al. 1963). Sensory functioning was indicated by a vision test (1 = the participant could see a break in a circle on a cardboard sheet and distinguish where the break was located, 2 = can not see) and by the interviewers' rating of the participant's ability to hear (1 = can hear, 2 = cannot hear). Third, self-rated health was measured using an item that is standard in the literature and that has often been shown to be a valid predictor of functioning and mortality among older people (for review, see Idler and Benyamini 1997): "How would you rate your health at present?" The response format ranged from 1 = very good to 5 = very bad.

The fourth set of correlates involved indicators of cognitive functioning. Two measures were used. An age-adjusted 23-item Chinese version of the Mini-Mental State Examination (MMSE; Folstein et al. 1975) assessed the facets orientation, registration, attention and calculation, recall, and language and movement. The second measure of cognitive functioning assessed verbal fluency. Participants were required to name as many kinds of food as possible within 60 s. This is a standard task in many intelligence tests.

The last fifth and sixth sets of correlates reflected social embeddedness and active engagement in life. Quantitative measures of social integration were available including whether the participant's spouse was alive and the number of children alive. Engagement in life was measured using a summed score of a list of eight activities. Participants were asked to indicate whether they performed the following eight activities regularly: Housework, grow vegetables and other field work, garden work, read newspapers or books, raise domestic animals, play cards and/or mah-jong, watch TV and/or listen to the radio, and religious activities. Due to the coding scheme (0 = never, 1 = sometimes, 2 = regularly), the maximum score was 16.

Mortality information was obtained in 2000 in interviews with relatives, caregivers, and community authorities. Mortality status was available for $N = 7,938$ and (partly) missing for $N = 867$. Of those for whom mortality status was recorded,

41 percent ($N = 3,247$) were deceased by 2000. Among the survivors ($N = 4691$), 85 percent ($N = 4,006$) provided valid data on resource indicators for well-being.

20.4 Results

Results are reported in three main sections. In a first section, we compare the profiles of psychological resources for positive and against negative well-being of octogenarians, nonagenarians, and centenarians in the 2-year survivor sample ($N = 4,006$). In a second step, selectivity analyses are carried out to examine the extent to which positive sample selectivity contributed to these findings. Here, we contrast the survivor sample against the attrition sample ($N = 4,799$). In a third step, multiple regression analyses are undertaken to examine a number of cross-disciplinary factors as potential individual difference correlates of subjective well-being among long-lived individuals.

20.4.1 Profiles of Well-Being in Advanced Old Age

Table contains descriptive statistics for the three age cohorts on the indicators of psychological resources for positive well-being and against negative well-being. Overall, participants in the three age cohorts had relatively high potential for well-being. For example, the mean for centenarians on the optimism item was 3.91, nonagenarians = 3.95, and octogenarians = 3.97. Statistically significant differences between the three age cohorts of long-lived individuals were found on personal control, ($F_{2,3,918} = 11.4$, $p < .000$), loneliness ($F_{2,3,953} = 6.3$, $p < .01$), and self-worth ($F_{2,3,940} = 14.8$, $p < .000$). On average, centenarians reported lower self-worth, greater loneliness, and less control over their lives than did octogenarians. As a result, centenarians had somewhat lower levels on the composite measures of resources for positive well-being ($F_{2,4,003} = 7.7$, $p < .000$) and against negative well-being ($F_{2,4,003} = 11.0$, $p < .000$). Although statistically significant, these differences were small amounting to 0.18 SD units for both sets of resources. The maximum difference between centenarians and octogenarians was on self-worth, but this reflected only 0.25 SD units. In sum, results from this 2-year survivor subsample of the CLHLS suggest that, on average, individuals are able to maintain relatively high well-being into advanced old age, and that there are cross-sectional differences between various cohorts of long-lived individuals, but they are small.

20.4.2 Sample Selectivity

To examine the effects of sample selectivity on the present findings about preserved well-being in very old age, participants from the 2-year survivor subsample of the CLHLS (s; $N = 4,006$) were contrasted against the total cross-sectional sample

Table 20.3 Descriptive statistics of resource indicators for positive well-being and for the lack of negative well-being across the three age cohorts at baseline assessment

Indicator	80–9 years N = 2,239	90–9 years (N = 1,216)	100–05 years (N = 551)
Resources for positive well-being			
1. I always look on the bright side of things.	3.97[a] (0.80)	3.95[a] (0.78)	3.91[a] (0.85)
2. I like to keep my belongings neat and clean.	4.11[a] (0.68)	4.07[a] (0.68)	4.03[a] (0.76)
3. I can make my own decisions concerning my personal affairs.	3.64[a] (1.02)	3.54[a,b] (0.99)	3.42[b] (1.05)
4. I am just as happy now as when I was younger.	3.40[a] (1.05)	3.36[a] (1.05)	3.32[a] (1.09)
Composite	3.78[a] (0.60)	3.73[a,b] (0.61)	3.67[b] (0.68)
Resources against negative well-being			
1. I often feel fearful or anxious.[+]	3.70[a] (0.82)	3.66[a] (0.82)	3.67[a] (0.84)
2. I often feel lonely or isolated.[+]	3.67[a] (0.84)	3.57[b] (0.85)	3.58[a,b] (0.89)
3. The older I get, the more useless I feel.[+]	3.10[a] (0.99)	3.00[a,b] (1.00)	2.85[b] (1.02)
Composite	3.49[a] (0.66)	3.40[b] (0.68)	3.37[b] (0.70)

N = 4006. Means and standard deviations shown in parentheses. [+] Scores were recoded so that high scores represent higher well-being-associated resources. Indices with different superscripts are significantly different between the age cohorts at $p < .01$ or below. Response format for the items ranges from 1 to 5.

($N = 8,805$). Following a procedure used by Lindenberger et al. (2002), effect sizes for sample selectivity were computed as the normed difference between the two nested samples: selectivity = $(M_s - M_{\text{total sample}})/SD_{\text{total sample}}$. Effects are expressed in SD units. It has to be noted that the effect size is a descriptive measure that is derived directly from the group level, so that there is no variance associated with it. For that reason, it is not possible to apply significance tests. Overall, only 15 percent of baseline participants who were eligible for repeated assessment were not willing or capable to do so. The majority of baseline participants who did not take part a second time were deceased. As a consequence, total selectivity was *not* separated into a mortality-associated component and an experimental component because the mortality component was the major source of drop-out.

In a first set of selectivity analyses, sample differences on the composite scores for the two sets of resources were determined. In a second set of analyses, sample differences in demographic, physical-functioning, and psychological characteristics were examined because these variables were used in a subsequent step as potential individual difference covariates of well-being.

Sample differences in psychological resources. The results of the selectivity analyses are shown in Fig. 20.1. From Panel A of Fig. 20.1, it can be seen that the

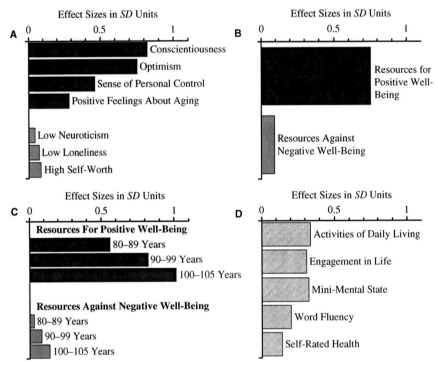

Fig. 20.1 Selectivity effects in the 2-year survivor sample of the CLHLS ($N = 4,006$) relative to the total CLHLS sample ($N = 8,805$) for variables assessed at baseline

Note: *Panel A*: Selectivity effects on the single psychological resource indicators for positive well-being (conscientiousness, optimism, sense of personal control, and positive feelings about aging) and against negative well-being (neuroticism, loneliness, and perceived loss of self-worth). *Panel B*: Selectivity effects on the composite measures of resources for positive well-being and resources against negative well-being. *Panel C*: Selectivity effects on resources for positive well-being and against negative well-being separately for the three age cohorts of octogenarians, nonagenarians, and centenarians. *Panel D*: Selectivity effects for individual difference correlates of well-being.

magnitude of sample selectivity was 0.82 SD units for conscientiousness (i.e., the 2-year survivor subsample was higher on this resource), 0.75 SD units for optimism, 0.46 SD units for control, 0.28 SD units for positive feelings about aging, 0.04 SD units for neuroticism, 0.07 SD units for loneliness, and 0.08 SD units for self-worth. Panel B of Fig. 20.1 illustrates that composite selectivity effects for resources linked to positive well-being (0.75 SD units) were much stronger than selectivity effects for lack of negative well-being (0.09 SD units). According to statistical convention (e.g., Cohen 1977), observed selectivity corresponds to medium effects for measures of positive well-being and to small effects for measures of lack of negative well-being.

Panel C of Fig. 20.1 displays sample selectivity effects for well-being separately for the three age cohorts. In both sets of resources, the positive selectivity effects tended to be stronger among the older cohorts. For resources linked to positive well-being, the magnitude of total sample selectivity was 0.56 SD units for octogenarians, 0.82 SD units for nonagenarians, and 1.01 SD units for centenarians. Again, based on Cohen's criteria 1977), the selectivity effects among octogenarians represent medium effects and among nonagenarians and centenarians correspond to large effects. For resources against negative well-being, effects were small: Sample selectivity was 0.03 SD units for octogenarians, 0.08 SD units for nonagenarians, and 0.14 SD units for centenarians.

Sample differences in individual difference correlates of well-being. An additional set of selectivity analyses examined sample differences in demographic, physical-functioning, and psychological characteristics. Panel D of Fig. 20.1 shows sample selectivity effects for those covariates that were available as continuous measures, which allowed the calculation of effect size estimates of selectivity analogous to those reported for well-being. The magnitude of total sample selectivity was 0.34 SD units for Activities of Daily Living, 0.31 SD units for engagement in life, 0.32 SD units for the Mini-Mental State Examination, 0.20 SD units for word fluency, and 0.14 SD units for subjective health. By convention (e.g., Cohen 1977), selectivity effects for the covariates of well-being examined here were small.

Table 20.1 provides additional information about all correlates between the CLHLS participants in the 2-year survivor subsample ($N = 4,006$) and the attrition subsample ($N = 4,799$), separately for the three age cohorts. There were no or only marginally significant sample differences in terms of gender distribution, place of residency, school education or not, availability of a spouse, and life satisfaction. For example, the lack of differences in life satisfaction across the samples as well as across the age cohorts indicates that life satisfaction represents but one and probably not the most sensitive indicator of successful aging (Kahneman et al. 1999). Substantive selectivity differences were found for the ratio of participants who were impaired in hearing and vision, and with regard to ADL, engagement in life, MMSE, word fluency, and self-rated health. These differences unequivocally indicate the positive selection of the 2-year survivor subsample. For example, among the centenarians who continued participation in the CLHLS, only 23 percent were found to be visually impaired as compared with 42 percent among those centenarians who were only available for testing once.

Table 20.1 also shows pronounced age-related differences. In both samples, the ratio of women over men increased drastically over the age cohorts, reflecting the higher mortality rates for men, chi-square $\chi^2_{2,N=4,006} = 169.4$, $p < .000$; $\chi^2_{2,N=4,799} = 357.6$, $p < .000$. Among the octogenarians, there was an almost equal distribution of gender, whereas the ratio was 4:1 for women among centenarians. In a similar vein, it was found that the older the CLHLS participants, the less likely they were to live in an urban area (chi-square $\chi^2_{2,N=4,006} = 61.8$, $p < .000$; $\chi^2_{2,N=4,799} = 126.3$, $p < .000$), to have had school education ($\chi^2_{2,N=4,006} = 140.6$, $p < .000$; $\chi^2_{2,N=4,799} = 360.3$, $p < .000$), to have a spouse ($\chi^2_{2,N=4,006} = 276.0$, $p < .000$; $\chi^2_{2,N=4,799} = 423.9$, $p < .000$), to have relatively well-preserved hearing ($\chi^2_{2,N=4,006} = 186.4$, $p < .000$; $\chi^2_{2,N=4,799} = 451.0$, $p < .000$) and vision ($\chi^2_{2,N=4,006} = 133.7$, $p < .000$; $\chi^2_{2,N=4,799} = 331.2$, $p < .000$), to be restricted in ADL ($N = 4,006$: $F_{2,4,003} = 122.7$, $p < .000$; $N = 4,799$: $F_{2,4,791} = 249.8$, $p < .000$) and engagement in life ($N = 4,006$: $F_{2,4,002} = 197.8$, $p < .000$; $N = 4,799$:

$F_{2,4,788} = 422.9$, $p < .000$), and to be cognitively fit on the MMSE ($N = 4,006$: $F_{2,3,991} = 261.8$, $p < .000$; $N = 4,799$: $F_{2,4,190} = 416.6$, $p < .000$) as well as in terms of word fluency ($N = 4,006$: $F_{2,3,371} = 76.9$, $p < .000$; $N = 4,799$: $F_{2,3,540} = 213.0$, $p < .000$). In contrast, there were no or only minimal differences between the three age cohorts in terms of life satisfaction ($N = 4,006$: $F_{2,3,994} = 0.7$, $p > .10$; $N = 4,799$: $F_{2,4,201} = 2.1$, $p > .10$) and self-rated health ($N = 4,006$: $F_{2,3,998} = 1.0$, $p > .10$; $N = 4,799$: $F_{2,4,201} = 3.0$, $p = .051$). The large majority of the sample was either living alone or living together with other household members, but not in nursing homes, and this was the case both for the total cross-sectional CLHLS sample ($N = 202$, 4.2 percent) and in the restricted positively selected sample of 2-year survivors ($N = 221$, 5.5 percent).

20.4.3 Regression Analyses

Multiple regression analyses were undertaken to examine a number of cross-disciplinary factors as potential correlates of subjective well-being among long-lived individuals. The effect of chronological age was covaried by entering age in a first step. This was followed by blockwise entry of the covariates. Intercorrelations among the constructs entered into the regression analyses are provided in the Appendix (see Table 20.5).

Results of the final age-partialed models of hierarchical regression analyses predicting the availability of resources for positive well-being and against negative well-being are displayed in Table 20.4. Analyses indicated that the linear

Table 20.4 Final age-partialed models from hierarchical regression analyses to predict resources for positive well-being and against negative well-being in the 2-wave sample of the Chinese Longitudinal Healthy Longevity Study

Unique predictors	Resources for positive well-being			Resources against negative well-being		
	B	SE	β	B	SE	β
Men/women	0.03	0.02	0.02	−0.11***	0.03	−0.08***
Urban/rural	−0.18***	0.02	0.15***	−0.07**	0.02	−0.05**
No school/school	−0.01	0.02	−0.01	0.03	0.03	0.02
Number of children alive	0.01**	0.00	0.05**	0.01*	0.00	0.03*
Poor vision/good vision	0.06	0.03	0.03	−0.05	0.04	−0.02
Poor hearing/good hearing	0.01	0.04	0.00	−0.10*	0.05	−0.04*
ADL	−0.02**	0.01	−0.05**	0.02*	0.01	0.05**
Engagement in life	0.03***	0.00	0.14***	0.00	0.01	0.01
Mini-Mental State	0.02**	0.00	0.16***	0.01*	0.00	0.05*
Word fluency	0.01***	0.00	0.07***	0.01***	0.00	0.08***
Self-rated health	0.21***	0.01	0.28***	0.15***	0.02	0.17***
R^2	0.21			0.10		
F	74.10			29.41		
df	12,3353			12,3353		
p <	0.000			0.000		

$N = 4006$.
* $p < .05$; ** $p < .01$; *** $p < .001$.

combination of the covariates accounted for 21 percent of the variance in resources for positive well-being, $F(12, 3,353) = 27.41$, $p < .000$, and 10 percent of the variance in resources against negative well-being, $F(12, 3,353) = 29.35$, $p < .000$. According to statistical convention (Cohen 1977), the effect of resources for positive well-being was of medium size and the effect of resources against negative well-being was small.

Unique positive predictors of resources for positive well-being were living in the city, being socially embedded (as indexed by the number of children alive), preserved functional health (Activities of Daily Living), engagement in life, cognitive fitness (Mini-Mental State, word fluency), and self-rated health. As expected, the largest standardized beta weight was found for subjective health (beta $\beta = 0.28$), but beta weights of the Mini-Mental State ($\beta = 0.16$), living in the city ($\beta = 0.15$), and engagement in life ($\beta = 0.14$) were also considerable. Being a man, living in the city, having higher social integration, preserved functioning (good hearing, ADL), cognitive fitness, and higher self-rated health uniquely predicted higher resources against negative well-being. Again, the beta weights for subjective health were largest ($\beta = 0.17$) for this dimension, whereas all other beta weights were rather small (ranging from $\beta = 0.08$ for word fluency to $\beta = -0.04$ for hearing).

20.5 Discussion

Our findings point to the existence of the well-being paradox in this Chinese sample of the oldest old. Despite constraints in objective life conditions (e.g., functional health), long-lived individuals showed reasonably high levels of psychological resources linked to the capacity of well-being. Differences between the three age cohorts of octogenarians, nonagenarians, and centenarians were present, but were relatively small. Selective mortality and individual differences in life-history and life-context factors accounted for substantial amounts of variance in the potential for well-being.

Overall, our analyses suggest that psychological processes associated with maximizing positive and minimizing negative well-being remain reasonably intact in advanced old age. Even among centenarians, the psychological system associated with sustaining well-being appears to be functional. At the same time, however, these results need to be interpreted in the context of the sample selectivity effects we report. Selectivity leads to an underestimation of actual losses associated with age-related change at the population level of the remaining portion of a given birth cohort in very old age. This is so because the less functional individuals in an age group are more likely to drop out. In addition, selectivity indicates that distance from death is associated with less efficacy of the psychological system linked to well-being. In future work, it will be important to more thoroughly attempt to separate the effects age-associated processes from mortality-associated processes.

We defined two sets of psychological resources that contribute to sustaining well-being by maximizing positive aspects of life (e.g., conscientiousness, optimism, personal control, and a positive perception of aging) and minimizing negative aspects (dealing with loneliness, maintaining self worth, and regulation of anxiety). The idea that personal well-being is multidimensional and that it reflects a positive balance of cognitive evaluations of life and emotional experience is well-established in the literature (e.g., Diener et al. 1999; Kahneman et al. 1999; Lawton 1991). Our definition of psychological resources and processes associated with well-being, however, was constrained by the items included in the CLHLS. Investigating well-being in the oldest old was not the primarily goal of the CLHLS.

Nevertheless, the seven items available in the survey do indicate key psychological resources that have been previously been highlighted in the gerontological literature as predictors of well-being and longevity, and our findings are theory consistent. Conscientiousness, for example, which was predictive of longevity in the Terman sample (Friedman et al. 1995) indicates a responsible well-organized lifestyle and a degree of self-discipline needed to deal with all the challenges of a long life. Optimism and a sense of personal control over the things that happen in one's life add positive future-oriented motivational resources to a conscientious lifestyle and probably ensure some aspects of variety to life routines and activities (see also Maruta et al. 2000). Minimization of anxiety, fear, and loneliness have been shown to have positive effects on the immune system (e.g., Kiecolt-Glaser et al., 2002), to reduce stress, and to foster well-being and health (e.g., Ryff and Singer 1998). Finally, the two items in the CLHLS which assess whether individuals perceive that, despite getting older, they have maintained a degree of happiness and sense of self-worth, were extracted from a PGCMS subscale (Lawton 1975) previously found to predict survival in the Berlin Aging Study (Maier and Smith 1999) and in the Ohio Longitudinal Study of Aging (Levy et al. 2002).

The limitations of the CLHLS data on psychological functioning are more than outweighed by its strengths. On the limitation side, one would have wished to have available a more comprehensive measurement battery to index not only psychological resources linked to well-being but also status on multiple components of well-being. Additional cognitive measures sensitive to average and higher levels of cognitive functioning (i.e., beyond the level of functioning assessed by the MMSE) would have allowed us to better differentiate individuals in terms of average and successful aging.

In terms of its strengths, the CLHLS offers a unique data set with large subgroup sizes at the oldest old ages to shed light on the differential impact of advanced age and selectivity. Furthermore, the CLHLS comprises a partially *representative* sample of octogenarians, nonagenarians, and centenarians in a non-Western country. As a consequence, the longitudinal data provide insight into the definition and mechanisms underlying the effects linked to social and economic status as well as technological and political changes. We also want to highlight the surprising similarity between our findings in these Chinese data about correlates of individual differences in the availability of resources for sustaining well-being and results from other countries (see Antonucci et al. 2002). This lends support to theories about successful aging and speaks to the universal nature of contributing factors including psychological resources, health, social integration, and engagement in life. The CLHLS provides a wealth of opportunities for future analyses to test hypotheses about determinants of longevity and life quality among the oldest old and also to distinguish sources of heterogeneity in this period of life (e.g., biological, social, age-related, and death-related). Such future work will further our understanding of how individuals age differently at the end of life.

Appendix

Table 20.5 Intercorrelations among the constructs entered INTO hierarchical regression analyses to predict resources for positive well-being and against negative well-being in the 2-wave sample of the Chinese Longitudinal Healthy Longevity Study

	1	2	3	4	5	6	7	8	9	10	11	12	13
1. Resources for positive well-being	–												
2. Resources against negative well-being	0.21[a]	–											
3. Age	−0.7	−0.9	–										
4. Men/women	−0.7	−1.4	0.17	–									
5. Urban/rural	−2.0	−0.8	0.13	0.3	–								
6. No school/school	0.12	0.13	−1.9	−5.4	−1.9	–							
7. Number of children alive	0.09	0.07	−1.2	−1.6	0.2	0.13	–						
8. Poor vision/good vision	−1.0	−1.1	0.18	0.13	0.06	−1.1	−0.7	–					
9. Poor hearing/good hearing	−1.0	−1.0	0.21	0.10	0.05	−1.2	−0.6	0.18	–				
10. ADL	0.08	0.13	−2.4	−1.3	.01	0.10	0.05	−2.5	−2.3	–			
11. Engagement in life	0.25	0.13	−3.3	−1.6	−1.4	0.28	0.12	−2.7	−1.6	0.28	–		
12. Mini-Mental State	0.27	0.18	−3.5	−2.7	−1.3	0.30	0.13	−.32	−.37	0.32	0.36	–	
13. Word fluency	0.19	0.16	−2.1	−1.4	−0.9	0.19	0.08	−1.3	−1.6	0.16	0.25	0.34	–
14. Self-rated health	0.34	0.23	−0.3	−1.0	−0.4	0.10	0.05	−1.2	−1.1	0.19	0.19	0.23	0.16

$N = 4006$.
[a] The correlation at the manifest level is somewhat lower than those reported from the LISREL models carried out at the latent level. Correlations that were not significantly different from zero at $p < .01$ or below, in italics.

References

Antonucci, T.C., C. Okorodudu, and H. Akiyama (2002), Well-being among older adults on different continents. *Journal of Social Issues* 58, pp. 617–626

Baltes, P.B. (1987), Theoretical propositions of life-span developmental psychology: On the dynamics between growth and decline. *Developmental Psychology* 23, pp. 611–626

Baltes, P.B. and M.M. Baltes (eds.) (1990), *Successful aging: Perspectives from the behavioral sciences*. New York, NY: Cambridge University Press

Baltes, P.B. and J. Smith (2003), New frontiers in the future of aging: From successful aging of the young old to the dilemmas of the fourth age. *Gerontology* 49, pp. 123–135

Berkman, L.F. (1988), The changing and heterogeneous nature of aging and longevity: A social and biomedical perspective. In: G.L. Maddox and M.P. Lawton (eds.): *Annual review of gerontology and geriatrics*. Vol. 8. New York: Springer, pp. 37–68

Coale, A. and S. Li (1991), The effect of age misreporting in China on the calculation of mortality rates at very high ages. *Demography* 28, pp. 290–301

Cohen, J. (1977), *Statistical power analysis for the behavioral sciences* (rev eds.). Hillsdale, NJ: Erlbaum

Diener, E., E.M. Suh, R.E. Lucas, and H.L. Smith (1999), Subjective well-being: Three decades of progress. *Psychological Bulletin* 125, pp. 276–302

Folstein, M.F., S.E. Folstein, and P.R. McHugh (1975), Mini Mental State: A practical method for grading the cognitive state of patients for the clinician. *Journal of Psychiatric Research* 12, pp. 189–198

Friedman, H.S., J.S. Tucker, J.E. Schwartz, C. Tomlinson-Keasey, L.R. Martin, D.L. Wingard, and M.H. Criqui (1995), Psychosocial and behavioral predictors of longevity: The aging and death of the "Termites". *American Psychologist* 50, pp. 69–78

George, L.K. (2000), Well-being and sense of self: what we know and what we need to know. In: K.W. Schaie and J. Hendricks (eds.): *Evolution of the aging self: The societal impact on the aging process*, New York: Springer, pp. 1–35

Idler, E.L. and Y. Benyamini (1997), Self-rated health and mortality: A review of twenty-seven community studies. *Journal of Health and Social Behavior* 38, pp. 21–37

Isaacowitz, D.M. and J. Smith (2003), Positive and negative affect in very old age. *Journals of Gerontology Series B: Psychological Sciences and Social Sciences* 58, pp. P143–P152

Kahneman, D., E. Diener, and N. Schwarz (eds.) (1999), *Well-being: The foundations of hedonic psychology*. New York: Sage

Katz, S., A.B. Ford, R.B. Moskowith, B.A. Jackson, and M.W. Jaffe (1963), Studies of illness in the aged. The index of ADL: A standardized measure of biological and psychosocial function. *Journal of the American Medical Association* 185, pp. 914–919

Kiecolt-Glaser, J.K., L. McGuire, T.F. Robles, and R. Glaser (2002), Emotions, morbidity, and mortality: new perspectives from psychoneuroimmunology. *Annual Review of Psychology* 53, pp. 83–107

Kunzmann, U., T.D. Little, and J. Smith (2000), Is age-related stability of subjective well-being a paradox? Cross-sectional and longitudinal evidence from the Berlin Aging Study. *Psychology and Aging* 15, pp. 511–526

Lawton, M.P. (1975), The Philadelphia geriatric center morale sale: a revision. *Journal of Gerontology* 30, pp. 85–89

Lawton, M.P. (1991), A multidimensional view of quality of life in frail elders. In: J.E. Birren, J.E. Lubben, J.C. Rowe, and D.E. Deutchman (eds.): *The concept and measurement of quality of life in the frail elderly*. San Diego, CA: Academic Press, pp. 3–27

Lennartsson, C. and M. Silverstein (2001), Does engagement with life enhance survival of elderly people in Sweden? The role of social and leisure activities. *Journals of Gerontology Series B-Psychological Sciences and Social Sciences* 56, pp. S335–S342

Levy, B.R., M.D. Slade, S.R. Kunkel, and S.V. Kasl (2002), Longevity increased by positive self-perceptions of aging. *Journal of Personality and Social Psychology* 83, pp. 261–270

Lindenberger, U., T. Singer, and P.B. Baltes (2002), Longitudinal selectivity in aging populations: separating mortality-associated versus experimental components in the Berlin Aging Study (BASE). *Journals of Gerontology Series B: Psychological Sciences and Social Sciences* 57, pp. P474–P482

Maier, H. and J. Smith (1999), Psychological predictors of mortality in old age. *Journal of Gerontology: Psychological Sciences* 54B, pp. P44–P54

Manton, K.G. (1990), Mortality and morbidity. In: R.H. Binstock and L.K. George (eds.): *Handbook of aging and the social sciences*. 3rd ed. San Diego, CA: Academic Press, pp. 64–90

Maruta, T., R.C. Colligan, M. Malinchoc, and K.P. Offord (2000), Optimists vs. pessimists: Survival rate among medical patients over a 30-year period. *Mayo Clinic Proceedings* 75, pp. 140–143

Menec, V.H. (2003), The relation between everyday activities and successful aging: A 6-year longitudinal study. *Journals of Gerontology Series B-Psychological Sciences and Social Sciences* 58, pp. S74–S82

Mroczek, D.K. and C.M. Kolarz (1998), The effect of age on positive and negative affect: A developmental perspective on happiness. *Journal of Personality and Social Psychology* 75, pp. 1333–1349

Rowe, J.W. and R.L. Kahn (1987), Human aging: usual and successful. *Science* 237, pp. 143–149

Rowe, J.W. and R.L. Kahn (1997), Successful aging. *Gerontologist* 37, pp. 433–440

Ryff, C.D. (1995), Psychological well-being in adult life. *Current Directions in Psychological Science* 4, pp. 99–104

Ryff, C.D. and B. Singer (1998), The contours of positive human health. *Psychological Inquiry* 9, pp. 1–28

Seeman, T.E., E. Crimmins, M.H. Huang, B. Singer, A. Bucur, T. Gruenewald, L.F. Berkman, and D.B. Reuben (2004), Cumulative biological risk and socio-economic differences in mortality: MacArthur studies of successful aging. *Social Science and Medicine* 58, pp. 1985–1997

Seeman, T.E., G.A. Kaplan, L. Knudsen, R. Cohen, and J. Guralnik (1987), Social network ties and mortality among the elderly in the Alameda County Study. *American Journal of Epidemiology* 126, pp. 714–723

Smith, J. (2001), Old age and centenarians. In: N. Smelser and P.B. Baltes (eds.): *International encyclopedia of the social and behavioral sciences*. Oxford: Elsevier, pp. 10843–10847

Smith, J., M. Borchelt, H. Maier, and D. Jopp (2002), Health and well-being in old age. *Journal of Social Issues* 58, pp. 715–732

Strawbridge, W.J., M.I. Wallhagen, and R.D. Cohen (2002), Successful aging and well-being: self-rated compared with Rowe and Kahn. *Gerontologist* 42, pp. 727–733

Suzman, R.M., K.G. Manton, and D.P. Willis(1992), Introducing the oldest old. In: R.M. Suzman, D.P. Willis, and K.G. Manton (eds.): *The oldest old*. New York, NY: Oxford University Press, pp. 3–14

Swan, G.E. and D. Carmelli (1996), Curiosity and mortality in aging adults: A 5-year follow-up of the Western Collaborative Group Study. *Psychology and Aging* 11, pp. 449–453

Taylor, S.E. (1991), Asymmetrical effects of positive and negative events: The mobilization-minimization hypothesis. *Psychological Bulletin* 110, pp. 67–85

Vaupel, J.W., J.R. Carey, K. Christensen, T.E. Johnson, A.I. Yashin, N.V. Holm, I.A. Iachine, V. Kanisto, A.A. Khazaeli, P. Liedo, V.D. Longo, Y. Zeng, K.G. Manton, and J.W. Curtsinger (1998), Biodemographic trajectories of longevity. *Science* 280, pp. 855–860

Zeng, Y. and J. W. Vaupel (2002), Functional capacity and self-evaluation of health and life of oldest old in China. *Journal of Social Issues* 58, pp. 733–748

Zeng, Y., J.W. Vaupel, Z. Xiao, C. Zhang, and Y. Liu (2002), Sociodemographic and health profiles of the oldest old in China. *Population and Development Review* 28, pp. 251–273

Chapter 21
An Exploration of the Subjective Well-Being of the Chinese Oldest-Old

Deming Li, Tianyong Chen and Zhenyun Wu

Abstract The research reported in this chapter focuses on subjective well-being, including life satisfaction and affective experience, among the oldest-old, and related factors, as compared to the young-old. A large sample was used from the Chinese Longitudinal Healthy Longevity Survey (CLHLS) conducted in 2002, which included 11,175 oldest-old (aged 80–120) and 4,845 young-old (aged 65–79). The results indicate that the average rating scores of life satisfaction and affective experience in the Chinese oldest-old were well above the neutral level. Self-reported life satisfaction remained constant or slightly increased with age, while scores of affective experience decreased with age. Subjective well-being was influenced by demographic variables (e.g., sex, education, residence, financial source, and living arrangement), and social supports from family members, friends or neighbors, and social workers. We discuss the implications of these results.

Keywords Affective experience, Age, ANOVA, Bivariate analysis, Chinese oldest-old, Demographic variables, Factor analysis, Life satisfaction, Oldest-old, Rating scores, Self-reported life satisfaction, Social support, Subjective well-being, Young-old

21.1 Introduction

Subjective well-being refers to people's cognitive and emotional evaluations of their lives (Andrews and Withey 1976; Diener 2000). That is, a happy man or woman should have relatively high life satisfaction and a pleasant affective experience (more positive and less negative affect). Previous studies have shown that subjective well-being (especially life satisfaction) remains stable throughout adulthood and does not show large negative age differences even at very old ages (e.g., Diener

---

D. Li
Key Laboratory of Mental Health, Institute of Psychology, Chinese Academy of Sciences, Beijing 100101, China
e-mail: lidm@psych.ac.cn

and Suh 1997; Smith et al. 1999). The absence of notable relationships between age and subjective well-being, despite social losses and an increase in physical and cognitive declines with age has been widely regarded as a paradox (Baltes and Baltes 1990). However, some studies have revealed that not all dimensions of subjective well-being demonstrate such age-related stability. For instance, some studies showed that only some dimensions of subjective well-being remained stable, while others declined with age (Cheng 2004; Kunzmann et al. 2000). In the first part of the research we report in this chapter, we examine the relationship between age and different dimensions of subjective well-being using a large sample from the Chinese Longitudinal Healthy Longevity Survey (CLHLS) conducted in 2002; this is the largest longitudinal study of oldest-old people in a developing country (Zeng et al. 2002).

Population aging is a serious issue in China, and it will be more serious with the rapid growth of the elderly population (for more discussion, see chapter 1 in this book by Poston and Zeng). The proportion of the population aged 60 and above increased from just over 6 percent in 1964 to more than 10 percent in 2000, and is expected to reach nearly 28 percent in 2050. Meanwhile, the population of the oldest-old increased more rapidly than that of the young-old. At present, the proportion of the oldest-old aged 80 and above is 10 percent of the total elderly population, but it could reach 20 percent in 2050 (Wu and Du 2002). In recent years, Chinese psychologists have paid more attention to subjective well-being and related factors in elderly people. However, most studies have focused on the young-old; very few have considered the subjective well-being of the Chinese oldest-old. There have been a few studies of the subjective well-being of the oldest-old in western nations (e.g., Kunzmann et al. 2000; Smith et al. 1999), but the results should not be extended to different cultures such as in China. Thus, in the second part of this chapter, we examine the life satisfaction and affective experience of the Chinese oldest-old and the factors related thereto, in comparison with the young-old.

21.2 Measures and Methods

21.2.1 Sample

We use data from the third wave of the CLHLS conducted in 2002, which included 11,175 oldest-old aged 80 and above, as well as 4,845 young-old aged 65–79. This survey collected a wide range of information about the oldest-old in China, such as their demographic data (presented in Table 21.1), social supports, and psychological characteristics. For the first time, this survey provides us with a complete profile of the oldest-old in China. In this chapter, we focus our analyses on oldest-old respondents aged 80 and above, as compared with young-old respondents.

Up to 2,511 respondents (15.7 percent of total participants) could not answer all eight questions on life satisfaction and affective experience. The proportion missing increased with age. That is, 11.5 and 28.3 percent of the oldest-old missed reporting

Table 21.1 Descriptive statistics, CLHLS 2002

Variable	Frequency	Percent
Age		
80–89	4,239	37.9
90–99	3,747	33.5
100+	3,189	28.5
% female	6,787	60.7
Education		
0 years	7,509	67.2
1–6 years	2,729	24.4
6+ years	836	7.5
Missing	101	0.9
Residence		
City	2,653	23.7
Town	2,584	23.1
Rural areas	5,938	53.1
Main financial source		
Own or spouse	1,728	15.5
Children or relatives	8,318	74.4
Government or others	1,129	10.1
Living arrangement		
Household member	8,979	80.3
In elderly home	633	5.7
Alone	1,563	14.0
% living with spouse	1,796	16.1

their life satisfaction and affective experience, respectively, compared with 0.6 and 11.9 percent of the young-old. Further analysis indicated that respondents could not answer questions mainly because of hearing loss, cognitive impairment, and/or other diseases (73.0 percent of missing values), and illiteracy (75.2 percent of the missing values). It is important to note that above two reasons (i.e., physical loss and illiteracy) overlapped each other, and account for 79.1 percent of the missing values.

21.2.2 Measures

Self-reported life satisfaction in the CLHLS was measured by a single question, namely, "how do you feel about your life?" Five levels of responses were given, i.e., very good, good, so so, bad, and very bad. Affective experiences were measured by seven questions, as follows: (1) "Do you always look on the bright side of things?" (2) "Do you like to keep your belongings neat and clean?" (3) "Do you often feel anxious or fearful?" (4) "Do you often feel lonely and isolated?" (5) "Can you make your own decisions concerning your personal affairs?" (6) "Do you feel useless with age?" and (7) "Are you as happy now as when you were when you were younger?" Five levels of responses were given, i.e., always, often, sometimes, seldom, and never. Respondents were asked to choose one of the five options. The scores of negative questions (i.e., question 3, 4, and 6) were reversed to positive scores (i.e., a higher score represents a less negative affect). Therefore, for all of the above eight questions of life satisfaction and affective experience, higher scores represent better subjective well-being.

Social support contained three questions, i.e., (1) "With whom do you chat with most?" (2) "With whom do you confide?" and (3) "Whose help do you ask for first when you are in trouble?" For the three questions, five response options were given, i.e., spouse, family members, friends or neighbors, social workers, and no one. Respondents were required to choose one of the five options.

21.2.3 Method

Our analysis consists of two parts. In the first part, exploratory factor analysis was employed to determine the dimensions of the measures of subjective well-being; then the relationship between age and the different dimensions of subjective well-being were explored. Second, a series of ANOVA analyses were presented to examine the effects of demographic variables and social support on life satisfaction and affective experience among the oldest-old, as compared to the young-old.

21.3 Results

21.3.1 Age and Different Dimensions of Subjective Well-Being

An exploratory factor analysis (with promax rotation to allow for correlated factors) was conducted on the seven affective variables to determine the distinct factors represented in the data. Two factors were identified with eigenvalues greater than one. The variables with the largest loadings on the first factor were questions 3, 4, and 6 (negative questions). Variables with the largest loadings on the second factor were questions 1, 2, 5, and 7 (positive questions). Therefore, it is reasonable to take the first factor as the negative affect of subjective well-being, and the second as the positive affect of subjective well-being.

Mean scores of life satisfaction, positive affect, and negative affect by age groups are displayed in Fig. 21.1. Consistent with previous studies, all means were above 3 (i.e., the neutral level). However, mean scores of life satisfaction remained constant or increased slightly with age, while the mean scores of positive affect and (lack of) negative affect decreased with age. That is, the older respondents reported constant (or a bit more) life satisfaction, less positive affect, and more negative affect, compared to the younger respondents.

The values of seven questions of affective experience in Table 21.2 also show that almost all the proportions of scores "1," "2," and "3" (representing worse and neutral affective experience) among the oldest-old were higher than among the young-old,

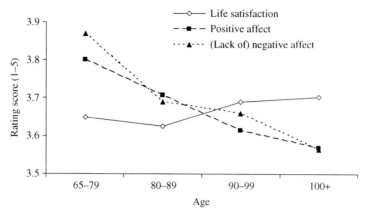

Fig. 21.1 Relation between age and mean scores of life satisfaction, positive affect, and negative affect. The scores of negative questions were transferred to positive scores (i.e., higher scores represent less negative affects)

while the proportions of score "4" and "5" (representing better affective experience) among the oldest-old were lower than among the young-old. A chi square test indicated that the differences between the oldest-old and the young-old on the seven questions of affective experience, as well as on the question of life satisfaction, were all significant. Because all the seven affect variables had a similar relationship with age, they were averaged to obtain one composite score in the following analysis, which represents affective experience.

Table 21.2 Comparison of life satisfaction and affective experience between oldest-old and young-old (%)

Variables	Oldest-old					Young-old					χ^2
	1	2	3	4	5	1	2	3	4	5	
Life satisfaction	1.1	6.3	31.3	47.3	14.0	0.7	5.6	35.4	44.0	14.2	31.36***
Affective experience											
Look on the bright side	0.4	4.7	18.2	66.4	10.2	0.5	4.2	17.1	66.6	11.6	9.88*
Keep neat and clean	0.1	1.9	24.1	60.6	13.3	0.1	1.0	22.2	61.8	14.9	27.24***
Feel anxious or fearful	1.2	4.2	23.7	35.7	35.2	1.0	3.4	19.9	36.2	39.5	42.94***
Feel lonely and isolated	2.3	7.1	27.2	30.2	33.2	1.8	4.5	20.1	31.2	42.4	179.34***
Make own decisions	5.8	12.9	25.8	25.4	30.0	3.6	7.6	17.9	26.2	44.8	394.94***
Feel useless with age	10.2	18.5	38.1	18.1	15.1	8.1	12.6	35.5	21.8	22.0	200.36***
Happy as younger	8.8	25.3	25.1	12.3	28.5	7.8	21.9	24.1	13.2	33.1	44.77***

The scores of negative questions were transferred to positive scores (i.e., higher scores represent less negative affect); the percentage for each variable should add up to 100, but some categories may not due to rounding
*$p < 0.05$, ***$p < 0.001$

21.3.2 Subjective Well-Being and Demographic Factors

Table 21.3 shows that a relatively higher proportion of the oldest-old were females and illiterate (49.8 and 48.4 percent, respectively) compared to the young-old. Moreover, the oldest-old also had a lower proportion with pensions and living with a spouse (47.4 and 59.9 percent, respectively) compared to the young-old. Meanwhile, slightly more oldest-old lived alone and in elderly homes (or nursing home) than the young-old (12.3 and 2.2 percent, respectively).

The effects of demographic variables on life satisfaction and affective experience were analyzed using ANOVA. The results were as follows. Life satisfaction was similar for oldest-old males and females, but scores for males' affective experience were significantly higher than for females. The life satisfaction and affective experience scores of the well-educated oldest-old were significantly higher than for the poorly educated oldest-old; life satisfaction and affective experience scores for illiterate individuals were the lowest. Life satisfaction and affective experience scores for the oldest-old were also different based on residence, i.e., they were highest in cities, lowest in rural areas and in between in towns. The oldest-old who received their main financial support from a spouse or who provided their own (main) financial support had the highest life satisfaction and affective experience scores.

Table 21.3 The effects of demographic variables on life satisfaction and affective experience in the oldest-old

Variables	Life satisfaction			Affective experience		
	M	SD	F	M	SD	F
Sex			1.91			151.04***
Male	3.68	0.82		3.75	0.58	
Female	3.66	0.84		3.59	0.59	
Education			34.19***			130.11***
0 years	3.64	0.84		3.60	0.58	
1–6 years	3.67	0.82		3.73	0.58	
6+ years	3.90	0.83		3.89	0.58	
Residence			61.67***			190.88***
City	3.82	0.85		3.86	0.59	
Town	3.69	0.83		3.67	0.58	
Rural areas	3.59	0.82		3.56	0.57	
Main financial source			79.63***			180.87***
Own or spouse	3.85	0.82		3.91	0.57	
Children or relatives	3.66	0.81		3.60	0.58	
Government or others	3.43	0.93		3.64	0.59	
Living arrangement			138.40***			37.81***
Household member	3.71	0.81		3.67	0.58	
In elderly home	3.85	0.77		3.75	0.59	
Alone	3.34	0.91		3.53	0.61	
Lived with spouse			1.59			139.66***
Yes	3.64	0.84		3.81	0.57	
No	3.67	0.83		3.62	0.59	

*** $p < 0.001$

Life satisfaction scores were in the middle for the oldest-old who received their main finances from children or relatives, but their affective experience scores were the lowest. Life satisfaction and affective experience scores also varied based on the living arrangements of the oldest-old. The lowest scores were for those who lived alone, the highest for those who lived in an elderly home (or nursing home), and they were in between for those who lived with a household member. Finally, life satisfaction and affective experience scores were higher for the oldest-old who lived with as opposed to without a spouse. The results, taken together, show that demographic factors have a significant effect on the life satisfaction and affective experiences of the oldest-old.

The effects of demographic factors on life satisfaction and affective experience among the young-old were similar to those among the oldest-old. However, the young-old who lived with household members had the highest life satisfaction and affective experience scores (3.69 and 3.85, respectively).

21.3.3 Subjective Well-Being and Social Support

As shown in Table 21.4, well-being was the highest among the oldest-old who asked household members for help. However, it was higher for the young-old who asked a spouse for help (44.2, 50.2, and 32.2 percent for questions 1, 2, and 3, respectively). In addition, the proportion who asked a social worker for help was a little higher for the oldest-old than for the young-old.

Table 21.4 The effects of social support on life satisfaction and affective experience in the oldest-old

Variables	Percent	Life satisfaction			Affective experience		
		M	SD	F	M	SD	F
Chat with whom				46.99***			52.05***
Spouse	11.6	3.65	0.84		3.82	0.57	
Household members	50.9	3.75	0.78		3.62	0.57	
Friends and neighbors	25.9	3.60	0.87		3.70	0.59	
Social workers	1.8	3.90	0.73		3.72	0.57	
No one	9.8	3.36	0.93		3.44	0.64	
Confide with whom				71.49***			42.55***
Spouse	12.5	3.64	0.84		3.83	0.57	
Household members	68.5	3.73	0.79		3.64	0.57	
Friends and neighbors	8.4	3.48	0.90		3.63	0.61	
Social workers	2.2	3.91	0.75		3.72	0.56	
No one	8.3	3.26	0.98		3.48	0.68	
Ask for help from whom				76.78***			29.31***
Spouse	6.7	3.60	0.86		3.85	0.58	
Household members	84.0	3.70	0.80		3.65	0.58	
Friends and neighbors	2.1	3.22	1.02		3.53	0.60	
Social workers	4.2	3.82	0.83		3.75	0.60	
No one	3.0	2.96	1.03		3.45	0.71	

*** $p < 0.001$

ANOVA analysis showed that life satisfaction and affective experience scores were higher for the oldest-old who often chatted with someone, confided in someone, and got help from someone when they were in trouble, while the scores were lower for the oldest-old who had no one to chat with, to confide in, or to get help from. It is important to note that the life satisfaction scores were the highest for the oldest-old who obtained support from social workers, while the scores for affective experience were the highest for the oldest-old who obtained support from a spouse. Ten percent of the oldest-old had no one with whom to chat or confide in, and 3 percent could not get help from anyone. These results as a whole suggest that the life satisfaction and affective experience among the oldest-old were affected significantly by social supports. Social workers play an important role in helping the oldest-old improve their life satisfaction and affective experience, in addition to family members, relations, friends, and neighbors.

The effects of social supports on life satisfaction and affective experience among the young-old were similar to those among the oldest-old. We note, however, that the scores of life satisfaction and affective experience were the highest among the young-old who got the support from a spouse.

21.4 Discussion

Data used in this article are from the third wave of the CLHLS conducted in 2002, the first year that young-old people aged 65–79 were included along with the oldest-old aged 80 and older. This expansion of the CLHLS made it possible for us to focus in this chapter on life satisfaction and affective experience and related factors among the oldest-old compared to the young-old. The large sample of 16,020 elderly people was collected from about half of the counties and cities in 22 of the 31 provinces in China. Eight questions represent a global judgment of life satisfaction and affective experience (i.e., positive and negative affect) which we used to explore life satisfaction and affective experience.

This chapter used bivariate analyses and is thus mainly descriptive. The relationships between variables might well be changed when the effects of other factors are introduced, as in a future multivariate analysis.

The results indicated that the average life satisfaction and affective experience scores among most of China's oldest-old were well above the neutral level. Life satisfaction remained constant or mildly increased with age, while affective experience apparently decreased with age (i.e., less positive affect and more negative affect). These results are very different from those found in western nations. For instance, Kunzmann et al. (2000) found that age was negatively related to positive affect and unrelated to negative affect. However, our previous studies have produced similar results to the study reported in this chapter. For instance, Li et al. (2006b) found that most elderly (2,225 respondents, aged 60–99) in seven Chinese cities, had a relatively high life satisfaction. Furthermore, material-life satisfaction scores tended to increase with age, and with increasing age mood trended towards quiet and contentment. On the other hand, spirit-life satisfaction scores decreased with age, and the influence of children also faded with age. These data, together with the results of the research reported here, suggest that the oldest-old in China are more satisfied with their lives (they have a higher material-life satisfaction) than the young-old, and have less positive affect and more negative affect (low spirit-life satisfaction).

In this research, we also analyzed possible factors which might explain the differences in life satisfaction and affective experience of the oldest-old. We found that life satisfaction and affective experience were influenced significantly by demographic variables. The oldest-old people with a relatively high life satisfaction and a pleasant affective experience were well-educated, lived in a city, were financially self sufficient (mainly supported by one's own pension or that of a spouse), lived with a spouse, and resided in an elderly home. However, the oldest-old who were illiterate, lived in rural areas, and were supported by the government or in some other way, had a relatively low life satisfaction and unpleasant affective experience. These results are consistent with our previous studies (Li et al. 2005, 2006a). It important to note that the oldest-old who lived in an elderly home, as well as the young-old who lived with household members, have the highest life satisfaction and affective experience. This result suggests that living in a nursing home (elderly home) rather than living alone might be a better choice for some oldest-old people. We also found that elderly people who lived in rural areas, and whose main source of support was from the government had a lower life satisfaction and affective experience. This indicates that some oldest-old people are unable to get sufficient financial support from the government or from other sources. Therefore, the government and society should pay more attention to the well-being of the elderly, especially the oldest-old who lack financial independence, particularly those who live in rural areas.

Finally, our results indicated that life satisfaction and affective experience among the oldest-old were also affected significantly by social supports. It should be emphasized that social workers played a very positive role in helping the oldest-old improve their life satisfaction and affective experience, in addition to family members, relatives, friends, and neighbors. Although social workers are presently not that popular in China, it is the duty of the government and society to set up a system of social work to help the elderly, especially the oldest-old, to achieve well-being and to be able to age successfully.

References

Andrews, F.M. and S.B. Withey (1976), *Social indicators of well-being: Americans' perception of life quality*. New York: Plenum.

Baltes, P.B. and M.M. Baltes (1990), Psychological perspectives on successful aging: The model of selective optimization with compensation. In: P.B. Baltes and M.M. Baltes (eds.): *Successful aging: Perspectives from the behavioral sciences*. Vol. 1. New York: Cambridge University Press, pp. 1–34

Cheng, S.-T. (2004), Age and subjective well-being revisited: A discrepancy perspective. *Psychology and Aging* 19 (3), pp. 409–415

Diener, E. (2000), Subjective well-being: The science of happiness and a proposal for a national index. *American Psychologist* 55 (1), pp. 34–43

Diener, E. and M.E. Suh (1997), Subjective well-being and age: An international analysis. In: K.W. Schaie and M.P. Lawton (eds.): *Annual review of gerontology and geriatrics*. Vol. 17. New York: Springer, pp. 304–324

Kunzmann, U., T.D. Little, and J. Smith (2000), Is age-related stability of subjective well-being a paradox? Cross-sectional and longitudinal evidence from the Berlin aging study. *Psychology and Aging* 15 (3), pp. 511–526

Li, D., T. Chen, Z. Wu, and G. Li (2005), Essential elements of healthy aging and their related factors. *Journal of Chinese Gerontology* 25 (9), pp. 1004–1006 (in Chinese)

Li, D., T. Chen, and G. Li (2006a), Life satisfaction of the elderly in Beijing and its related factors. *Chinese Journal of Clinical Psychology* 14 (1), pp. 58–60 (in Chinese)

Li, D., T. Chen, Z. Wu, J. Xiao, A. Fei, Y. Wang, L. Zhou, and F. Zhang (2006b), Life and mental status and age differences among urban elderly Chinese. *Journal of Chinese Gerontology* 26 (10), pp. 1314–1316 (in Chinese)

Smith, J., W. Fleeson, B. Geiselmann, R.A. Settersten, and U. Kunzmann (1999), Sources of well-being in very old age. In P.B. Baltes and K.U. Mayer (eds.): *The Berlin aging study: Aging from 70 to 100*. New York: Cambridge University Press, pp. 450–471

Wu, C. and P. Du (2002), Population aging in China: Present situation, trend prediction, and coping strategies. In: K. Chen (ed.): *Aging in China: problems and solutions*. Beijing: Chinese Union Medical University Press, pp. 1–12 (in Chinese).

Zeng, Y., J.W. Vaupel, Z. Xiao, and C. Zhang. (2002), Sociodemographic and health profiles of the oldest old in China. *Population and Development Review* 28 (2), pp. 251–273

Chapter 22
Social Support and Self-Reported Quality of Life: China's Oldest Old

Min Zhou and Zhenchao Qian

Abstract In this chapter, we explore how social support affects self-reported quality of life for the elderly (aged 80 and above) in China. We use data from the first wave of the Chinese Longitudinal Healthy Longevity Survey (CLHLS) conducted in 1998. Our results show clearly that all sources of social support are beneficial to quality of life. One's living arrangement, as one measure of social support, plays an important role. The elderly who live alone report the lowest quality of life. Paradoxically, the elderly who live in nursing homes report a higher quality of life than those living with children. We examined this paradox by comparing the differences in characteristics of those living in nursing homes and those living with children. Finally, we discuss the implications of these results for societies that are experiencing rapid aging as a result of sharp fertility declines.

Keywords Chinese oldest-old, Filial piety, Living arrangement, Nursing homes, Objective activities, Oldest-old, Perceived support, Rapid aging, Social support, Social structure, Self-reported quality of life, Type of social support

22.1 Introduction

Social support has long been known to affect an individual's emotional and physical health and general well-being (Dean et al. 1990). Social support can be a buffer against stress, protect people against developing illnesses, provide emotional support, and increase the life span (Ross and Mirowsky 2002). The elderly tend to have a better quality of life if they receive regular care and support from family members as well as from friends and peers (Matt and Dean 1993). Yet, social support has multiple dimensions; different sources of social support may have varying effects on the quality of life for the elderly.

M. Zhou
Department of Sociology, Ohio State University, 300 Bricker Hall, 190 N. Oval Mall, Columbus, OH 43210, USA
e-mail: zhou.144@sociology.osu.edu

Social support for the elderly involves both objective and subjective activities (Turner and Marino 1994). Elderly people living with children or other family members, for example, should have regular interactions with family members and receive physical and emotional support. These objective activities are expected to increase their quality of life. Meanwhile, some elderly individuals live independently, but the simple belief that family members would take care of them when they are ill is likely to shape their positive attitudes. In the research reported in this chapter, we examine both dimensions of social support on perceived quality of life for the elderly. Specifically, we pay special attention to the oldest old (aged 80 and above) and compare self-reported quality of life for those living with children and those living in elderly (nursing) homes[1]. Most research on social support and quality of life has focused on developed countries. Here we use data from the Chinese Longitudinal Healthy Longevity Survey (CLHLS) to examine this relationship in a different social setting. We address three main questions: (1) What is the self-reported quality of life like among China's oldest old? (2) How do objective and subjective dimensions of social support affect self-reported quality of life? And (3) Do elderly people who receive social support from families have a higher quality of life than those who receive social support from peers?

22.2 China's Oldest Old

The elderly population in China has been growing rapidly as a result of a sharp decline in fertility (Ogawa 1988; Zeng and George 2000; see also the discussion in Chap. 1 of this book by Poston and Zeng). The proportion of the population aged 60 or over increased from just over 7 percent in 1953 to more than 10 percent in 2000, and is projected to reach 27 percent in 2050 (Riley 2004). Whether the well-being of this rapidly increasing elderly population can be improved or even maintained has become a primary concern in Chinese society. In contrast to the situation in Western countries where retirees can rely in part on social security, the primary caretaker of the elderly in China is the family, and not the society. The lack of governmental support and an underdeveloped social security system mean that Chinese elderly must rely on their children's financial and social support to avoid economic hardships. In other words, the well-being of the Chinese elderly depends on whether they have children and how willing and able the children are to support their aging parents.

In China, filial piety is paramount. In a survey based in the city of Baoding, Whyte (1997) found that sentiments of filial obligation remain robustly intact among young people. They feel ashamed and are looked down on if their aging parents live alone or in nursing homes. Although stress and tension are common in extended families, the elderly parents often live with their children to conform to the social norms. The increase in life expectancy in recent decades means that families must live in such

---

[1] In the research reported here, most of the elderly who live in nursing homes do not have children.

arrangements for a very long time. Adult children, burdened by such arrangement especially when the aging parents become more dependent, may seek alternatives (Zeng et al. 2001). In reality, it is increasingly difficult to live in extended family settings because of the sharp decline in fertility. In Chinese cities where the one child per family policy has been in place for a long time, it is almost impossible for married couples (only children of their parents) to care for both sets of elderly parents. More elderly parents must choose to live alone or live in nursing homes.

22.3 Previous Research

Historically, and still today in many developing countries, an important goal of human reproduction has been old age support. Parents invest time and money to bear and raise children with the expectation that the children will, in turn, care for them when they are old. Whether the elderly later receive care tends to affect their quality of life. Scholars label this behavior as "reciprocity," that is, the normative obligation to assist people who have provided help to them (Finch 1989; Gouldner 1960; Horwitz et al. 1996). The elderly with no living children, however, cannot have such reciprocity. Thus, the elderly with living children should exhibit high levels of life quality as the investment to bear and rear children is paid back. We hypothesize that elderly with children are more likely to report a higher quality of life than childless elderly.

22.3.1 Social Support and Living Arrangement

Social support from children is essential to the well-being of elderly parents. Although the elderly in developed countries mostly live with their spouses or, if not married, live alone, it is customary for elderly parents in developing countries to reside with at least one child (Kramarow 1995). A study in Baoding, China finds that intergenerational support among families is the major source of old age security (Sun 2002). Living with children, an objective measurement of well-being, is expected to increase the quality of life (Hochschild 1973). However, a decline in fertility that has occurred in many parts of the world suggest that parents may no longer expect to rely fully on their children's support at older ages (Caldwell 1982). This is especially so in developed countries. Elderly parents may become increasingly unwilling to live in extended families. The elderly with sufficient economic means may in fact prefer to live alone so they can have their freedom and privacy and enjoy a better quality of life (Kramarow 1995).

Fertility decline has become widespread in developing countries, but living standards for the elderly and cultural norms about elderly living arrangements have not changed significantly. Living with children remains the most popular living arrangement. In such an arrangement, parents receive goods and services that they otherwise might have to purchase, while adult children may benefit from childcare and other household services (DaVanzo and Chan 1994). In addition, filial piety is the norm in many developing countries. As a result, most elderly parents in developing countries live with their children (Logan et al. 1998). It is unclear, however, whether changes in living arrangements are likely to occur in the future. In other words, it remains to be seen whether patterns will become closer to those observed in the United States and other developed countries where elderly parents with financial means are willing and increasingly likely to live on their own (Kramarow 1995). Why would elderly parents opt out to live on their own? One likely answer is that the quality of life may not be optimal among those living with children. Indeed, familial support does not always correspond with increasing quality of life among the elderly (Arling 1976; Wood and Robertson 1978). Some argue that voluntary interpersonal attachments such as friendships have a stronger positive association with self-reported quality of life compared to involuntary social ties, such as kinship bonds (Ellison 1990; Wood and Robertson 1978). In other words, elderly who are stuck with their children at home may not be happy with such an arrangement because family relationships can be potentially complicated and intergenerational conflicts may surface. This can be true as well in China under certain circumstances. A recent study shows that living arrangements among China's oldest old may change over time if the elderly suffer from health deteriorations or have lost their spouses (Zimmer 2005). A primary reason for such changes is that their children feel a tremendous burden and stress to deal with such difficult situations. Nursing homes or elderly homes, once thought to be a place for those without children, may become a viable option.

Although living in nursing homes is generally viewed negatively, this arrangement may strengthen family ties and renew closeness between the elderly parent and the adult child. If elderly parents and children can afford the cost, living in a nursing home may alleviate some of the strains and pressures caused by living together (Smith and Bengtson 1979). What matters most in this scenario is whether children pay regular visits to the elderly. Elderly parents living in nursing homes with frequent visits from children and grandchildren may report an increase in their quality of life because family visits raise their status and generate respect among their peers. Meanwhile, they also engage in frequent interactions with peers and friends to share common concerns and problems —another positive dimension that is shown to improve their quality of life (Matt and Dean 1993). Thus, living in nursing or elderly homes may lead to an increased sense of belonging and self-worth, which would promote positive attitudes toward life. We hypothesize that the elderly living in nursing or elderly homes may report a better quality of life than those residing with children when all other relevant factors are taken into account. In a society in which it is culturally unacceptable for elderly people with living children to live alone, living alone without strong financial stability may result in the lowest quality of life.

22.3.2 Perceived Support and Resources

In addition to objective measures of social support (family and peers), subjective appraisal of support is also an important dimension of social support (Turner and Marino 1994). Wethington and Kessler (1986) have presented evidence that perceived support is more important than received support in buffering the effect of stressful events. Knowing that support is available will reduce negative emotional and behavioral responses to stressful events and will promote health (Veiel and Baumann 1992). For the elderly, the thought of being cared for when sick can be an important predictor of life quality. But there is a relationship between receiving and perceiving support (Turner and Marino 1994). The elderly who live with families and peers or receive frequent visits from their children may form the strong perception that they will continue to receive support and care, which leads to better life quality.

22.3.3 Social Support and Social Structure

In many developing countries, such as China, there are pervasive urban and rural differences, and these usually reflect different access to educational and career opportunities, health care, financial support, and housing (Lin 1995). Urban residents usually have access to much better health care and public pensions than rural residents. However, fertility rates are lower in urban than in rural areas. This may increase the likelihood of rural elderly residents living with one of their children compared to urban elderly residents. Therefore, we expect that urban elderly may be more likely to live in nursing homes than their rural counterparts. The rural elderly living in nursing homes may be more selective and consist mostly of those with no living children. For these reasons it is important to control for place of residence.

Prior educational and occupational achievements measure a person's socioeconomic status, which is a direct indicator of how well an elderly person can support him or herself and, thus, how much he or she requires financial support from the children. With other variables such as health taken into account, the relationship between education and quality of life is indeed very strong. Education improves well-being because it provides access to paid work and economic resources that increase sense of control, as well as providing access to stable social relationships (Ross and Willigen 1997). In addition, well-educated people are more likely to have stable social relationships with their marriage partner and their children, both of which are positively associated with quality of life (Ross and Willigen 1997).

Occupation is expected to play a similar role. Those in professional jobs before retirement are likely to have a better quality of life compared to those in non-professional jobs.

22.3.4 Health and Quality of Life

Physical conditions may influence both social support and quality of life. Elderly with mental or physical illnesses are most likely to need assistance and support. These needs may increase the burden on family members and amplify intergenerational conflicts, which can lead to a lower quality of life. Moreover, they may lack access to daily activities such as entertainment and social interactions with friends that more active elderly people have, which decreases subjective well-being (Mutran and Reitzes 1984). Therefore, healthy elderly are expected to have a higher quality of life than their less healthy counterparts.

22.4 The Current Study

To summarize, our goal is to examine how different dimensions of social support affect the quality of life of China's oldest old. We argue that one's living arrangement is an important form of social support. Despite the social norm that adult children are responsible for caring for elderly parents, a growing percentage of the Chinese elderly live in nursing homes. We hypothesize that the elderly residing in nursing homes and reporting frequent visits from children report a better quality of life than those who live elsewhere, or who are not visited frequently by their children. We also hypothesize that elderly living alone should report the lowest quality of life.

In addition, perceived social support is an important factor predicting quality of life. We hypothesize that those with perceived support are more likely to report a better quality of life than those without such support. We recognize the importance of urban/rural residence in affecting the well-being of the Chinese elderly. Urban residents (who often have some level of educational attainment and well-paid jobs before retirement) are more likely to have financial resources and have been more "successful" in raising financially stable children compared to rural residents. Thus, we hypothesize that urban elderly will report a better quality of life than rural elderly.

The relationship between social support and quality of life has been well studied in developed countries. But to our knowledge, few studies have examined this relationship for the subpopulation of the oldest old in China. The research we report in this chapter is an attempt to fill this void. It will provide some implications for policy makers in their addressing of aging issues in a society that is aging at a fast pace.

22.5 Data and Methods

We use data from the first wave of the Chinese Longitudinal Healthy Longevity Survey (CLHLS) conducted in 1998. The survey was administered to 9,093 respondents from 631 randomly selected counties and cities from 22 of the 31 provinces in China. The questionnaire covers many aspects concerning the oldest-old Chinese, such as their socioeconomic status, living arrangements, family structure and support, daily activities, and health status. We restrict our analyses to respondents between the ages of 80 and 110. The final sample contains 7,871 elderly Chinese respondents.

The CLHLS over-sampled extremely old persons such as centenarians and nonagenarians, and over-sampled oldest old males, because there are fewer persons at the more advanced ages, and fewer males than females (Zeng et al. 2001). Therefore, appropriate weights will be used in the analyses.

22.5.1 Measuring Quality of Life

"Quality of life" is a term used loosely to indicate general well-being (Haug and Folmar 1986). Indicators of a good quality of life include health, sufficient funds, absence of psychological distress, and availability of supportive family and friends (Dowd and Bengtson 1978). One important aspect of quality of life is subjective attitudes and feelings (Schuessler and Fisher 1985). Subjective quality of life measures a person's sense of well-being, life satisfaction and happiness (Dalkey and Rourke 1973).

Self-reported quality of life is our dependent variable, and is measured by responses to one question, "How would you rate your quality of life?" There are six responses: very good, good, so so, bad, very bad, and not able to answer. The oldest old sample is very selective; they live longer than many of their peers (Christensen and Vaupel 1996; Zeng et al. 2001). It is not a surprise that many report a good quality of life. Only 3 percent of the respondents rate their quality of life as "bad" or "very bad," so we classify "so so," "bad," and "very bad" into one category "not good." Likewise, "very good" and "good" are grouped into one category "good." As shown in Table 22.1, 72 percent report their quality of life to be good while 28 percent report their quality of life to be not good.

22.5.2 Measuring Social Support

We include two dimensions of social support – objective behaviors and subjective perceptions. Objective behavior is measured by living arrangement, children's visits, and financial support. Living arrangement is divided into four categories: living with children and with or without spouse (67 percent), living with spouse only (13 percent), living in a nursing home (7 percent), and living alone (13 percent). We hypothesize that the elderly living with children are more likely to receive support from their children. However, intergenerational conflict and lack of privacy may lower the quality of life for the elderly.[2] Based on the prediction that peer networks will have a stronger effect on quality of life than family members, the elderly living in nursing homes[3] are likely to have a better quality of life than those living with children. We also include a variable that measures whether non-resident children

---

[2] Unfortunately, there are no data in the CLHLS allowing us to measure *intergenerational conflict*, *lack of privacy*, and *care for grandchildren*.

[3] Unlike the various forms of elderly homes and nursing homes in the US, the nursing homes in China are mainly run by the government and the quality of these nursing homes may not be as good as private nursing homes in developed countries.

Table 22.1 Descriptive statistics, weighted, CLHLS 1998

Variable	Percent
Total	7871
% reporting good quality of life	72.3
Demographic Characteristics:	
Age	
80–89	90.7
90–99	7.9
100–110	1.4
% Female	63.0
Education	
No schooling	62.9
1–6 Years of education	26.7
6+ Yeas of education	10.4
Occupation	
Professional or governmental	9.5
Worker or farmer	71.5
Housework or others	19.1
% Urban	37.2
No children ever born	5.0
At least one child alive	80.9
Physical conditions (ADL)	
Severe disability	5.9
Mild disability	12.2
Active	81.9
Social Support	
Living arrangement: (reference = alone)	
Child only	67.1
Spouse only	12.5
Nursing home	6.8
Alone	16.6
Financial support	
Own or spouse only	11.3
Children and relatives only	58.7
Government and others only	5.3
Own, spouse, and kins	16.7
All sources	8.0
Frequent visits	75.0
Caretaker when sick: (reference = nobody)	
Family members	89.8
Non-family member	8.2
Nobody	2.0

The percentage for each variable (more than two categories) should add up to 100, but may not due to rounding errors.

pay frequent visits[4] to their elderly parents. Three quarters of the elderly do receive visits from their non-resident children.

---

[4] This is measured by a question asking whether or not children pay frequent visits to the respondents. The answer is subjective and defined differently according to different respondents.

Furthermore, we include primary and secondary sources of financial support. We create a new variable based on these two sources that collapses respondents into mutually exclusive categories: (1) financial support only from self or spouse[5] (11 percent), (2) only from children or relatives (59 percent), (3) only from government or other sources (5 percent), (4) from self, spouse, children, or relatives (17 percent), and (5) from all sources, namely, self, spouse, children, relatives, government, and others (8 percent). The elderly who can support themselves are most likely to have a higher quality of life than those who depend on other sources of support.

The subjective perceptions of support are measured by whether the respondents think they will receive care if they get sick. This variable consists of support from family members, non-family members (friends and social workers), and no one. For the respondents in our data, 90 percent expect care from family members, 8 percent from non-family members, and 2 percent from no one. We expect those who would receive support from family members to have the highest quality of life. Respondents who think that nobody would care for them should have the lowest quality of life.

22.5.3 *Measuring Social Structure Variables*

Social structure variables include urban/rural residence, educational attainment, and occupation at retirement.[6] As discussed, urban/rural differences are expected to have a very strong effect. Educational attainment of the elderly is another important variable. However, most of the respondents in our sample obtained their education in the 1920s when the majority of Chinese received little formal education. Thus, education levels are grouped into three categories: no education (63 percent), 1–6 years of schooling (27 percent), and more than 6 years of schooling (10 percent). Primary occupation before age 60 is categorized into three groups: (1) professional and governmental (10 percent), (2) worker or farmer (71 percent), and (3) housework and others (19 percent).

22.5.4 *Measuring Health Status*

Physical conditions may influence both social support and quality of life. We hypothesize that the oldest-old Chinese with poor health will negatively assess their quality of life. The survey has questions on self-rated health status, activities of daily living (ADL), and interviewer-rated health status. Because of the strong association of the three variables, we rely on activities of daily living (ADL) as a measure of

---

[5] Retirement pensions are included in the first category (from self or spouse only) instead of the third category (from government or other sources only).

[6] Income is not measured owing to limitations in the data. However, education, occupation and region are good proxies since they are highly correlated with income.

an elderly person's health. We follow the work of Zeng and his colleagues (2001) and employ the Sullivan method of calculating active life expectancy. The ADL functional statuses include eating, dressing, transferring, using the toilet, bathing, and continence. If none of the six ADL activities are impaired, the person is coded "active" for ADL; if one or two activities are impaired, the person is coded "with mild disability"; if three or more activities are impaired, the person is coded "with severe disability." Cronbach's alpha for the ADL index shows high reliability (0.86). The sample is comprised of 82 percent of the elderly who are classified as active, 12 percent with a mild disability, and 6 percent with a severe disability.

We control for respondents' age and gender. Age is classified into three categories: ages 80–89, ages 90–99, and ages 100–110. As already noted, because the survey over-sampled the oldest individuals, it is necessary to incorporate sample weights. After weights are taken into account, our sample includes about 91 percent of the elderly aged 80–89, 8 percent of the elderly aged 90–99, and 1 percent of the elderly older than 100. Our weighted sample includes 63 percent of female elderly. We also control for whether the elderly have had children and whether any of the children are still alive.

22.5.5 Statistical Analysis

As described above, quality of life is treated as a dichotomous variable: good or very good (coded as 1), and so so, bad, or very bad (coded as 0).[7] Logistic regression is employed to predict the effects of the selected variables on the odds of the elderly reporting a good quality of life. We first introduce different measures of social support in the models (living arrangement, visits by children, perceived social support when sick) and then include other control variables (sources of financial support, ADL status, residence, education, occupation, age, sex, and whether the elderly person has children still alive). Our goal is to compare changes for different sources of social support after other control variables are taken into account.

22.6 Results

We begin our analyses by presenting the bivariate relationships between the independent variables and the dependent variable. Figure 22.1 presents percentages of the elderly reporting good quality of life by living arrangement. Statistical results show the significance of the differences among living arrangements. Eighty-seven percent of the elderly living in nursing homes report their quality of life to be good or very good. Those living with children or living only with their spouses are similar; close to three quarters report their quality of life to be good or very good. In sharp

---

[7] Preliminary analyses revealed that changing the coding scheme of this variable did not have much effect on the results. As such, we chose to code quality of life as a dichotomous variable.

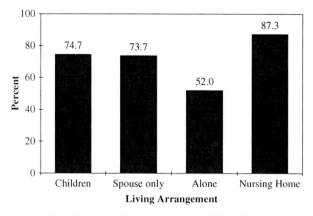

Fig. 22.1 Percentage of the elderly reporting good quality of life by living arrangement

contrast are the results for those living alone; approximately half of them report a good quality of life.

In addition, 73 percent of the elderly who are visited by children report good quality of life compared to 70 percent of the elderly who are not visited by children. The difference between the elderly with visits by non-resident children and those without is small, but is statistically significant. This result, however, is confounded by the number of elderly who do not have children and who live in nursing homes and who report a good quality of life. We will control for these factors in the multivariate analyses.

Perceived support is another important dimension of social support. It is measured by assessing who would take care of the respondent if he or she became sick. Seventy-seven percent of the elderly who perceive support from non-family members report good quality of life, while 73 percent of the elderly who perceive support from family members report good quality of life. Most of the elderly who perceive care from non-family members live in nursing homes. This unique group reports the highest quality of life. The elderly who perceive support from no one have the lowest quality of life: less than one third report their quality of life to be good.

The bivariate relationships between quality of life and the rest of the variables included in the multiple regression models are presented in Table 22.2. Sources of financial support are strongly associated with quality of life. Interestingly, the elderly who relied on their own financial means for support enjoyed the highest quality of life (78 percent). In contrast, elderly who relied on the government or other sources for support reported the lowest quality of life (69 percent). Elderly living in urban areas reported a better quality of life, seven percentage points higher than their rural counterparts (77 and 70 percent, respectively). The same difference exists between those with and without completed elementary education. Eighty-two percent of the elderly who had professional or government jobs reported a good quality of life compared to those in other occupations.

Table 22.2 Percentage of the elderly reporting good quality of life by selected variables, weighted

Select variables	% Reporting good quality of life
*Living arrangement****	
Child	74.6
Spouse Only	73.7
Alone	52.0
Nursing Home	87.3
*Frequent visits***	
No	70.2
Yes	73.0
*Perceived support when sick****	
Family Members	72.8
Nonfamily Members	77.2
Nobody	31.3
*Financial support****	
Own and Spouse only	77.6
Child and Relatives only	71.3
Government and Others only	68.8
Own, Spouse and Kins	73.6
All Sources	71.5
*ADL status****	
Active	72.9
Mild Disability	72.7
Severe Disability	63.1
*Place of residence****	
Rural	69.7
Urban	76.8
*Education****	
No Education	71.0
1–6 years	73.0
6+ years	78.5
*Occupation at retirement****	
Professional or governmental	82.4
Industrial or agricultural	70.6
Housework or others	73.8
*Age**	
80–89 years	72.0
90–99 years	75.5
100–110 years	77.1
*Having child****	
No	66.8
Yes	72.6
Child alive	
No	71.2
Yes	72.6
Sex	
Male	72.3
Female	72.3
Total	7871

*** $p < .01$; ** $p < .05$; * $p < .10$ (two-tailed).

As expected, physical condition is positively associated with quality of life. The elderly who have an active ADL status are no different from those who are mildly disabled (73 percent), but the severely disabled report a much lower quality of life on average (63 percent). In addition, whether they have ever had children affects quality of life. The elderly who have never had children report a lower quality of life compared to those who have, or have had children (67 and 73 percent, respectively).

22.6.1 Predicting Quality of Life

We now turn to logistic regression models to further explore the impact of social support on quality of life when other variables are taken into account. We focus on how different types of social support affect quality of life – objective activities (measured by living arrangement), support from children (whether non-resident children pay regular visits), and perceived support (who would take care of them if they became sick). Table 22.3 presents the results from the logistic regression equations. The first model includes living arrangement. We control for whether the

Table 22.3 Odds from logistic regression predicting effects of social support, health, and demographic characteristics on quality of life, 1998 ($n = 7,871$)

Independent variables	Model 1	Model 2	Model 3	Model 4	Model 5
Living arrangement (reference = children)					
Spouse only	0.76***	0.75***	0.76***	0.66***	0.70***
	(0.07)	(0.07)	(0.07)	(0.06)	(0.07)
Nursing home	2.00***	2.01***	2.25***	2.36***	2.41***
	(0.30)	(0.30)	(0.45)	(0.50)	(0.52)
Alone	0.40***	0.40***	0.44***	0.42***	0.44***
	(0.03)	(0.03)	(0.04)	(0.04)	(0.04)
No children ever born	0.80	0.80	0.86	0.83	0.87
	(0.12)	(0.11)	(0.13)	(0.12)	(0.13)
At least one child alive (Yes)	1.30***	0.97	0.97	0.94	0.90
	(0.09)	(0.11)	(0.11)	(0.11)	(0.10)
Regular visit of children (Yes)		1.40***	1.35***	1.38***	1.48***
		(0.14)	(0.14)	(0.14)	(0.15)
Caretaker when sick (reference = nobody)					
Family members			3.20***	3.52***	3.48***
			(0.75)	(0.84)	(0.84)
Nonfamily members			2.72***	2.96***	2.68***
			(0.73)	(0.81)	(0.74)
Financial support (reference = own and spouse only)					
Child and relatives only				0.62***	0.76***
				(0.07)	(0.09)
Government and others only				0.57***	0.67**
				(0.11)	(0.13)
Own, spouse, and kins				0.73*	0.81**
				(0.09)	(0.10)
All sources				0.72***	0.81
				(0.09)	(0.11)

Table 22.3 (Continued)

Independent variables	Model 1	Model 2	Model 3	Model 4	Model 5
ADL (reference = severe disability)					
Active				1.32***	1.48***
				(0.10)	(0.11)
Mild disability				1.43***	1.45***
				(0.13)	(0.13)
Urban (Yes)					1.27***
					(0.08)
Age (reference = 80–89)					
90–99					1.19***
					(0.08)
100 and higher					1.32***
					(0.10)
Female (Yes)					1.14**
					(0.08)
Education (reference = no schooling)					
1–6 years of schooling					1.14*
					(0.08)
6 and more years of schooling					1.33**
					(0.16)
Occupation (reference = worker or farmer)					
Professional					1.30**
					(0.17)
Housework					1.22***
					(0.09)
N	7871	7871	7871	7871	7871
Degree of freedom	5	6	8	14	22
LR chi2	203	214	240	284	348

Standard errors are shown in parentheses. *** $p < 0.01$; ** $p < 0.05$; * $p < 0.10$ (two-tailed)

elderly person ever had children and whether any of the children are still alive. This control is necessary because living arrangement is highly associated with whether an elderly person has children. Elderly living in nursing homes are twice as likely to report a good life quality as elderly residing with one of their children. Elderly living alone, however, are 60 percent less likely to report a good quality of life. These findings provide initial evidence that elderly living in nursing homes report the highest quality of life.

Whether or not non-resident children visit their elderly parents has a strong effect on quality of life for the elderly parents. The elderly are 40 percent more likely to report good quality of life if they report receiving regular visits from children. The variable measuring perceived support is added in Model 3. The effect of perceived support is strong compared to the effects of living arrangement and children's visits (objective activities). This is consistent with patterns found in developed countries (Turner and Marino 1994). In other words, elderly who perceive that their family members would take care of them if they were sick are 3.2 times as likely to report good quality of life, compared with those who perceive they would not receive any help. When living arrangement is taken into account, elderly who perceive support from family members have a higher quality of life than elderly who perceive support from non-family members, reversing the pattern seen in the bivariate relationship (see Fig. 22.2).

Sources of financial support and ADL status are added in Model 4. The elderly who completely rely on themselves, or their spouses, for financial security are most likely to report a good quality of life. The elderly who can only rely on the government or other sources for support are 43 percent less likely to report a good quality of life than the self-financed elderly. Clearly, for some elderly persons, social and financial capital accumulated over a lifetime pays off in terms of quality of life. Undoubtedly, health status affects quality of life. Interestingly, mild disability does not prevent an elderly person from enjoying a good quality of life. Quality of life only suffers when the elderly report severe disability.

Urban/rural residence, age, educational attainment, and occupational status at retirement are included in Model 5. Good quality of life is 27 percent more likely for urban than for rural elderly. The effect of education only becomes strong with regard to the difference between those who have completed elementary education and those who have not. Elderly with completed elementary education are 33 percent more likely to report a good quality of life than elderly with no schooling. As expected, former professional or government employees are 30 percent more likely to report a good quality of life compared to former farmers or workers. The social structure variables employed clearly demonstrate that the jobs and schooling the elderly had earlier in their lives and where they live continue to have important effects on their reported quality of life.

When other variables, such as education and occupation, are taken into account, the age effect remains strong. Elderly between the ages of 100 and 110 are 32 percent more likely, and elderly aged 90 to 99 are 19 percent more likely to report a good quality of life compared to elderly between the ages of 80 and 89. The positive effect of age is most likely reflective of the selective nature of the oldest old. Their positive attitudes towards life are likely to have played a role in their

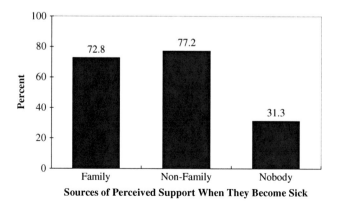

Fig. 22.2 Percentage of the elderly reporting good quality of life by sources of perceived support

healthy longevity. Gender has a weak but significant association with quality of life – elderly women are 13.7 percent more likely to report a good quality of life compared to elderly men.

When the variables are compared across models, it is clear that the regression coefficients remain stable. This means that the variables introduced in each model have additive effects on the predictions of quality of life. Elderly respondents living in nursing homes stand out as having the highest quality of life even when all other variables are taken into account. This significant finding suggests the importance of community support for the elderly. The elderly who reside with children may not feel so free to talk with their children; intergenerational gaps may prevent them from having good communication with their family members. In contrast, residence in nursing homes may provide an environment in which the elderly can share common interests and concerns. However, it is premature to conclude that nursing homes are an ideal living arrangement for the elderly. Selection effects may affect the findings. It is likely that elderly with certain personality traits (easy going, open, and so forth) are more likely to live in nursing homes – and live longer – than elderly without such traits.

Though the data prevent us from observing a clear picture of potential selectivity effects, Table 22.4 provides a glimpse of those living in nursing homes. The elderly

Table 22.4 Odds from logistic regression predicting whether living in nursing homes

Independent variables	Odds	Standard error
No child	2.24***	0.63
Child alive	1.37	0.54
Financial support (reference = own and spouse only)		
Child and relatives only	1.22	0.41
Government and others only	7.18***	2.41
Own, spouse and kins	1.42	0.48
All	2.24*	0.78
Visit	1.63	0.59
ADL (reference = severe disability)		
Active	0.60*	0.16
Mild disability	1.20	0.38
Caretaker when sick (reference = nobody)		
Family members	0.48	0.36
Nonfamily members	98.19***	73.21
Urban	5.25***	1.06
Age1	0.42***	0.09
Age2	0.32***	0.09
Female	1.10	0.25
1–6 years Education	1.49*	0.35
6+ years Education	0.96	0.33
Professional	0.46*	0.16
Housework or Others	0.51**	0.13
N	7871	
Degree of freedom	19	
LR chi2	2148.62	

***$p < 0.01$; **$p < 0.05$; *$p < 0.1$ (two-tailed).

living in a nursing home are more likely to have no children, depend on government and other sources for financial support, count on non-family members for support when they become ill, live in urban areas, and be between the ages of 80–89, compared to those who do not live in nursing homes. This profile suggests that the elderly living in nursing homes tend to be childless and depend on government support. Why do they report a better quality of life? This may be largely due to the impact of peers and friends who provide good company. It is also likely that their take on quality of life has a different reference point. In the Chinese society where elderly care is the responsibility of families and kinship, the elderly who have never had children may have been pessimistic about their old age support. However, nursing homes provide them with food, shelter, and social networks. The old age security they have in nursing homes may make them think more positively and report higher levels of social support.

In summary, we have found strong evidence of social support affecting quality of life among China's oldest old. Living arrangement, children's visits, and perceived social support are all strongly associated with quality of life. In addition, urban/rural residence and the socioeconomic status of the elderly also play important roles. Although the elderly living alone and the elderly living in a nursing home have many characteristics in common, their quality of life is at two extremes, which suggests the importance of community support and peer influences in shaping quality of life.

22.7 Discussion and Conclusion

Our objective has been to examine how various sources of social support affect quality of life for the elderly. We analyze data from the 1998 Chinese Longitudinal Healthy Longevity Survey (CLHLS). Although previous literature on social support is voluminous, research has mostly focused on developed countries. Our study fills this gap by providing new evidence from a developing country regarding the impact of social support on quality of life. This is an important endeavor given that population aging is rapidly increasing in many developing countries where social and financial security for the elderly is not yet fully in place. We center on how objective behaviors (living arrangement and children's visits) and subjective perceived social support (who would take care of them when they are sick) affect quality of life.

More elderly persons living on their own may indicate less availability of social support from family members. Our analyses recognize that social support is multi-dimensional and that the elderly benefit from all dimensions of social support. Where the elderly live indicates the kind of support, if any, they will receive. Unfortunately, elderly who live alone or those living only with a spouse report much lower levels of quality of life. In contrast, elderly persons living in nursing homes have the highest quality of life. Consistent with previous research, our study indicates that relationships with friends and peers are shown to be more positively associated with quality of life than relationships with family members (Ellison 1990; Wood and Robertson 1978). Indeed, community attachment and peer interactions are critical for fostering a good quality of life for the elderly.

For the elderly with children, our analyses show that quality of life does not so much depend on whether they have children of their own, but on whether their children are in their daily lives. For these elderly people, frequent visits by children and the perception that these children will be available when they are sick increase their quality of life. Family networks are important as well. Superficially, children's visits may indicate that children are in regular contact with their elderly parents. At a deeper level, these visits may indicate that the children care for their aging parents, the extended family is united and harmonious, and the elderly receive reverence from their descendants. The elderly value these things and thus they are highly associated with their quality of life.

In sum, our analysis shows the importance of social support on quality of life for China's oldest old. The fast pace of aging in China suggests that it will be difficult in the near future for the family to shoulder all the responsibilities of caring for aging parents. While a social security system needs to be in place to provide financial security for the elderly, nursing homes or retirement communities may be a viable option. The elderly in our study do indeed benefit from networks of peers and friends in their older years. These networks and communities can regularly facilitate interactions and communications among the elderly and provide a support system for them, which seem to be the most important key to a good quality of life for the Chinese elderly.

References

Arling, G. (1976), The elderly widow and her family, neighbors and friends. *Journal of Marriage and the Family* 38, pp. 757–768

Caldwell, J.C. (1982), *Theory of fertility decline*. London, New York: Academic Press

Christensen K. and J.W. Vaupel (1996), Determinants of longevity: Genetic, environmental and medical factors. *Journal of Internal Medicine* 240, pp. 333–341

Dalkey, N.C. and D.L. Rourke (1973), The Delphi procedure and rating quality of life factors. In: US Environment Protection Agency. *The quality of life concept*. Washington, DC: USGPO, pp. 209–221

DaVanzo, J. and A. Chan (1994), Living arrangements of older Malaysians: Who coresides with their adult children? *Demography* 31 (1), pp. 95–113

Dean, A., B. Kolody, and P. Wood (1990), Effects of social support from various sources on depression in elderly persons. *Journal of Health and Social Behavior* 31 (2), pp. 148–161

Dowd, J.J. and V.L. Bengtson (1978), Aging in a minority population: An examination of the double jeopardy hypotheses. *Journal of Gerontology* 33, pp. 427–436

Ellison, C.G. (1990), Family ties, friendships, and subjective well-being among black Americans. *Journal of Marriage and the Family* 52, pp. 298–310

Finch, J. (1989), *Family obligations and social change*. Cambridge, England: Polity Press.

Gouldner, A.W. (1960), The norm of reciprocity: A preliminary statement. *American Sociological Review* 25, pp. 161–178

Haug, M.R. and S.J. Folmar (1986), Longevity, gender, and life quality. *Journal of Health and Social Behavior* 27, pp. 332–345

Hochschild, A. (1973), *The unexpected community*. Englewood Cliffs, New Jersey: Prentice-Hall.

Horwitz, A.V., S.C. Reinhard, and S. Howell-White (1996), Caregiving as reciprocal exchange in families with seriously mentally ill members. *Journal of Health and Social Behavior* 37 (2), pp. 149–162

Kramarow, E.A. (1995), The elderly who live alone in the United State: Historical perspectives on household change. *Demography* 32 (3), pp. 335–352

Lin, J. (1995), Changing kinship structure and its implications for old-age support in urban and rural China. *Population Studies* 49 (1), pp. 127–145

Logan, J.R., F. Bian, and Y. Bian (1998), Tradition and change in the urban Chinese family: The case of living arrangements. *Social Forces* 76 (3), pp. 851–882

Matt, G.E. and A. Dean (1993), Social support from friends and psychological distress among elderly persons: moderator effects of age. *Journal of Health and Social Behavior* 34 (9), pp. 187–200

Mutran, E. and D.C. Reitzes (1984), Intergenerational support activities and well-being among the elderly: A convergence of exchange and symbolic interaction perspectives. *American Sociological Review* 49 (1), pp. 117–130

Ogawa, N. (1988), Aging in China: demographic alternatives. *Asian-Pacific Population Journal* 3 (3), pp. 21–64

Riley, N.E. (2004), China's population: new trends and challenges. *Population Bulletin* 59 (2).

Ross, C.E. and J. Mirowsky (2002), Social support and subjective life expectancy. *Journal of Health and Social Behavior* 43 (4), pp. 469–489

Ross, C.E. and M.V. Willigen (1997), Education and the subjective quality of life. *Journal of Health and Social Behavior* 38 (3), pp. 275–297

Schuessler, K.F. and G.A. Fisher (1985), Quality of life research and sociology. *Annual Review of Sociology* 11, pp. 129–149

Smith, K.F. and V.L. Bengtson (1979), Positive consequences of institutionalization: Solidarity between elderly patients and their middle-aged children. *The Gerontologist* 19, pp. 438–447

Sun, R. (2002), Old age support in contemporary urban China from both parents' and children's perspectives. *Research on Aging* 24 (3), pp. 337–359

Turner, J.R. and F. Marino (1994), Social support and social structure: a descriptive epidemiology. *Journal of Health and Social Behavior* 35 (9), pp. 193–212

Veiel, H.O.F. and U. Baumann (1992), The many meanings of social support. In: H.O.F. Veiel and U. Baumann (eds.): *The meaning and measurement of social support*. New York: Hemisphere Publishing, pp. 1–7

Wethington, E. and R.C. Kessler (1986), Perceived support, received support, and adjustment to stressful life events. *Journal of Health and Social Behavior* 27 (3), pp. 78–89

Whyte, M.K. (1997), The fate of filial obligations in urban China. *China* 38, pp. 1–31

Wood, V. and J.F. Robertson (1978), Friendship and kinship interaction: Differential effects on the morale of the elderly. *Journal of Marriage and the Family* 40, pp. 367–375

Zeng, Y. and L.K. George (2000), Family dynamics of 63 million (in 1990) to more than 330 million (in 2050) elders in China. *Demographic Research* 2 (5), pp. 1–48

Zeng, Y., J.W. Vaupel, Z. Xiao, C. Zhang, and Y. Liu (2001), The healthy longevity survey and the active life expectancy of the oldest old in China. *Population: An English Selection* 13 (1), pp. 95–116

Zimmer, Z. (2005), Health and living arrangement transitions among China's oldest-old. *Research on Aging* 27 (5), pp. 526–555

Chapter 23
Mortality Predictability of Self-Rated Health Among the Chinese Oldest Old: A Time-Varying Covariate Analysis

Qiang Li and Yuzhi Liu

Abstract Our research investigates the relationship between self-rated health (SRH) and mortality among the Chinese oldest old (age range 80–105) as well as risk patterns by gender. We use data for 7,783 respondents from the Chinese Longitudinal Healthy Longevity Study (CLHLS) conducted in 1998, 2000 and 2002. The results of a Cox Hazard Model show that self-rated health was significantly associated with mortality in the oldest old even when socio-demographic characteristics, physical and cognitive functioning, serious illness and engagement in activity were controlled. The association between SRH and the risk of mortality was stronger for males than for females. The modifying effects of some confounders on such relationships also differed by gender. Missing values on SRH predicted higher mortality, but such effects were explained by physical and cognitive functioning.

Keywords Chinese oldest old, Cognitive function, Cox proportional Hazard model, Engagement in Activities, Fourth age, Gender differences, Hazard model, Kaplan–Maier method, Missing values, Mortality, Physical function, Predictability, Self-rated health, Time-on-study, Time varying covariates, Types of time scales

23.1 Introduction

The significant and independent association between self-rated health (SRH) and mortality has been documented in previous studies (see reviews by Benyamini and Idler 1999; Idler and Benyamini 1997). However, the mechanisms that generate such a relationship are not fully understood. Early researchers have proposed that objective measures of health, such as disease burden, physical functioning, cognitive functioning and depression, only partly explain the relationship between SRH and mortality (e.g., Benyamini and Idler 1999; Idler and Benyamini, 1997;

Q. Li
Max-Plank Institute for Demographic Research, Konrad-Zuse-Str. 1, 18057 Rostock, Germany
e-mail: li@demogr.mpg.de

Bjorner et al. 1996; Strawbridge and Wallhagen 1999). SRH has a psychological component (Verbrugge and Balaban 1989), and is a broader measure of health than the presence or absence of disease and functional limitations (Idler and Benyamini 1997; Strawbridge and Wallhagen 1999). Research has suggested potential differences and similarities between SRH and objective measures of health, and that these differences are often associated with socio-demographic factors (Idler and Kasl 1995).

Gender differences in the relationship between SRH and mortality have been studied in an attempt to examine the mechanisms underlying this relationship (e.g., Bath 2003; Deeg and Bath 2003; Idler 2003). Although the association between SRH and mortality appears to be consistent, the predictive ability of SRH by gender differs by study. For instance, some studies have reported a significant relationship between SRH and mortality among men rather than women (e.g., Hays et al. 1996; Idler and Kasl 1991; Van Doorn and Kasl 1998), and others have proposed the reverse (e.g., Benyamini et al. 2003; McCallum et al. 1994; Simons et al. 1996). Further studies are warranted that explore gender differentials in risk factors, which may help explain the relationship between SRH and mortality.

Idler and Benyamini (1997) have suggested that SRH may likely change over time. Recently, scholars have pointed out that ignoring changes in SRH over time, as has been done in some studies, can lead to biased estimates of mortality risk associated with SRH (e.g., Fayers and Sprangers 2002; Ferraro and Kelly-Moore 2001; Han et al., 2003). Therefore, treating SRH as a time-varying variable may correct such problems and might shed light on understanding the link between SRH and mortality. The research we report in this chapter follows this strategy: it investigates the association between SRH and mortality, considering SRH and other health indicators as time-varying variables.

We explore this relationship among the Chinese oldest old, defined here as persons aged 80 years and older. Many studies have proposed that the oldest old (Fourth Age) are very different from the young old (Baltes and Smith 2003; Maier and Smith 1999; Suzman et al. 1992; Zeng et al. 2002). In general, oldest old populations are typically characterized by a higher proportion of women than men, higher levels of comorbidity, and a greater consumption of medical and other care services. Our research focuses on individuals born in China in 1893–1918; they comprised the oldest old observed in 1998 and possessed a set of characteristics that were cohort-specific: they have much lower levels of education and a much higher likelihood of widowhood. These cohort-specific characteristics may influence their self-assessment of health, and subsequently the relationship between SRH and mortality. The small sample size of the oldest old in previous studies impeded a meaningful analysis of this subpopulation. We thus know relatively little about the relationship between SRH and mortality risk in this subgroup. Further, some researchers have concluded that there is an accelerated decline of functional abilities among the oldest old (e.g., Baltes and Smith 2003; Fries et al. 2000). Thus, the SRH of the oldest old, which is related to changes in functional ability over time, is also more likely to change during follow-up (e.g., Fayers and Sprangers 2002; Ferraro and Kelly-Moore 2001; Han et al. 2003; Strawbridge and Wallhagen 1999). Some studies have found that older persons reported weaker associations between SRH and mortality compared to their younger counterparts (e.g., Benyamini et al. 2003; Idler and Angel 1990; Strawbridge and Wallhagen 1999). Considering that SRH is a dynamic rather than a static perception, the relationship between SRH and mortality may nevertheless have a different pattern in the oldest old. Our research explores this relationship between SRH and mortality using a large sample of the oldest old in China ($N = 7,783$) that may yield a more precise estimation of this relationship.

Our research also investigates gender specific patterns of SRH on mortality in the oldest old. Research has documented that life expectancy is higher for women than for men. Nevertheless, women report poorer health status and experience more years of disability and ill health than do men (e.g., Manton et al. 1995). This paradox may explain some of the gender differences in the SRH and mortality association. In addition, some studies have shown that women and men differ in the extent of their health problems, social conditions, and sensitivity to symptoms and signs of disease and impairment (Verbrugge 1990). These differences may affect the way that women and men rate their health, which then influences the relationship between SRH and mortality. Exploring the relationship between SRH and mortality separately for men and women in the oldest old may contribute to an understanding of the mechanisms that generate the effects of SRH on mortality.

Another question of interest in our research was the issue of missing values in the SRH data of the oldest old. Missing values are common in studies of the oldest old. A common way to handle missing data is the complete cases approach, which drops subjects with missing values from the analysis. However, some have indicated that the missing psychological function variables were related to mortality (Anstey et al. 2001; Maier et al. 2003) and that deleting cases where data were missing might bias the study results (Anstey et al. 2001). Considering that SRH may summarize both objective and subjective aspects of health (Maddox and Douglass 1973) we also examined whether the missing SRH data were predictive of mortality in the oldest old.

Based on the above, the research we report in this chapter examines the following research questions and hypotheses:

First, does SRH predict mortality in the Chinese oldest old? If indeed SRH is more likely to change during follow-up, which would indicate that the failure to consider these changes would underestimate the SRH–mortality association, we expect to find a significant and independent association between SRH and mortality, even when controlling for socio-demographic characteristics, physical and cognitive functioning, serious illness and engagement in activity.

Second, are there gender differences in the association between SRH and mortality? If oldest old women have a lower risk of mortality and report poorer health, we expect to detect gender differences in this relationship, as well as SRH having a stronger effect on mortality among females than among males. Moreover, it is expected that functional limitations and serious illnesses will play different roles in the relationship between SRH and mortality risk among women and men.

Third, do missing SRH values predict a higher risk of mortality compared to non-missing categories of SRH? If the oldest old with missing values on SRH have poorer health than do those with complete data, we expect missing values to be related to higher mortality.

23.2 Method

23.2.1 Sample

The sample in this research was taken from participants in the Chinese Longitudinal Healthy Longevity Survey (CLHLS). Detailed information about the sample, design, and assessment battery is published elsewhere (Zeng et al. 2002). This study comprises 3 waves of data collection: a baseline in 1998, and second and third waves in 2000 and 2002, respectively. The baseline survey was conducted in 631 randomly selected counties and cities of the 22 provinces in China. All centenarians in these counties and cities were interviewed on a voluntary basis. For each centenarian, one octogenarian (aged 80–89) living nearby and one nonagenarian (aged 90–99) living nearby, with pre-designated age and sex, were matched and interviewed. "Nearby" is loosely defined: it could be in the same village or street if such an individual was available, or in the same town or in the same sampled county or city (Zeng et al. 2002). The pre-designated age and sex were employed to have approximately equal numbers of male and female nonagenarians and octogenarians. For example, for a centenarian aged 102, a nearby octogenarian aged 82 and a nearby nonagenarian aged 92 were matched and interviewed. The sex of the octogenarians and nonagenarians to be interviewed was randomly determined with the goal of having approximately equal numbers of males and females at each age from 80 to 99. If such individuals could not be found, an alternative individual of the same sex and in the same 5-year age group (80–84, 85–89, 90–94, 95–99) was selected (Zeng et al. 2002). The total valid sample size of the baseline data is 8,805 elderly persons aged 80 to 105.

Our research uses three waves of respondent data. Mortality information was obtained in wave 2 and in wave 3. In the second wave in 2000, among the panel in the baseline survey in 1998, 4,690 participants survived, 3,263 died, and 851 were lost to follow up. In the third wave in 2002, among the 4,690 survivors, 2,562 individuals were alive, 1,563 died between the two waves, and 565 were lost to follow-up. Detailed information on sample size is given in Fig. 23.1.

Similar to previous research studying the oldest old using longitudinal data, we also encounter lost to follow-up and missing data in the covariates. With respect to the 851 lost to follow-up in the second wave, we do not know whether or not these individuals died during the study period; they did not provide critical survival information, and were thus deleted from our sample. With respect to the 565 lost to follow-up in the third wave, they survived to the 2000 survey and therefore provided useful information at the first and second waves. We kept them in our sample, and treated them as right censored at the second wave in 2000. Thus, the total sample with available survival data is 7,954 (3,127 survivors and 4,827 deceased) in total.

Missing values, except for gender and type of residence, have the potential of varying over time. The treatment of missing values in SRH, the MMSE (Mini-Mental State Examination), and in verbal fluency is described below. Some covariates with a very small percentage of missing values at baseline were deleted. Of the

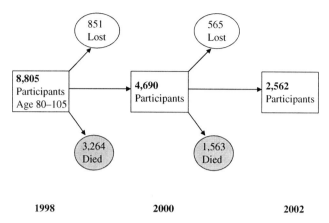

Fig. 23.1 Pattern of participants in the Chinese Longitudinal Healthy Longevity Study (CLHLS), 1998–2002

7,954 participants with full survival information, 107 (1.35 percent) of the deceased and 64 (0.82 percent) of the survivors had no information in these covariates at baseline.[1] The final sample was reduced to 7,783, 3,063 (39.36 percent) for those who survived and 4,720 (60.64 percent) for those who were deceased, respectively.

23.2.2 Measures

Self-rated health. SRH in the CLHLS was assessed using a single question, "How do you rate your health at present?" Participants' answers were coded as 1 through 5, representing: Very good, good, fair, poor and very poor, respectively. Because of the small number of respondents who evaluated their SRH as "very poor" ($n = 43$ in 1998 and $n = 61$ in 2000), we collapsed the last two categories. Thus, in our analysis, the responses of SRH were grouped into four levels: very good, good, fair, and poor.

Socio-demographic characteristics. Six measures of social-demographic characteristics were employed. Sex has two levels: female and male. Residence was defined as urban and rural. Years of schooling was divided into 3 groups: no schooling, 1–6 years of schooling, and 7 + years of schooling. Occupation before 60 years old was categorized into two levels: higher occupation and lower occupation. Professional and technical personnel, and governmental, institutional or managerial personnel were collapsed into the "higher occupation" category. Agri-

---

[1] Among the 107 dead, the following information is not available: years of schooling, 32; occupation before age 60, 2; marital status, 4; self-reported ADL,19; engagement in activity, 15; and suffered serious illness during last two years, 37. Among the 65 survivors, information is lacking on: years of schooling, 10; occupation before age 60, 2; marital status, 1; self-reported ADL, 13; engagement in activity, 9; and suffered serious illness during last two years, 23.

cultural, forest, animal husbandry, fishing, industrial worker, commercial or service worker, military personnel,[2] and housework and others were collapsed into the "lower occupation" category. Current marital status was composed of two groups: married and unmarried,[3] where the latter included widowed, separated, divorced and never married persons.

Physical functioning. Self-reported limitations in activities of daily living (ADL) were used to measure physical functioning. The respondents were asked whether they had difficulties in performing any of six ADLs: bathing, dressing, toileting, transferring, continence and feeding. Their answers were grouped into three levels: no functional limitation, one functional limitation, two or more functional limitations.

Cognitive functioning. Two measures were used to assess the cognitive functioning of the Chinese oldest old. The Mini-Mental State Examination (MMSE) is translated into the Chinese language and culturally adapted based on established international standards for the MMSE questionnaire, and carefully tested in a pilot survey (Zeng et al. 2003). It includes brief measures of orientation, registration, attention, and calculation, recall, and language, with scores ranging from 0 to 23.[4] MMSE scores were graded into four levels by quartiles, with a higher quartile indicating a higher level of cognitive functioning.

Verbal fluency was measured using a single question. Participants were asked to name in 60 seconds as many kinds of food as possible. This is a standard task in many intelligence tests. The number of foods named was grouped into two levels, using a median split (0–50 percent, 50–100 percent) with 50–100 percent indicating a higher level of verbal fluency.

Suffering from serious disease during the last two years. Due to the poor data quality of self-reported chronic illness by the Chinese oldest old (Zeng et al. 2002), we did not use these data, although they have often been employed in previous studies. Instead, we used whether or not a person has been suffering from a serious illness during the last two years, which provides some crude information on serious illnesses that might play a role in the relationship between SRH and mortality. This covariate was collapsed into two groups: not suffering from serious illness during the last two years, and suffering from one or more serious illnesses (including being permanently bedridden).

---

[2] Military personnel do not include officers.

[3] We did not categorize current marital status into three levels, "married," "widowed" and "separated, divorced and never married" because of the small number of "separated, divorced and never married" (only 3% in the whole sample and 1.5 percent in the female sample), which may result in zero exposures in the hazard models.

[4] Usually, MMSE has a score ranging from 0 to 30. For the sake of brevity, the CLHLS used 23 items to measure these five domains of cognitive functioning, which resulted in a score ranging from 0 to 23. The incomplete measures mainly occur on the assessment of orientation and language. The standard MMSE uses ten items to measure orientation, while the CLHLS uses five items. Also, the CLHLS does not measure reading and writing abilities.

Engagement in activity. Engagement in activity includes housework, growing vegetables and other field work, garden work, reading newspapers/books, raising domestic animals, playing cards and mah-jong, watching TV and/or listening to the radio, and religious activities. Participants had three response options: almost everyday, sometimes and never. We summed the answers with scores ranging from 8 to 24. Scores of engagement in activity were divided into two levels using a median split, with "50–100 percent" indicating a higher level of engagement.

23.2.3 Model

The survival analysis was formulated in the counting process framework. Age was employed as the time scale. The hazard model for the force of mortality in the oldest old is:

$$h(t) = h_0(t) \exp(\beta \mathbf{X}(t))$$

where $h(t)$ denotes the mortality risk of the oldest old at the attained age t, $h_0(t)$ is the baseline hazard, $\mathbf{X}(t)$ is the matrix of covariates, including SRH and other covariates, and β is the vector of the coefficients of covariates.

Using age as the time scale brings up the debate on the choice of time-scale in a mortality hazard model (e.g., Klein and Moeschberger 2002; Korn et al. 1997; Lamarca et al. 1998; Li 2005). Two types of time scales have been used in the literature: the age of studied subjects, and time-on-study with the baseline age treated as a covariate. We prefer to use the age of the subjects as the time scale on the following grounds. First, it indicates that the mortality risk at the attained age t depends on the sum of age at baseline and time-on-study. Secondly, it is consistent with the very idea of a baseline hazard, that is, the baseline intensity is the same for all individuals. Thirdly, it satisfies our expectation that the mortality hazard will change more as a function of age than as a function of time-on-study. Finally, it reflects that the interest is in the study of the aging process and the risk factors related to it, not in the use of time-on-study as the time scale. Employing time-on-study as the time scale will not satisfy these criteria.

Graphs of the survival function were plotted using the Kaplan–Maier method in the four year observation periods.

To estimate the effect of SRH on mortality controlling for covariates, we used a Cox Proportional Hazard Model with left truncation and right censoring, utilizing time-varying covariates. We first ran the bivariate models to check for the crude effects of SRH on mortality. Then the multivariate equations were estimated to examine the independent effects of SRH on mortality, controlling for sociodemographic characteristics, physical functioning, cognitive functioning, suffering from serious illness during the last two years, and engagement in activity. Separate models were developed for each gender. Treating self-rated health, current marital status, self-reported ADL, cognitive functioning, suffering from serious illness during the last two years, and engagement in activity as time-varying variables allowed us to use all of the information provided by the participants (baseline and follow-up), and should correct for the possible underestimation of the effect of lower level self-rated health on mortality.

23.2.4 Missing Values

In the present study, SRH, the MMSE, and verbal fluency included a large portion of missing data. For SRH, 6.76 percent (526 participants) have incomplete data. For the MMSE and verbal fluency, 7.23 percent (563) and 9.06 percent (705) of the participants failed to answer the questions, respectively. If we deleted the cases with missing values on these three variables, the results could well be biased. When dealing with missing values, researchers use different methods such as replacing missing values with the mean, imputing the values, or including an additional level comprising the persons with missing values (e.g., Allison 2001; Maier et al. 2003). For the purpose of obtaining an estimate of mortality risk associated with incomplete data on SRH, we applied the method of including an additional level of SRH comprised of persons with missing values. The same strategy was applied to replace missing values in the MMSE and verbal fluency covariates. Other covariates at the second wave had very small percentages of missing values (less than 0.1 percent), so we used their corresponding values at baseline for their missing values at follow-up.

23.3 Results

Results are presented in two main sections. In the first section, we address the relationship between SRH and mortality for the whole sample. In the second section, we estimate the hazard model separately for men and women.

23.3.1 SRH and Mortality Risk in the Oldest Old

Table 23.1 shows a breakdown of the sample according to factors at baseline (see the second column). It also displays the number and percent of people who died (see the third and fourth columns).

The cross tabulation of SRH and mortality (see Table 23.1) reveals their gross association. With a decline of SRH, the proportion of deceased persons increased. Participants with missing SRH scores had the highest proportion of deaths.

Table 23.2 presents the transitions between different levels of SRH as well as the transition to death for the oldest old. In addition to attrition due to death, significant changes were found between different levels of SRH. Although we observed some recovery from poor health to better health, a transition was more likely to occur in the opposite direction, that is, from better health to poor health. This may be the main reason why ignoring changes during follow-up tends to underestimate the effects of a lower level of SRH on mortality.

Table 23.1 Sample description at baseline in 1998, and number & percentage of participants who died before the third wave in 2002

	Total ($N = 7,783$)	Deceased ($N = 4,720$)	
	No.	No.	Percent[a]
Self-rated health			
Very good	907	419	46.20
Good	3,231	1,810	56.02
Fair	2,479	1,574	63.49
Poor	640	470	73.44
Missing	526	447	84.98
Socio-demographic characteristics			
Age groups			
80–89	3,000	1,191	39.70
90–99	2,710	1,839	67.86
10–105	2,073	1,690	81.52
Sex			
Male	3,114	1,823	58.54
Female	4,669	2,897	62.05
Type of residence			
Urban	2,742	1,503	54.81
Rural	5,041	3,217	63.82
Years of schooling			
0 year	5,332	3,428	64.29
1–6 year	1,874	1,021	54.48
7+ year	577	271	46.97
Occupation before age 60			
Higher	478	181	37.87
Lower	7,305	4,539	62.14
Current marital status			
Married	1,227	513	41.81
Unmarried	6,556	4,207	64.17
Self-reported ADL			
No functional limitation	4,905	2,470	50.36
1 functional limitation	1,032	689	66.76
2+ functional limitations	1,846	1,561	84.56
Cognitive functioning			
MMSE			
75–100%	2,080	869	41.78
50–75%	1,339	693	51.76
25–50%	2,093	1,305	62.35
0–25%	1,732	1,384	79.91
Missing	539	469	87.01
Word fluency			
50–100%	3,595	1,765	49.10
0–50%	3,483	2,349	67.44
Missing	705	606	85.96
Suffered from serious illness during last 2 years			
Suffering no serious illness	6,991	4,125	59.00
Suffering 1+ serious illness	792	595	75.13
Engagement in activity			
50–100%	3,282	1,458	44.42
0–50%	4,501	3,262	72.47

[a] percent = $\frac{\text{No. deceased}}{\text{No. total}} \times 100$

Table 23.2 Transition of self-rated health during the follow-up period for the oldest old (%)

Self-rated health at baseline in 1998	Self-rated health at the second wave in 2000 & decease						
	Very good	Good	Fair	Bad	Missing	Deceased	Total
Very good	15.33	30.76	15.99	5.51	4.74	27.67	907
Good	9.59	25.26	19.1	5.76	5.42	34.88	3,231
Fair	4.8	16.98	20.17	8.67	5.04	44.33	2,479
Poor	2.34	10.94	17.34	10.16	6.41	52.81	640
Missing	1.33	5.7	6.46	4.18	11.03	71.29	526
Total	7.58	20.76	18.08	6.91	5.68	40.99	7,783

We present percentage for the transition between different levels of self-rated health as well as attrition to death, and absolute number for Total in such way that frequency of transition can be calculated

The predictive ability of SRH on mortality risk may reflect the latent impact of other risk factors on mortality. Therefore, six sequential models were used to assess the relative impact of other adjustment variables on the relationship between SRH and mortality (see Table 23.3). The first model includes only SRH. In the second model, socio-demographic characteristics are added. The third model includes self-reported ADL, while the adjustment for cognitive functioning is performed in the fourth model. The fifth and the final models include suffering from serious illness during the last 2 years and the engagement in activity, respectively. This sequence allows assessing the extent to which SRH reflects socio-demographic characteristics, physical functioning, cognitive functioning, serious illness and engagement in activity by examining the relative impact of each subsequent set of variables on a previously computed relationship. For example, if SRH primarily reflects physical and cognitive functioning, then any observed relationship between SRH and mortality in the first two models may disappear when self-reported ADL and cognitive functioning are incorporated into the fourth model.

The graph in Fig. 23.2 showed the survival function for the oldest old. The survival probability decreases typically with increased age during the 4-year observation period.

Model 1 (see Table 23.3) shows that SRH was significantly associated with mortality. The predictive ability of SRH on mortality did not change with the inclusion of socio-demographic characteristics (see Model 2). The largest reduction occurs in model 3 and model 4 with the addition of physical and cognitive functioning. In these two models, the relative risks of each level of SRH related to the reference group are less pronounced compared with the relative risks in previous models. Different from our expectations, when cognitive functioning was entered into the model the significant effect of SRH missing values on mortality disappeared. This may suggest that the cognitive functioning of the oldest old is related to the SRH missing values. In other words, it is possible that due to poor cognition the oldest old could not answer the question. The addition of suffering from serious illness during the last two years into model 5 only slightly changes the pattern of the SRH–mortality association. The inclusion of engagement in activity reduces the relative risk of SRH on mortality in the final model.

Table 23.3 Mortality risk associated with SRH with control for socio-demographic characteristics, physical & cognitive function, serious illness and engagement in activities in Chinese oldest old

	Model 1 RR	95% CI	Model 2 RR	95% CI	Model 3 RR	95% CI	Model 4 RR	95% CI	Model 5 RR	95% CI	Model 6 RR	95% CI
Self-rated health (very good = 1)												
Good	1.12*	(1.01, 1.25)	1.13*	(1.01, 1.26)	1.08	(0.97, 1.20)	1.03	(0.93, 1.16)	1.04	(0.93, 1.16)	1.01	(0.91, 1.13)
Fair	1.50***	(1.35, 1.68)	1.53***	(1.37, 1.71)	1.34***	(1.20, 1.49)	1.23***	(1.10, 1.38)	1.23***	(1.10, 1.38)	1.19**	(1.06, 1.33)
Poor	2.01***	(1.77, 2.28)	2.06***	(1.81, 2.34)	1.57***	(1.38, 1.79)	1.37***	(1.20, 1.57)	1.35***	(1.18, 1.55)	1.30***	(1.13, 1.48)
Missing	2.30***	(2.03, 2.61)	2.35***	(2.07, 2.67)	1.76***	(1.54, 2.01)	1.17	(0.99, 1.39)	1.16	(0.98, 1.37)	1.11	(0.94, 1.32)
Sex (male = 1)												
Female			0.71***	(0.66, 0.76)	0.67***	(0.62, 0.72)	0.65***	(0.60, 0.70)	0.65***	(0.60, 0.70)	0.66***	(0.61, 0.71)
Type of residence (city = 1)												
Rural			1.01	(0.95, 1.08)	1.05	(0.99, 1.12)	1.03	(0.97, 1.10)	1.04	(0.97, 1.10)	1.03	(0.96, 1.10)
Years of schooling (0 years = 1)												
1–6 year			0.92	(0.85, 1.00)	0.93	(0.85, 1.01)	0.97	(0.89, 1.05)	0.97	(0.89, 1.05)	0.99	(0.91, 1.08)
7+ year			0.95	(0.82, 1.09)	0.92	(0.80, 1.06)	0.99	(0.86, 1.15)	0.99	(0.86, 1.15)	1.04	(0.90, 1.21)
Occupation before age 60 (higher = 1)												
Lower			1.46***	(1.23, 1.72)	1.50***	(1.27, 1.77)	1.46***	(1.24, 1.73)	1.46***	(1.24, 1.73)	1.41***	(1.19, 1.67)
Current marital status (Married=1)												
Unmarried			1.34***	(1.21, 1.48)	1.36***	(1.23, 1.50)	1.33***	(1.20, 1.47)	1.33***	(1.20, 1.47)	1.31***	(1.19, 1.45)
Self-reported ADL (no limitation = 1)												
1 functional limitation					1.39***	(1.27, 1.51)	1.33***	(1.22, 1.45)	1.33***	(1.21, 1.45)	1.27***	(1.17, 1.39)
2 + functional limitation					2.05***	(1.91, 2.19)	1.81***	(1.68, 1.94)	1.77***	(1.65, 1.91)	1.67***	(1.55, 1.80)
MMSE (75–100% = 1)												
50–75%							1.10	(0.99, 1.22)	1.10	(0.99, 1.22)	1.09	(0.98, 1.21)
25–50%							1.21***	(1.10, 1.33)	1.21***	(1.10, 1.33)	1.17**	(1.06, 1.29)
0–25%							1.51***	(1.36, 1.67)	1.50***	(1.35, 1.67)	1.42***	(1.28, 1.58)
Missing							1.84***	(1.55, 2.17)	1.82***	(1.54, 2.15)	1.71***	(1.45, 2.03)

Table 23.3 (Continued)

	Model 1		Model 2		Model 3		Model 4		Model 5		Model 6	
	RR	95% CI	RR	95% CI	RR	95% CI	RR	95% CI	RR	95% CI	RR	95% CI
Word fluency (50–100% = 1)												
0–50%							1.14***	(1.07, 1.23)	1.14***	(1.07, 1.23)	1.13***	(1.05, 1.21)
Missing							1.20**	(1.05, 1.36)	1.20**	(1.06, 1.37)	1.19**	(1.05, 1.36)
Suffered from serious illness during last 2 years (No serious illness = 1)												
1 + serious illness									1.12**	(1.03, 1.21)	1.11**	(1.03, 1.20)
Engagement in activity (50–100% = 1)												
0–50%											1.32***	(1.23, 1.43)

Relative risks (RR) are reported; 95% confidence intervals for RR are shown in parentheses; Reference group of each covariate is presented in parentheses; Model 1 is unadjusted model. Model 2 is adjusted model, controlling for socio-demographic characteristics; Model 3 added self-reported ADL; Model 4 added cognitive function; Model 5 added suffering from serious illness during the last 2 years. Model 6 added engagement in activity
* p<0.05, ** p<0.01, *** p<0.001;

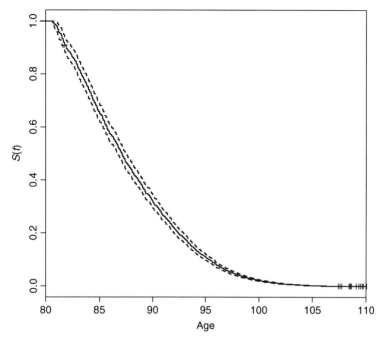

Fig. 23.2 Survival function for Chinese oldest old

23.3.2 Gender Difference in the Relationship Between SRH and Mortality

Before running the hazard model separately for men and women, we first performed descriptive analyses comparing SRH in the two gender groups (see Table 23.4). There are significant gender differences in SRH. Women report poorer health than the men, and the survival probability was higher among women than among men (see Fig. 23.3). These results are consistent with our hypothesis: we expected the gender differences in the relationship between SRH and mortality.

It was in this context of significant gender differences that we then conducted a hazard analysis separately for men and women (see Table 23.5). As we expected, gender differences in the relationship between SRH and mortality were observed.

Table 23.4 Self-rated health by gender, and comparison between men and women

	Male (%) (N = 3,114)	Female (%) (N = 4,669)	Significant gender differences
Self-rated health			χ^2 (1) = 10.25 P<0.001
Very good	14.61	9.68	
Good	44.06	39.82	
Fair	30.51	32.75	
Poor	7.00	9.04	
Missing	3.82	8.72	

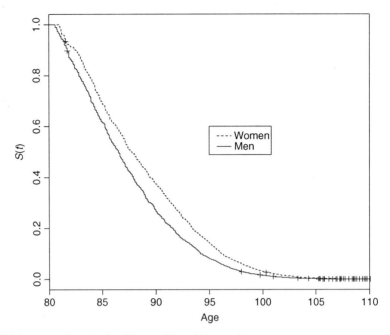

Fig. 23.3 Survival function for Chinese oldest old by gender

Although SRH was a significant predictor of mortality for both men and women, the effects were less pronounced among women than men. The relative mortality risks for men who rated their health as "fair" and "poor" were significantly higher than those with a "very good" rated health status, while the relative risk for women who rated their health as "bad" was only significantly higher compared to that of the reference group. In addition, the modifying effects of potential confounders were also different in men and women. Current marital status, and suffering from serious illness during the last 2 years modified the relationship between SRH and mortality among men, but not among women. The role of verbal fluency was more pronounced among women than among men.

23.4 Discussion

Our research examined the predictability of SRH for mortality in the Chinese oldest old as well as the risk pattern by gender. SRH was significantly associated with mortality when we controlled for socio-demographic characteristics, physical and cognitive functioning, serious illness and engagement in activity. The research results we reported in this chapter expand our knowledge of the association between SRH and the risk of death indicating that it predicts mortality risk not only among young adult people and old adults, but also among the oldest old. Moreover, there was a different predictive pattern of SRH on mortality by gender among the Chinese

Table 23.5 Mortality risk associated with SRH with control for socio-demographic characteristics, physical & cognitive function, serious illness and engagement in activity in Chinese oldest old by gender

	Men ($N = 3,114$)				Women ($N = 4,669$)			
	Model 1		Model 2		Model 1		Model 2	
	RR	95% CI	RR	95% CI	RR	95% CI	RR	95% CI
Self-rated health (very good = 1)								
Good	1.09	(0.93, 1.28)	0.98	(0.83, 1.15)	1.16	(1.00, 1.36)	1.04	(0.89, 1.21)
Fair	1.62***	(1.38, 1.89)	1.24*	(1.05, 1.46)	1.48***	(1.27, 1.72)	1.15	(0.99, 1.35)
Poor	2.51***	(2.08, 3.04)	1.46***	(1.19, 1.80)	1.84***	(1.55, 2.19)	1.21*	(1.01, 1.44)
Missing	2.50***	(2.02, 3.09)	1.12	(0.83, 1.53)	2.28***	(1.93, 2.70)	1.10	(0.89, 1.36)
Type of residence (city = 1)								
Rural			1.05	(0.95, 1.16)			(0.94, 1.10)	
Years of schooling (0 years = 1)							1.01	
1–6 year			1.01	(0.91, 1.12)			0.96	(0.83, 1.11)
7+ year			1.04	(0.88, 1.22)			1.12	(0.78, 1.61)
Occupation before age 60 (higher = 1)								
Lower			1.34**	(1.12, 1.61)			1.83**	(1.19, 2.83)
Current marital status (married = 1)								
Unmarried			1.39***	(1.24, 1.55)			1.05	(0.83, 1.33)
Self-reported ADL (No limitation = 1)								
1 functional limitation			1.26**	(1.09, 1.45)			1.27***	(1.13, 1.42)
2+ functional limitation			1.78***	(1.57, 2.02)			1.59***	(1.45, 1.75)

Table 23.5 Continued

	Men (N = 3,114)				Women (N = 4,669)			
	Model 1		Model 2		Model 1		Model 2	
	RR	95% CI	RR	95% CI	RR	95% CI	RR	95% CI
MMSE (75–100% = 1)								
50–75%			1.08	(0.94, 1.24)			1.10	(0.94, 1.29)
25–50%			1.18*	(1.03, 1.36)			1.16*	(1.01, 1.34)
0–25%			1.37***	(1.17, 1.61)			1.46***	(1.26, 1.69)
Missing			1.53**	(1.12, 2.07)			1.80***	(1.46, 2.23)
Word fluency (50–100% =)								
0–50%			1.10	(0.99, 1.22)			1.15**	(1.05, 1.27)
Missing			1.18	(0.93, 1.49)			1.21*	(1.04, 1.41)
Suffered from serious illness during last 2 years (No serious illness = 1)								
1 + serious illness			1.15*	(1.01, 1.30)			1.09	(0.99, 1.20)
Engagement in activity (50–100% = 1)								
0–50%			1.29***	(1.16, 1.44)			1.35***	(1.22, 1.49)

Relative risks (RR) are reported; 95% confidence intervals for RR are shown in parentheses; Reference group of each covariate is presented in parentheses; Model 1 is unadjusted model. Model 2 is adjusted model, controlling for socio-demographic characteristics, self-reported ADL, cognitive function, suffering from serious illness during the last 2 years and engagement in activity
* $p<0.05$, ** $p<0.01$, *** $p<0.001$;

oldest old. This indicates that the association of SRH with mortality risk is stronger for males than for females, and that the modifying effects of some covariates on such relationships also differ by gender.

Our results are consistent with the findings of other studies (e.g., Bjorner et al. 1996; Strawbridge and Wallhagen 1999; Wolinsky and Johnson 1992), indicating that SRH has significant and independent effects on mortality, and that other risk factors only partly explain the relationship between SRH and mortality risk. In line with earlier research, our analyses revealed that the role of socio-demographic characteristics in the association of SRH with mortality can be negligible, whereas physical and cognitive functioning, and serious illness are important in accounting for the relationship between SRH and mortality risk (e.g., Benyamini and Idler 1999; Idler and Benyamini 1997; Strawbridge and Wallhagen 1999; Wolinsky and Johnson 1992).

Interestingly, our study found that engagement in activity moderated the relationship between SRH and mortality risk in the oldest old. Controlling for engagement in activity reduced the ability of SRH to predict mortality, even though the adjustment for functional ability and serious illness had already been accounted for in the model. We may assume that to some extent SRH, per se, reflects ability and engagement in daily life, which can contribute to a better understanding of the definition of SRH in the oldest old. In other words, ability and engagement in daily life are components of the definition of health.

Our research showed gender differences in the association of SRH with mortality. Women had lower mortality risks than men; however, women also reported poorer health than did men. As expected, we did find a stronger effect of SRH on mortality risk among men than among women. In comparison with the reference category of "very good," among men, the adjusted relative risk for "poor" (RR = 1.46), and "fair" (RR = 1.24) SRH was significantly greater than among women (RR = 1.21 [poor] and RR = 1.15 [fair]). Poor rated health that included "poor" and "fair" health might be much more fatal among men than among women. This suggests many differences between men and women in the perception of health. Future studies on gender differences in health ratings are needed to further our understanding of the relationship between SRH and mortality among the oldest old.

Different confounding patterns of some covariates by gender revealed in the present study could help explain the gender differences in the relationship between SRH and mortality. Current marital status was a significant predictor of mortality in men but not in women. But it did not influence the relationship between SRH and mortality for men and women. Recalling the statement above about the negligible impact of socio-demographic characteristics on the relationship between SRH and mortality, we may conclude that gender-specific social conditions may have no influence on gender differences in the SRH–mortality association among the oldest old. Serious illness influenced the relationship between SRH in the male subgroup, but it did not in the female subgroup, whereas verbal fluency modified the association of SRH with mortality in women rather than in men. Our results support previous studies on gender differences in health problems. Men and women may well differ in their sensitivity to symptoms and signs of disease and impairments (Verbrugge 1990). Given a specific health problem, men are more likely to die compared to women, while women are more likely to have functional limitations (Arber and Ginn 1991). Suffering from a serious illness is more likely to result in death among men than among women, while it is more likely to lead to functioning limitations such as cognitive impairment among women. Thus, we may assume that serious diseases are more important for the relationship between SRH and mortality among the oldest old men, while functional limitations play a role in the association between SRH with mortality among the oldest old women.

Our research also evaluated the effect of SRH missing data on mortality. Having incomplete data on SRH significantly increased mortality risk in the bivariate model, and in the model that adjusted only for socio-demographic characteristics. This effect, however, was explained when self-reported ADL, MMSE and verbal fluency were added into the model. A possible explanation might be that the participants with missing data on SRH may have poorer physical and cognitive health compared to those with complete SRH data. However, we also took into account missing values on the MMSE and verbal fluency in the models, so that there may be some correlation between the missing values on SRH and the other two variables. This may introduce some complexity in understanding the link between missing values on SRH and mortality risk. For example, unobserved heterogeneity may speak to the relationship between the missing values on SRH and the risk of death. Future studies using logistic regression are needed to better understand the mechanisms that lead to missing values. Our research indicates that missing data on SRH did not occur completely at random and might reflect frailty in physical and cognitive functioning among the oldest old.

Admittedly, the present study has some limitations. Due to the poor quality of chronic disease data reported by participants, we do not have an exact understanding of the impact of disease burden on the relationship between SRH and mortality. Although we found that suffering from serious illness played different roles on the gender predictive pattern, because of lack of data on specific diseases we could not identify which illnesses in particular influenced the association.

In conclusion, the large sample of the oldest old and the good quality of the CLHLS data provided us with a wealth of opportunity to study the SRH and mortality association in this population subgroup, as well as gender differences in this relationship. Our research yielded precise estimations of the relationship between SRH and mortality risk in the oldest old. Statistical controls for confounding variables reveal the unique effects of SRH on mortality. Gender differences in the relationship between SRH and mortality in the oldest old may contribute to the extant literature. Treatment of missing values on SRH in mortality risk estimation is relevant to the discussion on incomplete data in survival analysis at old–old ages when missing values are more likely to occur.

References

Allison, P.D. (2001), *Missing data*. Thousand Oaks, CA: Sage.
Anstey, K.J., M.A. Luszcz, and L.C. Giles (2001), Demographic, health, cognitive, and sensory variables as predictors of mortality in very old adults. *Psychology and Aging* 1, pp. 3–11

Arber, S. and J. Ginn (1991), Gender, class, and health in later life. In: S. Arber and J. Ginn (eds.): *Gender and later life. A sociological analysis of resources and constraints.* London: Sage, pp. 107–128

Baltes, P.B. and J. Smith (2003), New frontiers in the future of aging: From successful aging of the young old to the dilemmas of the fourth age. *Gerontology* 49, pp. 123–135

Bath, P.A. (2003), Differences between older men and women in the self-rated health–mortality relationship. *The Gerontologist* 43 (3), pp. 387–395

Benyamini, Y. and E.L. Idler (1999), Community studies reporting association between self-rated health and mortality. *Research on Aging* 21 (3), pp. 392–401

Benyamini, Y., B. Tzvia, L. Ayala, and M. Baruch (2003), Gender differences in the self-rated health–mortality association: Is it poor self-rated health that predicts mortality or excellent self-rated health that predicts survival? *The Gerontologist* 43, pp. 396–405

Bjorner, J.B., T.S. Kristensen, K. Orth-Gomér, G. Tibblin, M. Sullivan, and P. Westerholm (1996), Self-rated health: A useful concept in research, prevention, and clinical medicine. Stockholm, Sweden: Ord and Form AB.

Deeg, D.J.H. and P.A. Bath (2003), Self-rated health, gender, and mortality in older persons: Introduction to a special section. *The Gerontologist* 43, pp. 369–371

Fayers, P.M. and M.A. Sprangers (2002), Understanding self-rated health. *The Lancet* 359, pp. 187–188

Ferraro, K.F. and J.A. Kelly-Moore (2001), Self-rated health and mortality among black and white adults: Examining the dynamic evaluation thesis. *Journal of Gerontology: Social Sciences* 56B, pp. S195–S205

Fries B.E., J.N. Morris, K.A. Skarupski, C.S. Blaum, A. Galecki, F. Bookstein, and M. Ribbe (2000), Accelerated dysfunction among the very oldest-old in nursing homes. *The Journal of Gerontology Series A: Biological Sciences and Medical Sciences* 55, pp. M336–M341

Han, B., C. Phillips, L. Ferrucci, K. Bandeen-Roche, M. Jylha, J. Kasper, and J.M. Guralnik (2003), Change in self-rated health and mortality among community-dwelling disabled older women. *The Gerontologist* 45 (2), pp. 216–221

Hays, J.C., D. Schoenfeld, D.G. Blazer, and D.T. Gold (1996), Global self-ratings of health and mortality: Hazard in the North Carolina Piedmont. *Journal of Clinical Epidemiology* 29, pp. 969–979

Idler, E.L. (2003), Discussion: Gender differences in self-rated health, in mortality, and in the relationship between two. *The Gerontologist* 43 (3), pp. 372–375

Idler, E.L. and R.J. Angel (1990), Self-rated health and mortality in the NHANES-I Epidemiologic Follow-up Study. *American Journal of Public Health* 80, pp. 446–452

Idler, E.L. and S.V. Kasl (1991), Health perception and survival: Do global evaluation of health status really predict mortality? *Journal of Gerontology: Social Sciences* 46, pp. S55–S65

Idler, E.L. and S.V. Kasl (1995), Self-ratings of health: Do they also predict change in functional ability? *Journal of Gerontology: Social Science* 50B, S344–S353

Idler, E.L. and Y. Benyamini (1997), Self-rated health and mortality: A review of twenty-seven community studies. *Journal of Health and Social Behavior* 38 (1), pp. 21–37

Klein, P.J. and M.L. Moeschberger (2002), *Survival analysis: Techniques for censored and truncated data.* 2nd ed. New York: Springer

Korn, E.L., B.L. Graubard, and D. Midthune (1997), Time-to-event analysis of longitudinal follow-up of a survey: Choice of the time-scale. *American Journal of Epidemiology* 145 (1), pp. 72–80

Lamarca, R., J. Alonso, G. Gomez, and A. Munoz (1998), Left-truncated data with age as time scale: An alternative for survival analysis in the elderly population. *The Journal of Gerontology Series A: Biological Sciences and Medical Sciences* 53 (5), pp. M337–M343

Li, Q. (2005), Subjective well-being and mortality in Chinese oldest old. Working paper in Max Planck Institute for Demographic Research.

Maddox, G.L. and E.B. Douglass (1973), Self-assessment of health: A longitudinal study of elderly subjects. *Journal of Health and Social Behavior* 14, pp. 87–93

Maier, H. and J. Smith (1999), Psychology predictors of mortality in old age. *Journal of Gerontology: Psychological sciences* 1, pp. 44–54

Maier, H., M. McGue, J.W. Vaupel, and K. Christensen (2003), Cognitive impairment and survival at older ages. In: C.E. Finch, Y. Christen, and J.M. Robine (eds.): *Brain and longevity*. Berlin: Springer, pp. 131–144

Manton, K.G., M.A. Woodbury, and E. Stallard (1995), Gender differences in human mortality and aging at late ages: The effect of mortality selection and state dynamics. *The Gerontologist* 35, pp. 597–608

McCallum, J., B. Shadbolt, and D. Wang (1994), Self-rated health and survival: a 7-year follow-up of Australian elderly. *American Journal of Public Health* 84, pp. 1100–1105

Simons, L.A., J. McCallum, Y. Friedlander, and J. Simons (1996), Predictors of mortality in the prospective Dubbo study of Australian elderly. *Australian and New Zealand Journal of Medicine* 26, pp. 40–48

Strawbridge, W.J. and M.I. Wallhagen (1999), Self-rated health and mortality over three decades. *Research on Aging* 21 (3), pp. 402–416

Suzman, R.M., D.P. Willis, and K.G. Manton (1992), *The oldest old*. New York: Oxford University Press.

Van Doorn, C. and S. V. Kasl (1998), Can parental longevity and self-rated life expectancy predict mortality among older persons? Results from an Australian cohort. *Journal of Gerontology: Social Sciences* 53B, pp. S28–S34

Verbrugge, L.M. (1990), Pathways of health and death. In: R.M. Apple (eds.): *Women, health, and medicine in America. a historical handbook,* New York: Garland, pp. 41–79

Verbrugge, L.M. and D.J. Balaban (1989), Patterns of change in disability and well-being. *Medical Care* 27, S128–S147

Wolinsky, F.D. and R.J. Johnson (1992), Perceived health status and mortality among older men and women. *Journal of Gerontology* 47 (6), pp. S304–S312

Zeng, Y., J.W. Vaupel, Z. Xiao, C. Zhang, and Y. Liu (2002), Sociodemograhic and health profiles of the oldest old in China. *Population and Development Review* 28 (2), pp. 251–273

Zeng, Y., Y. Liu, and L.K. George (2003), Gender differentials in the oldest old in China. *Research on Aging* 1, pp. 65–80

Chapter 24
Gender Differences in the Effects of Self-rated Health Status on Mortality Among the Oldest Old in China

Jiajian Chen and Zheng Wu

Abstract While self-rated health has been shown to be a powerful predictor of mortality among the old population, our understanding of how self-rated health interacts with gender and other health related conditions in predicting mortality remains inconclusive. Based on data from China's longitudinal survey of oldest-old population from 1998 to 2000, we examine gender differences in the effects of self-rated health, education and psychosocial factors on the 2-year mortality of respondents aged 80–105 in the baseline wave of the CLHLS. Self-rated health is found to be consistently related to longer life after controlling for the effects of socio-demographic and physical health conditions among oldest-old men and women. This finding suggests that self-rated health status is an important health indicator and mortality predictor for the oldest-old in China. As education and psychosocial factors are also found to be independently predictive of mortality, they appear to play an important role in modifying the longevity among the oldest-old in China.

Keywords Chinese oldest-old, Cox proportional hazards regression, education, gender difference, inexpensive instrument, interviewer-rated health, moderating effect, mortality, mortality converge, predictor, predictive power, psychosocial factors, self-rated health, subjective dimension

24.1 Introduction

According to numerous studies, self-rated health (SRH) is a robust indicator of overall health status. The literature consistently shows that this simple assessment of global health is an independent predictor of mortality, especially among the elderly (Benyamini and Idler, 1999; Idler and Benyamini 1997). Since the 1970s, investigations into self-perceptions of health have repeatedly demonstrated the high

---

J. Chen
East-West Center, Research Program, Population and Health Studies, 1601 East-West Road, Honolulu, HI 96848, USA
e-mail: chenj@eastwestcenter.org

face value of SRH measures, showing that the relationship between subjective health evaluations and short- and long-term mortality persists from country to country and among different populations (Helweg-Larsen et al. 2003). Moreover, the capacity for SRH to predict mortality and other subsequent health outcomes is not spurious, but is independent of conventional objective health risk factors, including sociodemographic characteristics, psychosocial resources, social support, and medical history. Indeed, Mossey and Shapiro (1982) suggest that SRH predicts long-term survival more accurately than do medical records.

These findings have justified further research into SRH, and are promising in several respects. Most obviously, the predictive power of SRH offers the health research community a cost-effective and uncomplicated means for estimating patterns of mortality, disease, and health care utilization, particularly in those cases where more refined health indicators are missing, inaccurate, or difficult to obtain (Kaplan and Baron-Epel, 2003). Moreover, perceived health ratings have the capacity to capture a highly inclusive range of health variables in a simple and comprehensive measure (Benyamini and Idler 1999). For example, as an assessment of global health status, SRH is more or less a cumulative proxy indicator for stressors, social support, health behaviors, personal attitudes, chronic conditions, functional limitations, and disease history (Benyamini et al. 2003a; Kaplan and Baron-Epel 2003). Available in many nationally representative surveys, SRH is, therefore, an inexpensive instrument for targeting at-risk populations for timely medical interventions to prevent health care crises and prolong life. As a targeting instrument, SRH represents the potential to add something more to medical diagnoses and medical screening while maximizing the cost benefits associated with preventative medicine and early detection of health problems.

However, as Kaplan and Baron-Epel (2003) remark, our understanding of the process through which SRH links up with specific health outcomes remains unclear, even though the literature is extensive. Helweg-Larsen et al. (2003) have suggested that there is a need to move in the direction of understanding why this relationship exists by examining variables that might moderate the relationship. Their research indicates that SRH may be less predictive of mortality when people age because their perception of health is increasingly relying on social comparison rather than actual physical health. Therefore, further assessment of the factors underlying the relationship between SRH and mortality among the elderly, especially the oldest-old, warrants further attention.

A broad range of variables affect personal assessments of global health, including medical history and current physical health status, as well as variables not usually considered in biomedical models of physical health (Idler et al. 1999). For example, having experienced a serious illness in the past often lowers SRH even if the illness is no longer present, for prior illness experiences are concrete reminders of health risks and the potential for illness (Benyamini et al. 1999). Multiple medical and non-medical variables tend to confound the formation of personal assessments of health, making SRH a fluid and dynamic health indicator. In other words, SRH reflects an introspective process involving diverse evaluation criteria and comparisons, and the subjective character of self-ratings is the main reason why health researchers have had limited success in determining why self-rated health influences mortality and other health outcomes. More importantly, people's health needs are "socially constructed phenomena located within a socially organized context" (Anderson 1986). There is a *subjective* dimension to how individuals experience illness, which means that health needs are not just neutral conditions related to corresponding medical problems. Rather, socially specific beliefs about illness define how individuals perceive their health and how these norms guide self-rated health. People's SRH is therefore believed to be influenced by both biomedicine and social and medical ideologies (Kaplan and Baron-Epel 2003).

Along with the recent debate on whether the effects of SES on mortality converge at old ages (Helweg-Larsen et al. 2003; House et al. 1990, 1994; House 2002; Ross and Wu 1996), recent research shows that SES inequalities in physical health may attenuate at old ages, but remain an important predictor of deteriorating emotional health and mortality among old women (Mishra et al. 2004). Emerging evidence suggests that SES inequalities in health begin from early life and accumulate over lifetime. Educational attainment is a preferable indicator of SES because it is universal and stable after young adulthood, and also because it structures other SES indicators (i.e., occupation and income) (Lynch 2003; Mishra et al., 2004; Ross and Wu, 1996). As such, the moderating effect of educational attainment on the relationship between SRH and mortality among the oldest-old should be addressed.

As also suggested, the impact of SRH on mortality may also be mediated through exposure to psychosocial factors at the interpersonal or intrapersonal level (Idler and Benyamini 1997; Mackenbach et al. 2002). A recent empirical study shows that social isolation is associated with mortality among elderly patients even after controlling for age and disease severity (Brummett et al. 2001). Assessment of possible moderating effects of psychosocial factors on the oldest-old SRH-mortality relationship is also needed.

Recent studies suggest that the overall predictive power of SRH could be overestimated because of gender-specific effects (Deeg and Kriegsman, 2003; Benyamini et al. 2003b). The physical, psychosocial, and societal differences between women and men suggest that gender is an important factor in determining how individuals perceive their health status. The gender-specific differences in the accuracy of SRH, which, combined with gender variation in health trajectories at old age, could also reflect how effectively gender-specific SRH predicts all-cause mortality. For example, on the one hand, women tend to have a greater understanding of their health symptoms and disease histories, and this knowledge may increase the accuracy of their SRH. This greater awareness may thus leave little information to be supplemented by SRH and lead to a weaker independent SRH-mortality relationship for women. On the other hand, women live longer than men, and also experience more years of non-life threatening but negative health problems. Therefore, a greater inclusiveness of minor health problems with less accuracy of SRH may also result in a weaker SRH-mortality relationship for women than for men.

These gender-related inconsistencies could be an artifact of differences in study design (Bath 2003), but enough evidence exists to warrant further consideration into whether gender-specific SRH has independent effects on all-cause mortality rates.

The recent literature includes several studies that focus on gender differences in the effects of SRH on mortality, but these only cover the developed world (e.g., Bath 2003; Benyamini et al. 2003b; Deeg and Kriegsman 2003). This leaves much work to be done in developing countries, where widespread poverty often exacerbates gender disparities (Frankenberg and Jones 2004).

Of special empirical interest here is gender variation in the effect of SRH on mortality among the oldest-old in Mainland China. Low mortality and fertility below the replacement level have initiated a transformation of China's population structure, which is now aging rapidly. As Myers et al. (1992) remark, "further aging of this enormous elderly population could produce massive numbers of persons aged 80 and over and strain society's capability to support and care for its oldest old." While the Chinese elderly population aged 65 and over has grown from 26 million in 1950 up to 87 million in 2000, growth in the proportionate numbers of the oldest old population aged 80 and over was especially intense, tripling during 1950–2000 (United Nations 2002; see more discussion of these trends in the first chapter of this volume by Poston and Zeng). As SRH could represent a highly inclusive and cost-effective health measure, it is important to assess its overall efficacy by examining whether SRH can accurately gauge health and mortality by gender, especially among women who account for a majority (66 percent) of the oldest old population in China (NBS 2002).

Recent studies demonstrate the relationship between SRH and mortality among the Chinese elderly population (Yu et al. 1998; Liang et al. 2000). These studies are noteworthy because they replicate the basic findings from North American and European studies, thus demonstrating the cross-cultural validity of the effects SRH has on all-cause mortality rates. However, gender-specific effects of SRH on mortality among the elderly in China have yet to be examined.

The research reported in this chapter aims to examine the gender differences in the SRH-mortality relationship and in the factors that mediate the relationship among the oldest old in China. Specifically, our research addresses the following three questions, among men and women separately, to further specify possible gender differences in underlying variables that moderate the relationship between SRH and oldest-old mortality:

(1) Is SRH predictive of oldest-old mortality in China?
(2) Do risk factors moderate the association between SRH and mortality?
(3) Is SRH a valid evaluation of health among the oldest-old in China?

24.2 Methods

24.2.1 Study Sample

Our data sources are the oldest-old male and female cohorts, aged 80 and above, participating in the China Longitudinal Healthy Longevity Survey (CLHLS) conducted in 1998 and 2000 (Zeng et al. 2001, 2002; Zeng et al. 2004; Gu and Zeng 2004).

The CLHLS is a large population-based sample of oldest-old people residing in private households or nursing homes from about half of the counties and cities in 22 of the country's 31 provinces, covering around 85 percent of the total population in 1998 and 2000. The CLHLS was conducted through an interview process using a structured questionnaire and included physical examinations by a medical doctor, a nurse, or a medical student. The baseline survey interviewed 9,073 seniors aged 80 or older in 1998, among them 8,805 respondents who were aged 80–105 whose reported age is considered reliable (Zeng et al. 2001, 2002). The survey participation rate was 88 percent, including those who were unavailable because of death, severe illness, and migration, and was 98 percent after excluding those unavailable for an interview (Zeng et al. 2001).

Of the original 8,805 respondents aged 80–105 years old who resided in private households or nursing homes in 1998, 4,691 respondents (53.3 percent) survived; 3,264 respondents (37.0 percent) died; and 850 respondents (9.7 percent) were untraceable in the follow-up survey in 2000. The follow-up survey collected information about date of death, cause of death, and included a comprehensive set of questions asked of close family members about the circumstances before death of persons who had been interviewed in 1998 but who had died before the 2000 survey. The follow-up death rates have been found to be accurate at ages 90 and over, and somewhat underestimated at ages 80–89 (Zeng et al. 2004; Gu and Zeng 2004). The sub-sample used in this research includes 7,938 respondents aged 80–105 after excluding 850 untraceable individuals and 17 individuals with unknown information on date of death. It is relevant to examine 2-year follow-up short-term mortality because the advanced age of the oldest old may dramatically increase the long-term risk of mortality regardless of self-rated health.

24.2.2 Statistical Analysis

A standard life-table method is used to estimate the cumulative mortality of the oldest old during a 2-year time interval since the baseline interview. Respondents alive at the follow-up are censored.

Cox proportional hazards regression models are used to estimate the effects of self-rated health status, education, psychosocial factors, and other socio-demographic and health-related covariates on mortality. We also use multinomial logistic regression models to examine the effects of the same set of independent variables on SRH and interviewer-rated health (IRH), respectively.

As Idler recommends (2003), we initially pool our data to conduct tests of the SRH-mortality relationship by gender interaction and then conduct data analyses for men and women separately to investigate gender-specific mediating effects of risk factors on mortality sequentially in a hierarchical approach. We supplement these analyses with an additional investigation that includes measures of IRH, for the reason that information from this evaluation (which includes a medical examination) could have more objectively shaped the measurement of general health status. Finally, we examine the effects of the same set of independent variables on SRH and

IRH, respectively, to further specify which factors underlie the relationship between SRH and mortality.

24.2.3 Measures

Self-rated health (SRH) is defined based on a single-item question, namely, "How do you rate your health at present—very good, good, so-so, bad, very bad?" The survey requires respondents to answer the question by themselves: no proxy responses were allowed (Zeng et al. 2002; Zeng and Vaupel 2004). A three-category self-rated health variable was created: good (including "very good" and "good"), fair ("so-so"), and poor (including "bad" and "very bad").

Interviewer-rated health (IRH) is based on a question answered by the interviewer: "The interviewee was: surprisingly healthy (almost no obvious ailments); relatively healthy (only minor ailments); moderately ill (moderate degree of major ailments or illnesses); or very ill (major ailments or diseases, bedridden, etc.)." Corresponding to the grouping of self-rated health (good, fair, and poor), interviewer-rated health is defined with three categories: sound health (i.e., "surprisingly healthy"), regular health (i.e., "relatively healthy"), and ill health (including "moderately ill" and "very ill").

Socio-demographic variables included education (coded by total number of years attending school), sex, age (by single year), rural/urban residence, and marital status (currently married, not married). A large proportion of the oldest-old did not go to school (68 percent) or had only 1–5 years of schooling (21 percent). Few respondents had 6 or more years of education (10 percent). Oldest-old women are more disadvantaged than oldest-old men in education. For example, the majority of oldest-old women (88 percent) did not attend school, and only a few had 1–5 years of school (8 percent) or 6 years or more (3 percent). The corresponding distribution for oldest-old men has more variation, with 38 percent having no schooling, 41 percent having 1–5 years of schooling, and 20 percent having 6 years of schooling or more. As such, a categorical variable for level of education for the oldest-old is not desirable.

A psychosocial scale variable, *feeling lonely* (House 2001; Brummet et al. 2001), is included in the analysis. Note that the respondents were required to answer this question by themselves. Based on a question of how similar the oldest-old respondents were to the people who "often feel lonely and isolated," the loneliness scale variables were coded 5 for very similar, 4 for similar, 3 for so-so, 2 for not similar, and 1 for not similar at all.

Functional and physical health conditions included *activities of daily living* (ADL) and various types of chronic illnesses. The ADL variable was categorized as *dependent* if the oldest-old respondents received assistance for bathing, dressing, toilet use, transfer to bed, continence, and feeding, or *partially dependent* if they only received some help in doing these, or otherwise *independent*. The chronic illnesses refer to hypertension and any other life-threatening chronic conditions including diabetes, heart disease, stroke, bronchitis, pulmonary tuberculosis, cancer, and Parkinson's disease (Verbrugge 1989).

24.3 Results

Table 24.1 shows the distributions of the selected variables in the study sample. On average, oldest-old men have more years of schooling than oldest-old women. While oldest-old men are less likely to be ADL dependent (20 percent) than oldest-old women (34 percent), oldest-old men are more likely to have life-threatening chronic conditions (24 percent) than women (19 percent). Nevertheless, the average feeling lonely scale is the same for both oldest-old men and women. Moreover, the distribution of SRH and IRH shows a clear gender difference; the percentage of good SRH or sound IRH among oldest-old men is always greater than that of oldest-old women. However, it is noteworthy that the percentage of good SRH is higher than the percentage of sound IRH for oldest-old men (59 vs. 48 percent) and women (49 vs. 35 percent). Apparently, oldest-old men and women tend to view their own health status more optimistically than do the interviewers.

Is SRH predictive of gender-specific oldest-old mortality in China? Table 24.2 shows that with both sexes pooled together, oldest-old men have a higher mortality risk than oldest-old women after controlling for age, other socio-demographic characteristics, physical health conditions, and psychosocial factors. Moreover, the gender-SRH interactions are always statistically significant when various risk factors are controlled hierarchically; this indicates that the SRH-mortality relation is stronger for oldest-old men than for oldest-old women (Model 1–4 in Table 24.2). Clearly, the effects of SRH on mortality are gender-specific, and separate analyses by gender are needed to test the mediating effects of risk factors on mortality among the oldest old in China.

Figure 24.1 shows a clear gradient in cumulative mortality by three levels of self-rated health in oldest-old men and women. The cumulative 2-year mortality is lowest in the self-rated good health group and is highest in the self-rated poor health group for oldest-old men and women, respectively. The gaps in cumulative mortality are wider for the men than for the women.

Further, gender-specific analyses also show a clear "dose-response" pattern where relative mortality risk is highest for poor SRH and next highest for fair SRH, as compared with good SRH for the men and women. Moreover, the relative mortality risks are greater for the men. For example, among oldest-old men, the relative risks of death among those with poor and fair SRH are 2.4 (95 percent confidence interval CI = 2.0–2.9) and 1.6 (95 percent CI = 1.4–1.8), times as much as those with good SRH. By contrast, the corresponding relative risks for oldest-old women are 1.6 (95 percent CI = 1.4–1.9) and 1.3 (95 percent CI = 1.1–1.4), times as much as those with good SRH (Model 1 in Table 24.2).

Do risk factors moderate the association between SRH and mortality? Model 2 of Table 24.2 shows that after controlling for the effects of education, marital status, and urban/rural residence, the inverse association between SRH and mortality is further enhanced rather than attenuated, for men and for women. There is also an inverse association between years of schooling and mortality among the oldest-old men and women when SRH and other socio-demographic variables are taken into account.

Table 24.1 Characteristics of variables used in the multivariate analyses

Variable	Men ($n = 3171$)	Women ($N = 4767$)
Age (years)	83.5	84.1
Self-rated health		
Good	62.1	56.7
Fair	29.8	33.0
Poor	6.9	8.6
Missing	1.2	1.7
Objective health		
Surprisingly healthy	54.6	48.0
Relatively healthy	37.3	43.9
Moderately or very ill	7.3	7.7
Missing	0.8	0.4
Education		
Years of schooling	3.4	0.8
Missing	–	–
Marital status		
Currently married	46.5	13.0
Not married	53.3	87.0
Missing	0.2	0.0
Place of residence		
Rural	66.2	64.7
Urban	33.8	35.3
ADL		
Independent	87.0	81.6
Partially dependent	3.4	5.9
Dependent	9.6	12.5
Hypertension		
Yes	15.3	18.0
No	84.7	82.0
Life threatening conditions		
Yes	26.1	21.5
No	73.9	78.5
Feeling lonely	2.3	2.5
Missing	–	–

Source: Longitudinal Healthy Longevity Survey in China, 2000–2002.
Note: The means and percentages are weighted based on sample in 1998.

Model 3 of Table 24.2 shows that after controlling for age, socio-demographic characteristics and functional and physical health conditions, including ADL and chronic conditions, the association between SRH and mortality attenuates substantially. Nevertheless, poorer health is still associated with greater mortality risks for both men and women. Education is still independently associated with reduced mortality among men and women, even when demographic and health-related factors are taken into account.

24 Gender Differences in the Effects of Self-rated Health Status

Table 24.2 Adjusted hazard ratios of mortality risks for self-rated health (SRH) among the oldest-old in China

Characteristics	Model 1 Risk ratio	95% CI		Model 2 Risk ratio	95% CI		Model 3 Risk ratio	95% CI		Model 4 Risk ratio	95% CI	
Both sexes												
Age	1.078*	1.073	1.084	1.073*	1.067	1.079	1.059*	1.052	1.065	1.059*	1.052	1.065
Gender (Women)												
Men	1.122*	1.007	1.250	1.283*	1.143	1.439	1.393*	1.241	1.564	1.387*	1.232	1.561
SRH (Good)												
Fair	1.268*	1.145	1.404	1.275*	1.150	1.412	1.179*	1.063	1.307	1.171*	1.051	1.305
Poor	1.606*	1.384	1.864	1.610*	1.394	1.861	1.345*	1.155	1.566	1.188*	1.001	1.411
Gender × SRH interaction												
Gender × Fair	1.262*	1.075	1.482	1.247*	1.061	1.465	1.231*	1.048	1.447	1.237*	1.047	1.461
Gender × Poor	1.525*	1.195	1.947	1.519*	1.188	1.941	1.321*	1.033	1.690	1.460*	1.119	1.903
Years of schooling				0.975*	0.960	0.989	0.970*	0.956	0.984	0.970*	0.956	0.985
Marital status (Not married)												
Currently married				0.708*	0.621	0.807	0.696*	0.610	0.793	0.695*	0.607	0.796
Place of residence (Urban)												
Rural				1.049	0.967	1.138	1.101*	1.015	1.195	1.095*	1.006	1.192
ADL (Independent)												
Dependent							1.892*	1.734	2.064	1.859*	1.707	2.025
Partially dependent							1.250*	1.087	1.438	1.233*	1.068	1.424
Hypertension (No)												
Yes							0.864*	0.767	0.972	0.864*	0.765	0.977
Life threatening conditions (No)												
Yes							1.150*	1.052	1.258	1.167*	1.063	1.281
Feeling lonely										1.067*	1.022	1.114
Men												
Age	1.075*	1.066	1.084	1.065*	1.056	1.075	1.052*	1.042	1.062	1.052*	1.042	1.062

Table 24.2 (Continued)

Characteristics	Model 1 Risk ratio	95% CI		Model 2 Risk ratio	95% CI		Model 3 Risk ratio	95% CI		Model 4 Risk ratio	95% CI	
SRH (good)												
Fair	1.594*	1.408	1.805	1.587*	1.402	1.798	1.431*	1.260	1.625	1.431*	1.255	1.632
Poor	2.430*	2.003	2.949	2.440*	2.009	2.965	1.704*	1.384	2.099	1.676*	1.344	2.089
Years of schooling				0.978*	0.962	0.995	0.973*	0.957	0.990	0.974*	0.958	0.991
Marital status (Not married)												
Currently married				0.674*	0.582	0.780	0.660*	0.571	0.764	0.660*	0.568	0.767
Place of residence (Urban)												
Rural				1.097	0.965	1.246	1.166*	1.025	1.327	1.167*	1.023	1.332
ADL (Independent)												
Dependent							2.012*	1.740	2.327	1.962*	1.688	2.281
Partially dependent							1.248	0.994	1.569	1.260	1.000	1.588
Hypertension (No)												
Yes							0.862	0.718	1.034	0.859	0.712	1.036
Life threatening conditions (No)												
Yes							1.196*	1.042	1.372	1.183*	1.026	1.363
Feeling lonely										1.083*	1.012	1.159
Women												
Age	1.081*	1.012	1.155	1.079*	1.071	1.087	1.064*	1.055	1.073	1.064*	1.055	1.073
SRH (good)												
Fair	1.268*	1.145	1.404	1.275*	1.150	1.413	1.188*	1.071	1.318	1.182*	1.060	1.318
Poor	1.607*	1.384	1.865	1.613*	1.389	1.874	1.370*	1.175	1.597	1.211*	1.017	1.441
Years of schooling				0.965*	0.935	0.996	0.961*	0.931	0.991	0.956*	0.925	0.988
Marital status (Not married)												
Currently married				0.831	0.608	1.134	0.854	0.626	1.167	0.861	0.625	1.184

Table 24.2 (Continued)

Characteristics	Model 1		Model 2		Model 3		Model 4	
	Risk ratio	95% CI	Risk ratio	95% CI	Risk ratio	95% CI	Risk ratio	95% CI
Place of residence (Urban)								
Rural			1.010	0.909 1.123	1.055	0.948 1.173	1.039	0.930 1.162
ADL (Independent)								
Dependent					1.824*	1.637 2.032	1.799*	1.606 2.016
Partially dependent					1.260*	1.054 1.506	1.226*	1.019 1.475
Hypertension (No)								
Yes					0.863	0.738 1.008	0.869	0.740 1.021
Life threatening conditions (No)								
Yes					1.114	0.989 1.255	1.153*	1.018 1.306
Feeling lonely							1.058*	1.001 1.119

Source: Longitudinal Healthy Longevity Survey in China, 2000–20; Reference group is in parentheses

* $p < 0.05$.

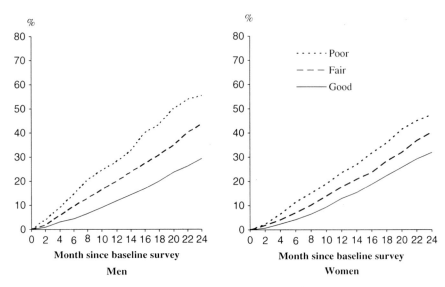

Fig. 24.1 Cumulative mortality by SRH among the oldest-old in China

Model 4 in Table 24.2 shows that after adjusting for "feeling lonely," the association between self-rated poor health and mortality further attenuates for both men and women, although the effects of SRH are still statistically significant for both groups. This finding provides new evidence that education and psychosocial factors play important causation roles in partially explaining the association between self-rated health and mortality.[1]

Is SRH a valid evaluation of health? Table 24.3 shows that after controlling for the strong effect of interviewer-rated health (IRH) on mortality, the effect of SRH is dramatically attenuated. For example, the relative risk of death for oldest-old men who self-rated their health as poor drops from 2.4 when only adjusting for age, to 1.5 when adjusting for age and interviewer-rated health (see Model 1 in Table 24.2 and Model 1 in Table 24.3). The result indicates that the association between SRH and oldest-old mortality is substantially explained by a more objectively evaluated IRH.

---

[1] Further analysis shows that after controlling for age, socio-demographic factors, physical health conditions, and psychosocial factors hierarchically, the age-SRH interaction and age-education interaction are not statistically significant for oldest-men and women (data not shown). This finding indicates that age does not moderate the association between SRH and mortality, nor the relationship between education and mortality among oldest-old men and women in China. In other words, the finding provides clear evidence that SRH and years of education are two important predictors of mortality among oldest-old men and women, regardless of age. Our results also show that after controlling for other risk factors, the interaction of age-feeling lonely is statistically significant for oldest-old men but not for oldest-old women. The result indicates that the independent effect of feeling lonely diminishes with advanced age among oldest-old men in China (data not shown). These findings indicate that the observed effects of SRH, education and feeling lonely are not age-dependent for oldest-old men and women, except that there is an "age-flattening-effect" of feeling lonely on mortality among oldest-old men.

Table 24.3 Adjusted hazard ratios of mortality risks for self-rated health (SRH) and interviewer-rated health (IRH) among the oldest-old men and women in China

Characteristics	Model 1 Risk ratio	95% CI		Model 2 Risk ratio	95% CI		Model 3 Risk ratio	95% CI		Model 4 Risk ratio	95% CI	
Men												
Age	1.067*	1.058	1.077	1.058*	1.048	1.068	1.050*	1.040	1.060	1.050*	1.040	1.060
SRH (Good)												
Fair	1.316*	1.150	1.507	1.319*	1.157	1.504	1.291*	1.132	1.472	1.298*	1.134	1.487
Poor	1.458*	1.168	1.821	1.448*	1.158	1.812	1.305*	1.042	1.634	1.304*	1.029	1.653
IRH (Surprisingly healthy)												
Relatively healthy	1.514*	1.324	1.731	1.501*	1.312	1.717	1.372*	1.194	1.577	1.368*	1.187	1.577
Moderately or very ill	2.745*	2.244	3.358	2.797*	2.284	3.424	2.111*	1.690	2.637	2.057*	1.632	2.593
Years of schooling				0.979*	0.962	0.995	0.976*	0.960	0.992	0.977*	0.960	0.994
Marital status (Not married)												
Currently married				0.658*	0.569	0.761	0.654*	0.565	0.756	0.651*	0.561	0.757
Place of residence (Urban)												
Rural				1.069	0.941	1.215	1.124	0.987	1.279	1.125	0.985	1.284
ADL (Independent)												
Dependent							1.681*	1.439	1.964	1.657*	1.411	1.945
Partially dependent							1.209	0.961	1.521	1.214	0.962	1.532
Hypertension (No)												
Yes							0.835	0.695	1.002	0.832	0.689	1.004
Life threatening conditions (No)												
Yes							1.092	0.949	1.256	1.083	0.937	1.251
Feeling lonely										1.066	0.995	1.142

Table 24.3 (Continued)

Characteristics	Model 1 Risk ratio	95% CI		Model 2 Risk ratio	95% CI		Model 3 Risk ratio	95% CI		Model 4 Risk ratio	95% CI	
Women												
Age	1.074*	1.066	1.082	1.072*	1.064	1.080	1.063*	1.054	1.071	1.063*	1.054	1.072
SRH (Good)												
Fair	1.076	0.967	1.198	1.084	0.973	1.207	1.079	0.968	1.202	1.075	0.960	1.204
Poor	1.192*	1.016	1.398	1.196*	1.018	1.404	1.161	0.988	1.364	1.028	0.857	1.232
IRH (Surprisingly healthy)												
Relatively healthy	1.371*	1.218	1.543	1.371*	1.217	1.544	1.254*	1.110	1.417	1.272*	1.121	1.444
Moderately or very ill	2.242*	1.937	2.595	2.245*	1.938	2.601	1.770*	1.507	2.078	1.785*	1.506	2.115
Years of schooling				0.970	0.940	1.001	0.966*	0.936	0.997	0.962*	0.931	0.994
Marital status (Not married)												
Currently married				0.848	0.621	1.158	0.867	0.635	1.185	0.870	0.632	1.197
Place of residence (Urban)												
Rural				0.995	0.895	1.107	1.032	0.927	1.149	1.021	0.913	1.142
ADL (Independent)												
Dependent							1.565*	1.393	1.759	1.548*	1.369	1.749
Partially dependent							1.216*	1.017	1.455	1.185	0.984	1.426
Hypertension (No)												
Yes							0.852*	0.729	0.997	0.860	0.732	1.011
Life threatening conditions (No)												
Yes							1.059	0.939	1.194	1.100	0.970	1.247
Feeling lonely										1.049	0.991	1.109

Source: Longitudinal Healthy Longevity Survey in China, 2000–20; Reference group is in parentheses
*$p < 0.05$.

Note that the association between SRH and mortality is still statistically significant among oldest-old men even when controlling for other sociodemographic factors, physical health conditions, psychosocial factors, and IRH (Model 2–4 in Table 24.3). Remarkably, SRH seems to reflect something more than the survey data have measured. For oldest-old women, the risk ratio of mortality for poor SRH is still statistically significant when controlling for IRH (Model 1–2 in Table 24.3), indicating that SRH reflects something independent of what IRH is measuring. However, when both IRH and physical health conditions are also controlled, poor SRH is no longer associated with increased mortality (Model 3 in Table 24.3). It appears that the SRH-mortality relationship can be well explained by more objective health measures for oldest-old women. One likely explanation for the result is that oldest-old women may evaluate their general health based on the more objective physical health conditions.

Table 24.4 shows the multinomial logistic regression estimated odds ratios of SRH and IRH for oldest-old men and women, respectively. The results show in detail whether or not the same control variables, which were used in the mortality analyses, are similarly associated with SRH and IRH. First, age shows the opposite effect on SRH and IRH: age is clearly associated with increased odds of interviewer-rated "ill" or "regular" health rather than "sound" health; by contrast, age is associated with decreased odds of self-rated poorer health rather than good health. The finding indicates a common subjective dimension of SRH among oldest-old men and women.

Second, education is consistently associated with reduced odds of poorer SRH and IRH for oldest-old men, whereas education is only associated with decreased odds of poorer IRH for oldest-old women, when controlling for other variables. For men, education is an important mediator linking the SRH-mortality relationship. For women, education is not linked with SRH as closely as with both IRH and mortality. This is likely attributable to the fact that women tend to experience more years of non-life threatening but negative minor health problems than men; hence, women's education differential in health may narrow more than men's, although education inequality in objective health and mortality may still widen.

Third, feeling lonely is consistently associated with increased odds of poorer SRH and IRH for both men and women. Moreover, the loneliness-SRH relationship is slightly stronger than the loneliness-IRH relationship. This finding suggests that psychosocial factors indeed play an important role in moderating the SRH-mortality relationship among men and women.

Last but not least, being ADL dependent and having life threatening chronic conditions are consistently associated with increased odds of poorer SRH and IRH for both oldest-old men and women. The associations of the two physical health condition variables with SRH are very strong, although not as strong as their associations with IRH. Being ADL dependent and having fatal conditions are also associated with increased risk of oldest-old mortality; this finding further indicates that the objective physical health conditions are also important underlying variables linking SRH with mortality among both oldest-old men and women.

Table 24.4 Multinomial logistic regressions (odds ratios) of SRH and IRH on selected characteristics among the oldest-old men and women in China

Characteristics	SRH				IRH			
	Poor vs. Good		Fair vs. Good		Ill vs. Sound		Regular vs. Sound	
	Risk ratio	95% CI	Risk ratio	95% CI	Risk Ratio	95% CI	Risk ratio	95% CI
Men								
Age	0.952*	0.926 0.977	0.989	0.975 1.002	1.010	0.984 1.036	1.018*	1.004 1.031
Years of schooling	0.940*	0.900 0.982	0.972*	0.951 0.994	0.936*	0.895 0.978	0.985	0.964 1.007
Marital status (No)								
Currently married	1.350	0.952 1.916	0.963	0.801 1.159	1.384	0.971 1.971	0.874	0.728 1.049
Place of residence (Urban)								
Rural	1.191	0.844 1.683	0.956	0.801 1.139	2.087*	1.467 2.970	1.331*	1.117 1.585
ADL (Independent)								
Partially dependent	2.252*	1.107 4.581	1.669*	1.205 2.313	3.437*	1.795 6.582	1.675*	1.204 2.331
Dependent	12.104*	8.314 17.620	2.566*	2.022 3.256	39.468*	26.269 59.299	4.757*	3.571 6.339
Hypertension (No)								
Yes	1.225	0.785 1.911	1.092	0.854 1.396	1.799*	1.156 2.799	1.449*	1.135 1.851
Life threatening conditions (No)								
Yes	4.016*	2.875 5.611	2.033*	1.675 2.468	6.458*	4.579 9.107	2.893*	2.358 3.550
Feeling lonely	1.876*	1.593 2.209	1.373*	1.247 1.511	1.754*	1.481 2.077	1.341*	1.216 1.479
Women								
Age	0.981*	0.964 0.999	0.994	0.934 1.057	1.038*	1.020 1.056	1.021*	1.011 1.032
Years of schooling	0.994	0.934 1.057	1.002	0.968 1.037	0.938*	0.881 0.999	0.964*	0.930 0.998
Marital status (No)								
Currently married	1.109	0.651 1.889	1.031	0.754 1.409	1.338	0.754 2.374	1.128	0.830 1.533
Place of residence(Urban)								
Rural	1.444*	1.118 1.865	1.222*	1.054 1.416	1.623*	1.274 2.068	1.183*	1.018 1.375

Table 24.4 (Continued)

	SRH						IRH					
	Poor vs. Good			Fair vs. Good			Ill vs. Sound			Regular vs. Sound		
Characteristics	Risk ratio	95% CI		Risk ratio	95% CI		Risk Ratio	95% CI		Risk ratio	95% CI	
ADL (Independent)												
Partially dependent	0.938	0.574	1.532	1.116	0.873	1.425	2.296*	1.490	3.536	1.449*	1.135	1.851
Dependent	3.271*	2.507	4.269	1.734*	1.469	2.046	20.011*	15.188	26.366	3.365*	2.755	4.110
Hypertension (No)												
Yes	0.991	0.704	1.395	1.219*	1.001	1.484	1.096	0.786	1.528	1.248*	1.017	1.532
Life threatening conditions (No)												
Yes	2.599*	2.008	3.364	1.489*	1.250	1.773	3.288*	2.514	4.301	2.072*	1.705	2.519
Feeling lonely	1.779*	1.579	2.003	1.450*	1.343	1.566	1.510*	1.342	1.699	1.220*	1.125	1.323

Source: Longitudinal Healthy Longevity Survey in China, 2000–20; Reference group is in parentheses.
*$p < 0.05$.

24.4 Limitations

There were 560 respondents who could not answer the SRH question. Respondents with missing information on SRH tend to have a greater mortality risk than any of the other respondents who have self-rated their health at different levels (data not shown). The majority of these respondents with missing information on SRH had no schooling, were ADL dependent, and not able to hear the interview. The percentage of missing information on self-rated health is slightly higher among oldest-old women (9 percent) than among oldest-old men (4 percent) (Table 24.1). As such, excluding these respondents from the analyses underestimates the strength of associations between risk factors and mortality, especially for oldest-old women. Nevertheless, we decided against imputing missing data with known information because hypothesis tests on the validity of SRH in predicting mortality may have been rendered less reliable.

Controls for physical health conditions and psychosocial factors resulted in a loss of 300 respondents with missing information. As these missing respondents are more likely to be ill and illiterate, the estimated associations between risk factors and mortality are also likely to be underestimated.

Our research focuses on the gender-specific mediating effects of demographic (as measured by age), social and SES (as measured by education), physical health conditions and psychosocial factors (as measured by diagnosed diseases, ADL, and feeling lonely) on the SRH-mortality relationship of the oldest-old in China. Health behavior is known to be an important predictor of mortality in the elderly population (Lee et al. 1997; Paffenbarger et al. 1993). Our data show that adding physical activity (as measured by regular exercise) changes the effect of SRH on mortality only slightly for oldest-old men and women when all the selected variables are controlled; the exception is when the psychosocial variable is also included, the effect of SRH on mortality is only marginally significant at the 0.05 level for oldest-old women (data not shown). The relationship between SRH and health behavior is known to be rather complex as SRH could influence health behavior (Idler and Benyamini 1997), and the addition of the health behaviors would not change very much in the relationship between SRH and overall mortality (Benjamins et al. 2004). The effect of health behavior on the SRH-mortality relationship will be addressed in a future study.

The follow-up period for our mortality analysis was only 2 years. It is desirable to have a longer-term follow-up. Nevertheless, the oldest-old mortality at advancing ages already approached 50 percent in 2 years. It is possible that we could reach the ceiling effect for follow-up studies in the near future.

The average education level of the oldest-old is very low, especially for women. This makes the grouping of education by level difficult. As such, the effect of education is modest.

24.5 Discussion

The research we presented in this chapter demonstrates that SRH is a powerful predictor of mortality among oldest-old men and women in China. The finding of a persistent "dose-response" pattern that poorer health is independently associated with a greater risk of mortality among the oldest-old in China is encouraging as SRH is indeed a comprehensive indicator for general health status and future mortality of the oldest-old men and women in China.

Consistent with findings from many other countries, our research observes certain gender differences in subsequent mortality among the oldest old in China. Self-rated poor health poses a greater risk of mortality for men than for women in our gender-SRH interaction analyses and gender-specific analyses after controlling for selected covariates. As the initial effect of SRH on mortality attenuates more for oldest-old women than for oldest-old men after sequentially adjusting for more objective health conditions in a hierarchical approach, it is tempting to attribute the observed gender differences to a gender difference in SRH sensitivity. That is, oldest-old men may evaluate their SRH based more on immeasurable pre-clinical and undiagnosed life-threatening conditions whereas oldest-old women's SRH evaluation may rely more on measurable disabling but nonfatal conditions (Deeg and Bath 2003; Idler 2003; Deeg and Kriegsman 2003; Idler and Benyamini 1997; Verbrugge and Ascine 1987).

As Zeng et al. (2004) have noted, self-reported medical conditions tend to be under-diagnosed and thus under-reported among the oldest-old in China. This is partially due to the fact that health care coverage and utilization are still at a developing stage in China, especially in the rural areas, despite recent efforts to improve health care insurance (Akin et al. 2004; Liu et al. 1999). With limited access to health care and medical diagnoses, it is amazing that SRH is still so predictive of mortality among both oldest-old men and women.

The present analyses reveal that despite certain gender differences in the predictability of mortality and SRH, the gender differences in associations between selected measurable mediators and SRH (as well as IRH) are actually quite small. Given the generally similar associations between measurable mediators and SRH and associations between SRH and mortality, our results suggest that gender differences in sensitivity to SRH may not be the case for the oldest-old in China. Further research is needed to improve our understanding of underlying variables that contribute to this finding.

It is remarkable that feeling lonely is persistently shown to be associated with both SRH and mortality among oldest-old men and women in China. This finding indicates that psychosocial factors do indeed play an important role in moderating the association between SRH and oldest-old mortality. It appears that SRH reflects not only the oldest-old persons' awareness of physical health conditions, but also their subjective appreciation of their mental health conditions (Kaplan and Baron-Epel 2003). The mechanism whereby excess mortality takes place among the oldest-old who have a high score of feeling lonely appears to be expressed through health care attention and utilization: individuals who are socially isolated are less likely to get access to medical care to take care of their health (Brummett et al. 2001).

It is also remarkable that the strength of the association between education and mortality persists at the oldest-old stage of life in China. The effect of education on mortality is sustained even when other socio-demographic characteristics and physical health conditions are controlled among men and women. Moreover, the effects of education on mortality are often age-independent for both oldest-old men and women.

The finding that mortality of the oldest-old in China is modifiable by education and the "feeling lonely" psychosocial factor has important public health implications for improving the "material, social and cultural resources of elderly people" as our finding supports the notion that "social causation plays an important role, even ongoing at old age" (Huisman et al. 2003). Recent research suggests that expectation of life is increasing owing to "the intricate interplay of advances in income, salubrity, nutrition, education, sanitation, and medicine," and that likely improvements in survival at advanced ages will substantially augment the need for pension, health-care and other social necessities (Oeppen and Vaupel 2002). As educational levels and the standard of living of the new oldest-old generations improve in the near future, policy makers and health care providers must prepare for increased needs for health care and social services due to likely further improvements in mortality in the oldest-old population in China.

Acknowledgments We thank Danan Gu, the editors, and two anonymous reviewers for their very helpful comments on an earlier draft of this chapter.

References

Akin, J.S., W.H. Dow, P.M. Lance, and C.P. Loh (2004), Did the distribution of health insurance in China continue to grow less equitable in the nineties? Results from a longitudinal survey. *Social Science and Medicine* 58 (2), pp. 293–304

Anderson, J.M. (1986), Ethnicity and illness experience: Ideological structures and the health care delivery system. *Social Science and Medicine* 22 (11), pp. 1277–1283

Bath, P.A. (2003), Differences between older men and women in the self-rated health-mortality relationship. *The Gerontologist* 43 (3), pp. 387–395

Benjamins, M.R., R.A. Hummer, I.W. Eberstein, and C.B. Nam (2004), Self-reported health and adult mortality risk: An analysis of cause-specific mortality. *Social Science and Medicine* 59 (6), pp. 1297–1306

Benyamini, Y. and E. L. Idler (1999), Community studies reporting association between self-rated health and mortality: additional studies, 1995–1998. *Research on Aging* 21 (3), pp. 392–401

Benyamini, Y., H. Leventhal, and E.A. Leventhal (1999), Self-assessments of health: what do people know that predicts their mortality? *Research on Aging* 21 (3), pp. 477–500

Benyamini, Y., E.A. Leventhal, and H. Leventhal (2003a), Elderly people's ratings of the importance of health-related factors to their self-assessments of health. *Social Science and Medicine* 56 (8), pp. 1661–1667

Benyamini, Y., T. Blumstein, A. Lusky, and B. Modan (2003b), Gender differences in the self-rated health-mortality association: Is it poor self-rated health that predicts mortality or excellent self-rated health that predicts survival? *The Gerontologist* 43 (3), pp. 396–405.

Brummett, B.H., J.C. Barefoot, I.C. Siegler, N.E. Clapp-Channing, B.L. Lytle, H.B. Bosworth, R.B. Williams, and D.B. Mark (2001), Characteristics of socially isolated patients with coronary artery disease who are at elevated risk for mortality. *Psychosomatic Medline* 63 (2), pp. 267–272

Deeg, D.J. and P.A. Bath (2003), Self-rated health, gender, and mortality in older persons: Introduction to a special section. *The Gerontologist* 43 (3), pp. 369–371

Deeg, D.J. and D.M. Kriegsman (2003), Concepts of self-rated health: Specifying the gender difference in mortality risk. *The Gerontologist* 43 (3), pp. 376–386

Frankenberg, E. and N.R. Jones (2004), Self-reported health and mortality: Does the relationship extend to a low income setting? *Journal of Health and Social Behavior* 45(4), pp. 441–452

Gu, D. and Y. Zeng, (2004), Sociodemographic effects on the onset and recovery of ADL disability among Chinese oldest-old. *Demographic Research* 11 (1), pp. 1–42

Helweg-Larsen, M., M. Kjøller, and H. Thoning (2003), Do age and social relations moderate the relationship between self-rated health and mortality among adult Danes? *Social Science and Medicine* 57 (7), pp. 1237–1247

House, J.S. (2002), Understanding social factors and inequalities in health: 20th century progress and 21st century prospects. *Journal of Health and Social Behavior* 43 (2), pp. 125–142

House, J.S. (2001), Social isolation kills, but how and why? *Psychosomatic Medline* 63 (2), pp. 273–274

House, J.S., R.C. Kessler, A.R. Herzog, R.P. Mero, A.M. Kinney, and M.J. Breslow (1990), Age, socioeconomic status, and health. *The Milbank Quarterly* 68 (3), 383–411

House, J.S., J.M. Lepkowski, R.P. Mero, R.C. Kessler, and A.R. Herzog (1994), The social stratification of aging and health. *Journal of Health and Social Behavior* 35 (3), pp. 213–234

Huisman, M., A.E. Kunst, and J.P. Mackenbach (2003), Socioeconomic inequalities in morbidity among the elderly: A European overview. *Social Science and Medicine* 57 (5), pp. 861–873

Idler, E.L. (2003), Discussion: Gender differences in self-rated health, in mortality, and in the relationship between the two. *The Gerontologist* 43 (3), pp. 372–275

Idler, E.L. and Y. Benyamini (1997), Self-rated health and mortality: A review of twenty-seven community studies. *Journal of Health and Social Behavior* 38 (1), pp. 21–37

Idler, E.L., S.V. Hudson, and H. Leventhal (1999), The meanings of self-ratings of health: A qualitative and quantitative approach. *Research on Aging* 21, pp. 458–476.

Kaplan, G. and O. Baron-Epel (2003), What lies behind the subjective evaluation of health status? *Social Science and Medicine* 56 (8), pp. 1669–1676

Kaplan, G.A., M.N. Haan, and R.B. Wallace (1999), Understanding changing risk factor associations with increasing age in adults. *Annual Review of Public Health* 20, pp. 89–108

Lee, I-M., R.S. Jr. Paffenbarger, and C.H. Hennekens (1997), Physical activity, physical fitness and longevity. *Aging: Clinical and Experimental Research* 9 (1–2), pp. 2–11

Liang, J., J.F. McCarthy, A. Jain, N. Krause, J.M. Bennett, and S. Gu (2000), Socioeconomic gradient in old age mortality in Wuhan, China. *Journal of Gerontology: Social Sciences* 55B (4), pp. S222–S233

Liu, Y., W.C. Hsiao, and K. Eggleston (1999), Equity in health and health care: the Chinese experience. *Social Science and Medicine* 49 (10), pp. 1349–1356

Lynch, S.M. (2003), Cohort and life-course patterns in the relationship between education and health: A hierarchical approach. *Demography* 40 (2), pp. 309–331

Mackenbach, J.P., J.G. Simon, C.W. Looman, and I.M. Joung (2002), Self-assessed health and mortality: Could psychosocial factors explain the association? *International Journal of Epidemiology* 31 (6), pp. 1162–1168

Mishra, G.D., K. Ball, A.J. Dobson, and J.E. Byles (2004), Do socioeconomic gradients in women's health widen over time and with age? *Social Science and Medicine* 58 (9), pp. 1585–1595

Mossey, J.M., and E. Shapiro (1982), Self-rated health: A predictor of mortality among the elderly. *American Journal of Public Health* 72 (8), pp. 800–808

Myers, G.C., B.B. Torrey, and K.G. Kinsella (1992), The paradox of the oldest old in the United States: An international comparison. In: R.M. Suzman, D.P. Willis, and K.G. Manton (eds.): *The oldest old.* New York: Oxford University Press, pp. 58–85

National Bureau of Statistics of China (NBS) (2002), *Tabulation on the 2000 Population Census of the People's Republic of China.* Beijing: China Statistics Press.

Oeppen, J. and J.W. Vaupel (2002), Broken limits to life expectancy. *Science* 296 (5570), pp. 1029–1031

Paffenbarger, R.S. Jr., R.T. Hyde, A.L. Wing, I.M. Lee, D.L. Jung, and J.B. Kampert (1993), The association of changes in physical-activity level and other lifestyle characteristics with mortality among men. *New England Journal of Medicine* 328 (8), pp. 538–545

Ross, C.E. and C. Wu (1996), Education, age, and the cumulative advantage in health. *Journal of Health and Social Behavior* 37 (1), pp. 104–120

United Nations (2002), *World population ageing 1950–2050*. New York: United Nations.

Verbrugge, L.M. (1989), The twain meet: Empirical explanations of sex differences in health and mortality. *Journal of Health and Social Behavior* 30 (3), pp. 282–304

Verbrugge, L.M. and F.J. Ascine (1987), Exploring the iceberg: Common symptoms and how people care for them. *Medical Care* 25 (6), pp. 539–569

Yu, E.S. Y.M. Kean, D.J. Slymen, W.T. Liu, M. Zhang, and R. Katzman (1998), Self-perceived health and 5-year mortality risks among the elderly in Shanghai, China. *American Journal of Epidemiology* 147 (9), pp. 880–890

Zeng, Y. and J.W. Vaupel (2004), Association of late childbearing with healthy longevity among the oldest-old in China. *Population Studies* 58 (1), pp. 37–53

Zeng, Y., J.W. Vaupel, Z. Xiao, C. Zhang, and Y. Liu (2001), The healthy longevity survey and the active life expectancy of the oldest old in China. *Population: An English Selection* 13 (1), pp. 95–116

Zeng, Y., J.W. Vaupel, Z. Xiao, C. Zhang, and Y. Liu (2002), Sociodemographic and health profiles of the oldest old in China. *Population and Development Review* 28 (2), pp. 251–273

Zeng, Y., D. Gu, and K.C. Land (2004), A new method for correcting the underestimation of disabled life expectancy and an application to the Chinese Oldest old. *Demography* 41 (2), pp. 335–361

Chapter 25
Epilogue: Future Agenda

Zeng Yi

To provide a more comprehensive profile of our work, in this Epilogue I summarize the major achievements, some selected findings published elsewhere, and the overall limitations of our Chinese Longitudinal Healthy Longevity Survey (CLHLS). At the end of this Epilogue, a future research agenda is outlined.

25.1 Publications and Research Reports Resulted from CLHLS Project

In addition to materials included in this book, as of April 2007, the CLHLS has resulted in the following publications and research reports:

- 24 papers written in English, published or accepted for publication in the U.S. or in European peer-reviewed journals.
- 4 chapters written in English in other books published in the U.S. or Europe.
- 80 papers written in English and presented at international conferences.
- 4 books published in China (two of them written in both English and Chinese).
- 91 papers written and published in peer-reviewed Chinese journals.
- 3 Ph.D. dissertations written in English and successfully defended in U.S. universities.
- 5 Ph.D. dissertations written in Chinese and successfully defended at Peking University.
- 13 M.A. theses written in Chinese and successfully defended at Peking University

---

Zeng Yi
Center for Study of Aging and Human Development, Medical School of Duke University, Durham, NC 27710, USA
Center for Healthy Aging and Family Studies/China Center for Economic Research, Peking University, Beijing, China
e-mail: zengyi68@gmail.com

- (We have not yet received information about graduate students' theses using the CLHLS data written and defended at other Chinese universities, but we know there are many.)
- 11 policy reports submitted to relevant government agencies.[1]

The lists of the published articles, books, book chapters and graduate student theses are available at the following websites.

PKU Websites http://www.pku.edu.cn/academic/ageing and
Duke Website: http://www.grei.duke.edu/china_study

Since 1998, when we started the baseline CLHLS survey, the media has tracked the progress of our project (training, field work, and research results) and has reported many results from the study to the public. About 60 Chinese and international media such as Xinhua *News Press*, the Voice of America, BBC, and the *Washington Post*, as well as *Science* magazine, have reported the progress and findings of the CLHLS project.

25.2 Summaries of Selected Striking Findings from the CLHLS Published Elsewhere

25.2.1 *Sociodemographic and Health Profiles of Oldest-Old in China*

Based on the CLHLS 1998 baseline survey data, we have documented demographic, socioeconomic and health portraits of the oldest-old aged 80–105 in China, a subpopulation growing very quickly and most likely in need of help. A large majority of the Chinese oldest-old live with their children, and rely mainly on their children for financial support and care. Most of the Chinese oldest-old have little or no education. Their functional capacities in daily living have declined quickly and their self-rated health has declined moderately across the oldest-old ages. Compared to their urban counterparts, extremely limited pension support is available to the rural oldest-old; the rural oldest-old are significantly less educated, and are more likely to be widowed and to rely on their children (Zeng et al. 2002). Compared to their male counterparts, oldest-old Chinese women are seriously disadvantaged

---

[1] These reports discussed the economic status, physiological and mental health, cognitive function, and marriage and families of the elderly, especially the oldest-old, and also proposed some related policy options. For example, based on data from the CLHLS and relevant studies, Prof. Hu Angang and Prof. Zeng Yi submitted a policy report on the oldest-old to the Chinese central government. Subsequently, the vice Prime Minister responded positively in writing, and the Whole-China Commission on Aging Work issued an special official policy note to the State Council (2003, No. 48) about the disadvantaged group of oldest-old, using research findings from the CLHLS report by Hu and Zeng.

not only socio-economically but also with respect to their health status as measured by activities of daily living, physical performance, cognitive functioning, and self-reported health; these gender differences are more marked with advancing age (Zeng et al. 2003). We have also discussed the need for policy related research suggested by these findings (Zeng et al. 2002, 2003).

25.2.2 The Active Life Expectancy of the Oldest-Old in China

Based on the 1998 CLHLS baseline survey data, we have found that after age 80, approximately 87, 8, and 5 percent of the remaining life of men living in rural areas are expected to be in active, mild disability and severe disability statuses, respectively. The corresponding figures for the urban male oldest-old are 80, 14, and 6 percent, respectively. The percentages of remaining life in the active, moderately disabled, or severely disabled statuses for the women living in rural areas after the age of 80 are 82, 10, and 8 percent, respectively. The figures for their urban counterparts are 75, 18, and 7 percent, respectively. Similar patterns of rural–urban and gender differentials are found for the life expectancies after age 90 and 100. It is clear that the percentage of remaining life in active status for the oldest-old in rural areas is substantially higher than for their urban counterparts. Men have a greater chance of being active at the oldest-old ages than women (Zeng et al. 2001).

We have proposed three potential explanations for why the oldest-old living in rural China are more likely to live active lives compared to those living in urban areas: (1) selection, i.e., those who survive to very advanced ages in rural areas are less likely to be frail and thus more likely to be active; (2) poorer facilities to assist the oldest-old in daily life may force the rural oldest-old to perform daily activities by themselves, and thus maintain their capacities for daily life for a longer time than their urban counterparts; (3) the physical environment is likely better in rural areas, and industrial pollution in the cities may likely worsen the health status of the oldest-old (Zeng et al. 2001).

25.2.3 Childhood Socioeconomic Status Is Associated with Health and Mortality at the Oldest-Old Ages

Using the unique data from the CLHLS, multivariate logistic regression analyses show that receiving adequate medical services during sickness, or not (or rarely) suffering from serious sickness in childhood, significantly reduces the risk of being ADL impaired, cognitively impaired, and of self-reporting poor health by 18 percent to 33 percent at the oldest-old ages. Estimates of the effects for five other indicators of childhood conditions are similarly positive, but most of the effects are not statistically significant. Multivariate survival analysis shows that better childhood socioeconomic conditions in general tend to reduce the 4-year period mortality risk among the oldest-old; but after controlling for fourteen covariates, the effects are not statistically significant. While we realize the limitations of lack of information about childhood illness, the oldest-olds' recall errors, and other data problems, we conclude that policies enhancing childhood health care and children's socioeconomic wellbeing can have large and long-lasting benefits up to the oldest-old ages (Zeng et al. 2007).

25.2.4 The Unique CLHLS Data on ADL Before Dying Have Led to Methodological Innovation

We found that the conventional multi-state life table approach without ADL data before dying tends to cause biases in estimating active and disabled life expectancies (see Fig. 25.1). This bias is due to the unreasonable assumption of no changes in ADL status from age x to death if a person dies during the age interval $(x, x + n)$; the biases are sizable and statistically significant. We extended the multi-state life table method and applied it to the CLHLS data to improve the ADL-status-based estimates of active and disabled life expectancies (Zeng et al. 2004).

25.2.5 Extent of Disability and Suffering Before Dying at Oldest-Old Ages

Based on information from 35 variables measuring the extent of disability and suffering before dying collected in the CLHLS, we found that the male oldest-old had a substantially higher chance of experiencing a non-suffering death than their female counterparts. The age differences in life table proportions of disability and suffering before dying are not substantial. We also found that ADL status reported in the

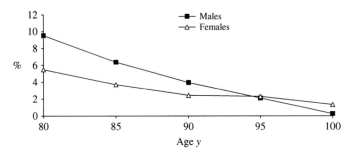

Fig. 25.1 Percentage distribution of the underestimation of disabled life expectancies based on the conventional method, which assumes no functional status changes between age x and death among those who die between age x and $x + 1$
(Source: Zeng, Gu and Land, 2004)

survey is a powerful predictor of the extent of subsequent disability and suffering before dying among the oldest-old (Zeng et al. 2004; see Fig. 25.2a, b).

25.2.6 Association of Late Childbearing and Healthy Longevity at Oldest-Old Ages

Statistical analysis of data from the CLHLS demonstrates that late childbearing after age 35 or 40 is significantly associated with survival and the healthy survival of very old Chinese women and men. The association is stronger for oldest-old women than for men. The estimates are adjusted for a variety of confounding factors pertaining to demographic characteristics, family support, social connections, health practices, and health conditions. Further analysis based on an extension of the Fixed Attribute Dynamics method shows that late childbearing is positively associated with long-term survival and healthy survival from ages 80–85 to 90–95 and to 100–105. This association exists among the oldest-old women and men, but, again, the effects are substantially stronger for women than for men. We have discussed four possible factors that may explain why late childbearing affects healthy longevity at the advanced ages: (1) social factors; (2) biological changes caused by late pregnancy and delivery; (3) genetic and other biological characteristics; and (4) selection (Zeng and Vaupel 2004).

25.2.7 Optimism Is One of the Secrets of Longevity

Our CLHLS data show that the percentages of being active in daily living, having good physical performance capacity and normal cognitive functioning drop dramatically from the ages 65–69 through 100–105. The percentage reporting good satisfaction in their current life, however, remains almost constant from ages 65–69 to 80–84 and declines slightly thereafter (Fig. 25.3). This suggests that being more positive in one's outlook on life is one of the secrets of longevity (Zeng and Vaupel 2002).

25.2.8 The Trajectories of Oldest-Old Mortality in China

We find that the Kannisto model, a two-parameter logistic formula, fits Han Chinese death rates at the oldest-old ages better than the classic Gompertz model and four other models. Chinese death rates appear to be roughly similar to Swedish and Japanese rates after age 97 for both males and females. Because reports of age in China seem to be serviceably reliable up to age 100, and even perhaps to age 105, we believe this convergence is mainly due to mortality selection in the heterogeneous Chinese population. We show that in China, as in developed countries, the rate of increase in mortality with age decelerates at the very old ages (Zeng and Vaupel 2003).

25.2.9 Characteristics of the Institutionalized Oldest-Old in China

Using data from the first three waves of the CLHLS, we have examined differences in institutionalization based on sociodemographic characteristics, family caregiving resources, health practices, religious activity, chronic conditions, and mortality risk. Our results indicate that the institutionalized oldest-old are more likely to be younger, to be male, to reside in urban areas, to have lower family-care resources, and to exhibit poorer health compared to those living in the community. We also find that the 2-year mortality risk for institutionalized elders is 1.35 times greater than for those residing in the community. However, the mortality differential is eliminated once the sociodemographic, family caregiving, and health characteristics of the oldest-old are taken into account. This study is the first to use nationally representative data to examine differences in the institutionalized and community-residing population of the oldest-old (age 80 and over) in China (Gu et al. 2007a).

25.2.10 Factors Affecting Place of Death and Evolutionary Theory

With data on 6,444 deceased respondents age 80 to 105 in the first three waves of the CLHLS, we used multilevel modeling to examine how community development, individual sociodemographic characteristics, health conditions, and health resources affect place of death. Our results indicate that 92 percent of the Chinese oldest-old die at home, with 7 percent dying in hospitals and 1 percent in institutions. Analyses indicate that residents from relatively developed communities in China and those who have pensions, and/or public and/or collective free (or subsidized) medical services, are more likely to die in hospitals and/or institutions. Individuals with higher socioeconomic status and worsening health are more likely to die in hospitals. We have proposed a theory about place of death consisting of three evolutionary stages, which might explain the disparity in the patterns of place of death in different societies (Gu et al. 2007b).

25.2.11 Sociodemographic Effects on the Dynamics of ADL Functioning at the Oldest-Old Ages

By pooling data from the three waves (1998, 2000, and 2002) of the CLHLS, we have shown that sociodemographic factors still play specific roles in determining disability dynamics and its task components at the very high ages, even after controlling for a rich set of confounding variables (Gu and Zeng 2004; Gu and Xu 2007). Our results also point out that the conventional method, which usually tends to exclude information about ADL changes before dying because of the unavailability of the data, yields some biases in estimating the effects of sociodemographic factors on disability transitions, when compared with the results from analyses including information of ADL changes before dying (Gu and Zeng 2004).

25.3 Limitations of the CLHLS Study

As discussed by Zeng et al. (2002) and in Chap. 3 of this volume by Danan Gu, we believe that the data quality of the CLHLS is acceptable. However, we also realize that there are some problems in the data. For example, we found that in the baseline survey a substantially higher proportion of the oldest-old, especially centenarians, were unable to answer the personality-related questions, compared to other types of questions. This may be because some illiterate oldest-old, especially centenarians, could not understand the questions about personality, which required them to compare themselves with a typical person of a specified disposition. We have subsequently revised these questions in the 2000, 2002, and 2005 follow-up surveys. However, although the personality questionnaire design and its data quality were substantially improved after the baseline survey, the reliability index of these particular data is still not ideal. Thus, one needs to pay special attention and to be cautious when using the personality data derived from the CLHLS, especially the 1998 baseline survey (for more discussion, see Chap. 3 of this volume).

Another weakness of the 1998 baseline survey and the 2000 follow-up survey was the lack of a comparative group of younger interviewees aged 65–79. This was due to funding constraints; thus we cannot make comparisons between the oldest-old and younger elders using the 1998–2000 data. We received additional financial support from NIA, UNFPA, and Chinese resources to include about 5,000 younger interviewees aged 65–79, and 10,000 oldest-old aged 80+ in our 2002 and 2005 surveys.

Similar to other studies focusing on the oldest-old, higher proxy use is a notable limitation. The higher proxy rate is related to older age, lower education, rural residence, lower cognitive functioning, and higher disability. Therefore, it would be better to add an indicator variable (i.e., either the presence or absence of a proxy) in the analyses when the aim of a proposed study is to examine the effects of these factors. Furthermore, we find that item incompleteness and sample attrition tend to be linked to age, gender, urban/rural residence, ethnicity, and health conditions. Although it is unlikely that these limitations will significantly affect the results, sufficient attention should be paid to them when verifying and reporting the outcomes of the related data analysis (for more information, see Chap. 3 of this volume).

While we are generally satisfied with our progress and achievements in the 10 years conducting the CLHLS study, we must admit that a major weakness is that the CLHLS study has been limited principally to social science research. We did collect blood dry-spot samples from 4,116 voluntary participants aged 80–110 in our 1998 baseline survey (45.4 percent of the total sample). Mainly due to limitations in research funds—the cost of genetic analysis has been very high (Brown 2005)—there has been no lab analysis work undertaken of these blood dry-spot samples[2]; also, no further DNA collections have been taken from the oldest-old and younger adults

---

[2] The blood dry-spot samples collected in 1998 are still well-preserved and could be used at a later date.

who subsequently and newly participated in the CLHLS study. However, it has been shown by various published studies that about one-fourth of an individuals' variation in longevity and health at advanced ages is attributable to genetic factors, and the rest, or three-fourths, is attributable to social, behavioral, and environmental factors, in combination with genetic interactions (Herskind et al. 1996; McGue et al. 1993). Nevertheless, the restriction so far of the CLHLS to social science research data is a major obstacle that prevents us from developing a complete understanding of the determinants of healthy longevity.

Recent literature has agreed that future research in this area must necessarily integrate socio-demographic and genetic analyses. This is a requirement for the development of true and complete scientific knowledge about the determinants of healthy longevity, and is the direction the research should go (Hobcraft 2006; Singer and Ryff 2001; Wise 2001). For example, Vaupel (2006) recently noted that knowledge about the genetics of human longevity is currently limited. Deeper understanding can be reached if geneticists work with demographers to use advanced demographic and statistical methods to study data on the genetics of long-lived individuals. One of the main reasons why people will live healthier and longer in the future is that medical interventions will be developed based on new genetic knowledge about healthy longevity. We draw inspiration from the above-cited and other important studies as well as the 296th Fragrant Hill Science Conference on Interdisciplinary Studies on Social, Behavior, Environmental and Genetic Determinants of Healthy Longevity organized by Professor Huanmin Yang and me, sponsored by the Chinese Academy of Sciences.[3] Indeed we plan to substantially expand our CLHLS study into a truly integrated research endeavor that includes demographic, social, behavior, environmental and genetic factors. The future agenda of our study is outlined in the next section.

25.4 Future Research Agenda

Researchers from the fields of demography, sociology, economics, genetics, and public health at Peking University (PKU) and the Chinese Academy of Sciences (CAS) have recently started a pilot research project supported by PKU and CAS. We expect that this pilot effort will lead to funded program projects for an integrated interdisciplinary study of the social, behavior, environmental and genetic determinants of healthy longevity. The integrated interdisciplinary investigations are expected to focus on analyzing already-collected, and to-be-collected, health, survival, socioeconomic, behavior, environmental and genetic data from centenarians, oldest-old sibling-pairs, longevity families, and other elders, as well as control groups of younger adults.

---

[3] We very much appreciate the support of the National institute of Aging (NIA) to cover the expense of air tickets for selected international participants of this conference through an Administrative Supplement Award (3 R01 AG023627-03S1; PI: Zeng Yi).

Besides adding the genetic component in our 2008 CLHLS nation-wide sample, which will be conducted in the 943 diversified counties and cities, we plan to conduct more in-depth studies in several typical areas where the density of centenarians is exceptionally. These in-depth studies will include the following data to be collected and analyzed by various experts: age validations, health examinations (including data collection of DNA and related bio-markers with informed consent), behavior, socioeconomics, nutrition, environmental conditions, and investigations of anthropological observations on family kinship structure and relations, and culture.

In our forthcoming research, we will also pay special attention to some targeted long-lived healthy individuals who are suspected to have healthy longevity related genes, based on demographic information and statistical analyses. For example, as mentioned earlier, Zeng and Vaupel (2004) found that those oldest-old female interviewees who had 2+ or 3+ late births after the age of 35 or 40 had a significantly higher probability of healthy survival, compared to other oldest-old women who were not late child-bearers, after controlling for various confounding factors. We plan to investigate and address research questions on whether these long-lived healthy late child-bearers may more likely to carry certain gene(s) related to healthy longevity and/or higher reproduction. This is just an example of the kinds of hypotheses that may be formulated using the CLHLS longitudinal data sets.

Integrated interdisciplinary analyses will certainly help us to better understand how genetic and non-genetic factors, and their interactions, may affect healthy longevity, and thus contribute to the human goal of living healthier and longer. There is no doubt that there are tremendous difficulties and challenges ahead of us. However, we are dedicated to continue to work hard and efficiently for science and for society and for those senior interviewees and their family members as well as other colleagues, friends and funding agencies that have supported our CLHLS study since 1997.

Acknowledgments The research reported here is supported by The National Institute on Aging grant (R01 AG023627-01) and National Natural Science Foundation of China key project grant (70533010).

References

Brown, S. (2005), Soul of the new gene machines. *Fortune* 151 (9), pp. 113–114
Gu, D. and Zeng, Y. (2004), Sociodemographic effects on the onset and recovery of ADL disability among the Chinese oldest-old. *Demographic Research* 11, pp. 1–41
Gu, D. and Q. Xu (2007), Sociodemographic effects on the dynamics of task-specific ADL functioning at the oldest-old ages: The case of China. *Journal of Cross-cultural Gerontology* 22, pp. 61–81
Gu, D., M.E. Dupre, and G. Liu (2007a), Characteristics of the institutionalized and community-residing oldest-old in China. *Social Science and Medicine* 64, pp. 871–883
Gu, D., G. Liu, D.A. Vlosky, and Zeng, Y. (2007b), Factors associated with place of death among the Chinese oldest-old. *Journal of Applied Gerontology* 26 (1), pp. 34–57

Herskind, A.M., M. McGue, N.V. Holm, I.T. Sorensen, B. Harvald, and J.W. Vaupel (1996), The heritability of human longevity, a population-based study of 2872 Danish twin pairs born 1870–1900. *Human Genetics* 97, pp. 319–323

Hobcraft, J. (2006). The ABC of demographic behavior: How the interplays of alleles, brains, and contexts over the life course should shape research aimed at understanding population processes. *Population Studies* 60 (2), pp. 153–187

McGue, M., J.W. Vaupel, N. Holm, and B. Harvald (1993), Longevity is moderately heritable in a sample of Danish twins born 1870–1880. *Journal of Gerontology* 48, pp. B237–B244

Singer, B. and C.D. Ryff (2001), *New horizons in health: An integrative approach*. Washington, DC: National Academy Press

Vaupel, J.W. (2006), *Human longevity: Plasticity, prospects and genetics*. Keynote speech presented at the International Conference on "Interdisciplinary Studies on Social, Behavior, Environmental and Genetic Determinants of Healthy Longevity", December 19–21, 2006, Beijing

Wise, D.A. (2001), *Themes in the economics of aging*. Chicago: University of Chicago Press

Zeng, Y. and J.W. Vaupel (2002), Functional capacity and self-evaluation of health and life of the oldest-old in China. *Journal of Social Issues* 58, pp. 733–748

Zeng, Y. and J.W. Vaupel (2003), Oldest-old mortality in China. *Demographic Research* 8 (7), pp. 215–244

Zeng, Y. and J.W. Vaupel (2004), Association of late childbearing with healthy longevity among the oldest-old in China. *Population Studies* 58 (1), pp. 37–53

Zeng, Y., J.W. Vaupel, Z. Xiao, C. Zhang, and Y. Liu (2001), The healthy longevity survey and the active life expectancy of the oldest-old in China. *Population: An English Selection* 13 (1), pp. 95–116

Zeng, Y., J.W. Vaupel, Z. Xiao, C. Zhang, and Y. Liu (2002), Sociodemographic and health profiles of oldest-old in China. *Population and Development Review* 28 (2), pp. 251–273

Zeng, Y., Y. Liu, and L.K. George (2003), Gender differentials of oldest-old in China. *Research on Aging*, 25, pp. 65–80

Zeng, Y., D. Gu, and K.C. Land (2004), A new method for correcting underestimation of disabled life expectancy and application to Chinese oldest-old. *Demography* 41 (2), pp. 335–361

Zeng, Y., Y. Liu and L.K. George (2005), Female disadvantages among the elderly in China. In: Y. Zeng, E. Crimmins, Y. Carrière, J. Robine (eds.): *Longer life and healthy aging*. Dordrecht: Springer

Zeng, Y., D. Gu, and K.C. Land (forthcoming), The association of childhood socioeconomic conditions with healthy longevity at the oldest-old ages in China. *Demography*

Index

Accessibility of health services, 277, 279
Adult children, 25, 27–28, 31, 33–34, 179, 193–195, 200, 210, 219, 236–237, 243, 246, 251–265, 359, 362
Affective experience, 292, 347–355
Aged dependency
  aged dependency burden, 2, 13
  total dependency ratio, 10
  youth dependency ratio, 10
Age difference, 84, 87, 347, 422
Age exaggeration
  age heaping, 63, 64–66, 71, 86, 95
  age misreporting, 71, 79–80, 81, 83–84, 87–89, 91–92, 94–95, 100, 273
  age pattern, 4, 47, 71, 100, 103, 108, 261, 264, 265
  age reporting, 16, 20–22, 25, 33, 40, 61–74, 79–97, 105, 108, 134, 160–161, 180
  age validation, 81, 93, 427
  animal year, 62, 81, 100, 182
  lunar calendar, 62, 100
  Myer's index, 64, 65
  Western calendar, 62
  Whipple's index, 64, 65
Altruism, 231
American oldest-old, 289, 305–306
Anhui Province, 194, 235, 239
ANOVA, 253, 350, 352, 354
Australia, 20–21, 63–64, 66–72, 74, 79, 105, 269, 291, 294–295, 300–301

Barefoot doctors, 275
Beneficial effect, 119, 179, 185, 187, 189, 210, 251
Biases, 20–22, 31–32, 41–43, 45–47, 71, 73–74, 82, 84, 95, 100, 102, 104–105, 108, 113, 159, 166–167, 170, 173, 200, 224, 252, 378–379, 384, 422, 424

Bivariate analysis, 260, 261, 318
Blood dry-spot samples, 425

Canada, 21, 24, 63–64, 66–72, 74, 105, 197
Caregiving, 113, 194–195, 251–266, 424
Care provision, 13, 195, 199, 252, 254–256, 258, 261, 262, 264–265
Cause of death, 31, 43, 106–107, 109, 113, 270, 401
Cause-specific death rates, 106
Census data, 14, 21, 64, 70, 103, 153, 193, 283
Census micro data, 151
Centenarian
  age distribution, 11, 20, 35, 63–64, 66–68, 74, 82–83, 86–87, 91, 95, 422
  age ratios, 63, 66–68, 86–87, 91–92, 95–96
  density, 21, 63, 68, 70, 72, 74, 427
  proportion, 33, 85, 86, 89, 91, 93, 284, 425
Center for Medicare and Medicaid Services, 31
Childhood, 19, 28, 40, 117–119, 139–140, 144, 146, 149–151, 160, 163, 165–167, 169–170, 173–174, 307, 421–422
Childhood socioeconomic status, 139, 421
Chile, 21, 63–64, 66–71, 74
China, 1–16, 19–21, 236, 32–34, 36, 39–40, 53, 61–64, 66, 70–75, 79–81, 84–86, 89, 91, 94, 99, 101–103, 106–107, 110–111, 113, 118, 121–124, 126–128, 130, 133–146, 149, 151–154, 157–160, 177–189, 193–195, 197–212, 215–232, 235–247, 251–253, 256, 258–262, 264, 269–285, 290–292, 293–301, 305, 307, 309–313, 315–326, 329, 333, 347–348, 354–355, 357–374, 377–380, 397–416, 419–421, 423–424, 427
China National Disease Surveillance Point System, 106

429

Index

Chinese elderly, 19, 53, 71, 110, 122–123, 149–152, 189, 193, 242, 290, 295, 300, 307, 315, 317, 358, 362, 374, 400
Chinese oldest-old
  active life expectancy, 316, 366, 421
  ADL before dying, 50, 422
  dynamics of ADL functioning, 424
  extent of suffering before dying, 31–34, 50, 424–425
  gender differentials, 378, 421
  health profile, 420–421
  institutionalized, 424
  late childbearing and healthy longevity, 423
  mortality trajectories, 423
  personality, 50–56, 141, 143–145, 318, 331, 372, 425
  place of death, 424
  rural-urban differentials, 421
  secrets of longevity, 423
  socio-demographic profile, 420–421
Chinese oldest-old, 28, 33, 43, 74, 111, 119, 290, 292, 297, 300–301, 308, 348, 420, 424
Chronic conditions, 100, 109, 113, 118, 135–136, 204, 208, 398, 402–404, 411, 424
CLHLS
  accuracy of imputation, 46
  accuracy of mortality data, 105
  achievements, 419, 425
  background, 16, 23
  contacts, 34
  data collection, 19, 25–26, 34, 62, 113, 151, 380, 427
  future research agenda, 16, 419, 426
  limitations, 325–326, 342, 425–426
  media reports, 420
  methodological innovation, 422
  objectives, 256
  organizational framework, 19, 25–26
  over-sampling, 24, 333
  policy reports, 420
  publications, 419–420
  refusal rate, 28, 31
  research opportunities, 16, 25, 31
  sample distribution, 16, 19, 25–27
  striking findings, 16, 420
  student thesis, 420
  study design, 16, 19, 25–26, 80
  unique features, 31, 33, 50
  websites, 420

weight, 20, 27, 34–35, 46, 81, 93, 101, 104, 141, 163–164, 166, 205–206, 253, 277, 318, 363–364, 368, 404
Cluster sample, 80
Cognition
  cognitive function, 20, 28, 40–42, 45, 50, 52, 54, 56, 105, 220, 290, 294–296, 308, 335, 342, 377, 379, 382–383, 385–388, 390–394, 420–421, 423, 425
  mental health, 178, 206–207, 211, 215, 217, 330, 415, 420
  Mini-Mental State Exam, 28, 50–52, 55, 182, 220, 296, 333, 335, 339, 380, 382
Cohort, 3, 9, 15, 33, 67–68, 72, 83, 91–92, 102, 118, 124, 149, 151–152, 154, 232, 289–291, 294, 295, 301, 307, 323, 331–333, 336–341, 378, 400
Cohort difference, 291, 332
Comparison, 7, 11, 25, 40, 42, 47, 50, 53, 56, 63, 66–69, 71, 74, 78, 79, 82–86, 90–93, 101, 103–105, 107, 110–112, 198, 226, 230, 253, 274, 282, 283, 289, 295, 297, 306, 312–313, 322, 348, 351, 389, 393, 398, 425
Complex sampling design, 101
Compression of morbidity, 15–16
Convergent validity, 54–55
Convoy network model, 187
Cooperative Medical System, 270
Coresidence, 194, 216–217, 228, 230–232, 238–239
Cronbach's alpha, 20, 50–52, 203, 366
Cross-sectional sample, 330, 336

Data assessment
  attrition, 333, 336
  data quality, 16, 19–22, 25, 35–56, 62–64, 67, 70, 94, 110, 382, 425
  digit preference, 64, 82, 84, 86
  discriminant validity, 54, 55
  Don't know answer, 44
  first year death, 102
  first-year mortality rate, 22, 108, 113
  full proxy response, 42
  impacts of age exaggeration, 72–73
  imputation, 20, 46
  inaccuracy, 63
  inconsistent responses, 49
  internal consistency, 20, 50
  item nonresponse, 20, 41, 43, 45–47
  item-total correlations, 51–53
  knowledgeable proxy, 41
  lost to follow-up, 47–48, 73, 108, 113, 161, 218, 222–224, 300, 380

Index

minimum reliability coefficient, 50
missing completely at random, 45
missing item, 46, 48, 56
missing value, 46, 48, 141, 173, 181, 253, 290, 349, 379–380, 384, 386, 394
nonresponse, 20, 40–41, 47–48
nonresponse rate, 20, 39, 43, 46, 62
proxy, 20, 28, 31–32, 39–44, 48–50, 54, 56, 62, 106, 110, 113, 136, 139, 257, 398, 402, 425
proxy reporter, 41
proxy response, 32, 42, 402
proxy use, 20, 39, 40–43, 54, 56, 62, 425
response pattern, 403, 414
response structure, 46
sample attrition, 20, 39, 44, 46–49, 54, 56, 62, 100, 425
second year death, 102
sources of error, 20, 40
underestimation, 22, 62, 102–105, 107–109, 112–113, 341, 384, 422
underestimation of mortality, 62, 102, 104–105, 108–109, 113
unit nonresponse, 43, 46
validity, 20, 39, 41, 50–51, 53–55, 62–64, 93–94, 110, 134, 220, 400, 414
Data source, 25, 63, 95, 100, 108, 113, 400
Decreasing degrees of independence, 161
Demographic variables, 46, 75, 204–205, 208, 209, 292, 350, 352, 355
Denture
 dental care, 315, 321, 324–325
 dentate status, 317, 323–324
 edentulism, 317, 326
 tooth loss, 290, 316–321, 324–326
Determinant, 217, 227–228, 279
Disability
 active life expectancy, 316, 366, 421
 activities of daily living, 24, 28, 50, 118, 135, 158, 161, 163, 181, 199, 203–205, 210, 219, 256, 294, 296, 305–306, 308, 316, 333, 335, 338–339, 341, 365, 382, 402, 421
 disability dynamics, 424
 disability paradox, 198
 "Instrumental activities of daily living" after "disability paradox", 28, 40, 50, 256, 306, 308
 Katz Index of ADL, 161
 task components, 424
Disengagement theory, 178
Disparity, 282, 424
DNA, 425, 427

Dyadic data, 195, 253–254
Dynamic equilibrium, 15–16

Education, 19–20, 28, 42, 45, 56, 75, 118, 122–128, 130, 139–141, 144, 149–154, 158–160, 163, 165–167, 169–170, 172–173, 179, 182, 184, 203–204, 208–209, 221, 225, 229, 237–238, 240, 242–245, 256, 290–292, 296–298, 300–301, 317, 321, 323, 333, 335, 339, 349, 352, 361–362, 364–368, 370–372, 378, 399, 401–404, 408, 411, 414–416, 420, 425
Elderly population, 2–3, 13, 15, 23, 35, 118, 211, 273, 291, 295, 300, 305–307, 317, 322, 348, 358, 389, 400, 414
Elder parents, 13, 258, 262
Emotional support, 13–14, 188, 194, 201, 210–211, 237, 239–246, 252, 265, 357–358
England and Wales, 20–21, 63–64, 66–68, 70–72, 84–85, 97, 107, 271
Environmental variables, 166, 167
Epidemiological transition, 100, 270–272, 281, 284
Ethnic minority
 Yao, 65, 89, 94, 97
 Zhuang, 65, 89, 94, 134, 273
Explanatory variables, 164–166, 203
Extent of disability and suffering before dying, 31, 32–34, 50, 422
External resources, 236
Extrapolated, 102

Factor analysis, 54, 137–138, 350
Family planning, 8–9, 158
Family relation, 32–33, 180, 266, 360
Family support, 19, 28, 32–33, 40, 74–75, 118, 200, 246, 252, 307, 423
Fertility decline, 3, 8, 89, 122, 193, 359
Filial piety, 200, 216, 264–265, 358–359
Financial support, 97, 194, 203–204, 208–209, 211, 237–238, 240–241, 243–246, 265, 275, 352, 361, 363–369, 371–373, 420, 425
Fixed-attributes dynamics, 150, 151
Fixed-effects, 206
Fourth age, 330, 332, 378
Functional classification, 295–296, 297, 299–300
Functional limitations, 53, 117, 135, 194, 198, 201, 207, 210–211, 227, 230–231, 242, 256, 294, 307, 378–379, 382, 385, 394, 398

Gender
  gender difference, 180, 217, 245, 290, 320, 323, 325–326, 378–379, 389–390, 393–394, 397–416, 421
Genetic factors, 426
Gini index, 275
Global health, 135, 139, 144, 146, 312, 397–398
Grossman Model, 118, 157–160, 173
Guangxi, 21, 26–27, 40, 86–87, 89, 90, 93–95, 97, 283–284

Han Chinese
  Han-dominated provinces, 21, 73
  Han majority, 62, 85, 89, 95
Health and Retirement Study, 108
Health capital, 117–119, 157
Health care
  Government Insurance Scheme, 274
  health care coverage, 274, 277, 283, 415
  health care system, 32, 195, 270, 272, 274, 276, 281
  health services, 31, 106, 159, 261, 274–279
  Labour Insurance Scheme, 274
Health disorders, 201, 215, 231
Health endowments, 166
Health expenditure, 274
Health outcome, 117–118, 150, 161, 164, 170, 198, 201, 398–399
Health profile, 420–421
Health status, 24–25, 28, 32, 74, 117–119, 122, 125, 127, 134–139, 142, 144–145, 149, 158–160, 162, 164–167, 181, 184–185, 187, 194, 202–204, 207–210, 217, 224, 235–240, 242–247, 254, 258, 263–266, 273, 290, 308–310, 312, 316, 318, 321, 326, 362, 365–366, 371, 379, 390, 397–399, 401, 403, 415, 421
Health workers, 275
Healthy aging, 1, 15–16, 23, 26, 34, 61, 193, 198–199, 211, 419
Healthy longevity, 1, 15–16, 19, 22, 24–28, 31–35, 39–40, 43, 62, 74, 80–81, 100, 117–119, 122, 124, 134, 139, 141, 151, 158, 160, 180, 194–195, 198, 202, 218, 253, 270, 281, 283–285, 295, 301, 307, 312, 317, 330–332, 340, 343, 348, 358, 362, 372–373, 380–381, 400, 404, 407, 410, 413, 419, 423, 426–427
HIV/AIDS, 277, 282
Homogeneity, 20, 53

Identity accumulation hypothesis, 178, 188
Impairment, 41, 135, 146, 207, 210–211, 220, 246, 252, 256, 258, 260, 289–290, 305–308, 312–313, 316, 331, 333, 349, 379, 393–394
Income distribution, 270, 275, 282
Inequality, 195, 211, 270, 278, 282, 285, 411
Inexpensive instrument, 398
Influential factor, 118, 124, 321
Institutionalization, 330, 424
Instrumental support, 194, 210–211, 237–241, 243–246
Instrumental variables, 170, 171
Interaction, 178, 183, 185, 189, 194, 198, 201, 211, 218, 222, 228–230, 231, 358, 360, 362, 373–374, 401, 403, 405, 408, 415, 426–427
Interdisciplinary analyses
  interdisciplinary studies, 426
Intergenerational relations
  intergenerational support, 195, 200, 236–238, 240–241, 243–246, 252, 265, 359
Internal resources, 236
Interviewer's assessment, 136
Interviewer-rated health, 365, 401–402, 408, 409
Intraclass correlation, 253

Japan, 4, 20–21, 24, 33, 43, 63–64, 66–72, 74, 84–85, 87, 91, 96–97, 271, 273, 284, 316–317, 324, 423
Jiangsu, 26–27, 40, 86, 89–90, 94, 97
Joint model, 195, 253–255, 262–264, 265

Kannisto-Thatcher Database, 64
Kannsito model, 104
Kaplan-Maier method
  Kaplan-Meier Curve, 124–126, 129

Large cities, 89, 274, 276–281
Large sample size, 80, 290
Late childbearing, 15, 72–75, 94, 105, 260, 423
Less developed areas, 276–277, 279–280, 282–284
Life expectancy, 1, 4, 16, 63, 66, 70, 85, 91, 108, 123, 158, 195, 269–273, 280–281, 283–284, 316, 325, 358, 366, 379, 421
Life satisfaction, 19, 28, 40–41, 198, 256, 292, 307, 331, 333, 339–340, 347–355, 363
Life style, 118–119, 145, 150, 160, 166–167, 173–174, 318

Living arrangement
  living alone, 48, 182, 184, 186, 194, 201–203, 206–210, 217, 219, 223–224, 226–227, 291, 340, 355, 360, 362–363, 367, 370, 373
  living with children, 203, 206, 210, 217, 219, 222–225, 227–228, 230–231, 240, 242, 244–245, 358–360, 363, 366
  living with others, 186, 219, 222–224, 227–228, 231
  living with spouse, 182, 184, 186, 219, 223, 226–227, 229–230, 349, 363
Logistic regression, 74–75, 194, 222, 225, 229, 242, 243–245, 296–298, 309, 311, 322–323, 366, 369, 372, 394, 401, 411–412, 421
Longevity, 1, 9, 79–80, 82, 85–86, 93–95, 100, 129, 140–141, 149–154, 232, 270, 281, 283–285, 290–291, 330, 342–343, 423, 426
Longitudinal analysis, 224
Longitudinal study, 24, 32–33, 47, 291, 294, 313, 332, 342, 348

Marital status, 31, 45, 119, 122–123, 125, 128–130, 163, 166, 169, 172, 180–181, 183–185, 187, 205, 208–209, 217–219, 221, 225, 228–230, 237, 240, 242, 244, 256–257, 259–260, 263, 291, 296, 298, 300, 381–383, 385, 387, 390–391, 393, 402–406, 409–410, 412
Mean age at childbearing, 87–90, 93
Medical conditions, 107, 295, 297–298, 301, 308, 415
Medical health, 219–220
Minimal functional limitations, 294
Missing values, 48, 141, 173, 181, 253, 290, 349, 379–380, 384, 386, 394
Mixed models procedure, 206
Moderating effect, 399
Morbidity
  prevalence of morbidity, 15
Mortality
  all-cause mortality, 100–102, 108, 197, 399–400
  mortality converge, 399
  mortality decline, 15–16, 270, 272, 282
  mortality differential, 108
  trajectory of mortality, 159, 423
Multicollinearity, 125, 127, 242
Multidimensional concept, 134, 146
Multinomial regression, 222, 225, 229, 401, 411, 412

Multiple imputation, 46
Multi-state life table, 422
Multivariate analysis, 20, 222, 261, 312, 319, 354

National Long-Term Care Survey, 49, 108
Natural teeth, 319, 321, 324
Need-based support, 252
Next-of-kin, 102, 106–107, 109–110, 112–113
Non-suffering death, 422
Normal ageing
  normal aging, 178
Nursing homes, 31, 161, 182, 184, 194, 203, 206–209, 218, 222–224, 259, 257, 291, 340, 352–353, 355, 358–364, 366, 370, 372–374, 401

Objective activities, 358, 369–370
Objective life conditions, 291, 341
Objective question, 41–42
Oldest-old
  nonagenarian, 20, 26–28, 33–34, 40, 66, 81–82, 108, 113, 199, 202, 210, 260–261, 291, 330, 332–333, 336, 338–339, 341–342, 362, 380
  octogenarian, 20, 26–28, 33–34, 40, 66, 81–82, 108, 113, 202, 207, 210, 260–261, 291, 330, 332–333, 336, 338–339, 341–342, 380
  oldest-old population, 134, 158, 160, 163–166, 189, 306–308, 316, 416
  young-old, 289, 291–292, 316, 348–355
OLS models, 205
One-child policy, 158
One per thousand fertility survey, 89, 92–93
Optimism, 200, 232, 331, 334, 336, 338, 341–342, 423
Oral disease
  oral hygiene, 290, 317, 325–326

Panel data, 118, 160–161, 164, 166, 173, 218
Parent support ratio, 11–12
Perceived support, 360–362, 367–371
Personality, 50–56, 141, 143–145, 318, 331, 372, 425
Physical function, 119, 178, 255, 300, 333, 337, 339, 377, 382–383, 386
Physical health measures, 135
Place of death, 424
Population aging
  ageing, 34, 79–81, 270–271, 291, 293, 296, 300, 305, 420
  aging, 158, 160–161, 200, 313

population growth, 5–7, 199
rapid aging, 15
Population health, 197–198, 201, 282
Population projection, 2–3
Positive psychological disposition, 199, 201
Pre-designed, 66
Predictability
  predictive power, 398–399
  predictor, 22, 48, 105, 108, 123, 134, 167, 173, 244, 255, 260, 290, 296, 331–332, 335, 340–342, 361, 390, 393, 397, 399, 408, 414, 423
Probit model, 166–168, 170, 254
Protection, 145, 154, 195, 251, 254–255, 260–262, 264
Psychological well-being
  psychological disposition, 194, 198–204, 206–211
  psychological factors, 166, 330
  psychological hardiness, 199
  psychological resources, 291, 330–332, 334–337, 341–342
  robust psychological disposition, 200–201
Psychosocial factors, 290, 399, 401, 403, 408, 411, 414–415

Quality of life, 15, 118, 135–138, 141–142, 144–146, 162, 164–168, 171, 173, 194, 289–291, 294–295, 301, 316–317, 325, 357–374

Random effects model, 205–206, 208–209
Rating scores, 351
Ratio index, 66
Ratio of survivorship, 76, 151
Re-assessment, 80, 84
Regional differences
  regional variation, 84, 89, 272, 285
Relative bias, 104
Reliability, 20, 22, 39, 50–56, 62, 83, 93, 95, 100, 109–110, 113, 201, 241–242, 366, 425
Residence difference, 325
Restricted Maximum Likelihood, 206
Rural areas
  rural China, 200, 236–237, 245–246, 261, 264, 283, 320, 421

Salutogenesis, 199
Sample design, 21, 46, 80–81, 82, 87, 93–94, 202, 380
Sanitary conditions, 276–277
Second National Epidemiological Survey on Oral Health, 318, 324

Second National Health Service Survey, 110
Secrets of longevity, 423
Selection, 21, 31–32, 61, 71–72, 74, 82, 100, 108–109, 140–141, 178, 195, 202, 207, 242, 246, 251–253, 255, 256, 258, 261–262, 264–265, 332, 339, 372, 421, 423
Self-perceived health status, 308–309
Self-rated health
  perceived health status, 235, 305, 308–309
  self reported health, 22, 73–74, 105, 135, 137–138, 141, 144, 162, 164–165, 203–204, 208–209, 308–310, 312
Self-reported chronic conditions, 109, 113
Self-reported life satisfaction, 349–350
Self-reported quality of life, 135, 136, 137, 138, 144, 146, 162, 164, 165, 167, 168, 171, 173, 358, 360, 363
Semi-standardized hazard ratio, 125–126, 128–129
Shanghai, 26–27, 40, 86, 89–90, 94, 97
Significant others, 41
Simulation, 105–106
Social change, 217
Social engagement
  engagement in activities, 379, 381, 383, 385, 386, 387, 388, 390, 391, 392, 393
  engagement in life, 331, 333, 335, 338–343
Social network, 179, 211, 216, 373
  social structure, 361, 365, 371
  social support, 118, 198, 200–201, 207, 210–211, 236–237, 239, 246, 252, 256, 261, 291–292, 348, 350, 353, 355, 357–367, 369, 373–374, 398
  type of social support, 246, 369
Sociodemographic factors
  sociodemographic profile, 420–421
  socioeconomic status, 72, 117, 124, 139, 146, 149, 157–159, 166, 170, 173, 194, 203–204, 207–209, 218, 238, 361–362, 373, 421–422, 424
Socioemotional selectivity theory, 178
Standards of living, 273, 276–277, 284
Strong predictor, 260
Structural equation, 334
Subjective dimension, 358, 399, 411
Subjective health status
  subjective measures of health, 134–135
Subjective well-being, 289, 292, 330, 336, 340, 347, 348, 349, 350, 351, 352, 353, 362
Successful aging, 291, 330, 339, 342
Survey of Family Dynamics of the Elderly's Children, 253

Survey of Welfare of Elderly in Anhui
  Province, 194
Survival analysis
  Cox proportional hazard model, 128,
    129, 383
  Cox proportional hazards regression, 401
  hazard model, 103–105, 108, 122, 124,
    128–130, 183, 290, 382–384, 389
  hazard of dying, 122, 126–130
  hazard ratio, 126, 128–129, 186, 405, 409
  survival curve, 122, 124–126, 129, 323–324
  survival status, 100, 164–165, 184
  survivor sample, 336, 338
  survivorship, 76–77, 117, 118, 151, 153, 272
  time-on-study, 383
  time varying covariates, 383

Weibull hazard models, 103, 183
Sweden, 11, 20–21, 61, 63–64, 66–68, 70–72,
  74, 83–85, 91, 189, 271, 284

Traditional family, 15
Transitional model, 194
Two-level hierarchical linear model, 206
Types of time scales, 383

U.S.A., 24
Univariate analysis, 154, 312
Unobserved heterogeneity, 255, 257–258, 394
Unobserved propensity, 257–258, 262
Urban/rural difference, 140, 203, 320–321,
  325–326, 361, 365

Printed in the United States
215015BV00003B/1/P